DRAFTING:

TIPS AND TRICKS ON DRAWING AND DESIGNING HOUSE PLANS

BOB SYVANEN

LIBRARY OF CONGRESS CATALOGING IN PUBLICATION DATA

SYVANEN, BOB, 1928-
DRAFTING: TIPS AND TRICKS ON DRAWING AND DESIGNING HOUSE PLANS.

1. ARCHITECTURE, DOMESTIC.. DESIGNS AND PLANS.
2. ARCHTECTURAL DRAWING. I. TITLE.
NA 7115.S97 720'.28'4 81-17344
ISBN 0-914788-48-5 AACR2

PRINTED IN THE UNITED STATES OF AMERICA

THIS BOOK WAS FIRST PRINTED BY THE AUTHOR IN 1981.
IF YOU HAVE ANY DRAFTING QUESTIONS, INCLUDE A STAMPED, SELF-ADDRESSED ENVELOPE AND SEND THEM TO:

BOB SYVANEN
179 UNDERPASS ROAD
BREWSTER, MASSACHUSETTS 02631

AN EAST WOODS PRESS BOOK
FAST AND McMILLAN PUBLISHERS, INC.
820 EAST BOULEVARD
CHARLOTTE, NC 28203

INTRODUCTION

IF YOU PLAN TO BUILD YOUR OWN HOUSE, THERE IS NO BETTER WAY TO START THAN BY DRAWING THE PLANS. THERE IS NO WAY TO SUCCEED AT A TASK WITHOUT GIVING IT A TRY. IT'S BEEN SAID THAT MOST OF US USE ABOUT FIVE PERCENT OF BODY-MIND POTENTIAL SO WITH A LITTLE EFFORT, SURELY WE CAN TAP THAT NINTY FIVE PERCENT. I HAVE FELT FOR SOME TIME NOW THAT ANYONE CAN DO ANYTHING, AND DESIGNING AND DRAWING YOUR OWN HOUSE PLANS IS JUST ANOTHER ONE OF THOSE THINGS ANYONE CAN DO.

WITH THE HIGH COST OF EVERYTHING, HERE IS AN AREA WHERE A SUBSTANTIAL SAVING CAN BE MADE. A $40,000.00 HOUSE, DESIGNED AND DRAWN BY AN ARCHITECT WOULD COST FROM $1600.00 ON UP, DEPENDING ON DESIGN. (COMPLICATED DESIGNS COST MORE). IF A DRAFTSMAN DRAWS THE PLANS, THEY WOULD COST FROM $800.00 ON UP. MOST ARCHITECTS WONT TOUCH THE SMALL HOUSE. PLANS FOR A $90,000.00 HOUSE, ARCHITECT DESIGNED AND DRAWN, WOULD COST FROM $3200.00 ON UP AND DRAFTSMAN DRAWN, $1600.00 ON UP.

MOST PROBLEMS ARE OR SHOULD BE IRONED OUT ON PAPER BEFORE THE FIRST STICK IS CUT; IT'S CHEAPER THAT WAY. THE MORE KNOWLEGE YOU HAVE ON THE TOTAL SUBJECT, THE BETTER THE END PRODUCT WILL BE. IN FACT, I GOT INTO CARPENTRY THIRTY YEARS AGO SO THAT I WOULD BE A BETTER ARCHITECT. MY CARPENTRY BOOKS WOULD BE A HELP IN BOTH DESIGNING AND BUILDING. I ALWAYS KEEP THE CARPENTER IN MIND WHEN I DESIGN A HOUSE, BUT MY PRIMARY INTEREST IS THE CLIENT'S IDEAS. I TRY TO PUT ON PAPER WHAT THE CLIENT WANTS, NOT WHAT I WANT. I MERELY SUGGEST AND GUIDE. YOU, AS DESIGNER, CAN DO THE SAME WITH THAT INNER WISDOM WE ALL HAVE. VERY FEW HOUSES THAT ARE BUILT ARE ARCHITECT-DESIGNED, SO COME ON IN, THE WATERS FINE!

CONTENTS

CONTENTS

1 - EQUIPMENT

YOU CAN DRAW THE SAME SET OF PLANS WITH A T-SQUARE, TWO WOOD PENCILS A KITCHEN TABLE, AND A TRIANGLE AS YOU CAN WITH A FANCY TABLE, SEVEN TRIANGLES AND FIVE PENCILS. ONE SET WILL COST $15.00, THE OTHER $2500.00, BUT IT ALL BOILS DOWN TO HAND AND HEAD. THE BARE BONES EQUIPMENT AND A DESIRE ARE ALL YOU NEED FOR DRAWING YOUR OWN HOUSE PLANS.

EQUIPMENT

THE BARE BONES EQUIPMENT AND A DESIRE ARE ALL YOU REALLY NEED FOR DRAWING YOUR OWN HOUSE PLANS.

A TABLE WITH A STRAIGHT EDGE WILL WORK FINE FOR A DRAFTING BOARD.

1¼"

TABLE THICKNESS

¾"

2½"

ANY TABLE WILL WORK FINE, EVEN ONE WITHOUT A STRAIGHT EDGE. JUST MAKE UP A STRAIGHT EDGE FROM TWO PIECES OF WOOD, STRAIGHT PIECES WOULD HELP.

CLAMP IT TO THE END OF THE TABLE

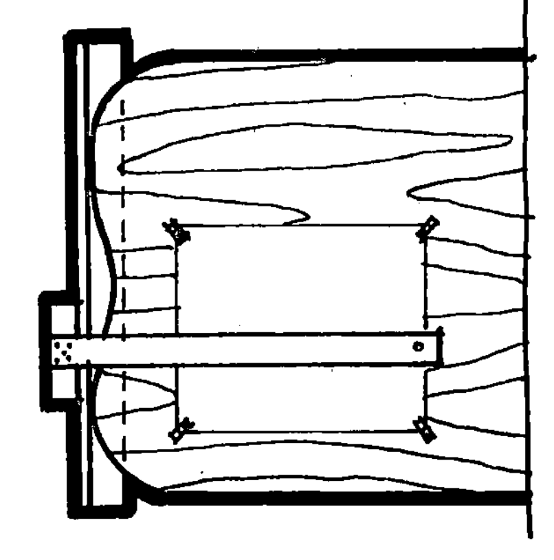

EQUIPMENT

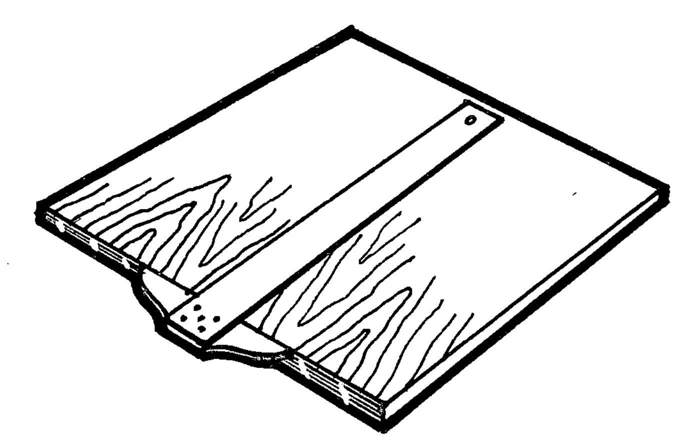

YOU CAN BUY A SMALL BOARD FOR ABOUT $25.00

OR YOU CAN MAKE ONE FROM 3/4" AC PLYWOOD, BOUGHT OR SCROUNGED. IF A T-SQUARE IS TO BE USED, ONE END HAS TO BE STRAIGHT AND SMOOTH. TO GET A NICE SLICK END, CEMENT A PIECE OF FORMICA (KITCHEN COUNTER TOP MATERIAL) TO THE EDGE. IT IS TOO HARD TO USE FOR THE SURFACE. A SHEET OF HARD SURFACE PAPER OR $15.00 A SQUARE YARD VINYL (AVAILABLE AT DRAFTING SUPPLY STORES) MAKES A NICE WORK SURFACE. THE VINYL SURFACE IS SOFT ENOUGH SO THAT PENCIL WORK WONT TEAR EVEN THIN TRACING PAPER. IT'S EASY TO CLEAN AND LASTS FOREVER. ATTACH IT TO THE BOARD WITH 1" DOUBLE COATED TAPE.

A SOLID CORE DOOR MAKES A BEAUTIFUL TABLE (VERY HEAVY, BUT A NICE SURFACE). THERE ARE PLENTY OF USED DOORS AT BUILDING SALVAGE YARDS. A HOLLOW CORE DOOR WILL DO IF YOU CHOOSE CAREFULLY. THE OLDER HOLLOW CORE DOORS HAVE A BETTER SURFACE THAN THE NEW ONES. THE SMOOTHER THE SURFACE THE FEWER PROBLEMS YOU WILL HAVE IN KEEPING THE DRAWINGS CLEAN. THE LITTLE BUMPS GET RUBBED BY THE TOOLS, SMUDGING THE DRAWING. DOORS FOR TOPS WILL BE ABOUT 3 FEET BY 7 FEET. I WOULDN'T GO OVER 2 FEET BY 3 FEET WITH 3/4" PLYWOOD UNLESS IT IS REINFORCED UNDERNEATH TO PREVENT SAG AND WARP. SAND THE SURFACE AND BRUSH IT CLEAN BEFORE PUTTING PAPER OR VINYL ON. ANY SMALL PARTICLES WILL CAUSE BUMPS ON THE SURFACE AND THESE WILL CAUSE SMUDGES ON THE DRAWINGS.

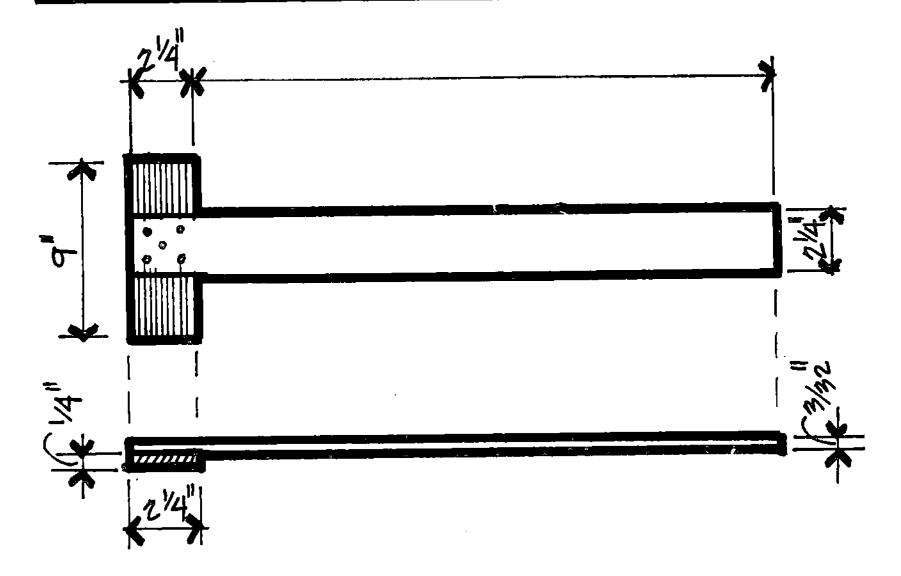

SINCE THE PARALLEL STRAIGHT EDGE WAS REFINED, THE T-SQUARE ISN'T USED MUCH THESE DAYS, BUT IT STILL WILL DO A GOOD JOB. YOU CAN MAKE ONE, BUT IT IS HARDLY WORTH WHILE WHEN A SMALL WOOD ONE COSTS BUT A FEW BUCKS, EVEN A GOOD ONE A FEW DOLLARS MORE. IF YOU MAKE ONE, TRY TO MAKE THE BLADE PERPENDICULAR TO THE HEAD, BUT DON'T BE CONCERNED IF IT IS NOT; IT WILL STILL SLIDE UP AND DOWN PARALLEL. JUST BE SURE THE HEAD IS FIRMLY ANCHORED TO THE BLADE WITH 5 WOOD SCREWS.

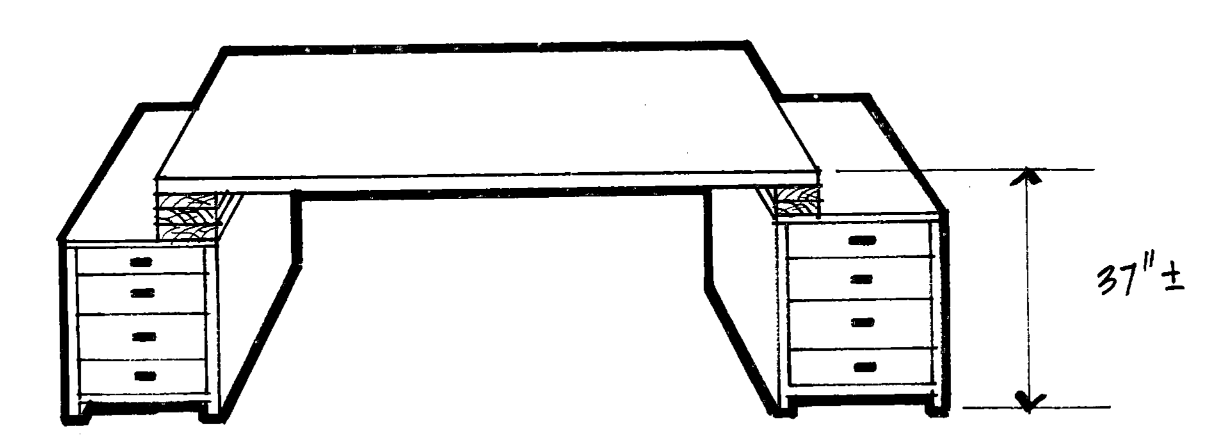

THE HEIGHT OF THE BOARD IS A MATTER OF PERSONAL PREFERENCE; REMEMBER YOU MIGHT WANT TO WORK BOTH SITTING AND STANDING. MY BOARD IS 37" HIGH AND FLAT, I DON'T LIKE TO CHASE PENCILS. MOST PEOPLE LIKE THE BOARD RAISED A FEW INCHES ON THE FAR EDGE. IF THE BOARD IS HIGH (37"), THEN YOU MUST SIT ON A HIGH STOOL, 30" IN MY CASE. TABLE HEIGHT (29") IS FINE TO WORK AT, BUT ONLY SITTING. THE BOARD CAN BE SUPPORTED ON HORSES MADE FOR THE JOB, OR BLOCKED UP ON CABINETS, DRESSERS, OR BOXES.

IF PLYWOOD IS USED, A WOOD SUB-FRAME SHOULD BE BUILT AND THEN THE WHOLE BUSINESS SUPPORTED.

THERE IS A NICE LITTLE 18"x24" BOARD WITH A PARALLEL STRAIGHT EDGE FOR ABOUT $35.00. THE LARGEST ONE I KNOW OF IS 23"x31" AND THE SMALLEST IS 12"x17".

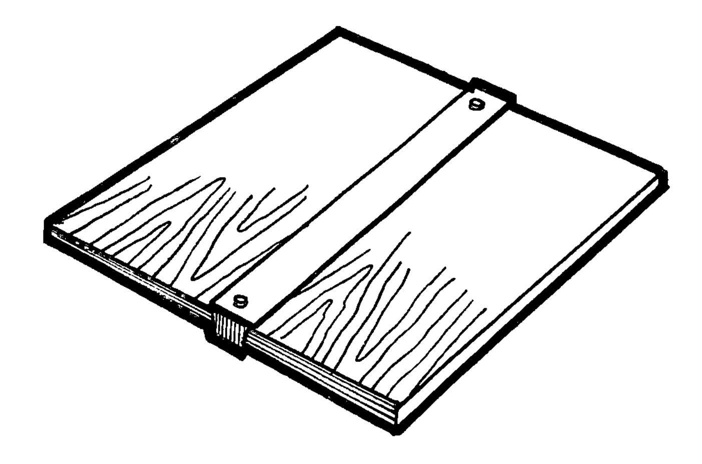

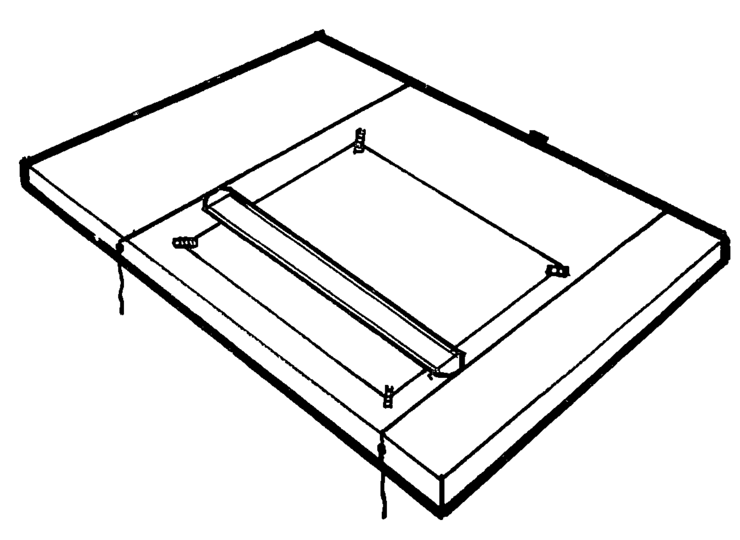

PARALLEL STRAIGHT EDGES COME IN SIZES 36", 48", 54", 60", 72". I LIKE THE 48" SIZE, A NICE IN-BETWEEN LENGTH. WITH THE PARALLEL STRAIGHT EDGE YOU DON'T HAVE TO HOLD THE STRAIGHT EDGE AND TRIANGLE AS YOU WOULD WITH THE T-SQUARE. AND IT IS ALWAYS PARALLEL NO MATTER WHERE YOU PUSH IT ON THE BOARD. BE SURE THE BOARD IS LONGER THAN THE STRAIGHT EDGE.

EQUIPMENT

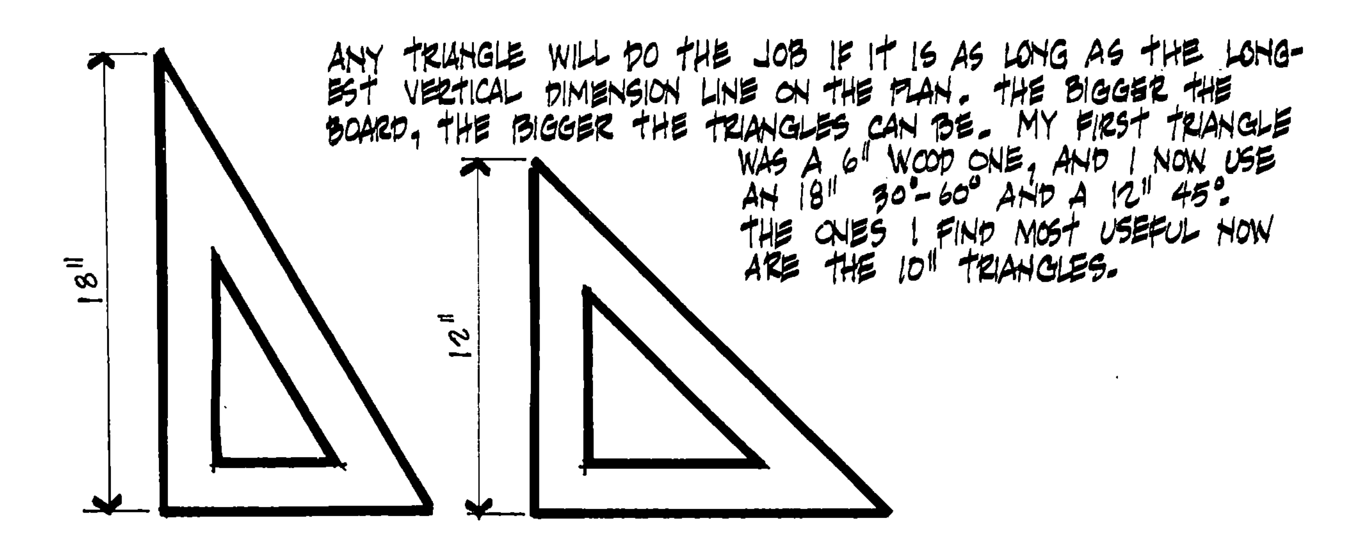

ANY TRIANGLE WILL DO THE JOB IF IT IS AS LONG AS THE LONGEST VERTICAL DIMENSION LINE ON THE PLAN. THE BIGGER THE BOARD, THE BIGGER THE TRIANGLES CAN BE. MY FIRST TRIANGLE WAS A 6" WOOD ONE, AND I NOW USE AN 18" 30°-60° AND A 12" 45°. THE ONES I FIND MOST USEFUL NOW ARE THE 10" TRIANGLES.

A REALLY GREAT TRIANGLE IS THE 10" ADJUSTABLE THAT COMES IN TWO STYLES. ONE MEASURES DEGREES AND THE OTHER ROOF PITCHES; EITHER WILL DO, BUT SINCE MY ROOF PITCH ONE VANISHED I CANNOT LOCATE ANOTHER. A GOOD REASON TO SCRATCH YOUR INITIALS ON TOOLS!

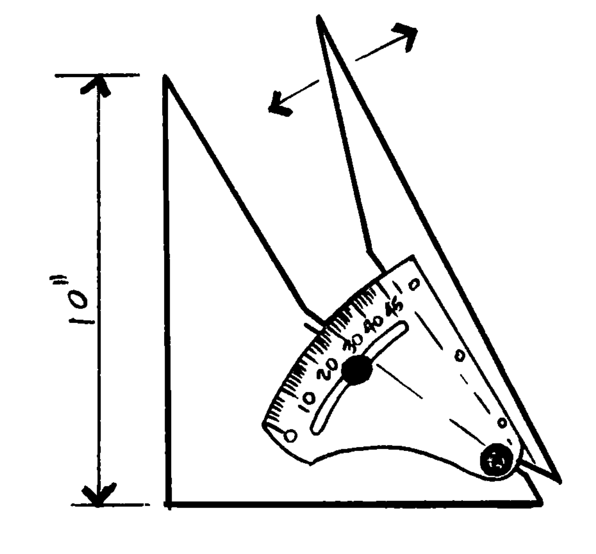

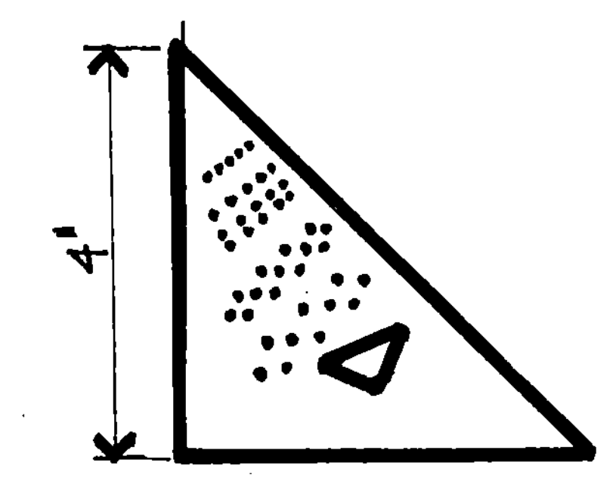

ANOTHER GOOD TRIANGLE IS THE 45 LETTERING GUIDE. THERE IS A 6" MODEL, BUT I FIND THE 4" K&E 1859-C-4 MODEL GOOD ENOUGH.

THE AMES LETTERING GUIDE IS PROBABLY THE MOST COMMON GUIDE AROUND, BUT TAKE YOUR PICK.

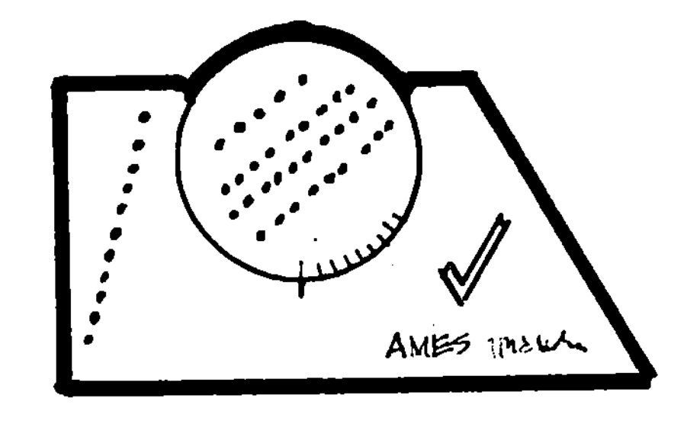

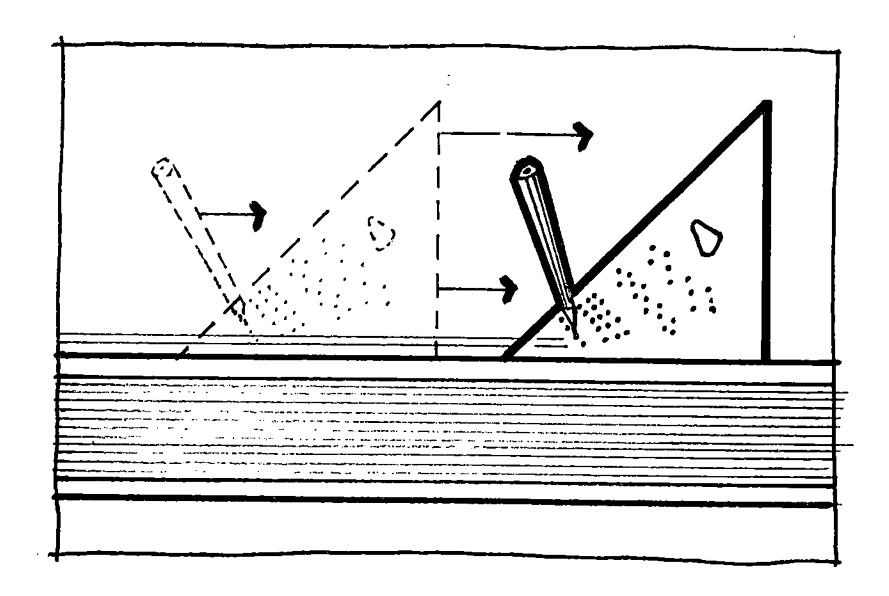

IT IS PRETTY OBVIOUS HOW THE LETTERING GUIDES ARE USED - THE PENCIL POINT IS PUT INTO A HOLE OF YOUR CHOICE AND THEN YOU SLIDE THE TRIANGLE, ALONG THE STRAIGHT EDGE, WITH THE PENCIL. BRING THE WHOLE BUSINESS BACK, PUT THE POINT IN A HOLE THAT WILL GIVE THE SPACING DESIRED, AND THEN SLIDE AGAIN. PESTO, TWO PARALLEL LINES! ONE CAUTION; THE GUIDES GET DIRTY. KEEP THEM CLEAN AND USE ONLY FOR LETTERING WORK OR ELSE YOU WILL FIND THE DRAWING WILL GET MESSY.

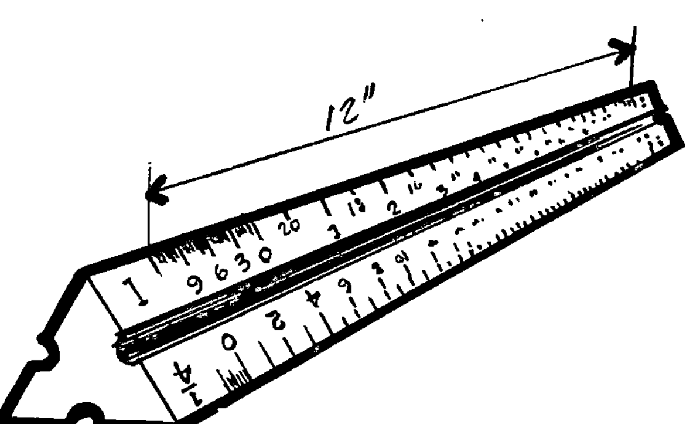

HOUSE PLANS DON'T HAVE TO BE DRAWN SUPER ACCURATELY AS LONG AS THE NUMBERS ARE CORRECT. I USED A PLAIN WOOD RULER WITH 1/16" INCREMENTS WHEN I STARTED. THE 12 INCH TRIANGULAR SCALE IS THE MOST FAMILIAR AND AN INEXPENSIVE WOOD SCALE WILL MEASURE THE SAME AS AN EXPENSIVE ONE, BUT NOT AS EASY ON THE EYES.

THIS 6" FLAT SCALE HAS 8 GRADUATIONS, 4 ON THE TOP AND 4 ON THE BOTTOM.

THIS FLAT SCALE HAS 4 GRADUATIONS AND THEY ARE ALL VISIBLE ON THE TOP. THE TRIANGLE SCALES ALWAYS SEEM TO HAVE THE SCALE YOU NEED ON THE BOTTOM.

EQUIPMENT

IN THE EARLY DAYS EVERYTHING WAS INKED AFTER FIRST DRAWING IN PENCIL. WHAT A DRAG! NOW ALL WE NEED IS GOOD PENCIL WORK AND WOOD PENCILS WORK FINE.

USE ANY SHARP TOOL TO CUT AWAY THE WOOD LEAVING ABOUT 3/4" OF LEAD EXPOSED.

THE LEAD IS THEN SHARPENED BY RUBBING THE TIP ON A PIECE OF SANDPAPER. WIPE THE DUST OFF WHEN FINISHED.

THE MECHANICAL PENCIL SAVES A LOT OF TIME. THE ONE MOST USED IS A PENCIL SHAPED METAL TUBE WITH A CHUCK AT THE POINT END THAT HOLDS ONE LONG PIECE OF LEAD. THE LEAD IS ADVANCED BY PRESSING THE REAR OF THE PENCIL WITH THE THUMB, OPENING THE CHUCK. THE LEAD IS FREE TO SLIDE OUT AS FAR AS NEEDED. THE TIP IS INSERTED INTO A PENCIL SHARPENER OR RUBBED ON A PIECE OF SANDPAPER. I USE THIS STYLE PENCIL WITH AN "H" OR "F" LEAD FOR HEAVY LINE WORK AND HEAVY LETTERING.

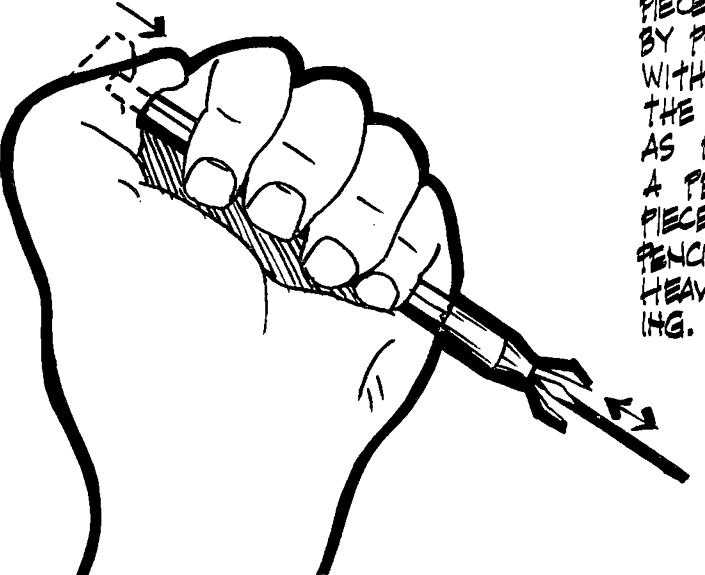

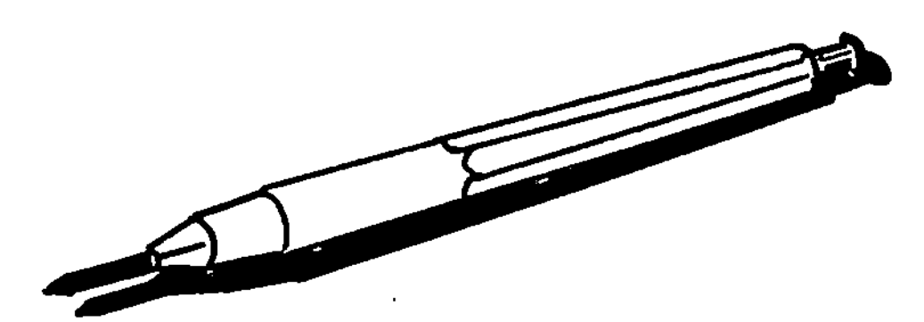

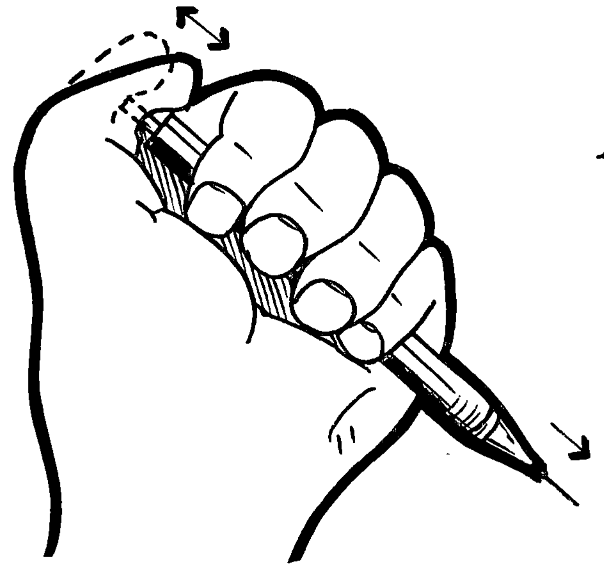

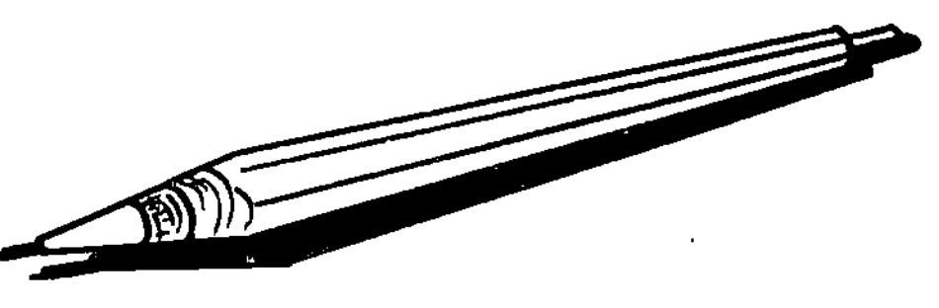

I THINK THE BEST PENCIL TO COME DOWN THE PIKE IS THE 5 MM MICRO LEAD. NOT EVERYONE CAN USE THIS PENCIL WITHOUT CONSTANTLY BREAKING THE LEADS. THIS PENCIL MUST BE HELD NEARLY VERTICAL AND WITH A GOOD TOUCH. THE BARREL IS LOADED WITH A DOZEN LEADS AND A CLICK OF THE THUMB, SIMILAR TO A BALL POINT PEN, WILL ADVANCE THE LEAD AND YOU ARE READY FOR WORK. IT'S CLEAN AND IT'S QUICK. YOU CAN'T BEAT IT FOR LAYOUT WORK WHERE YOU NEED A LIGHT TOUCH. THE LEADS COME IN ALL THE STANDARD DEGREES. I FIND THE 2H GOOD FOR LAYOUT AND THE H OR F GOOD FOR LETTERING.

THE MICRO LEADS NEED NO SHARPENING, ALL OTHERS NEED AT LEAST A PIECE OF SANDPAPER. THERE IS A 4" WOOD STICK WITH A DOZEN PIECES OF SANDPAPER GLUED ON MADE JUST FOR SHARPENING LEADS.

THERE IS A NEAT LITTLE K&E SHARPENER. MIGHT BE HARD TO FIND, BUT THERE ARE SIMILAR ONES.

JUST INSERT THE TIP AND TWIST

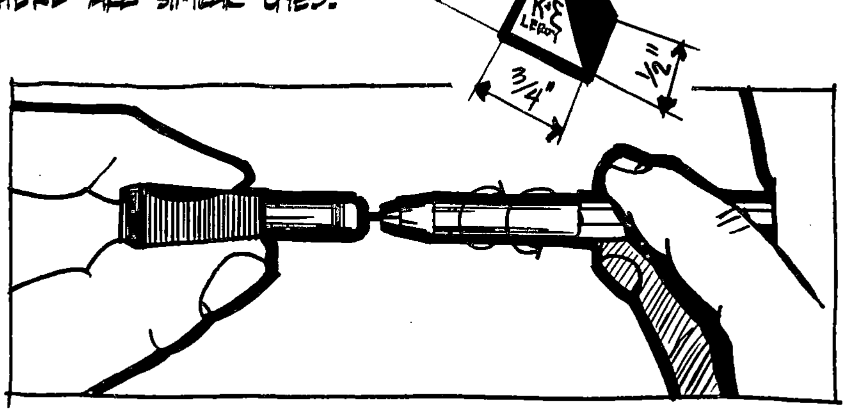

YOU CAN GO TO THE BIGGER METAL SHARPENERS.

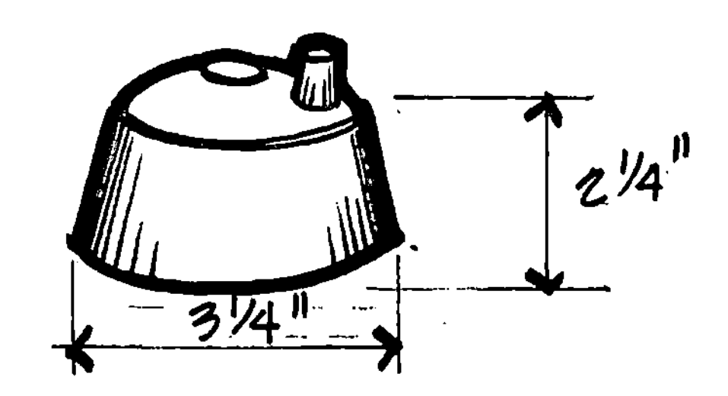

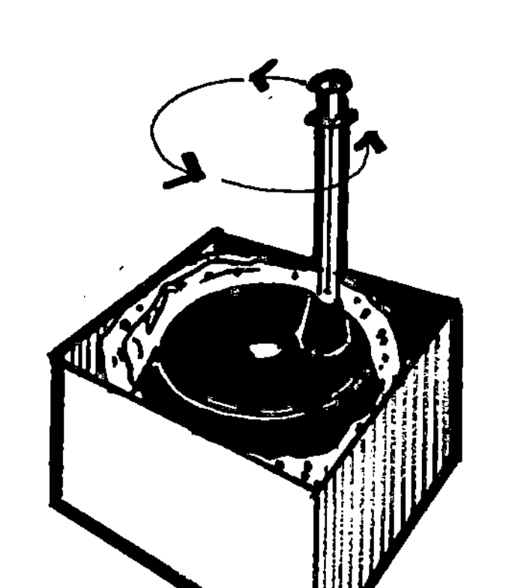

I KEEP THE ONE I HAVE IN ITS BOX. SURROUND IT WITH FOAM TO KEEP IT STEADY AND TO CLEAN THE POINT AFTER SHARPENING.

KOH-I-NOR MAKES A NICE PLASTIC SHARPENER THAT CLAMPS TO THE BOARD.

ALL OF THESE SHARPENERS CREATE A LOT OF DUST, SO BE CAREFUL. WIPE THE LEAD POINTS AND BRUSH OFF THE DRAWING FREQUENTLY, ESPECIALLY AFTER ERASING. ERASER PARTICLES ARE DIRTY AND THEY WILL SMUDGE THE DRAWINGS.

THERE IS A 4" CLOTH BAG, FILLED WITH CLEAN GROUND-UP ERASER PARTICLES, THAT YOU PAT ON THE DRAWING. THIS LEAVES SALT GRAIN SIZE BITS OF ERASER THAT THE TOOLS RIDE ON.

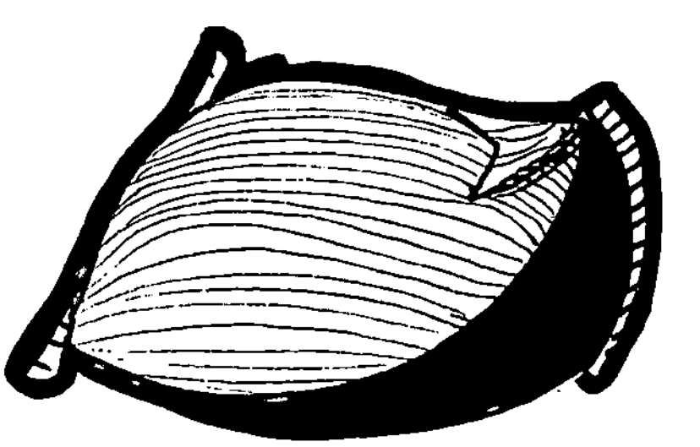

A GOOD WAY TO KEEP DRAWINGS CLEAN IS TO COVER EVERYTHING EXCEPT THE AREA BEING WORKED ON. ANY CLEAN PAPER WILL DO, BUT DO TAPE IT DOWN. THIS IS PARTICULARLY GOOD FOR LARGE DRAWING WITH MUCH WORK ON IT.

A THIN METAL ERASING SHIELD IS AN INEXPENSIVE ITEM WORTH HAVING.

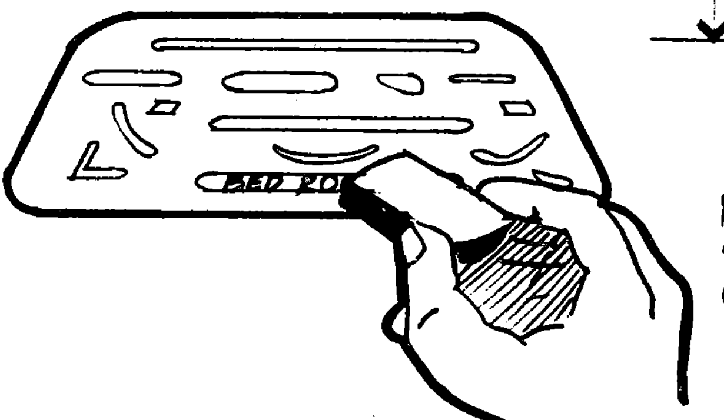

IT LETS YOU ERASE AN AREA WITHOUT DISTURBING WHAT IS AROUND IT. IT ALSO CLEANS THE ERASER.

THERE ARE THREE BASIC ERASERS: THE VINYL, THE GUM, AND THE RUBBER. THE OLD FASHIONED "SOAP" OR GUM ERASER IS STILL A GOOD ONE, BUT THE VINYL IS THE MOST POPULAR NOW. THE "PINK PEARL" RUBBER ERASER HAS BEEN AROUND A LONG TIME AND IT TOO IS A GOOD ERASER.

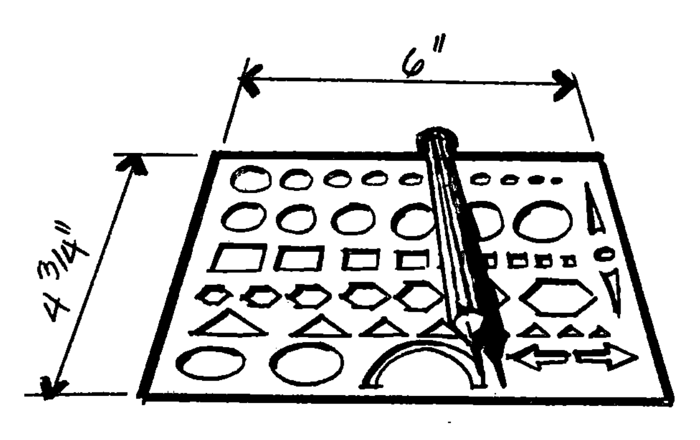

I HAVE A COUPLE OF FANCY COMPASSES THAT I RARELY USE SINCE ACQUIRING CIRCLE TEMPLATES. THE ONE MOST USED IS A 4"x6" COMBINATION CIRCLE, SQUARE, HEX, TRIANGLE TEMPLATE, A GOOD BASIC TEMPLATE. THERE IS ALSO A LARGE CIRCLE TEMPLATE WITH RADIUSES FROM 1/16" TO 3".

PLUMBING FIXTURE TEMPLATES, SOMETIMES AVAILABLE FROM PLUMBING SUPPLY STORES, BUT MORE EASILY PURCHASED AT DRAFTING SUPPLY STORES, ARE HANDY TOOLS. THERE ARE TEMPLATES FOR JUST ABOUT ANYTHING YOU NEED TO DRAW AND THEY ARE VERY HELPFUL, BUT NOT NECESSARY.

EQUIPMENT

3/4" WIDE MASKING TAPE IS A NICE SIZE TO USE FOR HOLDING DRAWINGS ON THE BOARD. BE SURE TO USE DRAFTING MASKING TAPE; IT'S NOT AS STICKY AS THE OTHER TAPES. YOU WILL HAVE LESS TROUBLE PEELING IT OFF. 'SCOTCH TAPE NO. 230 IS A DRAFTING MASKING TAPE.

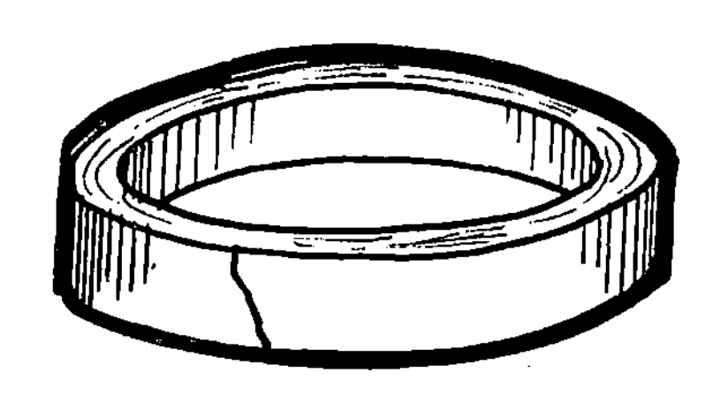

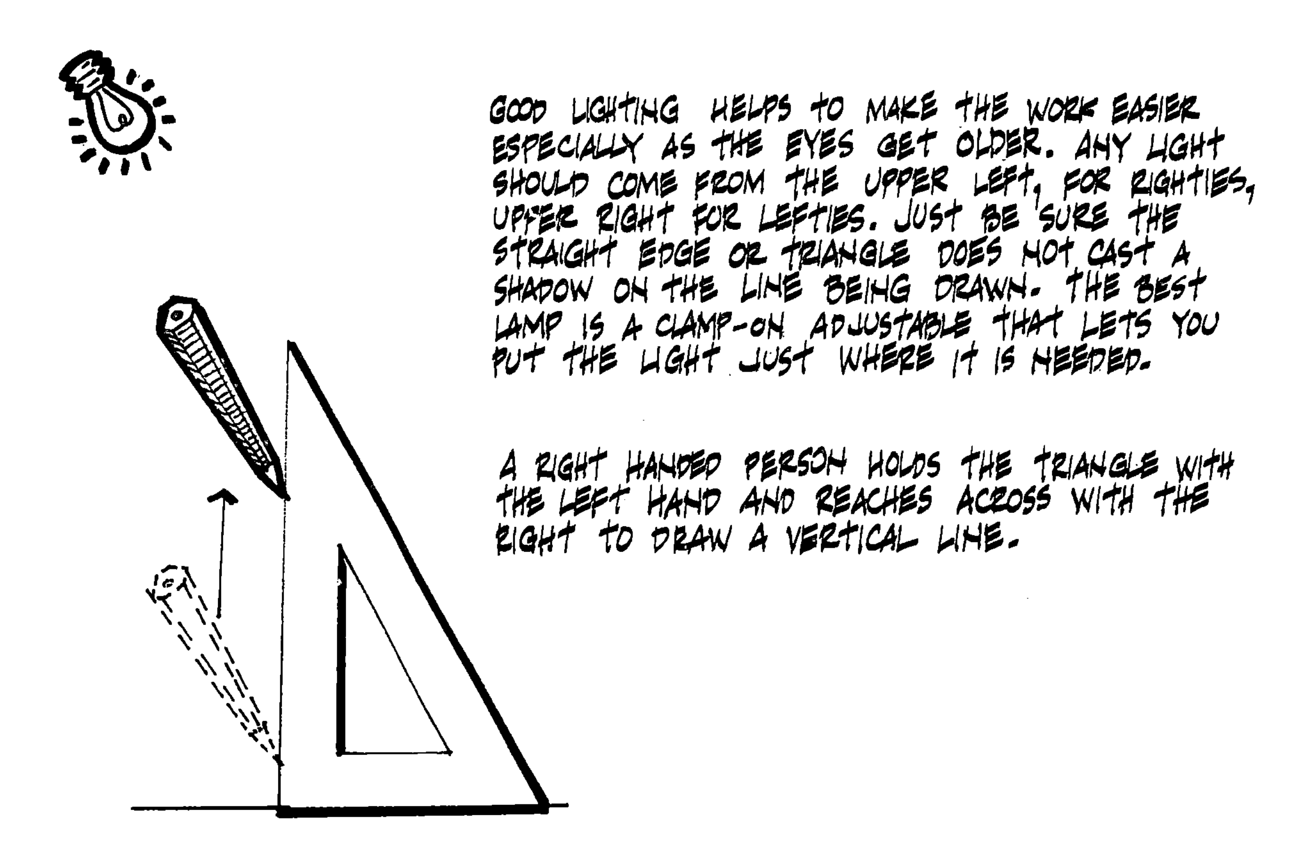

GOOD LIGHTING HELPS TO MAKE THE WORK EASIER ESPECIALLY AS THE EYES GET OLDER. ANY LIGHT SHOULD COME FROM THE UPPER LEFT, FOR RIGHTIES, UPPER RIGHT FOR LEFTIES. JUST BE SURE THE STRAIGHT EDGE OR TRIANGLE DOES NOT CAST A SHADOW ON THE LINE BEING DRAWN. THE BEST LAMP IS A CLAMP-ON ADJUSTABLE THAT LETS YOU PUT THE LIGHT JUST WHERE IT IS NEEDED.

A RIGHT HANDED PERSON HOLDS THE TRIANGLE WITH THE LEFT HAND AND REACHES ACROSS WITH THE RIGHT TO DRAW A VERTICAL LINE.

A POCKET CALCULATOR IS A TOOL THAT IS NOT NECESSARY, BUT IT SURE HELPS ME. IT NOT ONLY SAVES TIME AND HELPS AVOID ERRORS, IT MAKES MATH ALMOST FUN.

TRACING PAPER IS PRETTY MUCH A MATTER OF PREFERENCE. ANY TRACING PAPER WILL BE FINE, BUT SOME ARE BETTER THAN OTHERS. FOR THE SERIOUS DRAFTSMAN I LIKE CLEARPRINT 1000H, K&E CRYSTALENE OR K&E ALBANENE MEDIUM WEIGHT. ALL CAN BE PURCHASED BY THE ROLL OR SHEET. THESE PAPERS HAVE A "TOOTH" OR TEXTURE THAT SUITS MY HAND AND PENCILS. IF XEROX IS USED FOR PRINTS, THEN ANY KIND OF PAPER WILL DO, BUT TRACING PAPER IS PREFERRED FOR EASE OF TRACING THE DIFFERENT FLOOR PLANS. IF YOU HAVE A LIGHT TABLE THEN ANYTHING GOES.

I WOULD NOT BE WITHOUT A ROLL OF THIN YELLOW TRACING PAPER (THE TALKING PAPER) AND FELT TIP PENS. 14" AND 18" WIDE BY 50 YARDS I FIND THE MOST USEFUL AND AT $5.00 A ROLL - WELL WORTH IT! THE CHOICE OF FELT TIP DEPENDS ON WHAT WORKS BEST FOR YOU. SOME PEOPLE LIKE WORKING WITH FINE LINES WHILE OTHERS PREFER HEAVY LINES.

A LIGHT TABLE CAN BE VERY EXPENSIVE OR YOU CAN MAKE A WOOD FRAME BOX WITH A FEW LIGHT SOCKETS INSIDE AND A PIECE OF 3/16" OR 1/4" FROSTED GLASS FOR A WORK SURFACE. A PIECE OF 1/4" PLATE GLASS ON A COUPLE OF BOOKS AT EACH CORNER AND A LIGHT DIRECTED UNDER IS AN EFFECTIVE LIGHT TABLE. THERE IS A LOT OF HEAT GENERATED PARTICULARLY IF THE LIGHTS ARE ENCLOSED, SO VENTILATE.

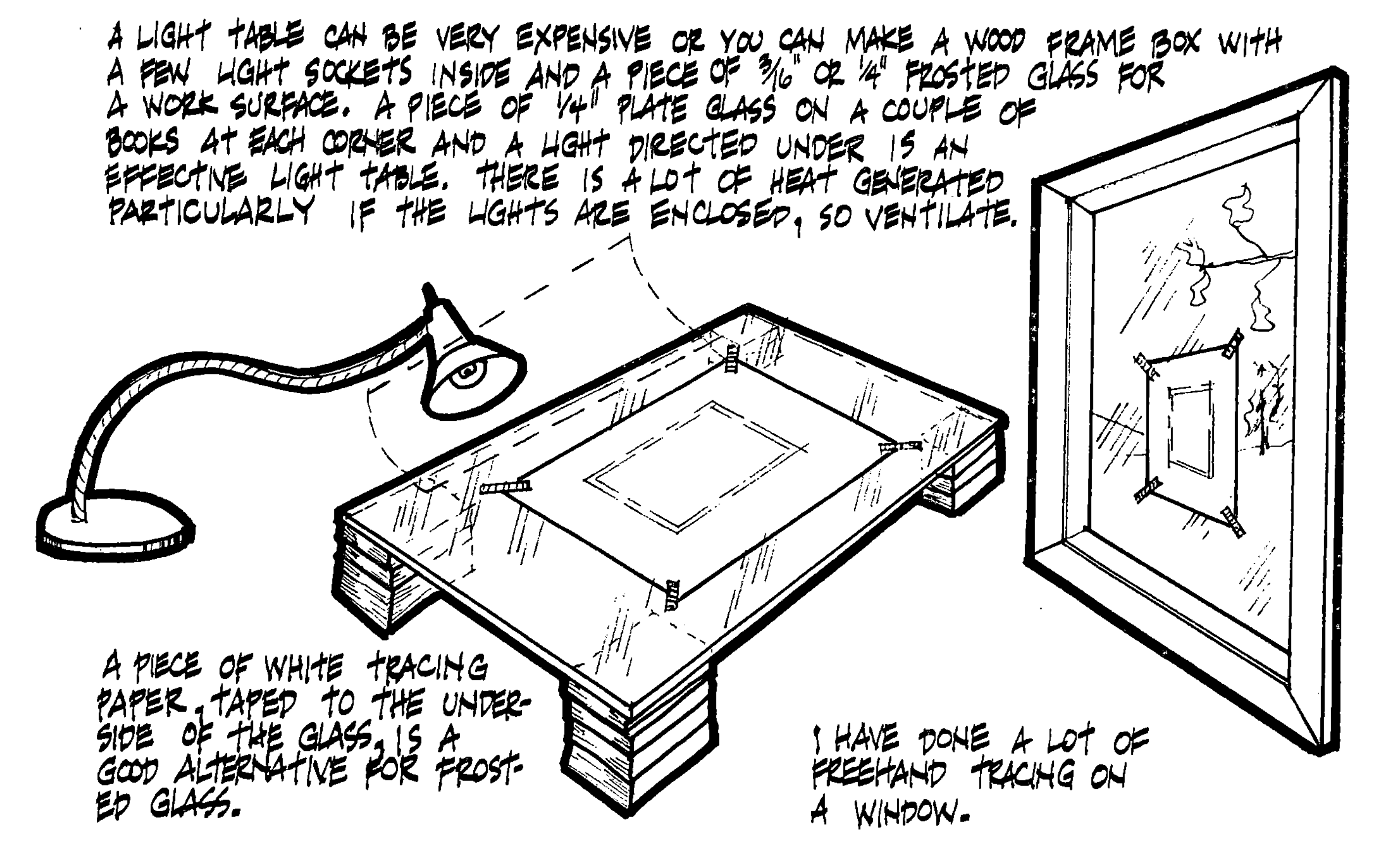

A PIECE OF WHITE TRACING PAPER, TAPED TO THE UNDERSIDE OF THE GLASS, IS A GOOD ALTERNATIVE FOR FROSTED GLASS.

I HAVE DONE A LOT OF FREEHAND TRACING ON A WINDOW.

EQUIPMENT

CLEAN ALL TRIANGLES, STRAIGHT EDGES, SCALES, TEMPLATES AND TABLE TOPS BEFORE THEY GET TOO DIRTY. IT WILL GO A LONG WAY TOWARDS KEEPING DRAWINGS CLEAN. CLEAN DRAWINGS REPRODUCE SHARPER AND THEREFORE ARE EASIER TO READ.

THESE ARE SOME BASIC TOOLS. THE DRAFTING BOARD IS A 24"X 30"X 3/4" PIECE OF PINE FACED PLYWOOD COVERED WITH VINYL. THE SURFACE UNDER THE VINYL SHOULD BE SMOOTH OR EVERY BUMP WILL SHOW.

A FEW MORE TOOLS TO MAKE THE WORK EASIER.

THE NEXT STEP UP. THIS BOARD IS A 3'-0" X 6'-8" SOLID CORE DOOR. THE VINYL COVERING OVER THE BIRCH DOOR MAKES FOR A NICE SMOOTH SURFACE.

2-DESIGN

MOST HOUSES ARE OWNER OR BUILDER-DESIGNED, SO IF YOU ARE AN OWNER OR BUILDER YOU CAN DESIGN YOUR HOUSE. WE ALL KNOW A GOOD HOUSE WHEN WE SEE IT. WE ALSO KNOW OUR LIMITATIONS, OR DO WE SET OUR LIMITS?

AS WE ENTER THE 80'S KEEP IN MIND WHAT THE "EXPERTS" ARE SAYING IN LIGHT OF THE HIGH COST OF CONSTRUCTION AND FUEL:

1. MORE ENERGY-EFFICIENT CONSTRUCTION.

2. FEWER AND SMALLER BEDROOMS.

3. NO FAMILY ROOMS.

4. SUMMER AND WINTER LIVING (CLOSE OFF PART OF THE HOUSE TO CONSERVE FUEL).

5. TWO GENERATIONS UNDER ONE ROOF.

THE DESIGN INFORMATION THAT FOLLOWS IS NOT VERY SOPHISTICATED, JUST VERY BASIC CONSIDERATIONS. THINGS LIKE FEELINGS, OBSERVATION, STAIRWAYS, CHIMNEYS, ETC. YOU CAN EXPAND FROM THESE BASICS AS YOUR NEEDS AND DESIRES DICTATE.

DESIGN

THE FIRST CONSIDERATION IN DESIGNING YOUR HOUSE IS TO REALIZE THAT YOU CAN DO IT. AFTER ALL, YOU DESIGN EVERY DAY WHEN DRESSING....

ARRANGING FURNITURE, FLOWERS, AND DRAPES.

YOU LOOK AT A HOUSE AND SOMETHING CLICKS- IT'S GOOD LOOKING OR IT'S AWFUL. THE ROOF ANGLE IS WRONG, THE CHIMNEY IS TOO THIN. TALK TO YOURSELF, CALL ON THAT INNER WISDOM WE ALL HAVE. TAKE PLENTY OF NOTES.

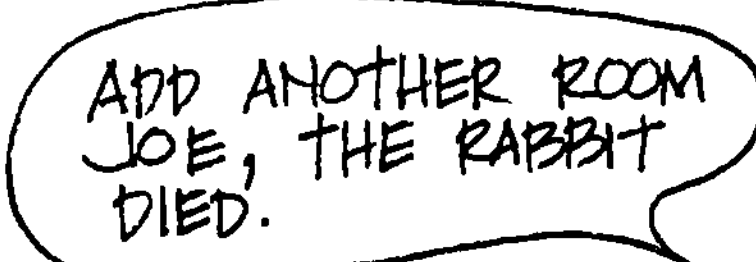

NEXT, CONSIDER THE NEED, WANTS AND LIFE STYLE. TO START SMALL BECAUSE POCKET-BOOK AND NEED ARE SMALL AND ADD AS THEY CHANGE IS A GOOD WAY TO GO. WHAT WE THOUGHT TODAY FREQUENTLY CHANGES TOMORROW.

NO MATTER WHAT YOU DESIGN, HALF THE PEOPLE WILL LOVE IT AND THE OTHER HALF HATE IT.

IF YOU SEE A HOUSE YOU LIKE DON'T BE AFRAID TO COPY IT OR USE PARTS IN YOUR OWN DESIGN. ALL THE EARLY CAVES WERE PRETTY MUCH ALIKE.

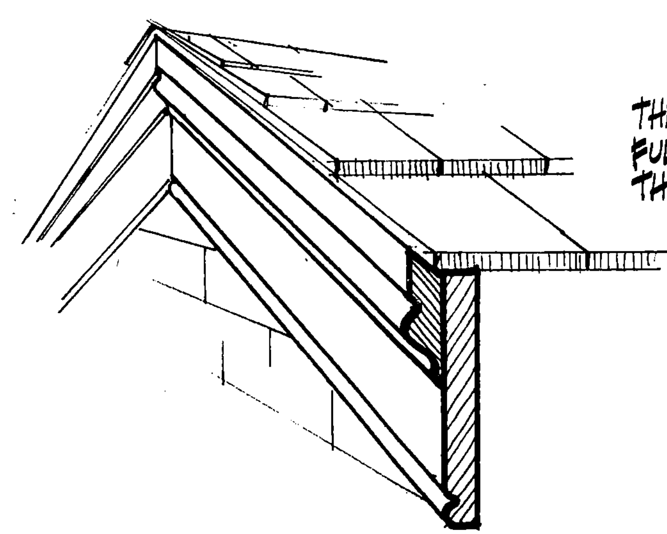

THE EARLY CAPE COD HOUSES WERE EITHER FULL CAPE, HALF CAPE, OR SALT BOX. THE VARIATIONS WERE IN THE ADDITIONS AND THE TRIM.

WHEN YOU SEE A NICE HOUSE, TAKE PICTURES AND MEASUREMENTS. TALK TO THE OWNERS. THEY WILL BE FLATTERED THAT YOU LIKE THEIR HOUSE.

I'M BOB SMITH AND WE'RE BUILDING A HOUSE. WE LOVE YOURS, MAY WE TAKE SOME PICTURES AND MEASUREMENTS?

WONDERFUL. WOULD YOU LIKE TO SEE THE INSIDE TOO?

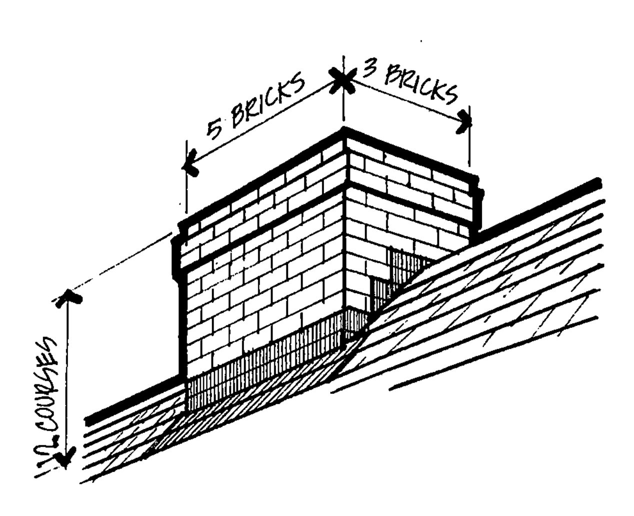

COUNT THE BRICKS OF A CHIMNEY YOU LIKE TO DETERMINE ITS SIZE. A BRICK IS 2" THICK BY 4" WIDE BY 8" LONG. MAKE NOTE OF THE ROOF SIZE AND PITCH. THAT SAME CHIMNEY MIGHT NOT LOOK GOOD ON A DIFFERENT ROOF.

TRY TO CARRY A TAPE MEASURE AND POCKET CALCULATOR.

YOU MIGHT SEE A PIECE OF TRIM YOU LIKE. OR MAYBE IT WOULD LOOK BETTER THINNER. TAKE NOTES; YOU WILL FILL A BOOK BEFORE YOU KNOW IT!

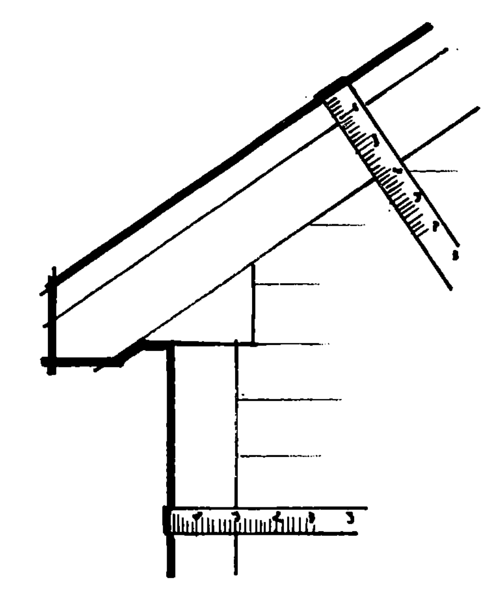

IF YOU DON'T HAVE A TAPE MEASURE WITH YOU, USE WHAT YOU DO HAVE, YOUR BODY. KNOW YOUR STRIDE (ABOUT 36")....

POTENTIAL BASKETBALL STAR.

BIG DOESN'T ALWAYS MEAN A LONG STRIDE.

SOME WOMEN HAVE VERY LONG LEGS.

DON'T STRETCH IT, THE STRIDE WILL BE INCONSISTENT.

DESIGN

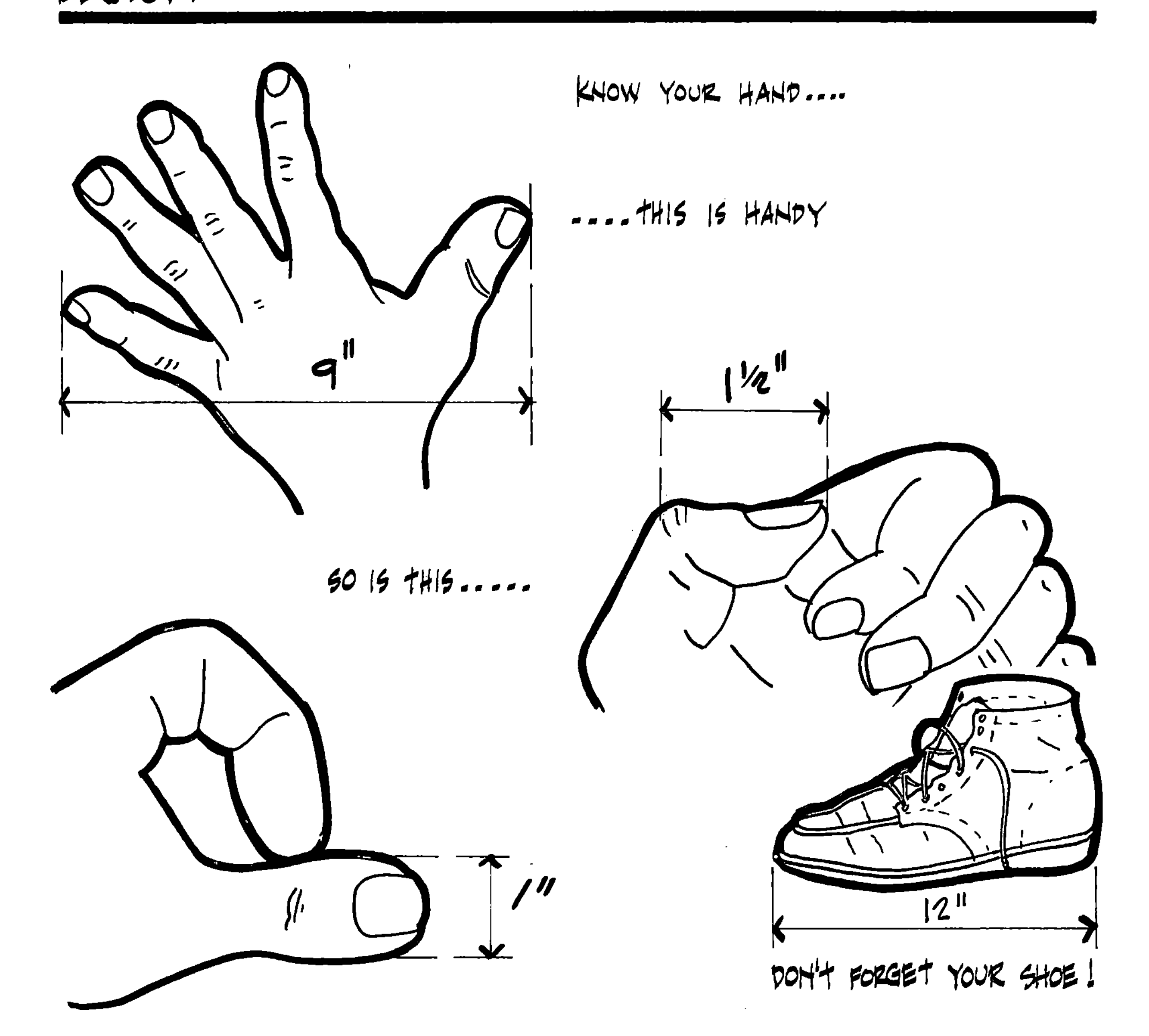

WHEN IN A ROOM THAT FEELS GOOD AND COMFORTABLE, TRY TO FIGURE WHY. IS IT THE CEILING HEIGHT, LENGTH TO WIDTH RATIO OF THE ROOM, WINDOW SIZES AND SPACES? MEASURE ALL THE CONTRIBUTING FACTORS. HOW FAR IS YOUR HAND FROM THE CEILING WHEN RAISED? COUNT FLOOR TILES AND HERE IS A GOOD TIME TO USE YOUR HAND SPAN. TILES ARE EITHER 9" OR 12". TAKE NOTES ABOUT THINGS THAT DON'T PLEASE YOU TOO; WE LEARN FROM GOOD AND BAD.

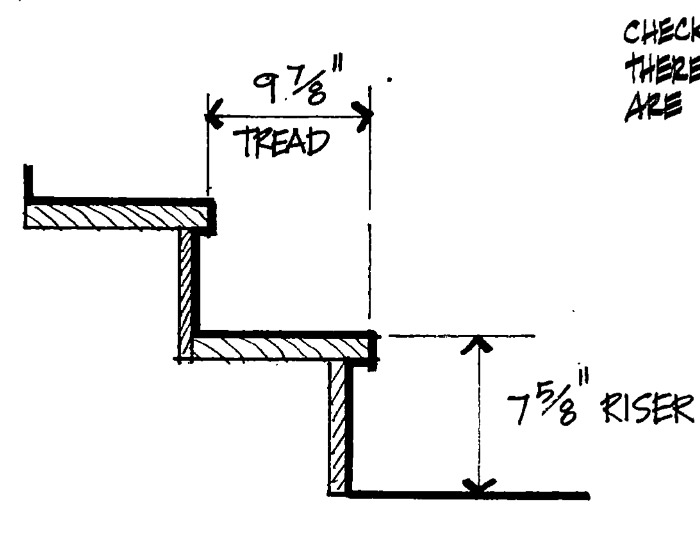

CHECK THE RISE AND TREAD ON STAIRS. THERE IS AN IDEAL OR AVERAGE, BUT ARE YOU AVERAGE?

$$9\tfrac{7}{8}'' + 7\tfrac{5}{8}'' = 17\tfrac{1}{2}''$$

THIS IS THE IDEAL TOTAL OF RISE AND TREAD.

AVAILABLE SPACE CONTROLS SHAPE OF STAIRWAYS....

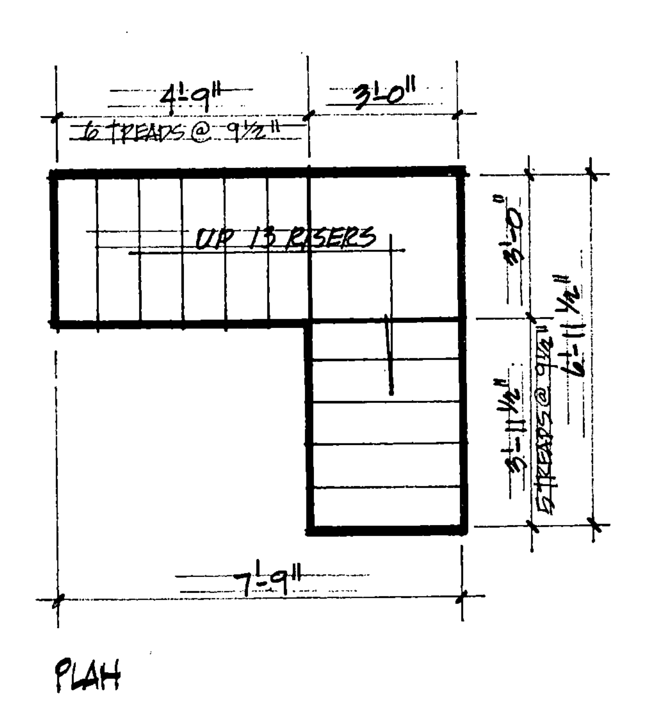

PLAN

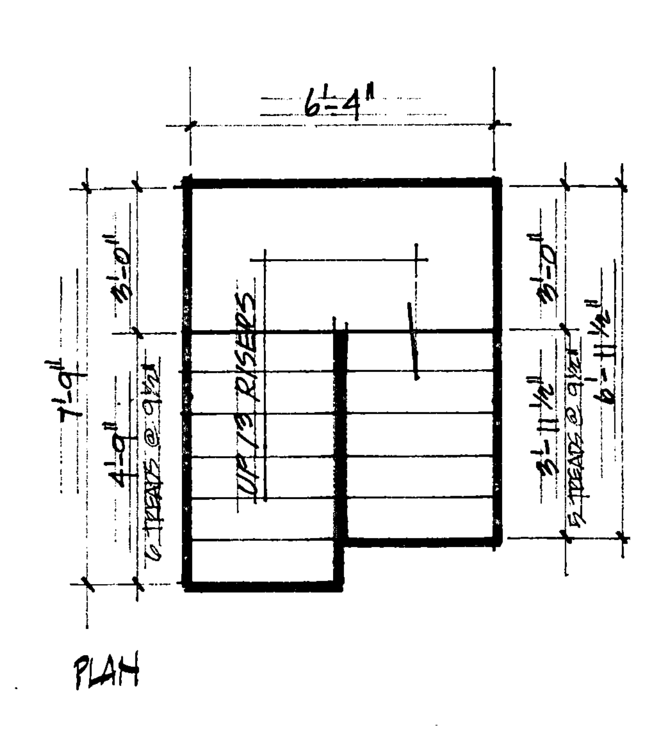

PLAN

CODE AND COMFORT DETERMINE SIZE.

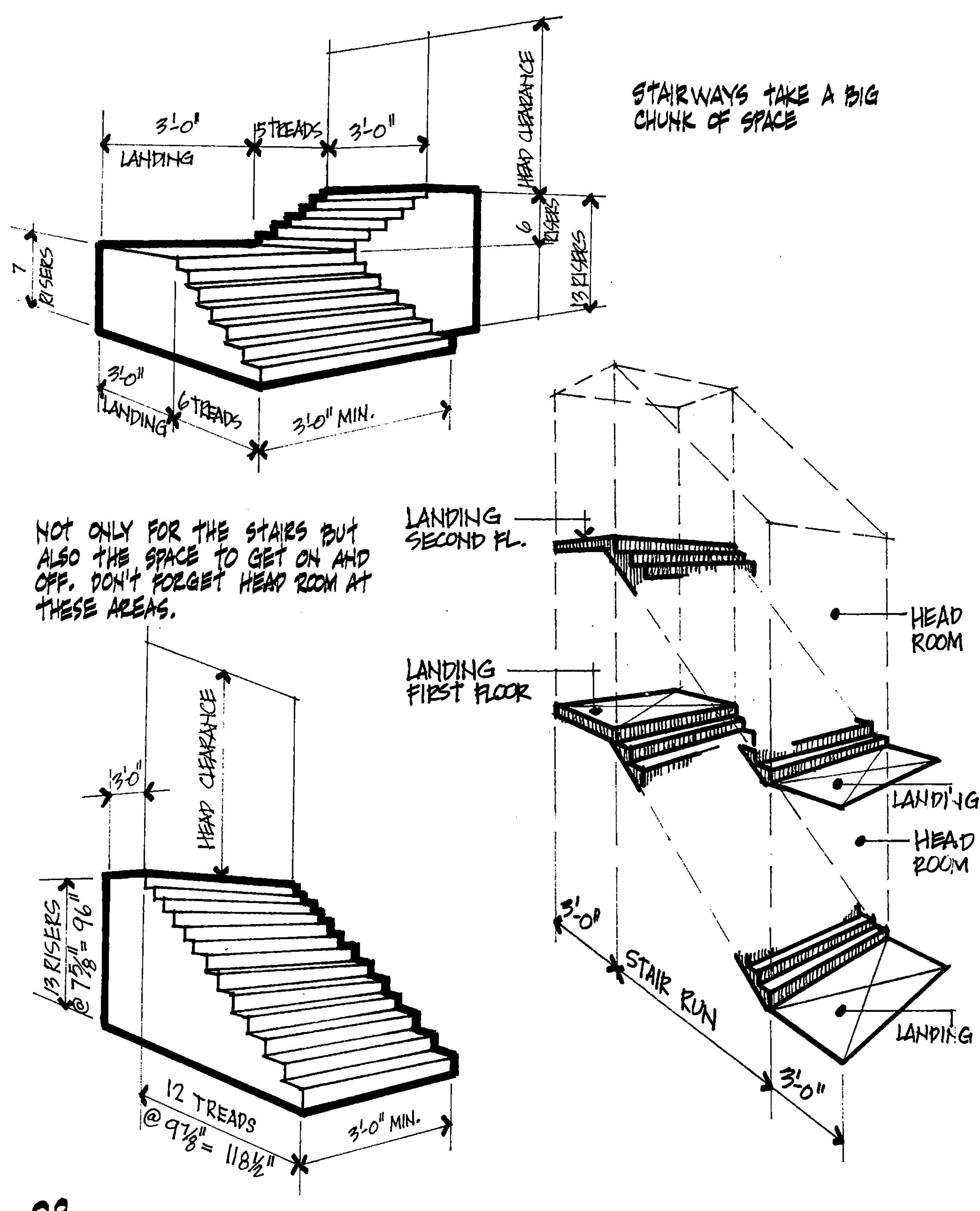
STAIRWAYS TAKE A BIG CHUNK OF SPACE
3'-0"
LANDING
5 TREADS
3'-0"
HEAD CLEARANCE
6 RISERS
13 RISERS
7 RISERS
3'-0"
LANDING
6 TREADS
3'-0" MIN.
NOT ONLY FOR THE STAIRS BUT ALSO THE SPACE TO GET ON AND OFF. DON'T FORGET HEAD ROOM AT THESE AREAS.
LANDING SECOND FL.
LANDING FIRST FLOOR
HEAD ROOM
LANDING
HEAD ROOM
3'-0"
STAIR RUN
LANDING
3'-0"
3'-0"
HEAD CLEARANCE
13 RISERS @ 7 5/8" = 96"
12 TREADS @ 9 7/8" = 118 1/2"
3'-0" MIN.

BATHROOMS HAVE SPACES TO CONSIDER- TUB, SINK, SHOWER, JOHN.

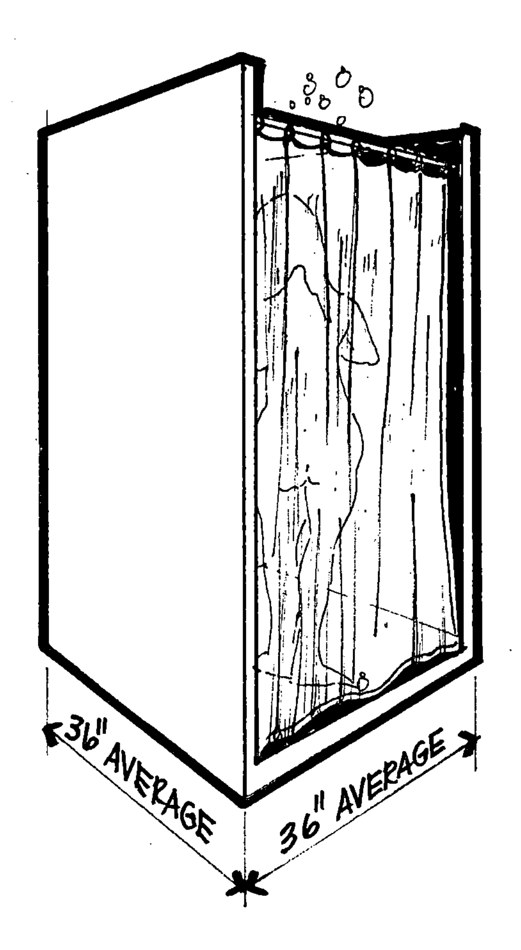

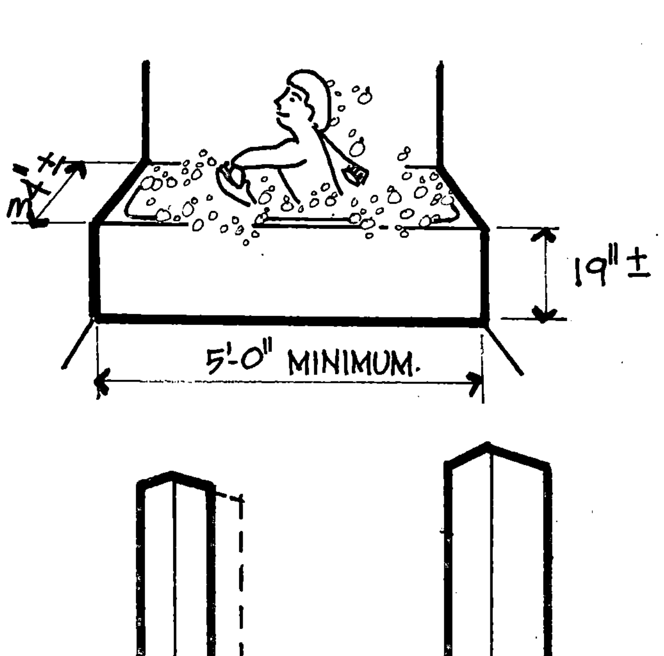

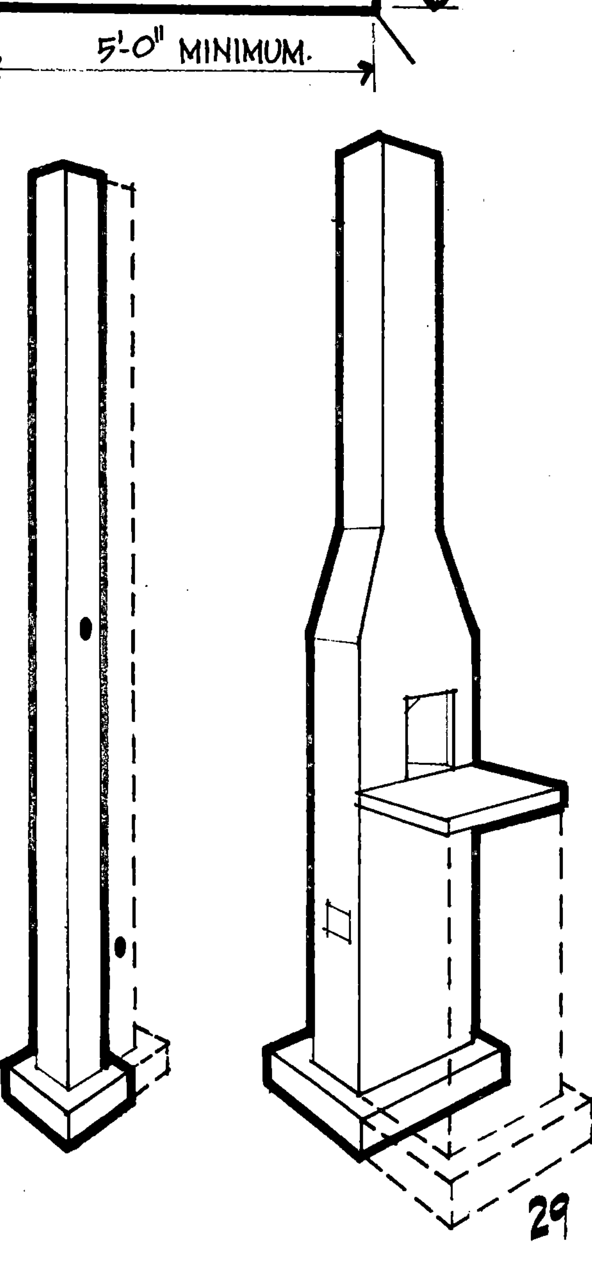

THE FIREPLACE IS ANOTHER BIG CHUNK OF SPACE, AND VERY CENTRAL TO THE THEME OF THE INTERIOR. WHEN YOU SEE A GOOD ONE USE IT IN YOUR DESIGN, BUT BE SURE IT WILL LOOK GOOD IN THE SIZE ROOM YOU PLAN. A BIG FIREPLACE IN A SMALL ROOM IS OVERPOWERING AND A SMALL ONE IN A BIG ROOM IS LOST. REMEMBER THE BRICK WORK RUNS FROM BASEMENT TO ROOF AND THE MORE FLUES IT HAS THE BIGGER THE CHIMNEY. EVEN A WOOD STOVE CHIMNEY WILL TAKE A LOT OF SPACE FROM BASEMENT TO ROOF.

DESIGN

THE KITCHEN HAS THE REFRIGERATOR, RANGE, SINK AND COUNTER SPACE AS MINIMUM REQUIREMENTS. THEN THERE COULD BE A WALL OVEN, DISH WASHER, BROOM CLOSET. AND TABLE OR COUNTER. TAKE NOTES AS YOU SEE LAYOUTS THAT ARE GOOD AND BAD FOR YOU. YOU WON'T FORGET IF YOU WRITE IT DOWN.

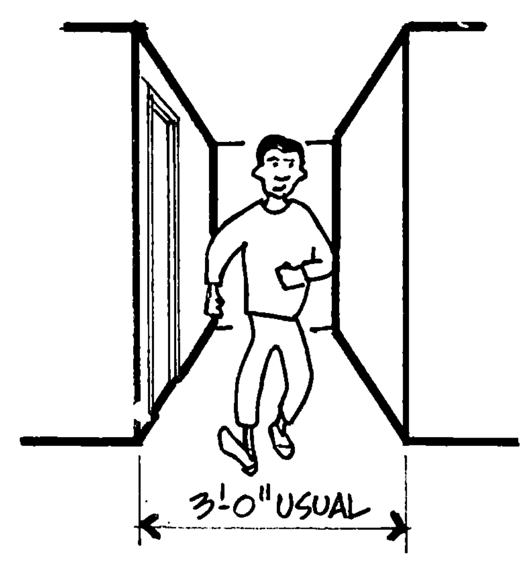

HALLS TAKE A LOT OF SPACE. THEY SHOULD BE 3'-0" FOR COMFORT, BUT CAN BE NARROWER IF THE LOCAL CODE WILL ALLOW.

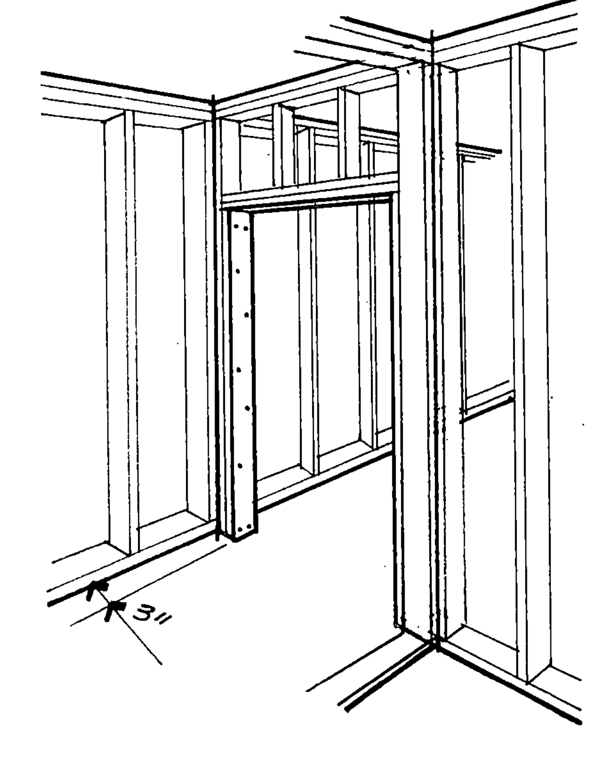

A NARROW HALL MAKES FOR A NARROW DOOR AT THE END BECAUSE OF THE FRAMING REQUIRED FOR THE DOOR. MAKING THE TRIM THINNER WILL REQUIRE LESS FRAMING MATERIAL AND THEN A WIDER DOOR.

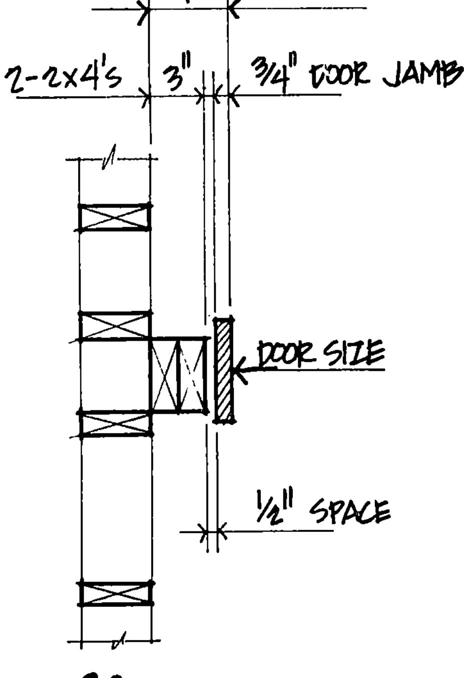

A CLOTHING CLOSET SHOULD BE 2'-0" CLEAR FOR HANGING GARMENTS....

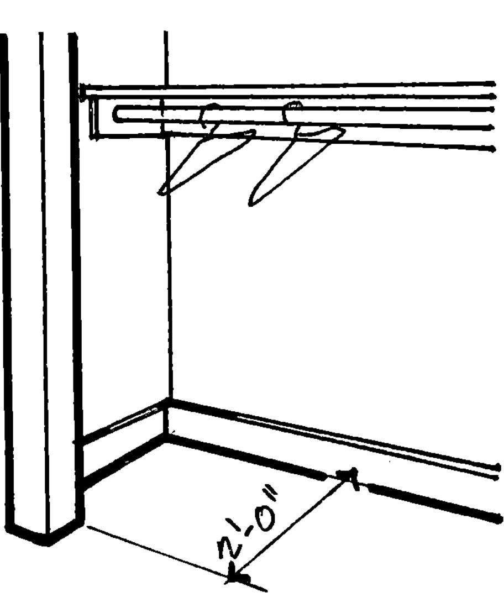

OTHER, AND VERY IMPORTANT, CONSIDERATIONS IN DESIGN ARE: GROUND SLOPE, SUN, AND TREES. WITH THE COST OF FUEL THE ENERGY EFFICIENT HOUSE LOOKS MIGHTY GOOD. WE CAN, AT LEAST, TAKE ADVANTAGE OF SUN, SHADE AND WIND BUFFERS.

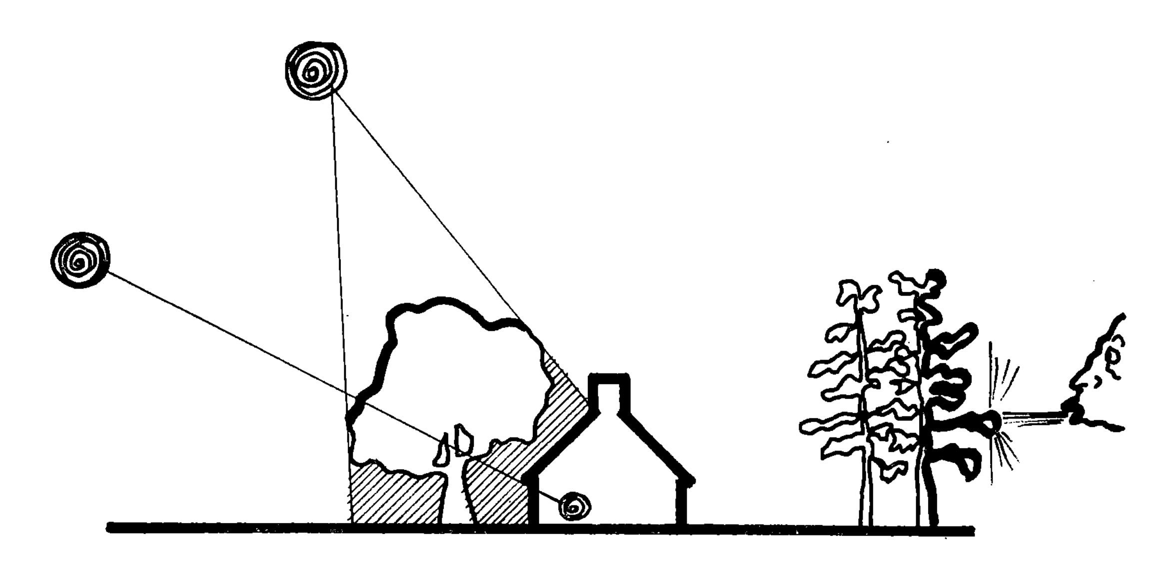

IDEAL CONDITIONS

1. SUMMER SUN SHADED BY LEAFY TREE.
2. WINTER SUN IN HOUSE WHEN LEAVES FALL.
3. NORTH WINTER WINDS BLOCKED BY PINE TREES.

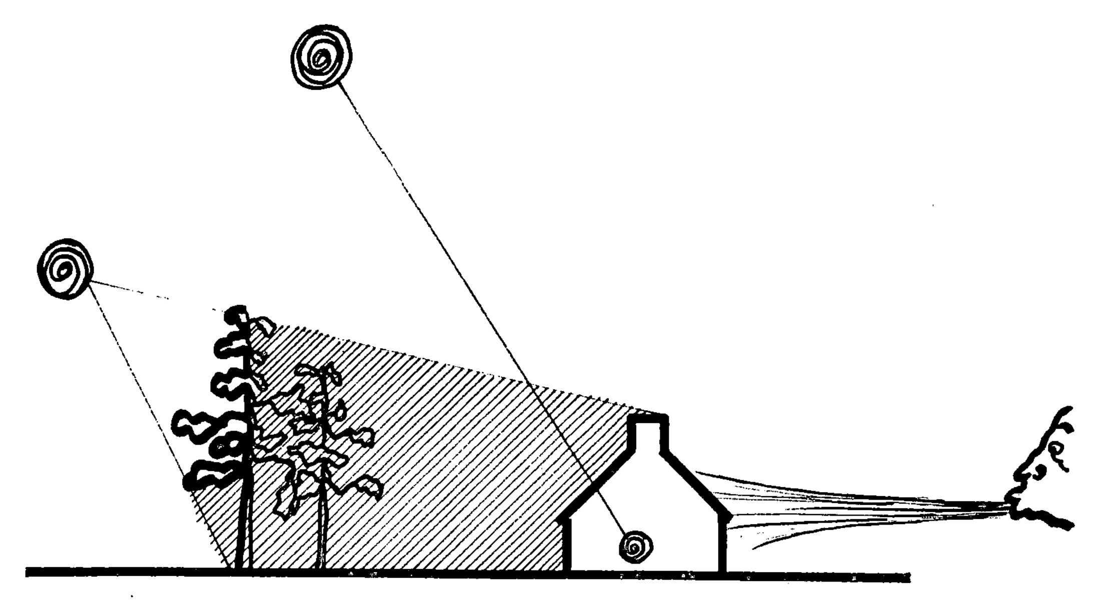

BAD CONDITIONS

1. SUMMER SUN COOKS HOUSE.
2. WINTER SUN BLOCKED BY PINE TREES.
3. NORTH WINTER WINDS CHILL HOUSE.

DESIGN

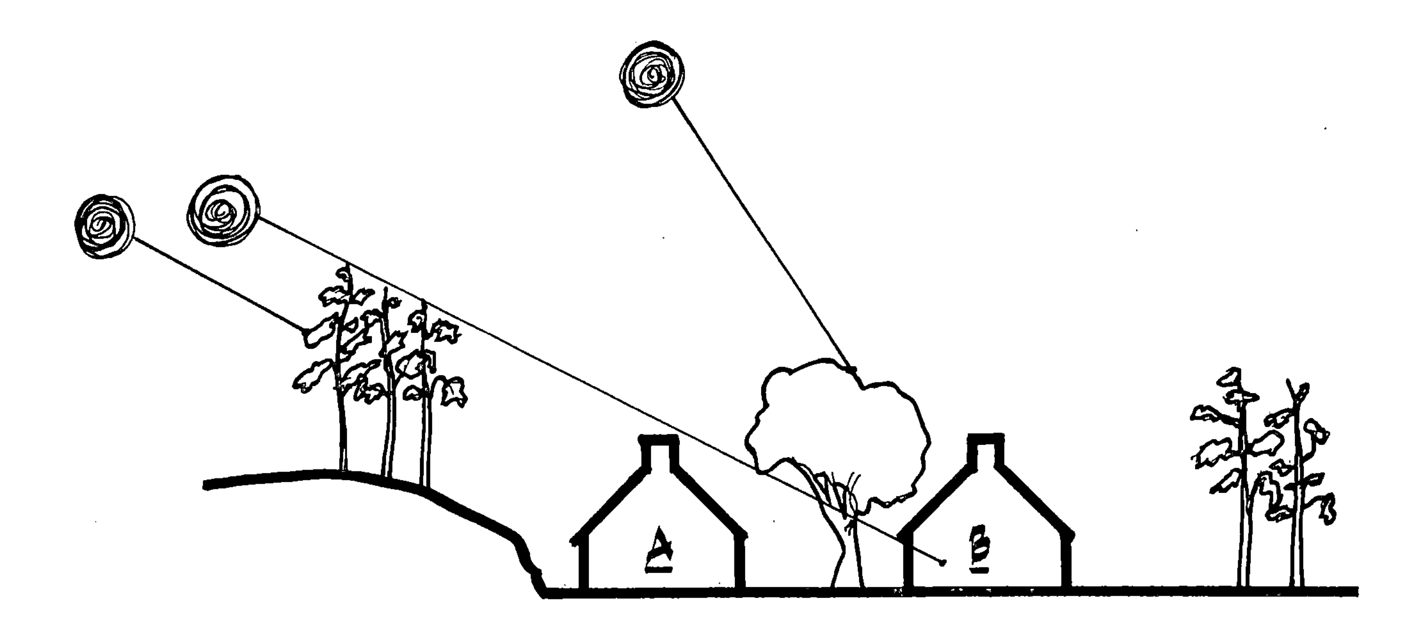

<u>HOUSE "B"</u> LOCATED BETTER THAN HOUSE "A".

<u>HOUSE "A"</u>
1. WINTER SUN BLOCKED BY PINES
2. WATER RUN OFF FROM HILL COULD CAUSE DAMP BASEMENT
3. SUMMER SUN COOKS HOUSE.

<u>HOUSE "B"</u>
1. SUMMER SUN BLOCKED BY LEAFY TREE.
2. WINTER SUN IN HOUSE WHEN LEAVES FALL.
3. NORTH WINTER WINDS BLOCKED BY PINE TREES. BOTH HOUSES BENEFIT BY THIS.

IF THE LAND SLOPES ENOUGH TO THE SOUTH AND THE WINTER SUN IS NOT BLOCKED, THEN AN UNDERGROUND HOUSE SHOULD BE CONSIDERED. TRY TO KEEP SOME LEAFY SHADE TREES TO BLOCK THE SUMMER SUN.

A GOOD QUICK SURVEY METHOD FOR FINDING THE SLOPE OF THE GROUND IS TO USE A SIMPLE CARPENTER'S LEVEL, A FEW WOOD STAKES AND A GOOD EYEBALL.

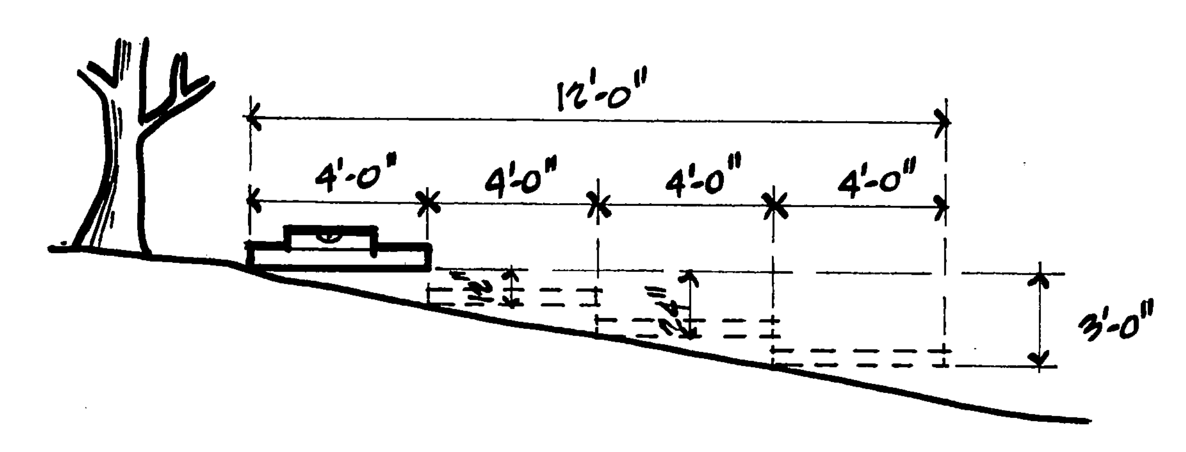

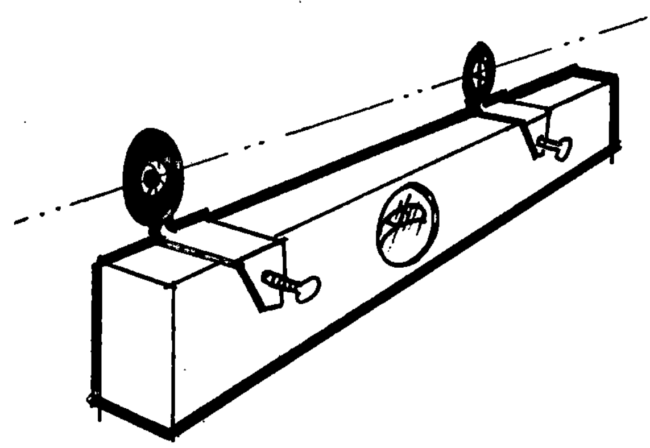

STANLEY TOOLS MAKES AN INEXPENSIVE SET OF LEVEL SIGHTS THAT CLAMP ONTO A WOOD CARPENTERS LEVEL. GOOD ACCURACY CAN BE ACHIEVED WITH EITHER SYSTEM, CERTAINLY GOOD ENOUGH TO DESIGN WITH, IF YOU CAN BORROW A FRIEND'S BUILDERS LEVEL (TRANSIT), SO MUCH THE BETTER, BUT NOT NECESSARY.

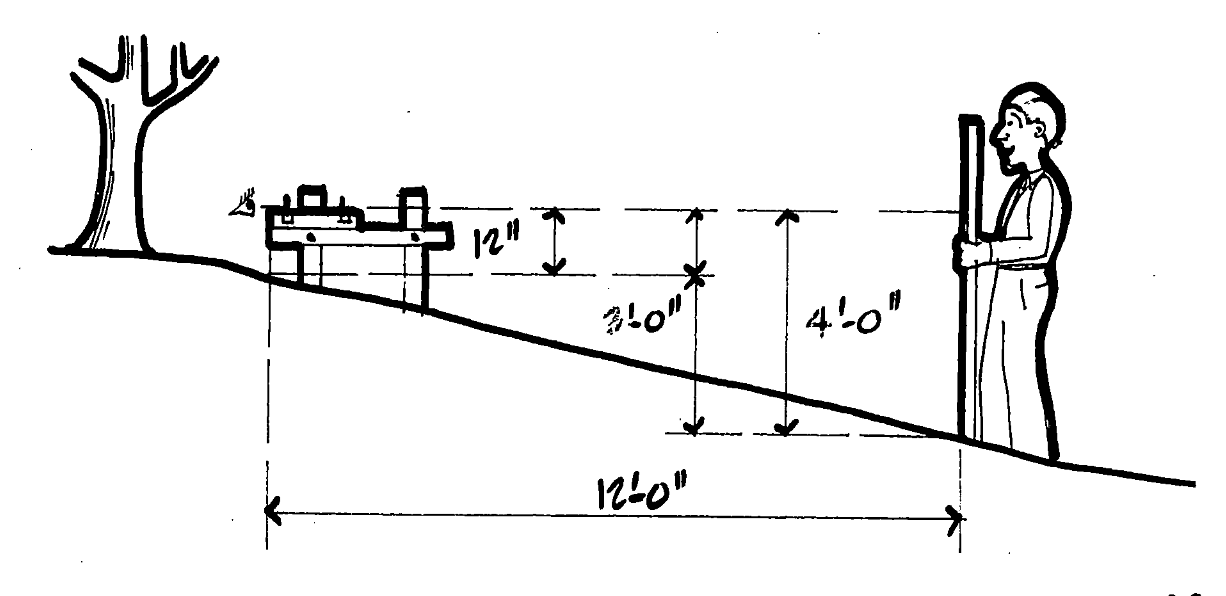

DESIGN

AS FOR STRUCTURAL DESIGN, CHECK THE HOUSES YOU VISIT. GO DOWN TO THE BASEMENT AND PACE OFF THE SPANS OF THE FLOOR JOISTS AND BEAMS.

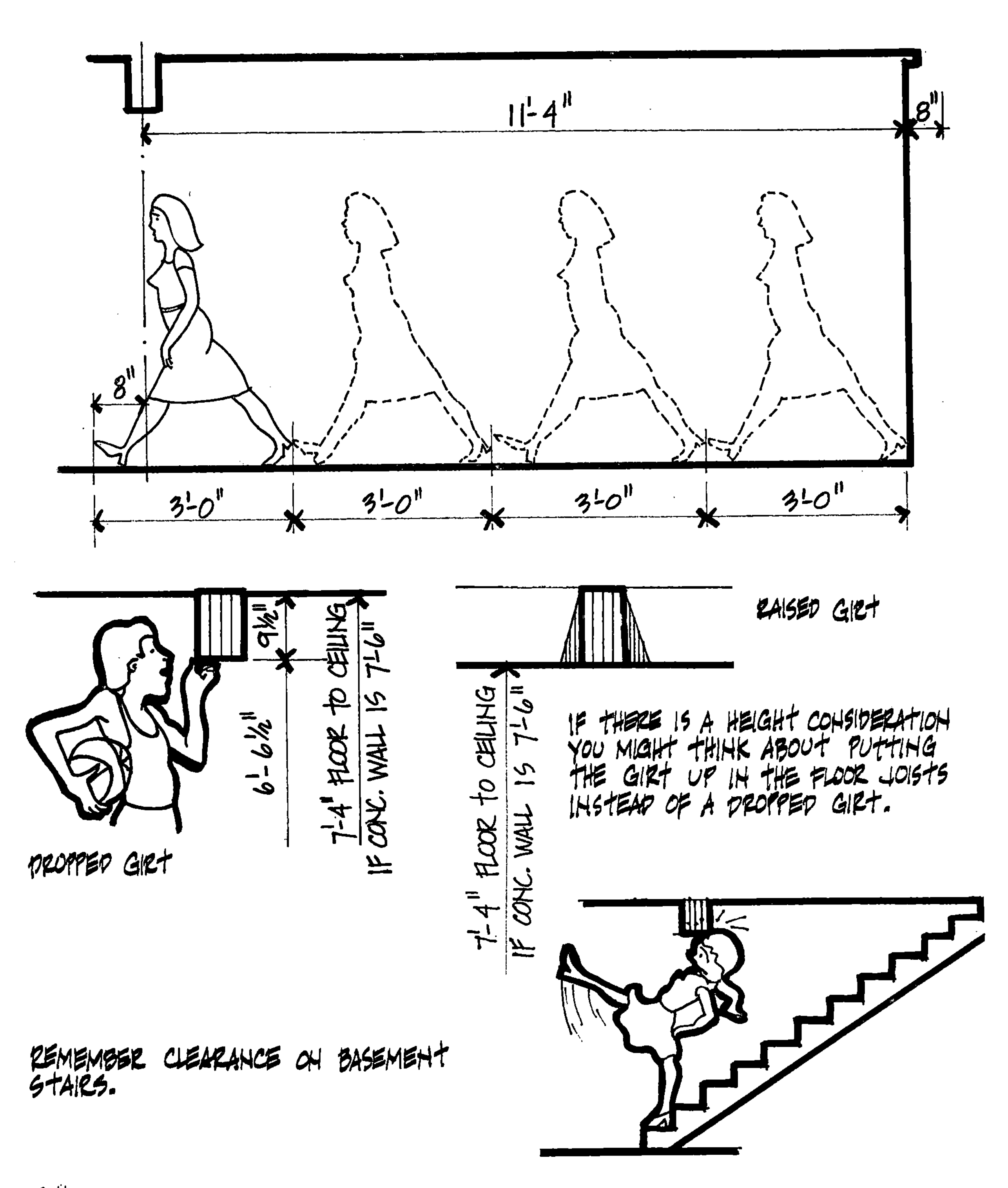

COUNT THE JOISTS FOR SPACINGS. LOOK IN THE ATTIC FOR RAFTER SIZES AND SPACING. CHECK A SAGGING ROOF TOO; IT TELLS YOU WHAT NOT TO DO. IF THE DISHES RATTLE WHEN YOU WALK ACROSS THE FLOOR, THAT TELLS YOU SOMETHING TOO. CHECK NEW CONSTRUCTION.

USE THE CODE BOOKS; THEY WILL TELL YOU THE SIZE AND SPANS FOR THE VARIOUS FRAMING CONDITIONS.

TABLE 1

MAXIMUM SPANS FOR GIRDERS

SIZE	1 STORY
4x6	6'-0"

TABLE 2

MAXIMUM SPANS FOR FLOOR JOISTS

SIZE	NO 1.	NO2.
2x6	9'-1"	8'-6"

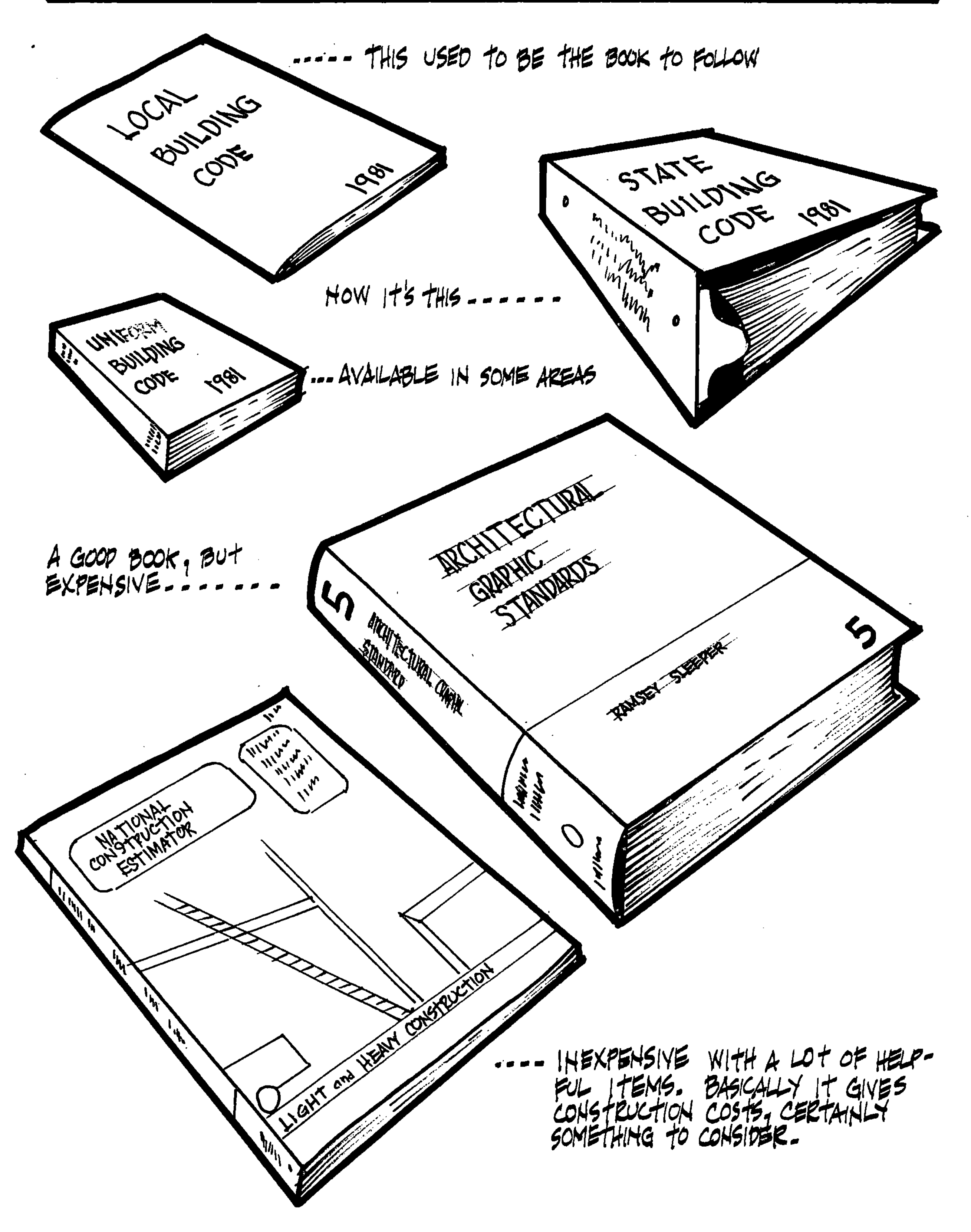
THIS USED TO BE THE BOOK TO FOLLOW
LOCAL BUILDING CODE 1981
STATE BUILDING CODE 1981
NOW IT'S THIS
UNIFORM BUILDING CODE 1981
AVAILABLE IN SOME AREAS
A GOOD BOOK, BUT EXPENSIVE
ARCHITECTURAL GRAPHIC STANDARDS
5
RAMSEY SLEEPER
NATIONAL CONSTRUCTION ESTIMATOR
LIGHT and HEAVY CONSTRUCTION
INEXPENSIVE WITH A LOT OF HELPFUL ITEMS. BASICALLY IT GIVES CONSTRUCTION COSTS. CERTAINLY SOMETHING TO CONSIDER.

ANY LARGE DEPARTMENT STORE CATALOG HAS A WEALTH OF INFORMATION. ANY ITEM IN HOUSE, GARAGE, GARDEN OR SHOP IS COVERED BY SIZE, SHAPE, COLOR AND WEIGHT.

LOCAL LUMBER YARDS SOMETIMES HAVE FREE OR INEXPENSIVE DESIGN CATALOGS.

CHECK WITH THE BUILDING INSPECTOR; HE IS THERE TO HELP YOU. DON'T TRY TO SLIP ANYTHING BY HIM—HE OR SHE HAS WAYS OF MAKING UP FOR IT TEN TIMES OVER. ONE OTHER CAUTION: CHECK WITH BUILDERS OR CARPENTERS ABOUT THE INSPECTOR'S ABILITY TO DEAL WITH THE OPPOSITE SEX BE IT MALE OR FEMALE. I HAVE RUN ACROSS SOME MALE INSPECTORS THAT GIVE WOMEN A HARD TIME. DON'T FIGHT HIM, TRY USING A MALE FRIEND TO DEAL THROUGH. OF COURSE FOLLOW THE CODES.

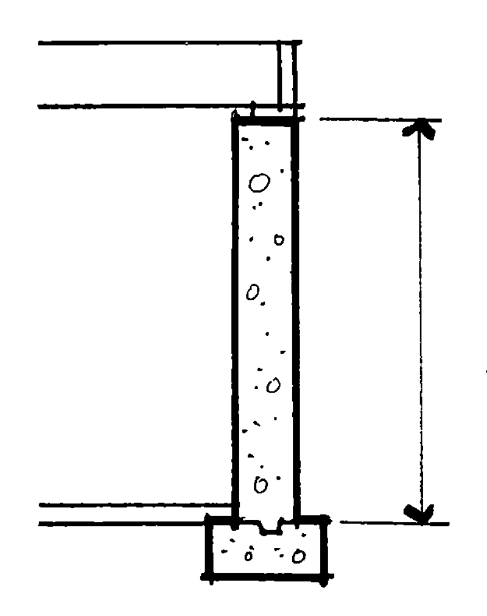

FOUNDATION WALL HEIGHTS ARE PRETTY MUCH CONTROLLED BY THE LOCAL CONCRETE CONTRACTOR. HIS FORMS ARE JUST SO HIGH, LIMITING THE HEIGHT OF THE CONCRETE POUR. IN MY AREA, 7'-6" IS THE COMMON POUR, BUT ANY HEIGHT CAN BE POURED IF THE FORMS ARE MODIFIED. IT ALWAYS COSTS MORE IF YOU STRAY FROM THE NORMAL CONDITIONS.

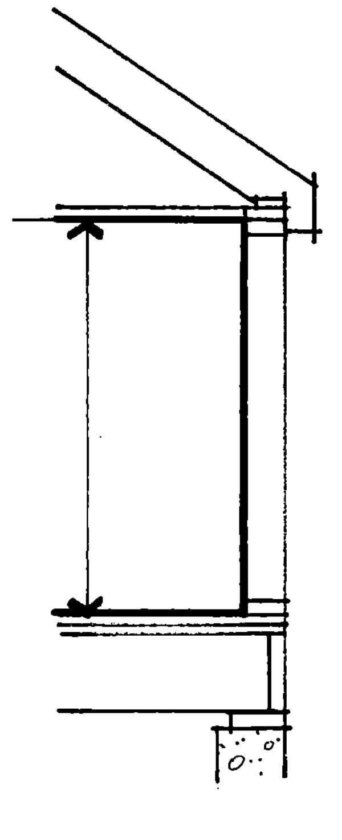

FIRST AND SECOND FLOOR MINIMUM CEILING HEIGHTS ARE SOMETIMES SET BY CODE, AND IF NOT, USE YOUR OWN FEELINGS. TRY TO KEEP STUD LENGTHS UNDER 8' FOR ECONOMY.

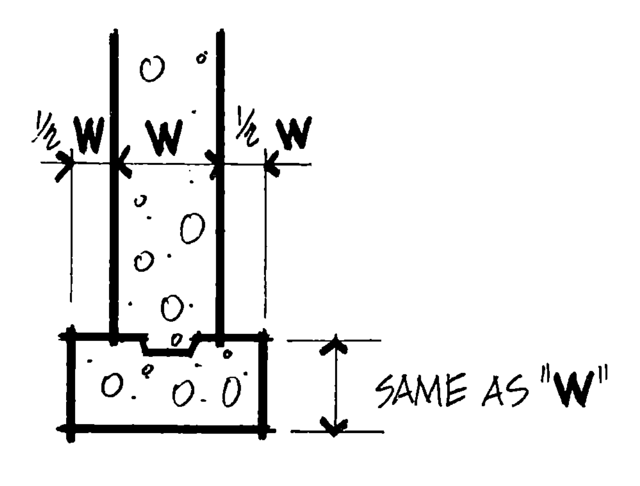

CONCRETE WALL THICKNESS IS GENERALLY SET BY CODE, AND THE FOOTINGS ARE TAKEN OFF THAT DIMENSION.

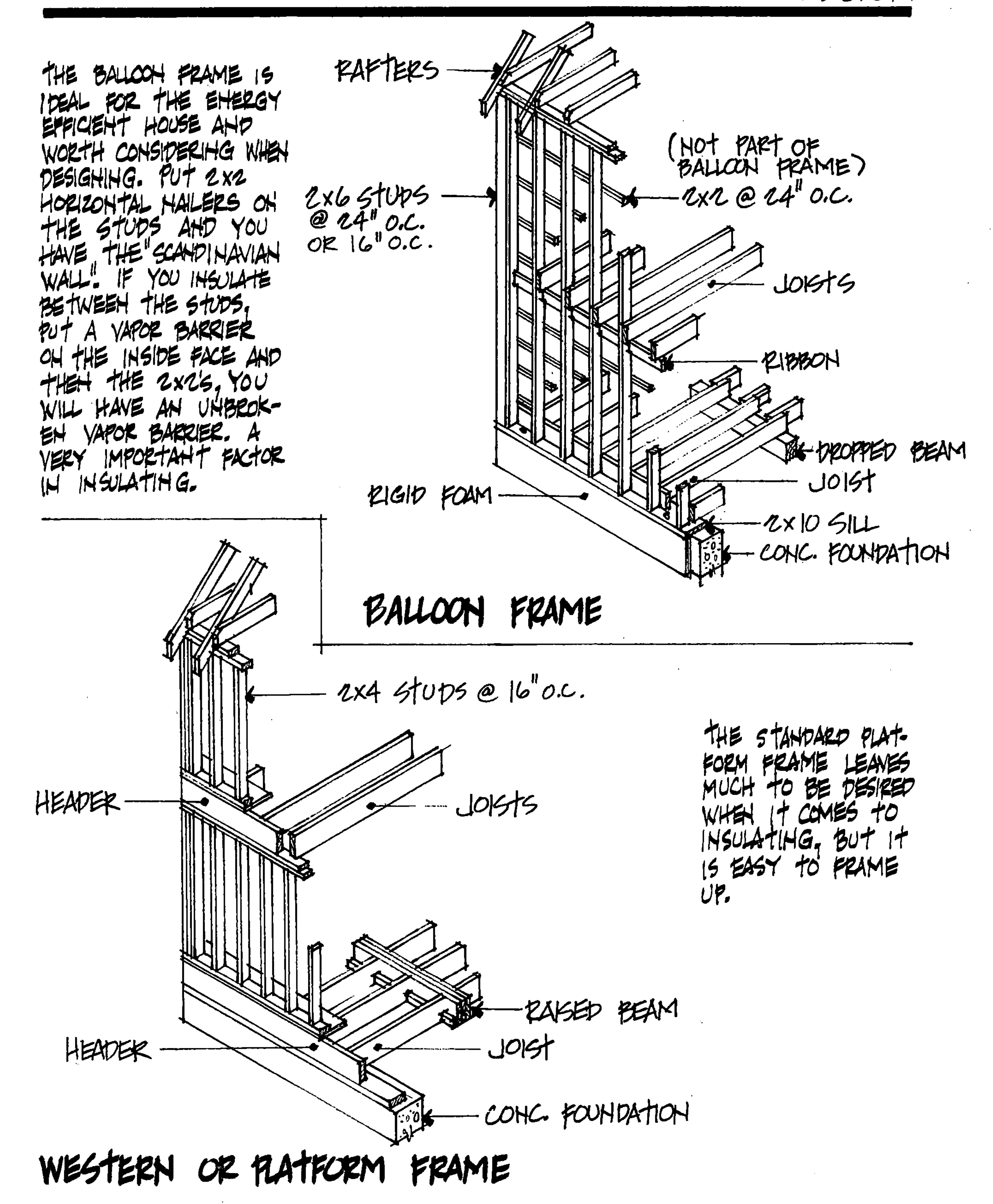

THE BALLOON FRAME IS IDEAL FOR THE ENERGY EFFICIENT HOUSE AND WORTH CONSIDERING WHEN DESIGNING. PUT 2x2 HORIZONTAL NAILERS ON THE STUDS AND YOU HAVE THE "SCANDINAVIAN WALL". IF YOU INSULATE BETWEEN THE STUDS, PUT A VAPOR BARRIER ON THE INSIDE FACE AND THEN THE 2x2'S, YOU WILL HAVE AN UNBROKEN VAPOR BARRIER. A VERY IMPORTANT FACTOR IN INSULATING.

BALLOON FRAME

THE STANDARD PLATFORM FRAME LEAVES MUCH TO BE DESIRED WHEN IT COMES TO INSULATING, BUT IT IS EASY TO FRAME UP.

WESTERN OR PLATFORM FRAME

DESIGN

ONCE THE BASIC HOUSE IS DECIDED ON MENTALLY, IT IS TIME TO SKETCH. USE A FELT TIP PEN ON YELLOW SKETCH PAPER (OR ANY CHEAP PAPER). YOU MIGHT WANT TO PUT A PIECE OF GRAPH PAPER UNDER TO HELP WITH THE HORIZONTAL AND VERTICAL LINE WORK. JUST LET YOUR INNER WISDOM GO TO TOWN AND CHECK YOUR NOTES. USE FLAT PAPER FURNITURE CUT OUTS, THE SAME SCALE AS THE FLOOR PLANS FOR ROOM DESIGNING.

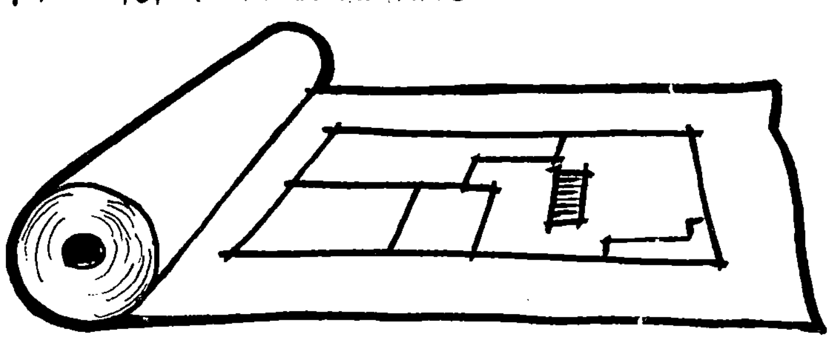

KEEP AN EMPTY WASTE BASKET WITHIN "SHOOTING" RANGE.

WHEN YOU MAKE A GOOD SKETCH, BUT IT NEEDS A CHANGE, JUST LAY THE YELLOW TRACE OVER AND MAKE A NEW SKETCH WITH THE CHANGE. IT IS A FAST WAY TO WORK AND YOU WILL FIND YOUR MIND WORKING FAST TOO.

THIS IS DONE WITH THE EXTERIOR AND INTERIOR DESIGN.

TO FIRM IT UP, DRAW IT FREE HAND ON GRAPH PAPER. THIS WILL GIVE YOU AN IDEA OF SIZES.

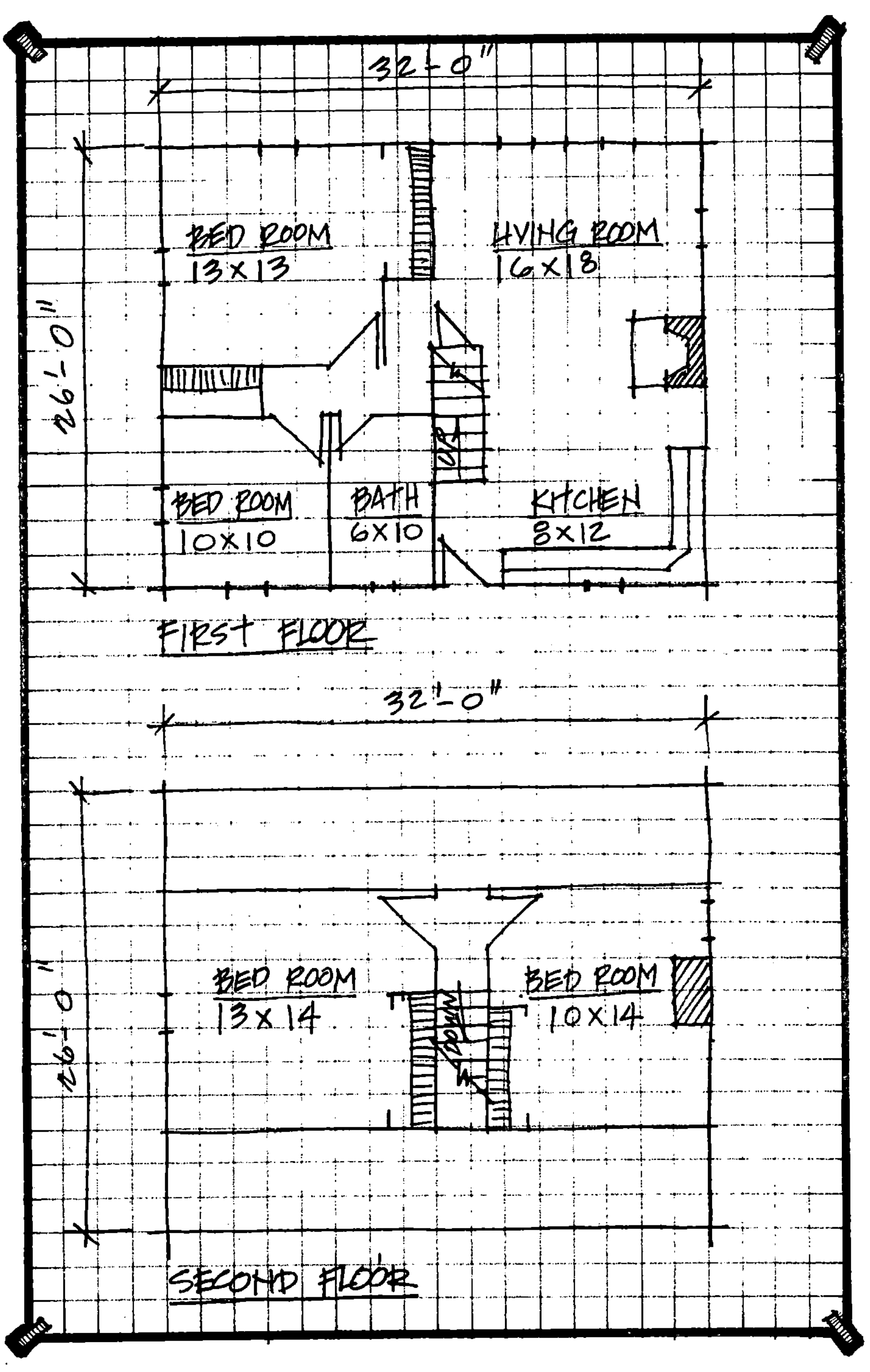

ONE SQUARE = 2 FEET

DO THE SAME WITH THE EXTERIOR.

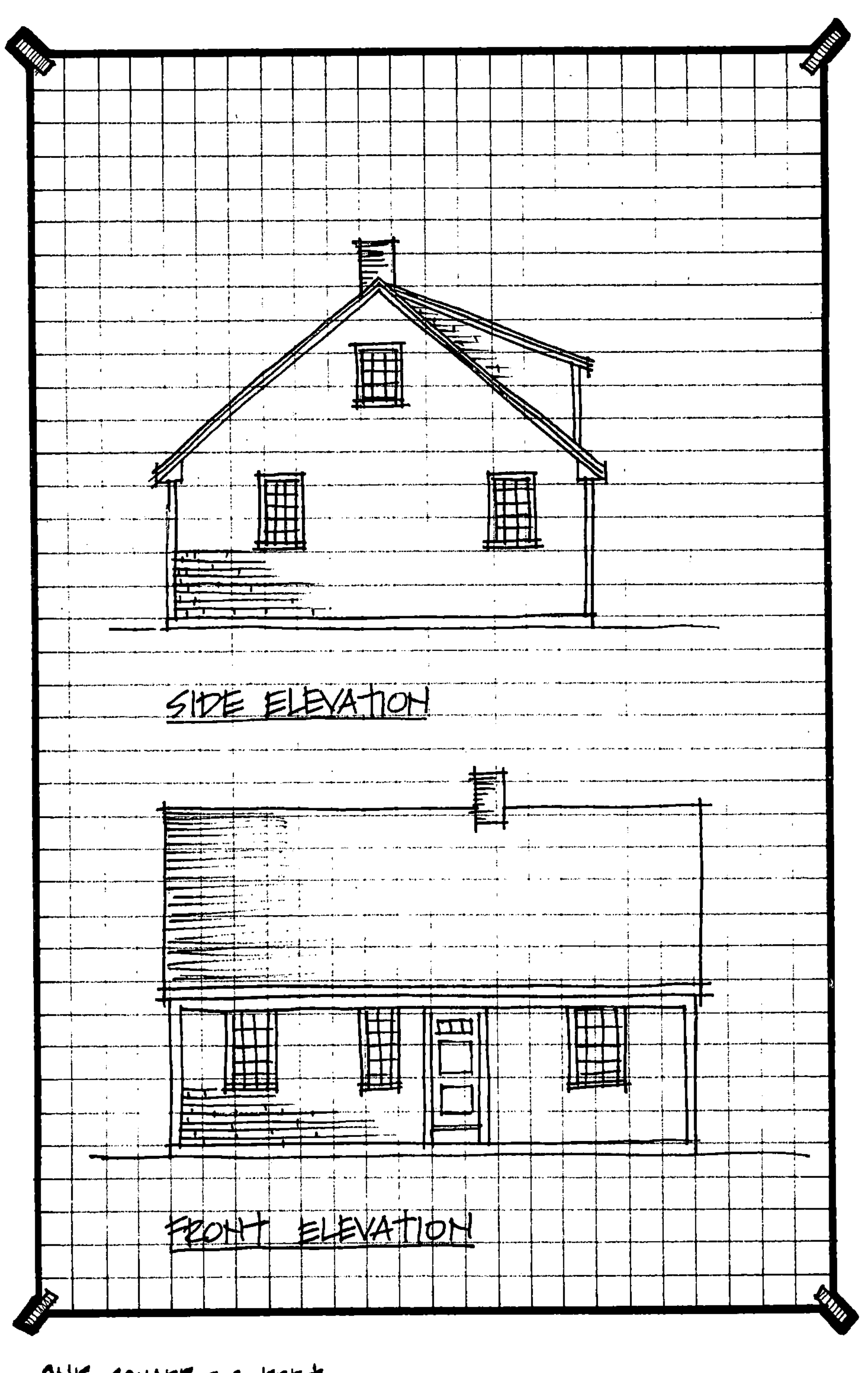

ONE SQUARE = 2 FEET

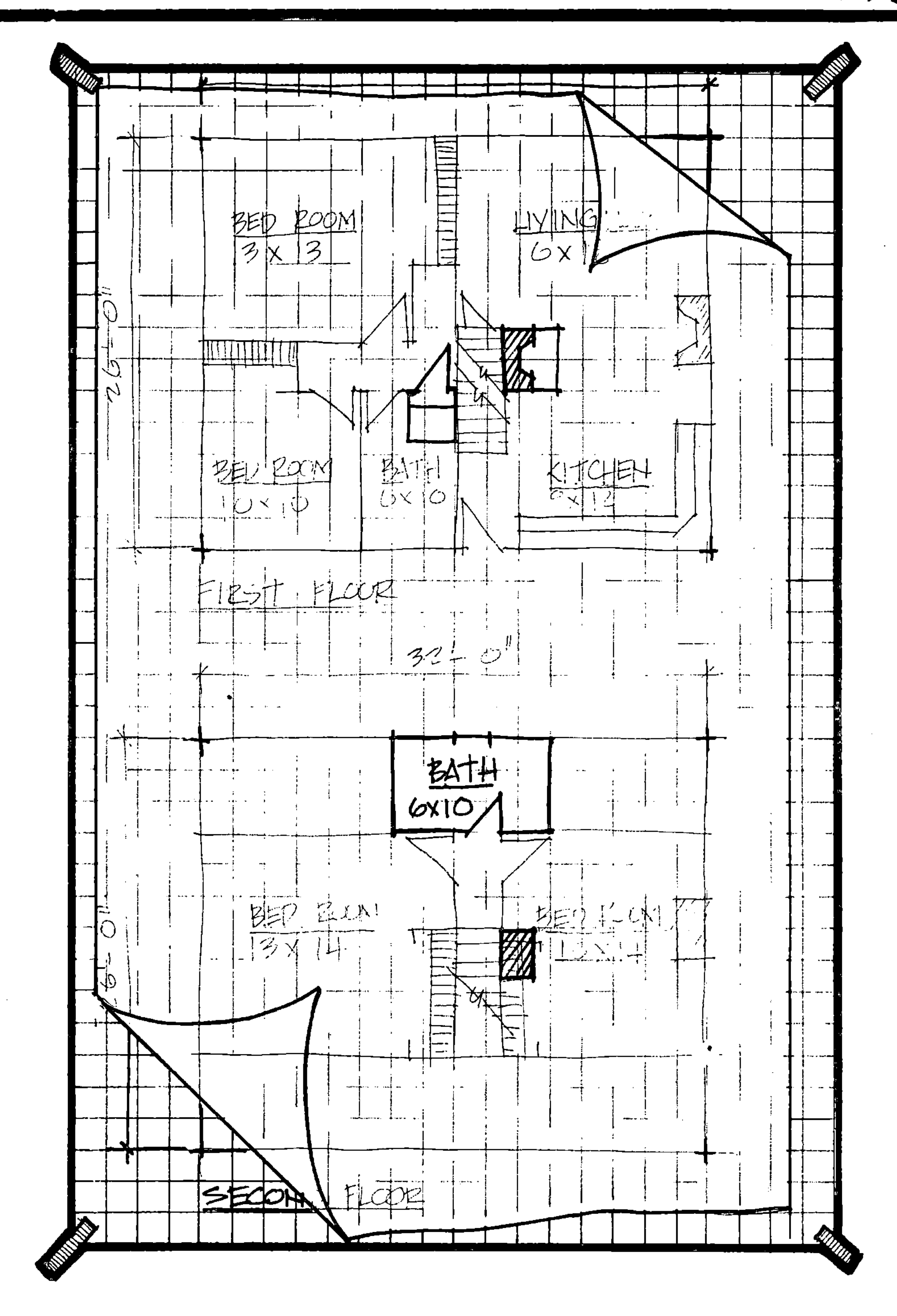

YOU MIGHT WANT TO MAKE MORE CHANGES. USE THE YELLOW TRACING PAPER. YOU CAN GO FROM THE YELLOW TRACING OR BACK TO THE GRAPH PAPER TO START THE WORKING DRAWINGS. REMEMBER WHEN USING LINE SKETCHES THE WALL THICKNESS IS NOT SHOWN; KEEP THAT IN MIND WHEN DIMENSIONING.

YOU NEVER KNOW WHEN YOUR TAPE MEASURE WILL COME IN HANDY.

PLUMBING VENTS LOOK BAD ON THE ENTRANCE SIDE OF THE ROOF.

THINK OF THE FUTURE.

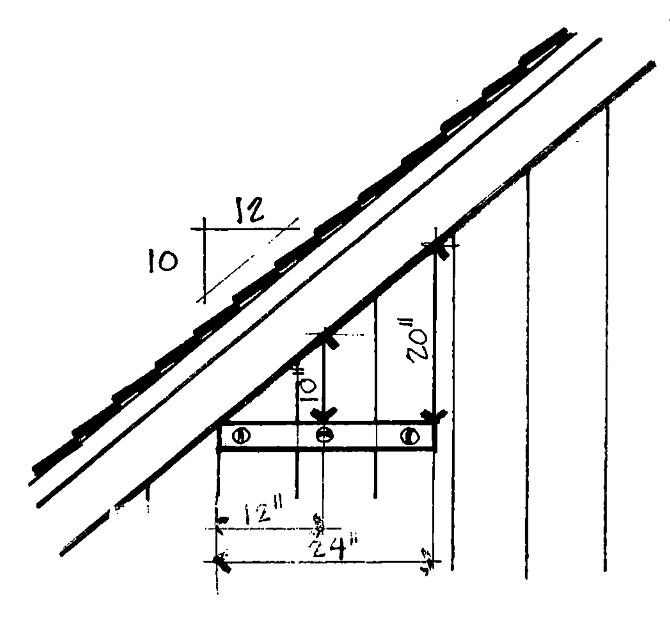

USE A CARPENTER'S LEVEL AND A TAPE MEASURE TO FIND THE ROOF PITCH.

3 - DRAFTING

YOU DON'T HAVE TO BE A TERRIFIC ARTIST TO BE A GOOD DRAFTSMAN. WHAT IT DOES TAKE IS OBSERVATION AND PRACTICE. TRY TO LEARN WHAT MAKES FOR A GOOD-LOOKING JOB AND THEN PRACTICE THE TECHNIQUES REQUIRED. THE PLANS DON'T HAVE TO BE A FANTASTIC WORK; AS LONG AS THE BUILDER KNOWS WHATS GOING ON THEY WILL BE FINE. WHAT CAN BE DIFFICULT ABOUT RUNNING A PENCIL ALONG A STRAIGHT EDGE OR MEASURING WITH A SCALE? WITH THE INEXPENSIVE CALCULATORS, MATHAMATICS BECOMES FUN (ALMOST). THE ONLY DIFFICULT THING IS THE LETTERING AND IT IS ONLY DIFFICULT (AT FIRST) TO MAKE IT STYLISH, BUT NOT DIFFICULT TO MAKE IT LEGIBLE.

DRAFTING

THE FIRST THINGS TO DECIDE ARE WHAT TO DRAW ON AND WHAT SIZE THE SHEETS WILL BE. BECAUSE FLOOR PLANS ARE TRACED ONE FROM THE OTHER, TRACING PAPER WOULD BE A GOOD CHOICE. THE SHEET SIZE IS DETERMINED BY THE BUILDING SIZE AND THE PRINTING PROCESS. A 2" SPACE ON EACH SIDE AND TOP WITH 3" ON THE BOTTOM ADDED TO THE LENGTH AND WIDTH OF THE HOUSE WILL GIVE THE SIZE SHEET TO USE. THE 2" AND 3" DIMENSIONS ALLOW SPACE FOR DIMENSIONS AND TITLES. A HOUSE 26'X32', DRAWN AT 1/4" SCALE, WILL USE A SHEET 11 1/2"X12". IF THE PRINTS ARE TO BE XEROX, THEN THE MAXIMUM SIZE SHEET WILL BE 11"X17" (CHECK WITH YOUR LOCAL PRINTER). THE 1/2" LOST ON HEIGHT CAN BE FUDGED WITH THE 3" BOTTOM. THE EXTRA 5" CAN BE USED FOR DETAILS. TRY TO KEEP THE DETAILS THAT ARE RELATIVE TO THE PLAN ON THE SAME SHEET. IF THE PRINTS ARE TO BE THE OZALID PROCESS, THEN A MUCH LARGER SHEET CAN BE USED, BUT THE 2" AND 3" DIMENSIONS SHOULD BE USED IN FIGURING THE SHEET SIZE.

THE NUMBER OF SHEETS IS DETERMINED BY THE SHEET SIZE, AND BUILDING INSPECTOR'S REQUIREMENTS. A LARGE SHEET COULD HAVE ALL THE PLANS ON ONE, A SMALL SHEET REQUIRES MANY SHEETS.

A GOOD SET OF WORKING DRAWINGS WOULD BE:

FOUNDATION PLAN _ _ _ _ _ _ _ _ 1/4"=1'-0"
FLOOR PLANS _ _ _ _ _ _ _ _ _ _ 1/4"=1'-0"
ELEVATIONS _ _ _ _ _ _ _ _ _ _ 1/4"=1'-0" OR 1/8"=1'-0"
SECTION THROUGH BUILDING _ _ 3/8"=1'-0" (SHOWS CONSTRUCTION)
DETAILS _ _ _ _ _ _ _ _ _ _ _ _ 1 1/2"=1'-0" (BETTER UNDERSTANDING OF CONSTRUCTION)
CABINET ELEVATIONS _ _ _ _ _ 1/4"=1'-0"
FIREPLACE ELEVATION _ _ _ _ _ 1/16"=1'-0"
FLOOR FRAMING _ _ _ _ _ _ _ _ 1/16"=1'-0" OR 1/4"=1'-0"
ELECTRICAL PLAN _ _ _ _ _ _ _ 1/4"=1'-0" (CAN BE ON FLOOR PLAN)
PLOT PLAN _ _ _ _ _ _ _ _ _ _ _ 1"=20'

THE FIRST FOUR WOULD BE THE MINIMUM REQUIREMENTS.
ELEVATIONS CAN BE AT 1/8" SCALE TO CUT DOWN ON THE NUMBER OF SHEETS.

WHEN THE NUMBER OF SHEETS IS DECIDED, CUT THEM ALL AT THE SAME TIME OR USE PRECUT SHEETS.

DETERMINE SHEET SIZE AND CUT AS MANY AS NEEDED.

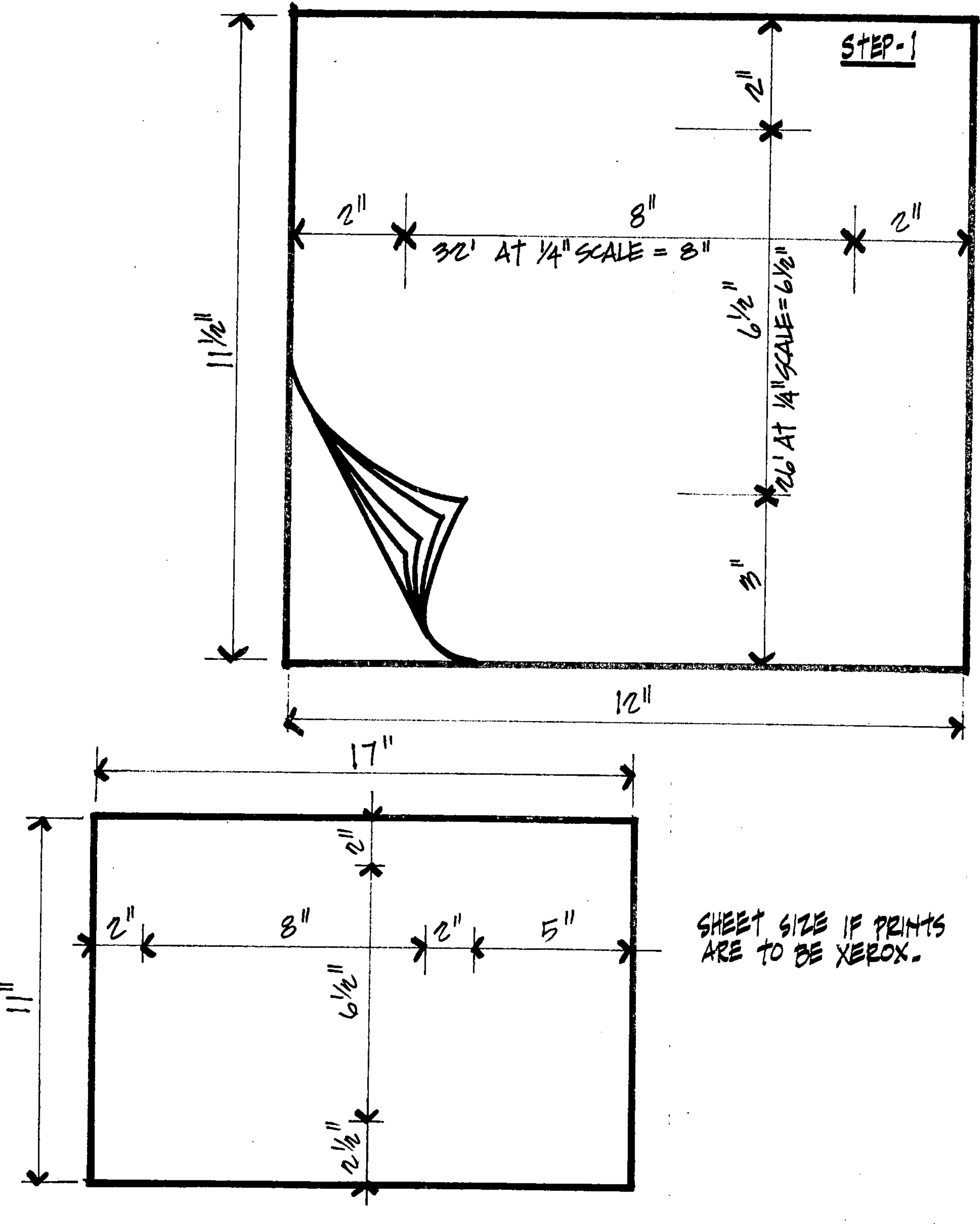

SHEET SIZE IF PRINTS ARE TO BE XEROX.

TAPE THE FIRST SHEET DOWN AND YOU ARE READY TO LAY OUT THE FIRST FLOOR PLAN. USE A 2H LEAD WITH A LIGHT TOUCH, JUST DARK ENOUGH TO SEE WITHOUT TOO MUCH TROUBLE.

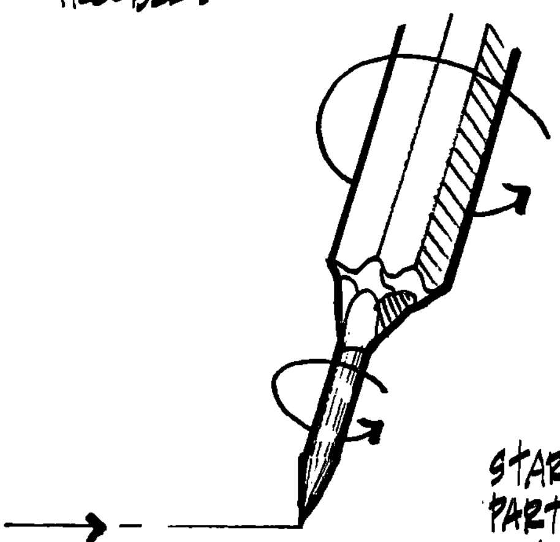

ALWAYS TWIRL THE PENCIL WITH THE FINGERS WHEN DRAWING HORIZONTAL AND VERTICAL LINES. THIS WILL KEEP THE WEAR EVEN ON THE POINT AND THE LINE WILL BE A CONSISTENT THICKNESS. IT TAKES A LITTLE GETTING USED TO, BUT IS A MUST.

START WITH THE PERIMETER OF THE HOUSE AND THEN THE PARTITIONS; DON'T WORRY ABOUT DOORS AND WINDOWS YET. DRAW THE EXTERIOR WALLS 5" (AT 1/4" SCALE CLOSE IS GOOD ENOUGH) IF STUDS ARE TO BE 2X4 AND 7" IF 2X6. THE INTERIOR PARTITIONS WILL BE 2X4 SO 4½" FOR THEM. THESE DIMENSIONS TAKE INTO ACCOUNT THE WALL MATERIAL, STUD, SIDING, SHEATHING, SHEETROCK.

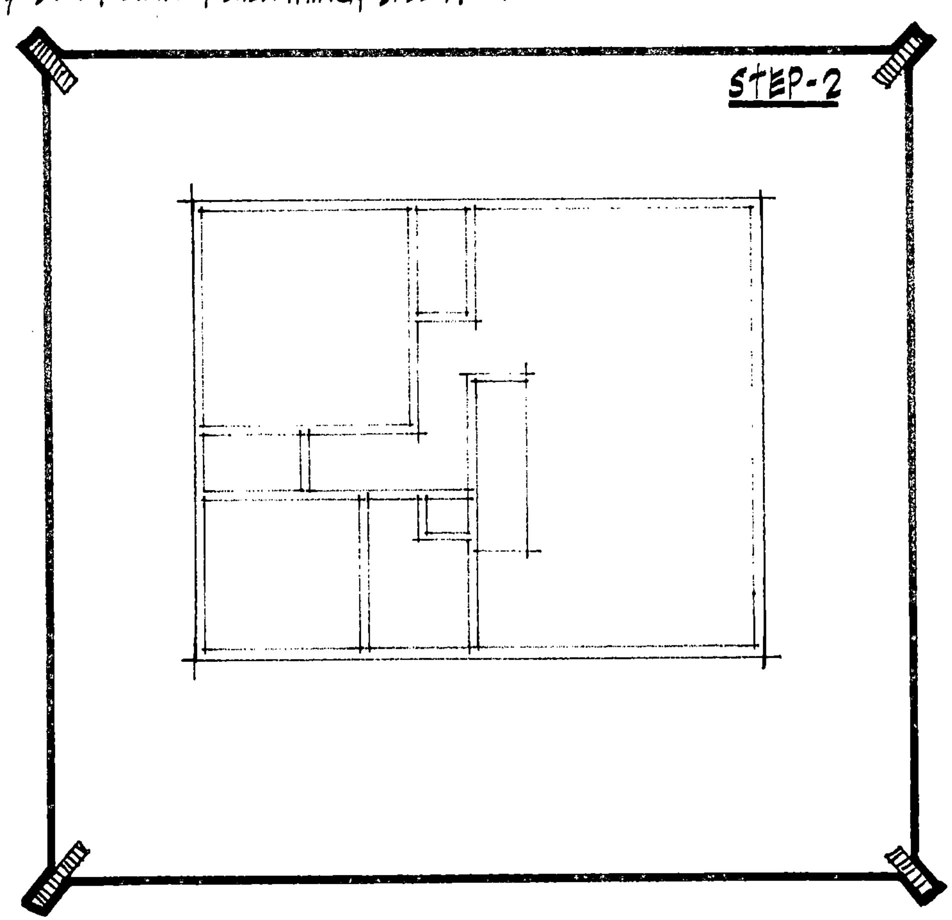

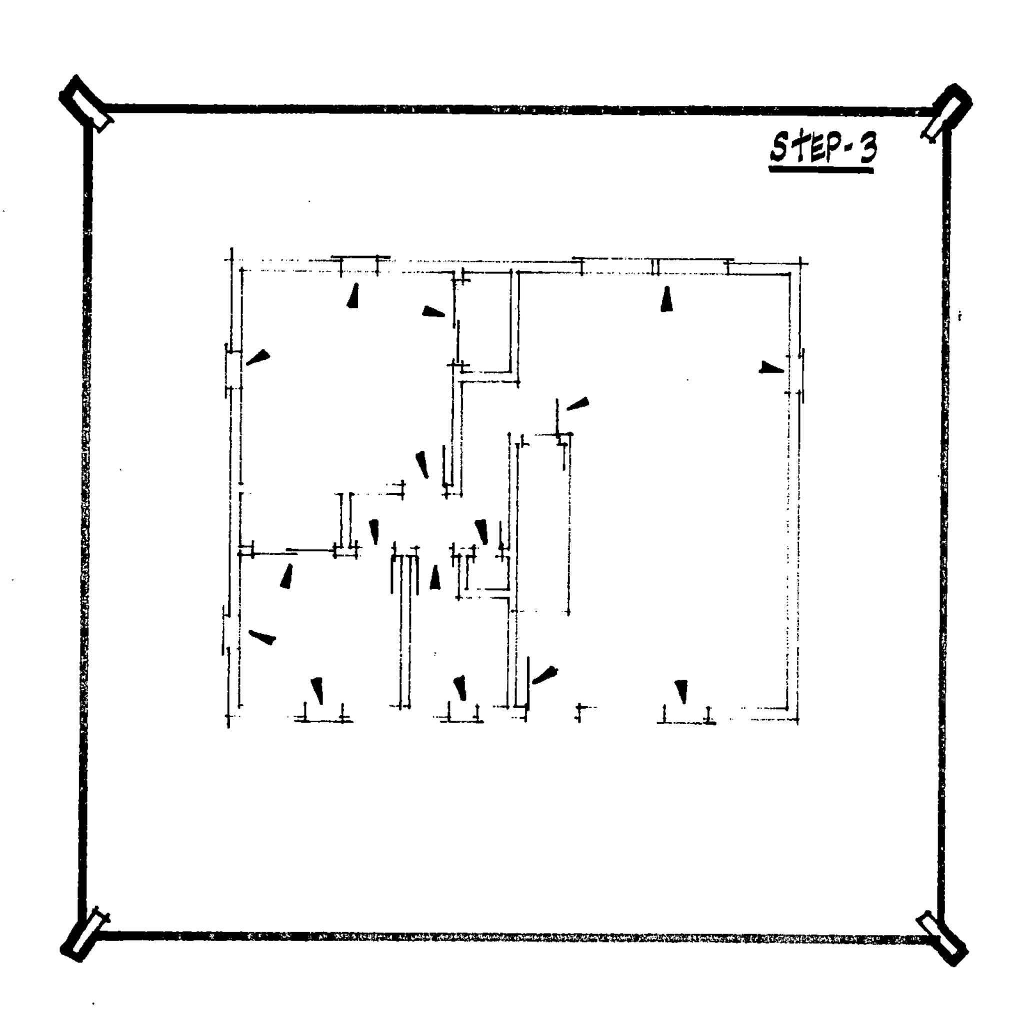

LOCATE DOORS AND WINDOWS.

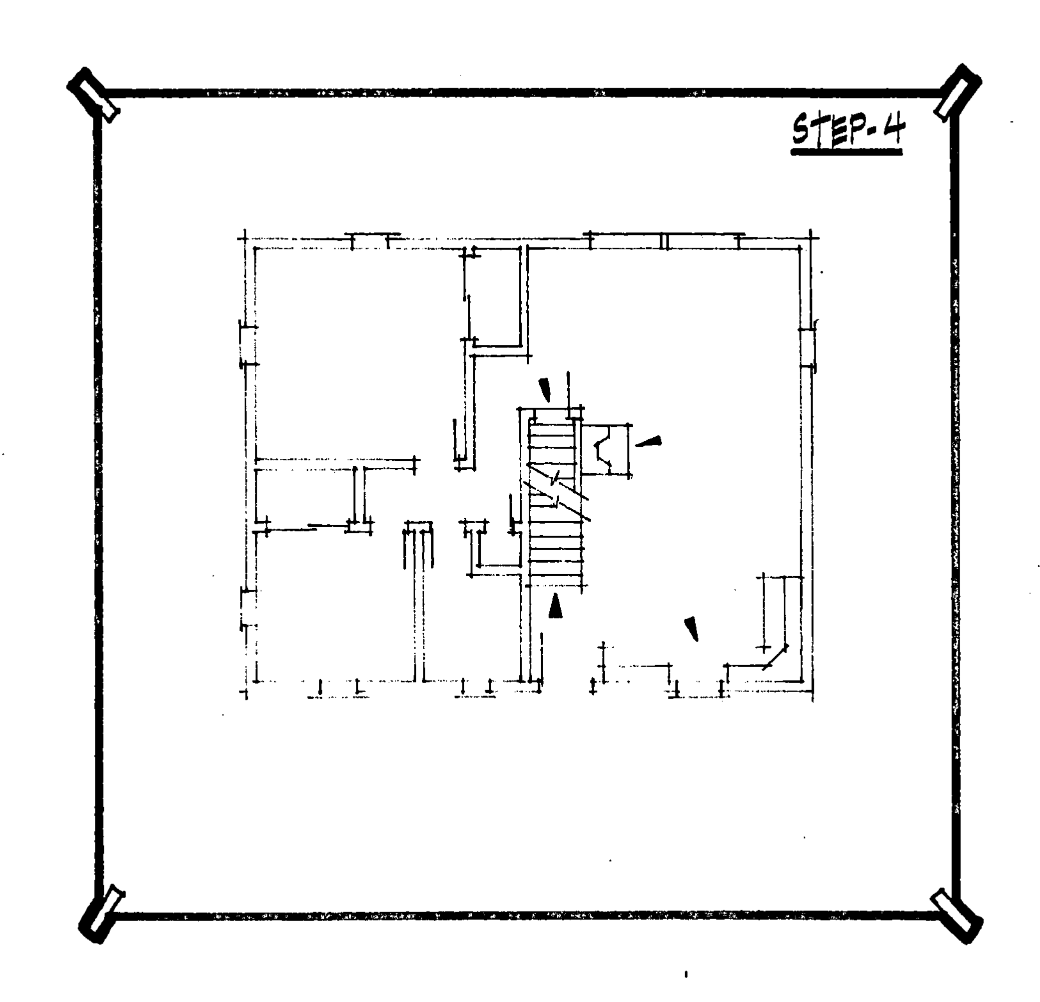

LOCATE STAIRS, FIREPLACE AND KITCHEN CABINETS.

LOCATE BATHROOM FIXTURES, KITCHEN SINK, RANGE AND REFRIGERATOR. WHEN SATISFIED THAT ALL IS WELL, DARKEN THE LINE WORK TO MAKE THINGS CLEARER. SAVE THE FINAL PUNCHING UP (DARK AND HEAVY) UNTIL AFTER THE DIMENSIONS ARE ON. IF LINE WORK IS HEAVIED UP AND WORK CONTINUES, THE SHEET WILL GET DIRTY.

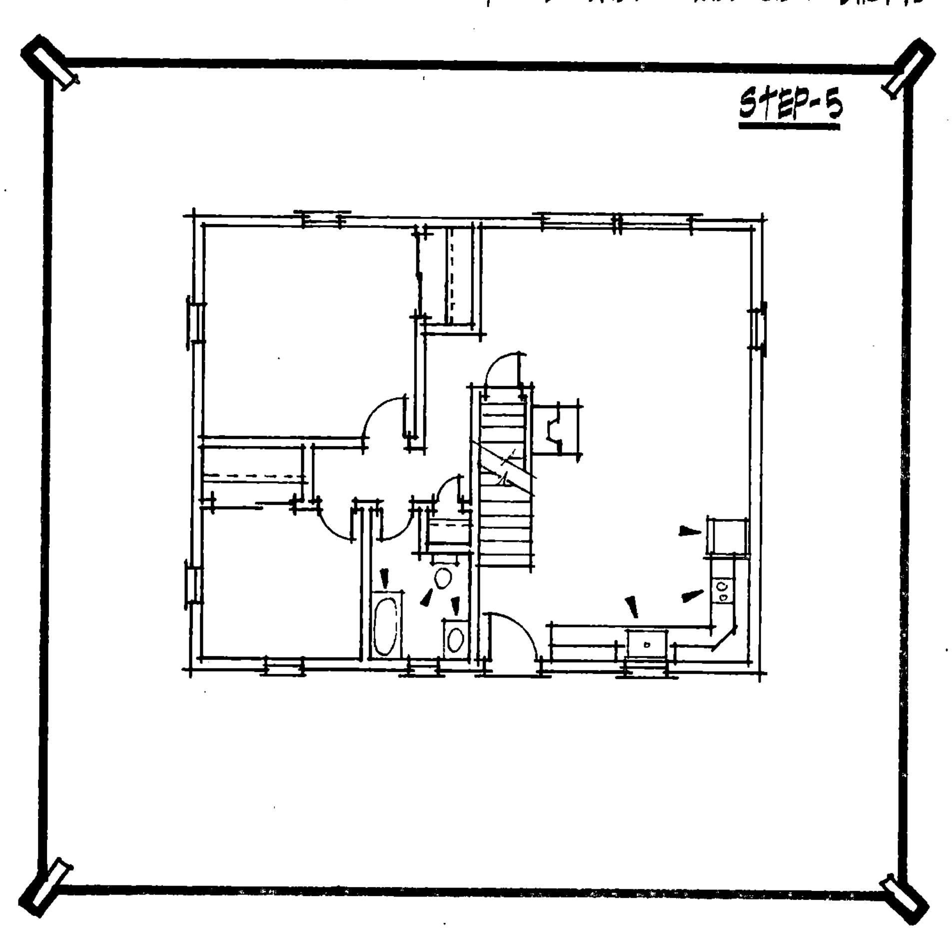

WHERE LINES CROSS AT CORNERS, EXTEND THEM PAST EACH OTHER.

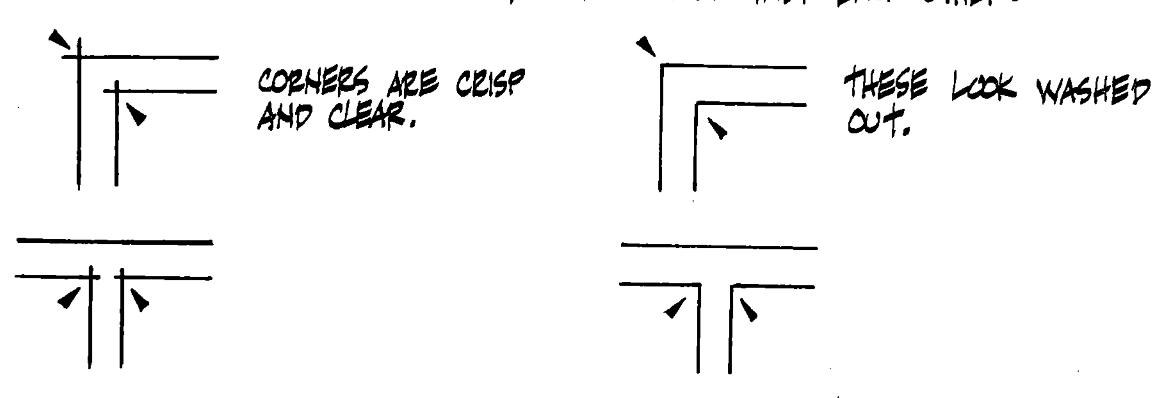

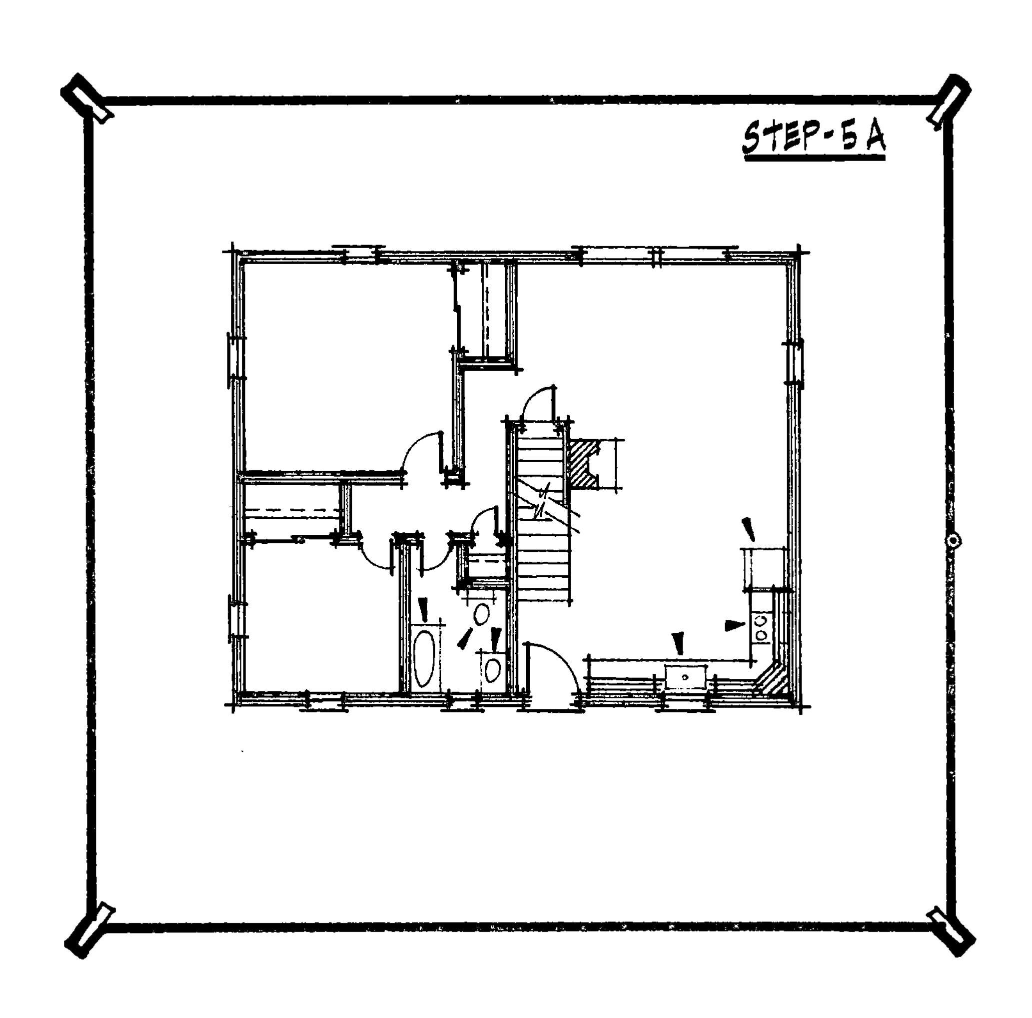

FILLING IN THE PARTITIONS WITH LIGHTER PARALLEL LINES HELPS TO SHOW THEM UP, HOWEVER DIMENSION LINES ARE NOT QUITE AS CLEAR. IT IS NOT NECESSARY TO DO, IT JUST DRESSES THE DRAWING UP.

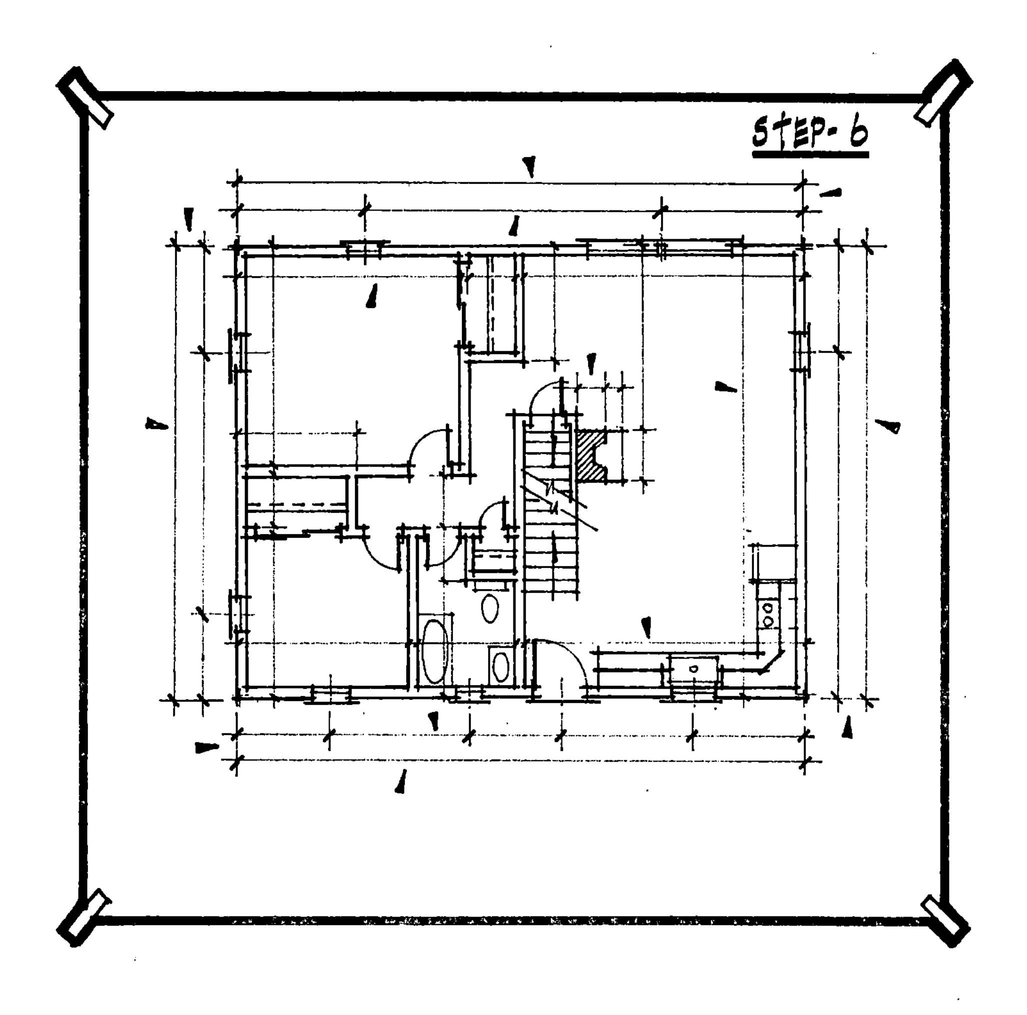

DIMENSION LINES COME NEXT AND THEY SHOULD BE LIGHTER THAN THE PARTITION LINES, BUT DARK ENOUGH TO BE POSITIVE. THE WALLS WANT TO JUMP OUT AT YOU.

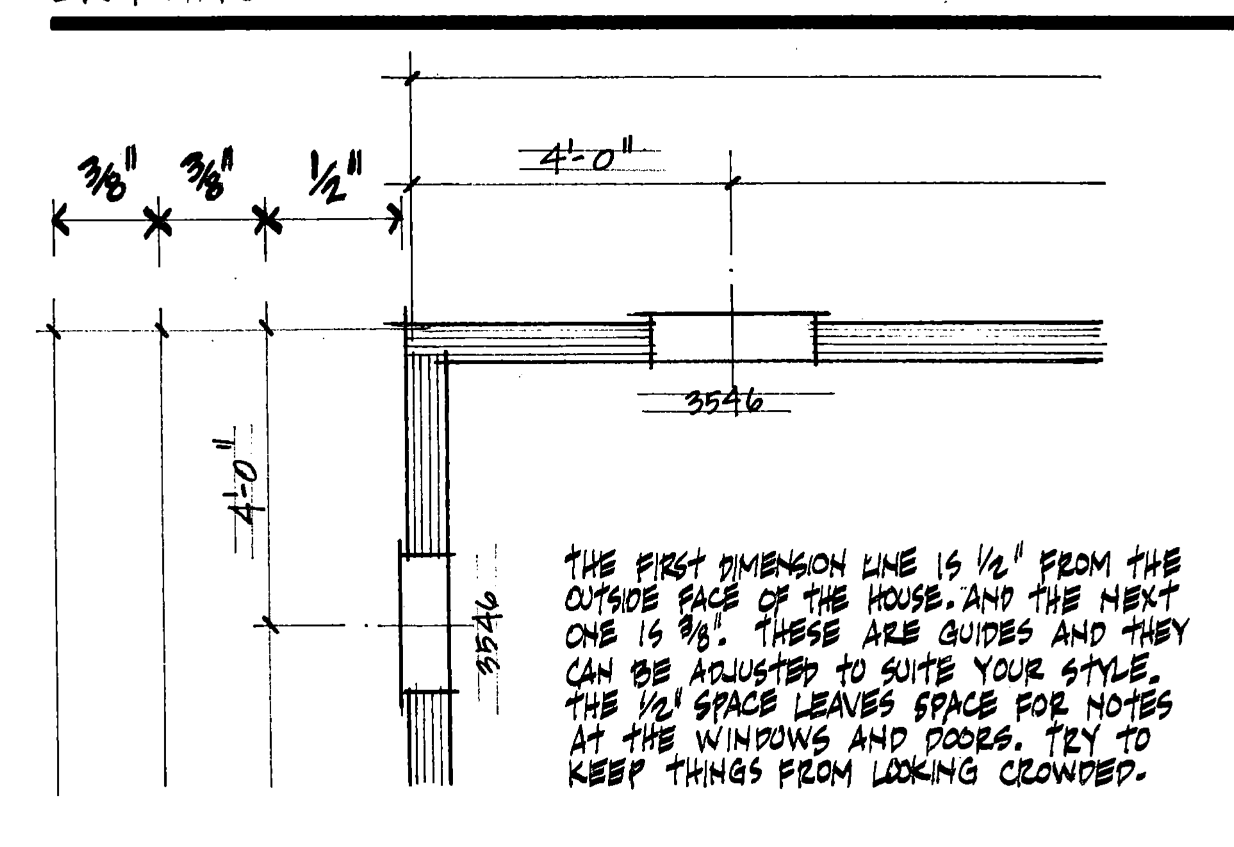

THE FIRST DIMENSION LINE IS 1/2" FROM THE OUTSIDE FACE OF THE HOUSE. AND THE NEXT ONE IS 3/8". THESE ARE GUIDES AND THEY CAN BE ADJUSTED TO SUITE YOUR STYLE. THE 1/2" SPACE LEAVES SPACE FOR NOTES AT THE WINDOWS AND DOORS. TRY TO KEEP THINGS FROM LOOKING CROWDED.

THERE ARE THREE BASIC SYMBOLS FOR TERMINATING DIMENSION LINES:

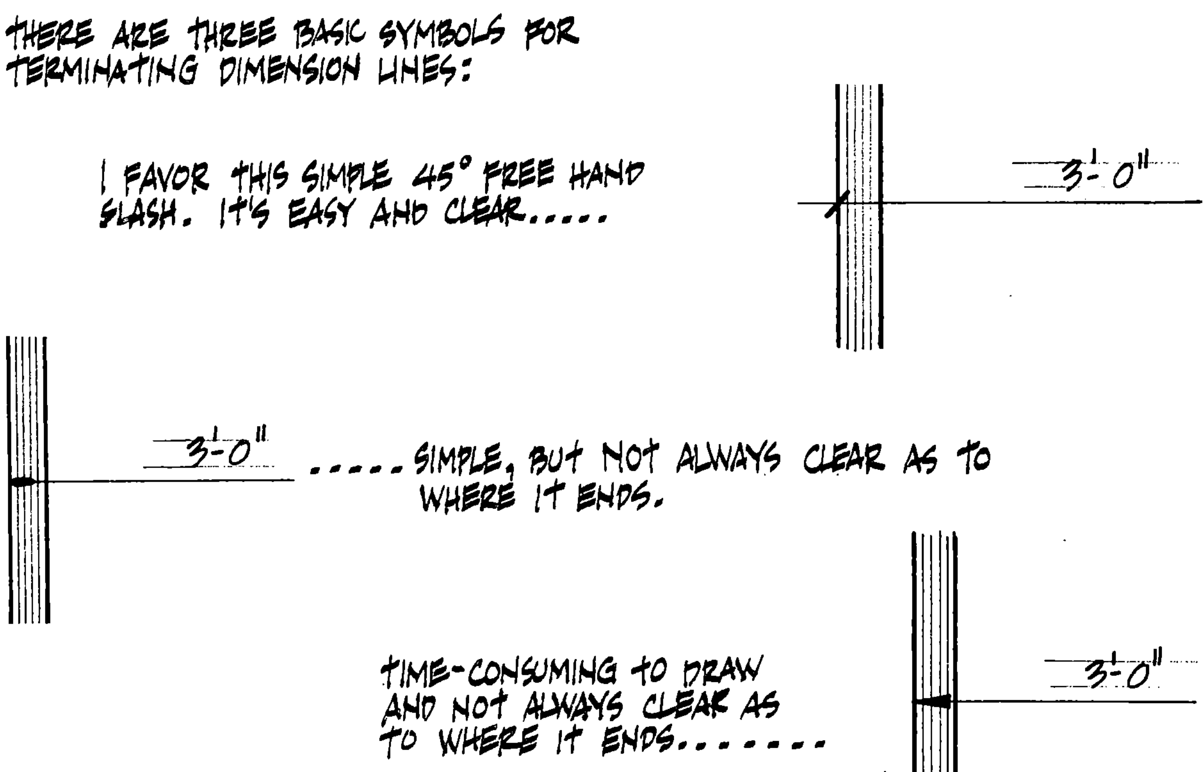

I FAVOR THIS SIMPLE 45° FREE HAND SLASH. IT'S EASY AND CLEAR.....

..... SIMPLE, BUT NOT ALWAYS CLEAR AS TO WHERE IT ENDS.

TIME-CONSUMING TO DRAW AND NOT ALWAYS CLEAR AS TO WHERE IT ENDS.......

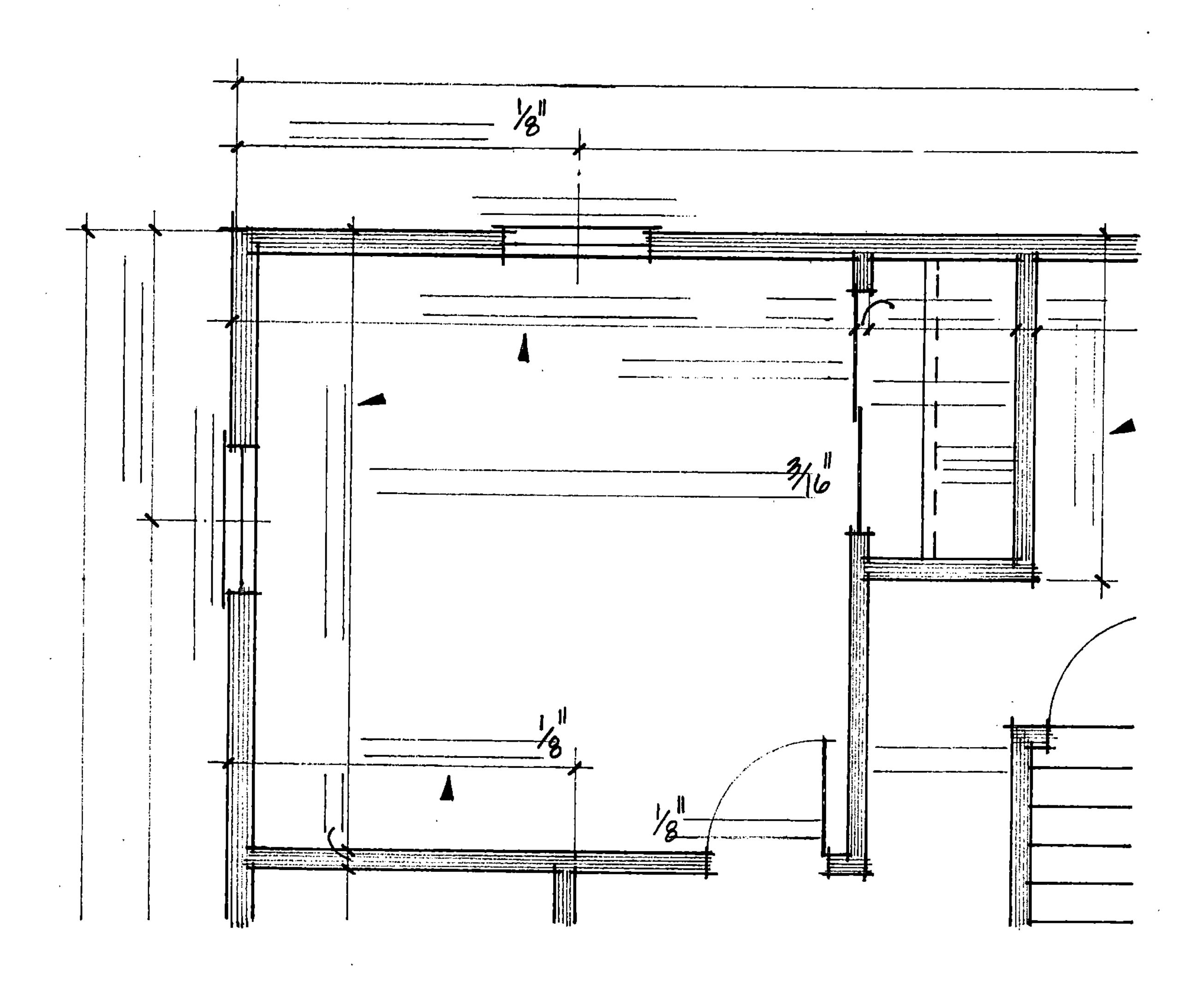

INSIDE DIMENSION LINES ARE KEPT TOWARD THE WALLS TO ALLOW FOR ROOM LABELING AND UNINTERRUPTED DIMENSION NUMBERS. THE DIMENSION SHOULD BE AT THE MID POINT OF WHAT IT IS DESCRIBING. CLARITY IS IMPAIRED IF A DIMENSION RUNS THROUGH IT. SOMETIMES IT CAN'T BE HELPED. DOORS AND WINDOWS ARE LOCATED ON THEIR CENTER LINE.

THERE IS A CHOICE TO BE MADE WHEN DIMENSIONING INSIDE PARTITIONS AND EACH HAS IT'S GOOD AND BAD POINTS.

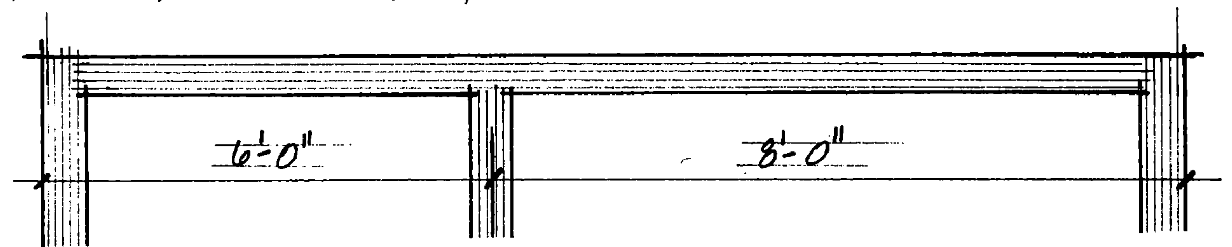

USING THE CENTER LINE OF PARTITIONS IS GOOD, BUT MAKE SURE TO BE CONSISTENT. THE CARPENTER THEN KNOWS THAT HE MUST LOCATE EACH PARTITION FACE OFF THE CENTER LINE ON THE PLANS.

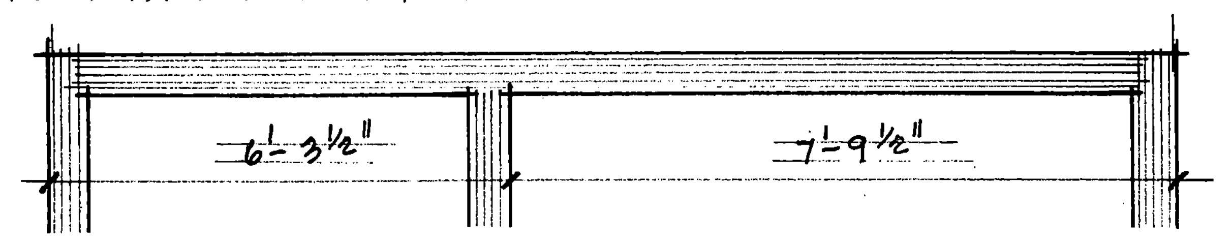

DIMENSION LINES TO ONE FACE OF A PARTITION (EXCLUDING THE EXTERIOR WHICH IS ALWAYS TO THE OUTSIDE FACE OF STUD) CAN BE CONFUSING. MANY MISTAKES ARE MADE BY PUTTING THE PARTITION ON THE WRONG SIDE OF THE LINE.

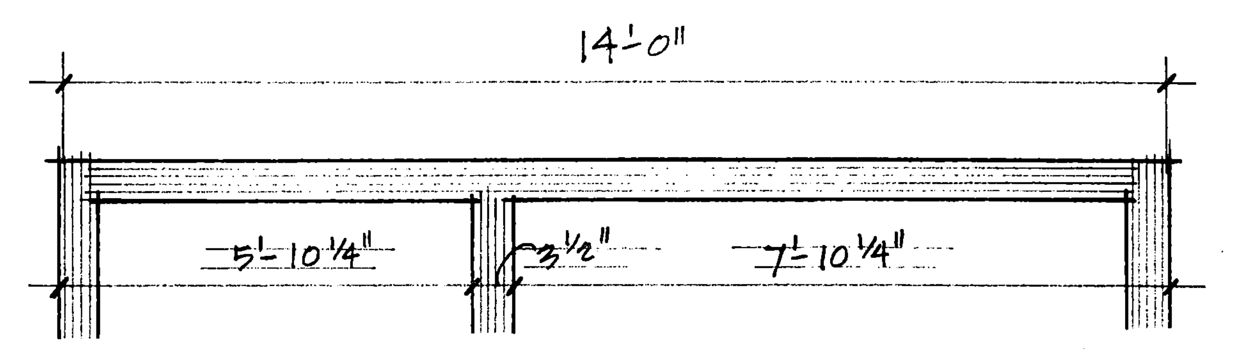

TO DIMENSION BOTH SIDES OF THE PARTITION IS A LOT MORE WORK FOR THE DRAFTSMAN, BUT VERY NICE FOR THE CARPENTER. MOST INTERIOR PARTITIONS ARE 3½" STUDS WHICH WILL CAUSE QUITE A FEW FRACTIONS TO COME UP IN THE DIMENSIONS; THAT IS WHY SOME DRAFTSMEN USE 4" FOR THIS DIMENSION. THE CARPENTER LAYS OUT THE ROOMS, USING THE 4" AND THEN ADJUSTS FOR THE ACTUAL 3½" WALL.

WHATEVER SYSTEM IS USED, THE SUM OF THE INSIDE DIMENSIONS MUST EQUAL THE EXTERIOR DIMENSION; ALWAYS CHECK THIS OUT. NEVER USE A DIMENSION LESS THAN ⅛" ANYWHERE, IT'S A HOUSE NOT A PIANO. STAY WITH ¼" MIN. FOR DIMENSIONS.

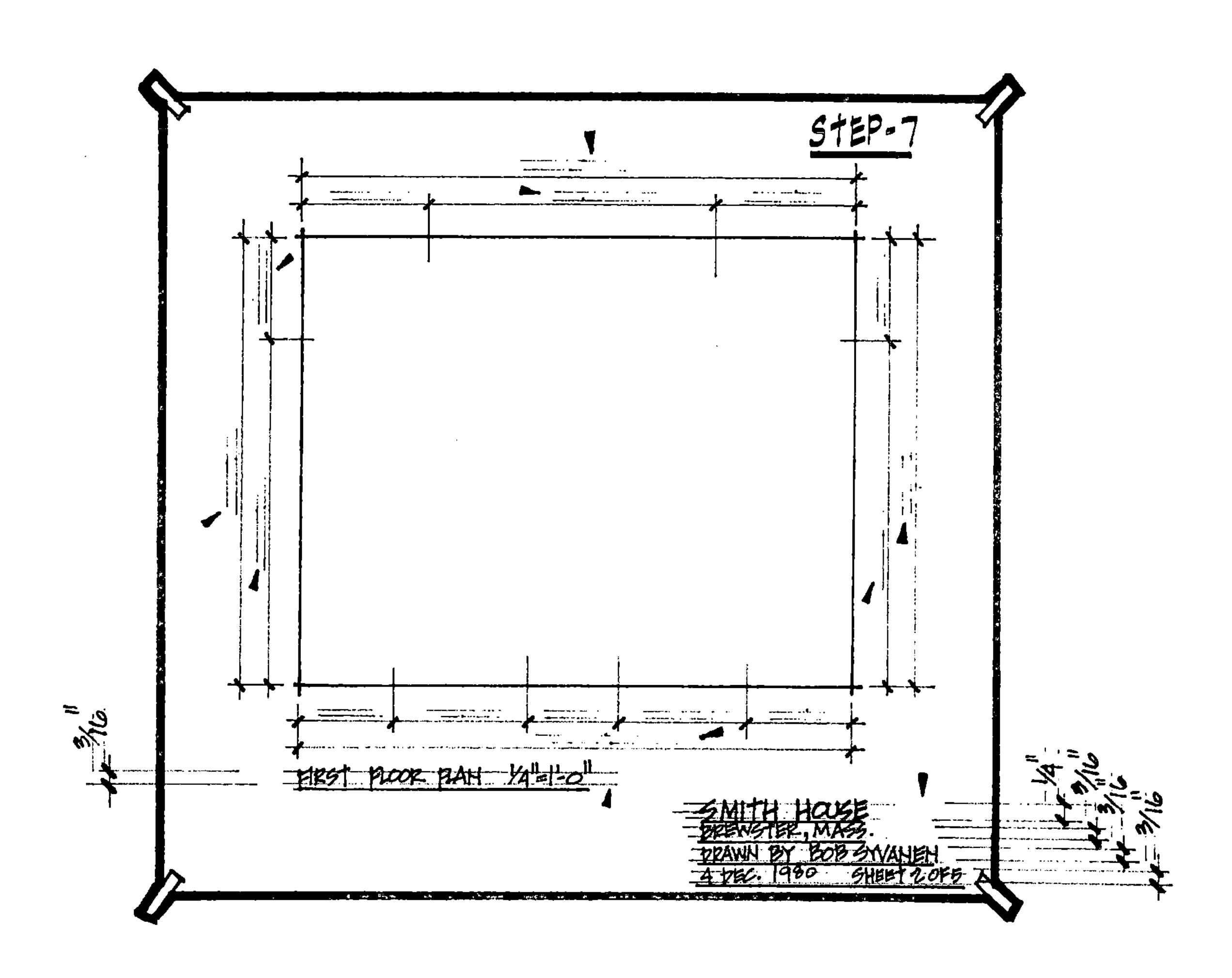

PUT IN ALL THE LETTERING GUIDE LINES AND DON'T BE AFRAID IF THEY SHOW UP ON THE PRINTS, I THINK THEY ADD TO THE LOOKS OF THE DRAWING. THE SIZES I USE MIGHT NOT SUIT YOU; USE THEM AS A GUIDE. PUT THE DIMENSIONS IN AFTER ALL THE GUIDE LINES ARE IN. DON'T WORRY IF THE PLAN DOES NOT MEASURE WHAT THE DIMENSION READS UNLESS IT IS VERY MUCH OUT OF SCALE. IT'S GOOD IF IT IS DIMENSIONED AS YOU WANT AND THE TOTALS ADD UP. I LIKE TO SAVE ALL THE LARGE LETTERING FOR LAST, AND DO ALL THE SHEETS AT THE SAME TIME, THE SHEETS STAY CLEANER AND THE LETTERING IS MORE CONSISTENT.

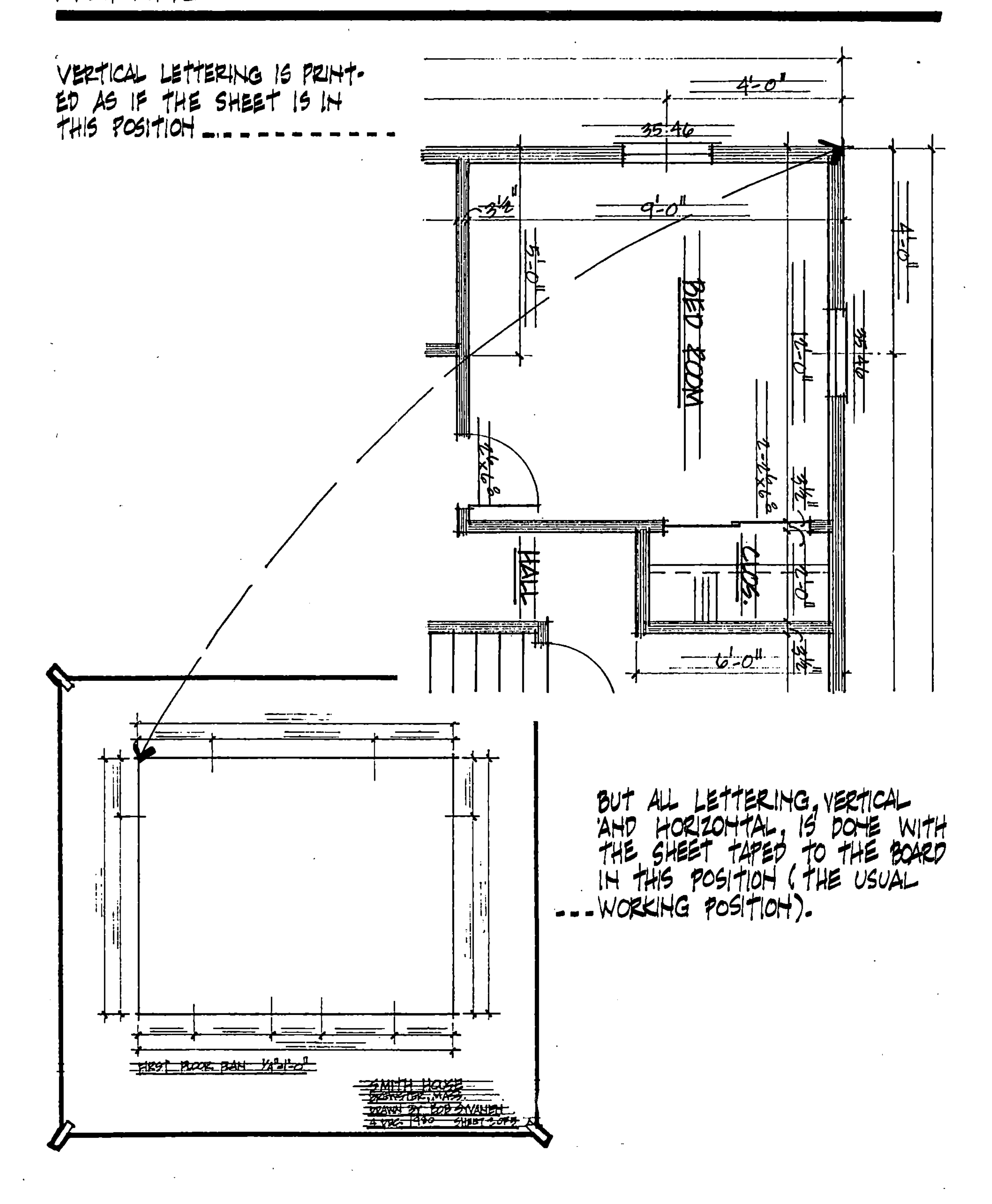
VERTICAL LETTERING IS PRINTED AS IF THE SHEET IS IN THIS POSITION
BUT ALL LETTERING, VERTICAL AND HORIZONTAL, IS DONE WITH THE SHEET TAPED TO THE BOARD IN THIS POSITION (THE USUAL WORKING POSITION).
4'-0"
35.46
9'-0"
5'-0"
BED ROOM
HALL
CLOS.
6'-0"

THE FOUNDATION PLAN IS TRACED FROM THE FIRST FLOOR PLAN.

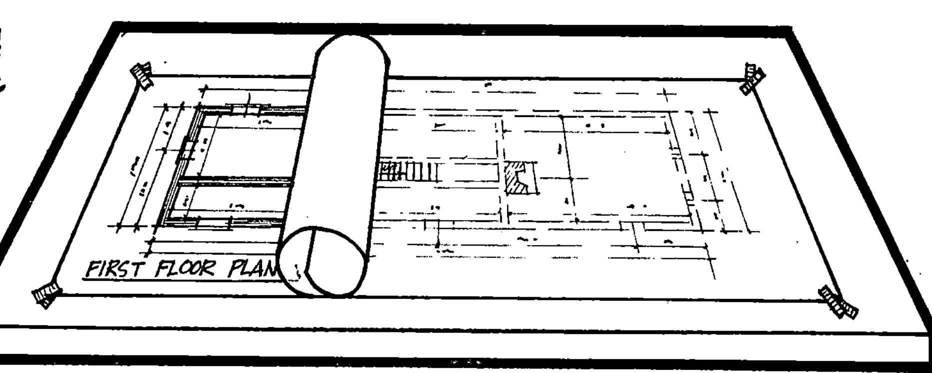

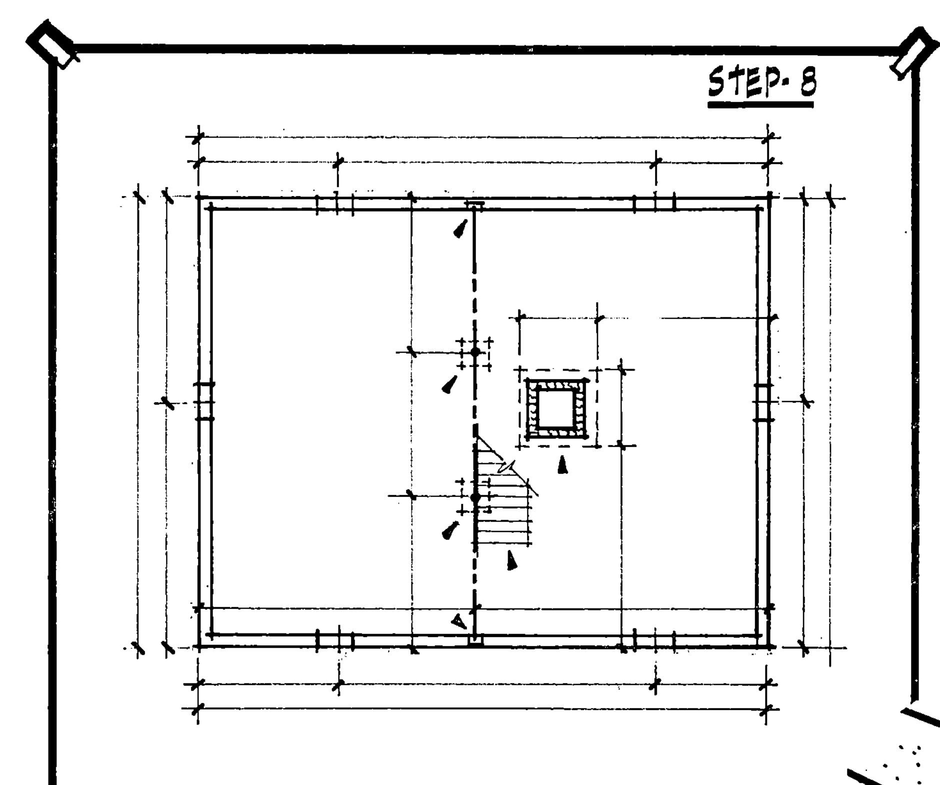

LOCATE CHIMNEY FOOTINGS, COLUMNS, BEAM POCKETS, WINDOWS, SEPTIC LINE, WATER LINE AND ANY SLEEVES GOING THROUGH THE FOUNDATION WALL. THE DRAWING STEPS ARE THE SAME AS THE FIRST FLOOR PLAN.

IF THE PRINTS ARE TO BE OF THE OZALID PROCESS, COLOR THE CONCRETE WALLS, WITH A BLUE PENCIL, ON THE BACK FACE OF THE DRAWINGS. MAKE IT MEDIUM DARK AND IT WILL SHOW A NICE SHADING ON THE PRINTS. DON'T COLOR THE WINDOW OPENINGS.

BEAM POCKET

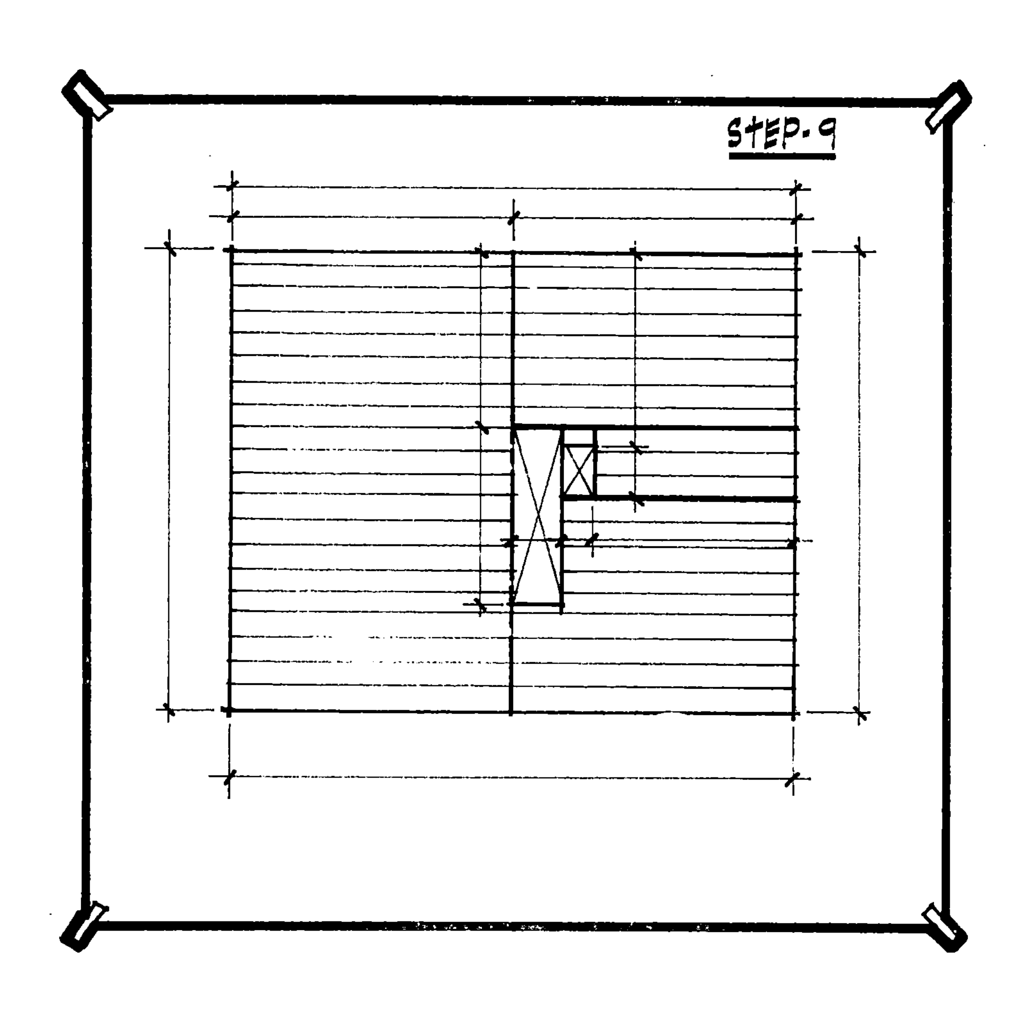

LAY ANOTHER SHEET OVER THE FOUNDATION PLAN AND LAY OUT THE FLOOR FRAMING PLAN. IT IS A GOOD SAFE WAY TO DRAW THE FRAMING BECAUSE THE STAIRS, CHIMNEY AND FLOOR BEAM ARE RIGHT THERE TO SEE. A SIMPLE WAY TO DRAW THE JOISTS IS TO TAKE THE LENGTH X 12" TO GET THE TOTAL INCHES. THEN DIVIDE BY 16" TO GET THE NUMBER OF JOIST SPACES. THE CALCULATOR DOES A GREAT JOB HERE. FIND A SCALE THAT IS CLOSE ENOUGH TO ANGLE FROM ONE SIDE TO THE OTHER (SEE ILLUSTRATIONS FOR SIDING SPACING ON PAGE 62) AND MARK THE SPACES. USE SINGLE LINES FOR ALL MEMBERS, JOISTS, HEADERS, BRIDGING, BLOCKING.

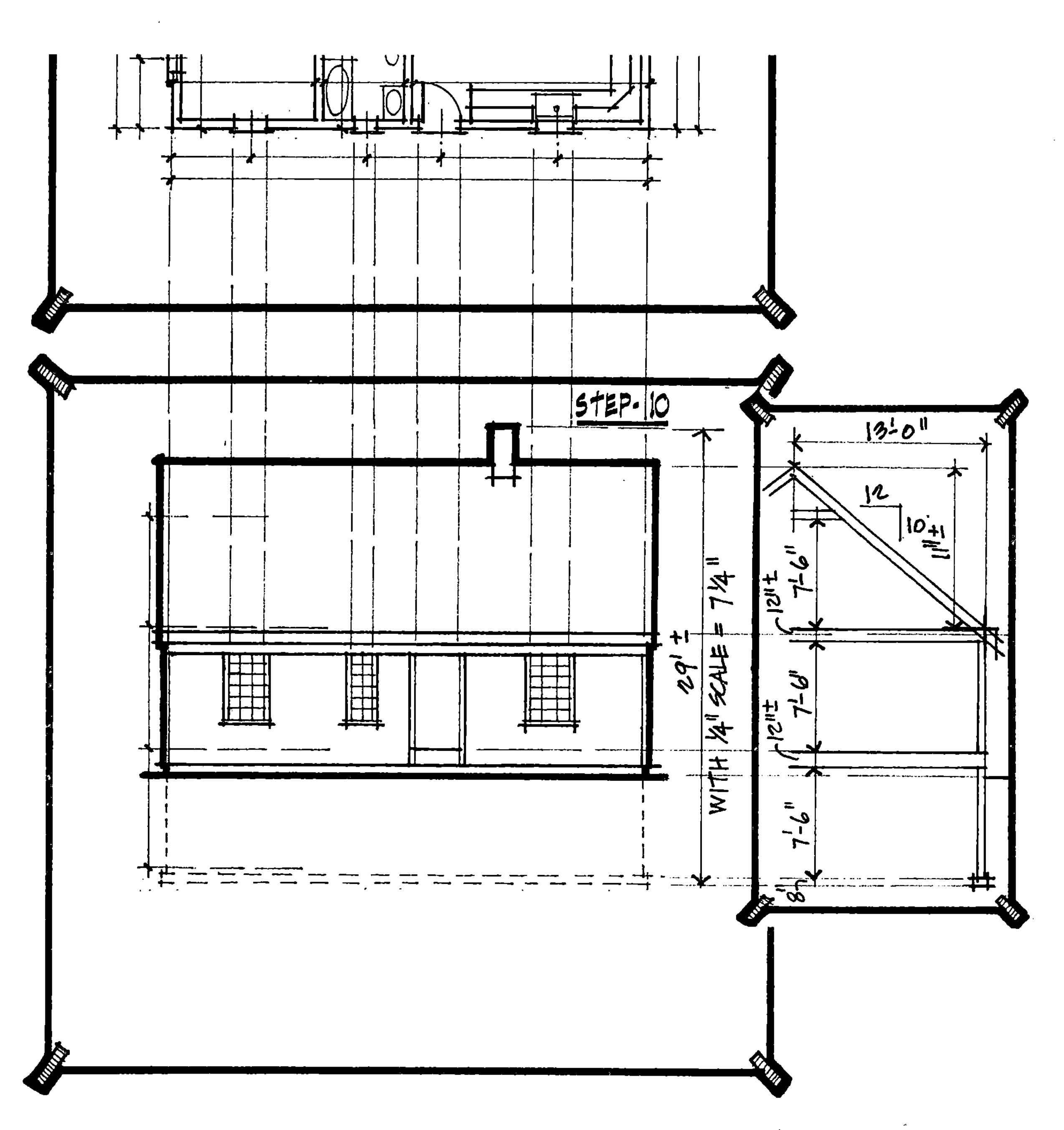

THE ELEVATIONS ARE PROJECTED OFF THE FLOOR PLAN AND A 1/4" SCALE SECTION. IF THE ELEVATIONS ARE DRAWN AT 1/8" SCALE, EVERYTHING MUST BE SCALED.

TEXTURES ARE DRAWN TO SUIT YOUR ARTISTIC SENSE. THEY SHOULD RESEMBLE THE MATERIAL USED.

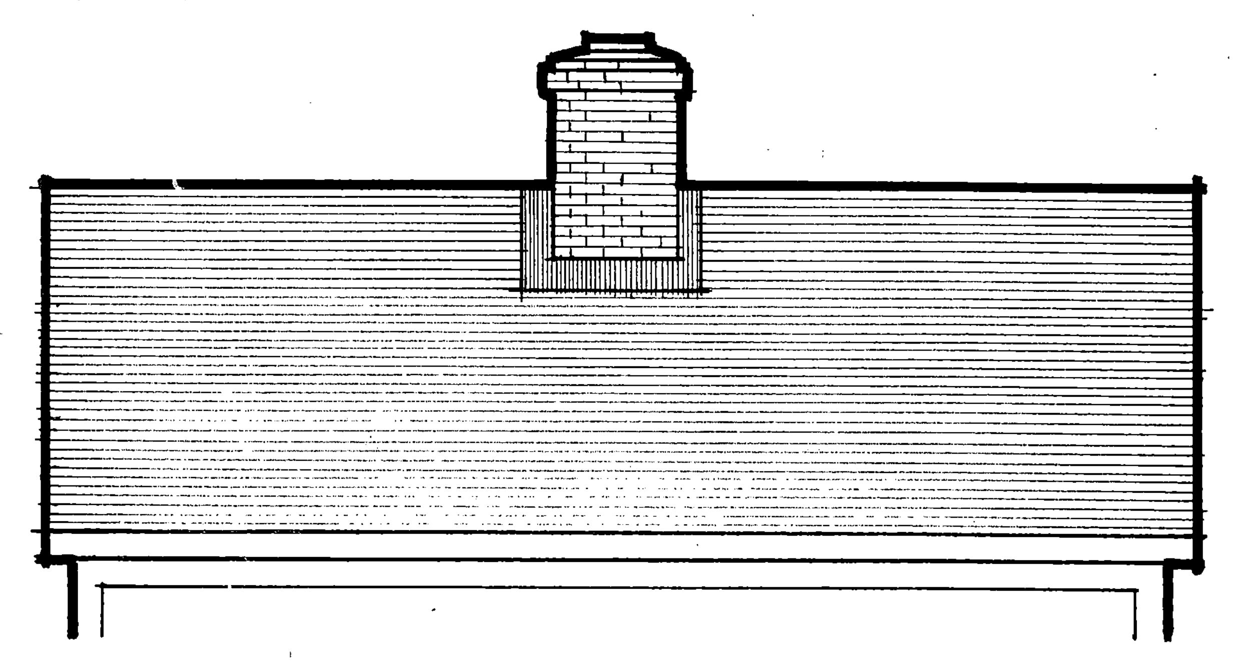

ROOF WITH CLOSE PARALLEL LINES SIMULATE SHINGLE COURSES.

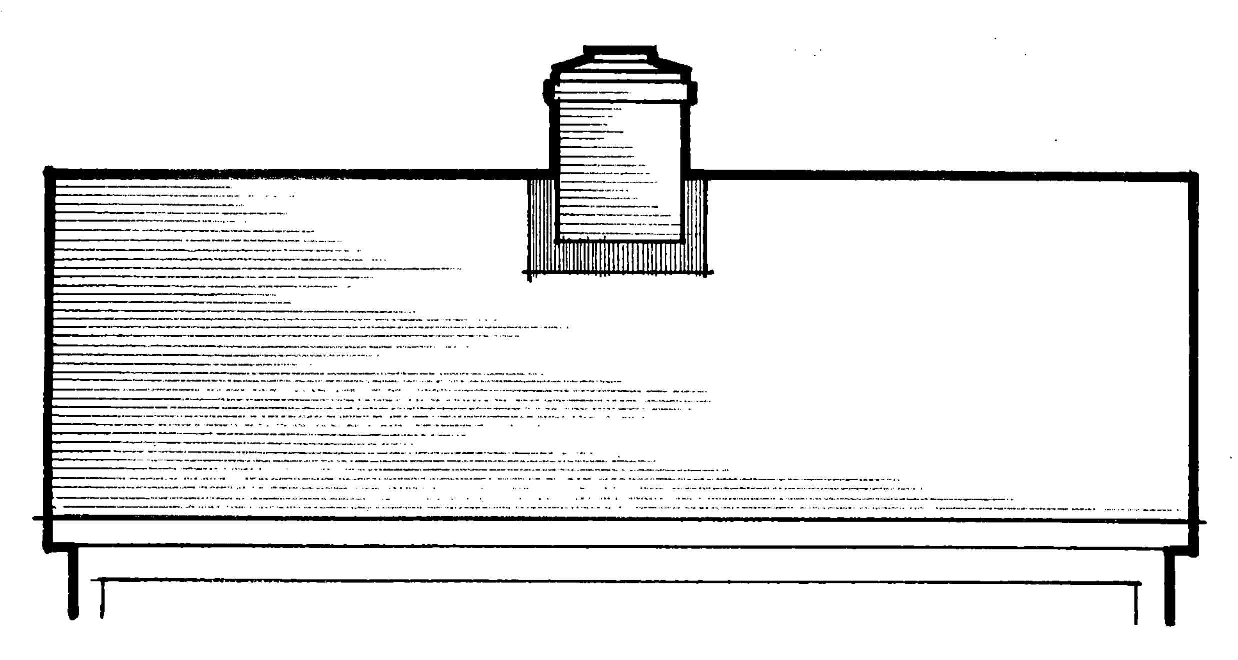

ROOF TEXTURE BROKEN BACK TO SIMULATE SUN ON ROOF.
THE SAME FOR BRICK WORK.

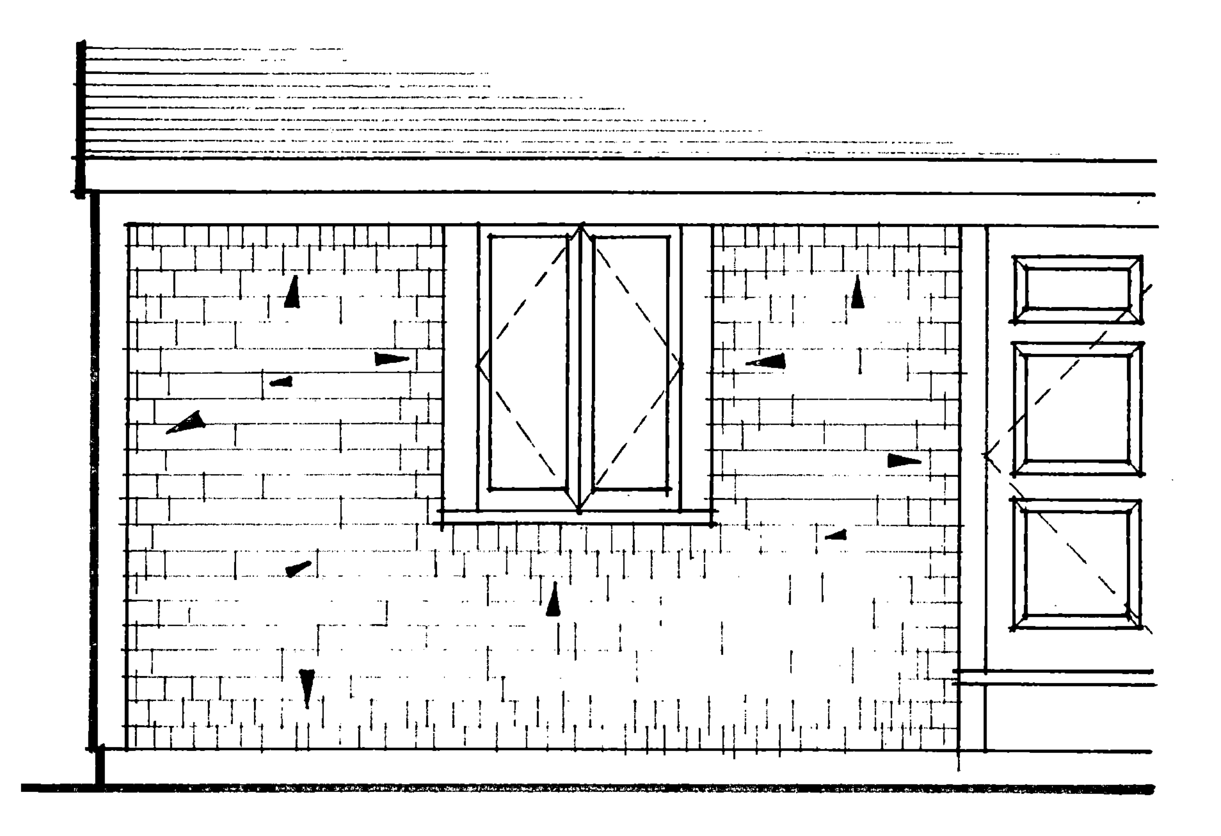

SHINGLES CAN BE PLAIN HORIZONTAL LINES, BROKEN BACK LIKE THE ROOF, OR USE CONTINUOUS WITH VERTICAL SHINGLE LINES. ACCENT THE CORNERS, WINDOWS, DOORS, EAVES, AND THE BOTTOM COURSE OF SHINGLES. BY USING MORE VERTICAL LINES AT THESE AREAS. USE A TRIANGLE AGAINST A PARALLEL STRAIGHT EDGE TO DO THIS. SPOT A FEW VERTICAL LINES IN THE EMPTY SPACES.

DRAW THE ROOF PITCH WITH THE ADJUSTABLE TRIANGLE. LAY OUT, ON PAPER, A HORIZONTAL 12" LINE AND A VERTICAL 10" LINE AND CONNECT. THESE POINTS TO GET THE PITCH ANGLE. SET TRIANGLE TO THIS.

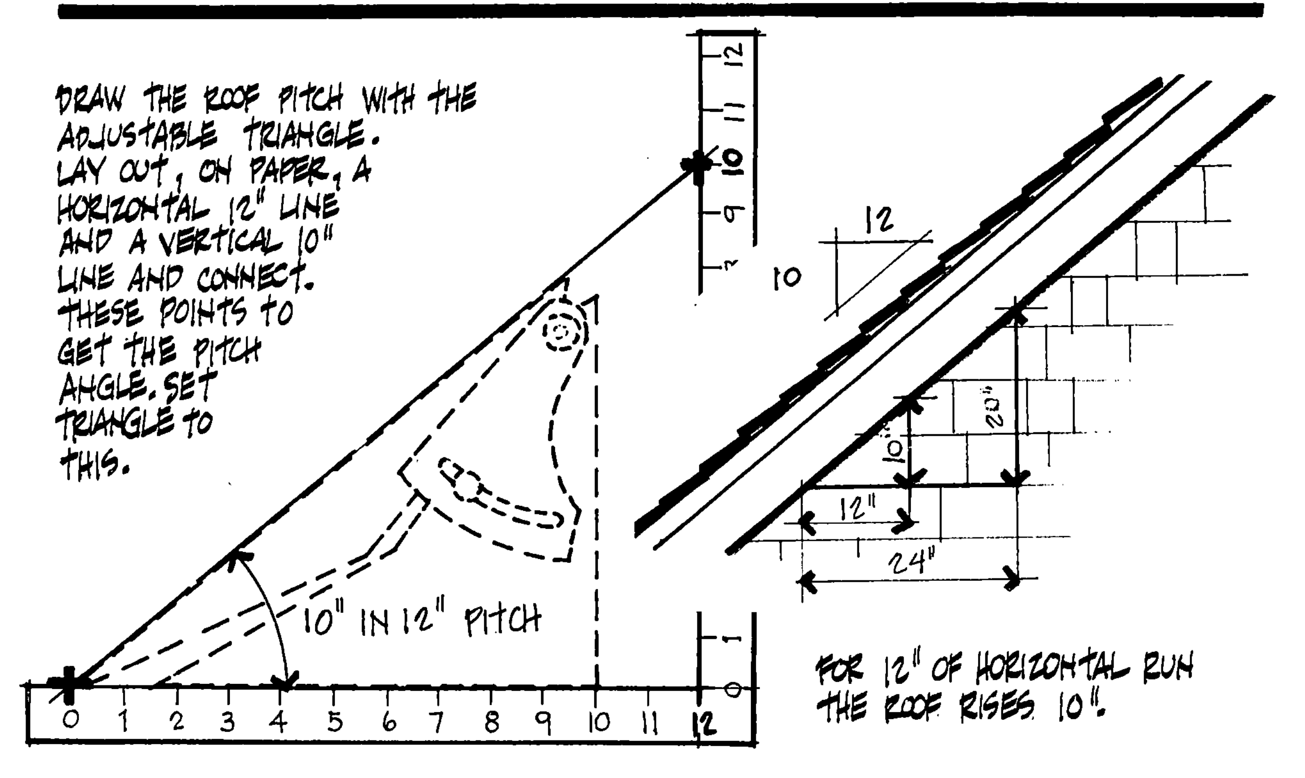

FOR 12" OF HORIZONTAL RUN THE ROOF RISES 10".

SIDING IS DRAWN USING THE DIMENSION OF THE MATERIAL, 5" SHINGLE COURSES, 4", 5", OR 10" CLAPBOARDS AND 4" TO 10" VERTICAL AND DIAGONAL SIDING.

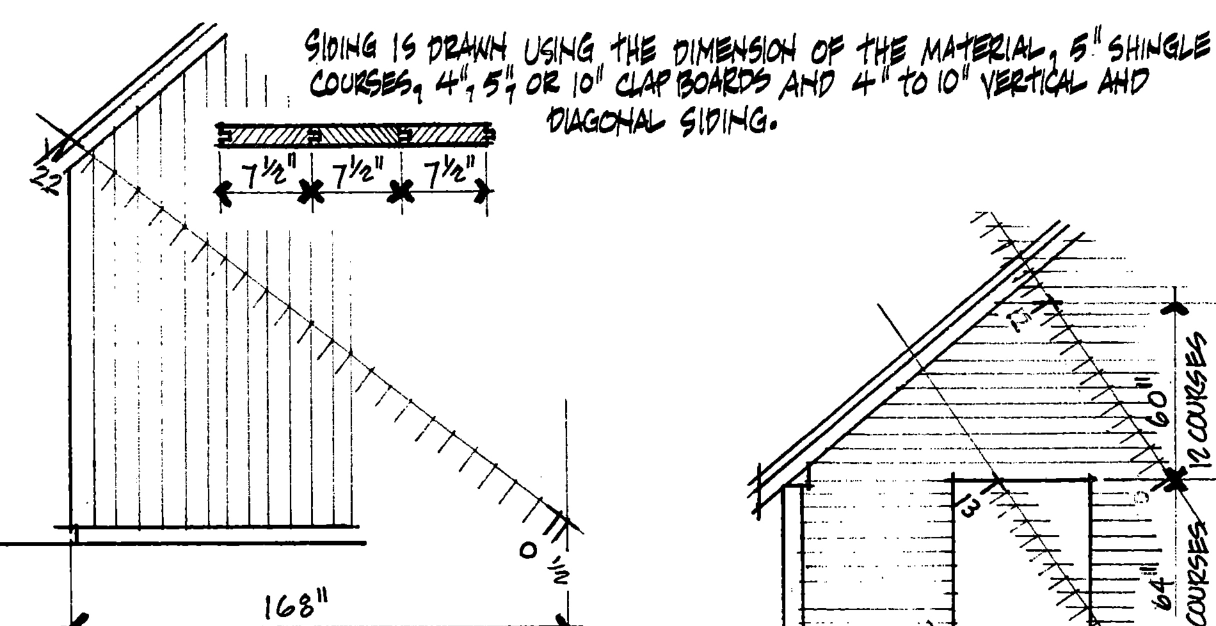

PUT THE SHINGLE COURSES ON AS THEY WOULD BE DONE BY THE CARPENTER. ANGLE THE SCALE TO GET THE RIGHT NUMBER OF COURSES FOR VERTICAL AND HORIZONTAL SIDING.

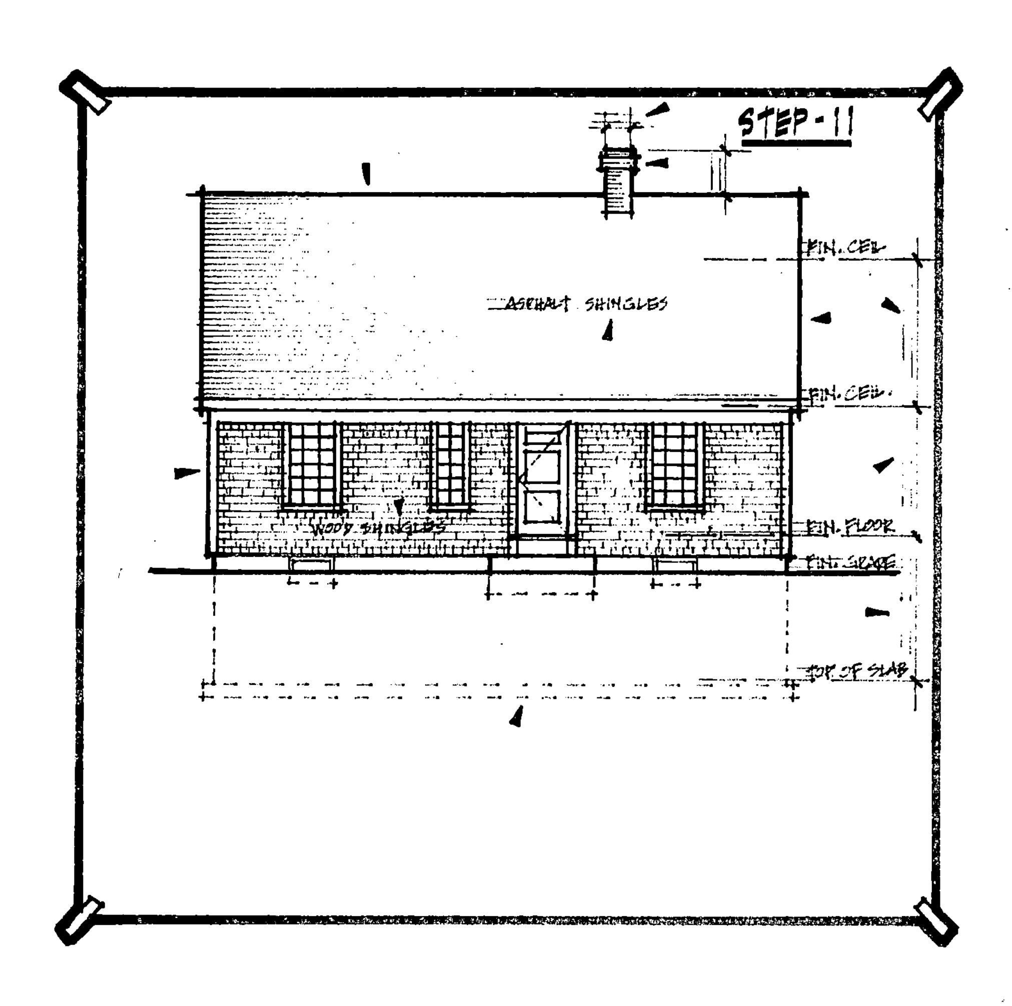

THIS IS A GOOD PLACE TO SHOW THE RELATIONSHIP BETWEEN THE FOUNDATION AND THE FINISHED GRADE. DIMENSION THE FLOOR HEIGHTS AND CHIMNEY SIZE. HEAVY UP THE WHOLE DRAWING AND IN PARTICULAR, THE FINISH GRADE AND THE OUTLINE OF THE HOUSE.
THE MAIN STEPS, 1 THROUGH 11, ARE FOLLOWED BY THE DETAILS. THERE SHOULD BE AT LEAST ON SECTION THROUGH THE BUILDING SHOWING FOUNDATION, EXTERIOR WALL, AND ROOF.

DETAILS CAN BE DRAWN AT ANY TIME. MAKE SKETCHES OF THEM AS YOU GO ALONG. THESE DETAILS WILL MAKE THINGS EASIER FOR THE BUILDER AND SHOW THE BUILDING INSPECTOR HOW THE HOUSE IS TO BE BUILT. HE CAN TELL YOU BEFORE IT'S BUILT IF IT IS ACCEPTABLE.

WHEN DRAWING DETAILS, PUNCH UP THE SECTIONS THAT THE DRAWING CUTS THROUGH. THINGS LIKE 2X4 PLATES, JOISTS, TRIM AND FOUNDATION.

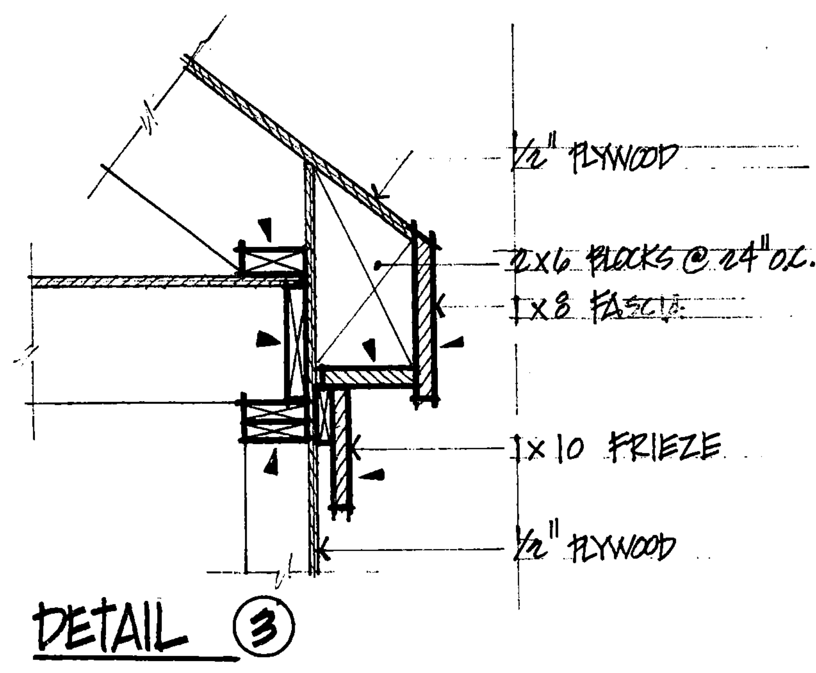

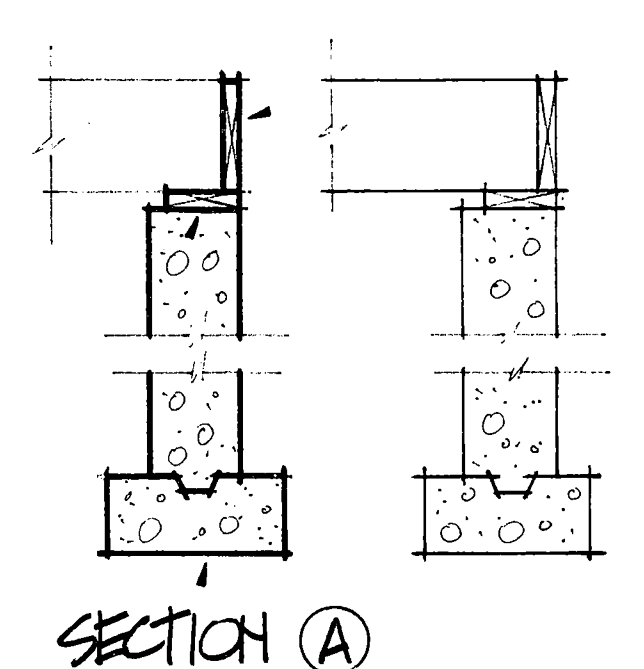

DRAW ALL THE DETAILS THAT FIT ON A SHEET BEFORE LETTERING. THE PROCEDURE IS SIMILAR TO DRAWING FLOOR PLANS: LIGHT LAYOUT, THEN HEAVIER FOR CLARITY, DIMENSION LINES, LETTERING, FINAL PUNCHING UP OF THE LINE WORK, AND LAST THE HEAVY TITLE.

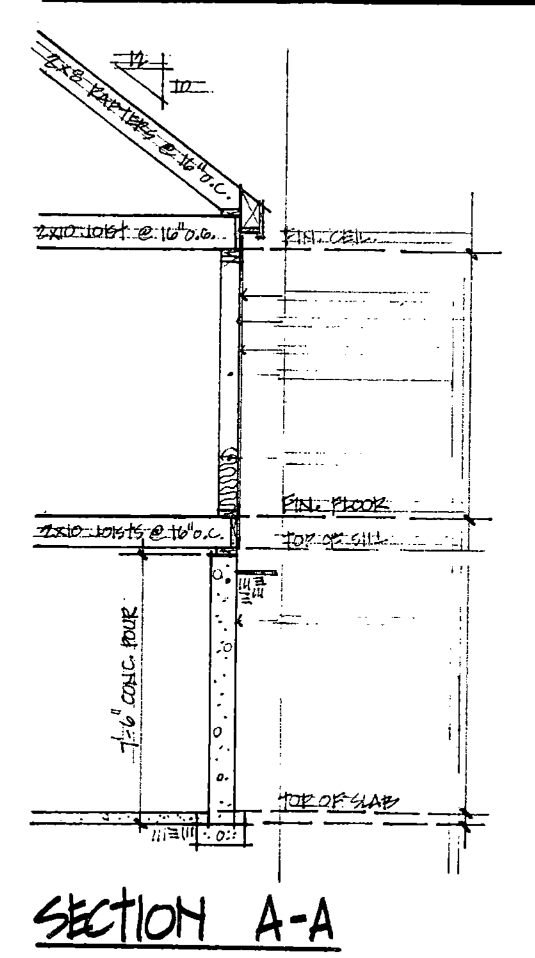

SECTION A-A

THE SECTION THROUGH THE BUILDING IS REALLY A LARGE DETAIL AT 3/8" SCALE. YOU CAN'T SHOW MUCH DETAIL AT A SCALE SMALLER THAN THAT, BUT IF THE SECTION IS SIMPLE THEN 1/4" SCALE WILL WORK.

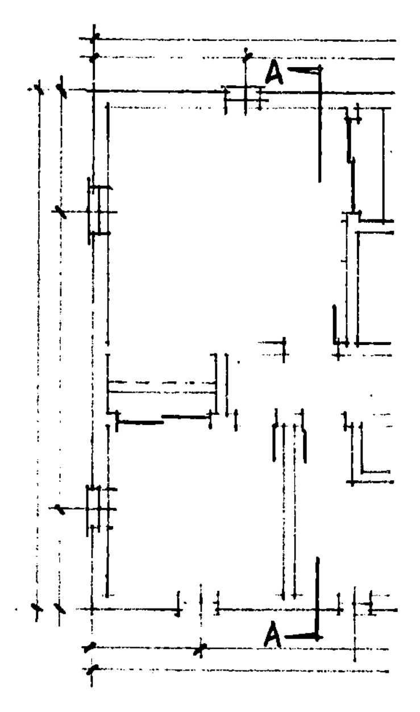

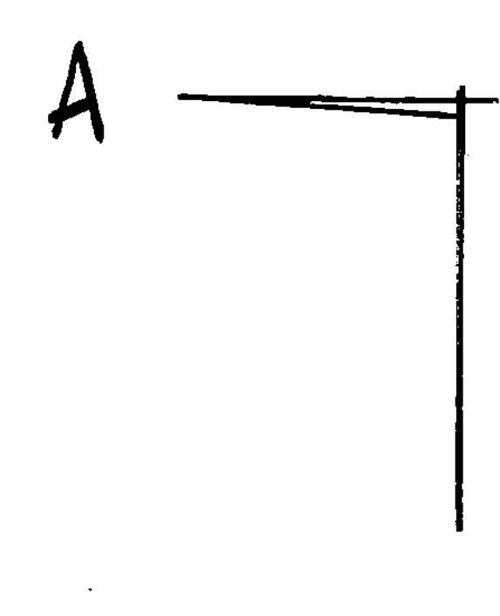

THIS IS A NICE SIMPLE WAY TO SHOW WHERE THE SECTION CUTS THROUGH ON THE PLAN.

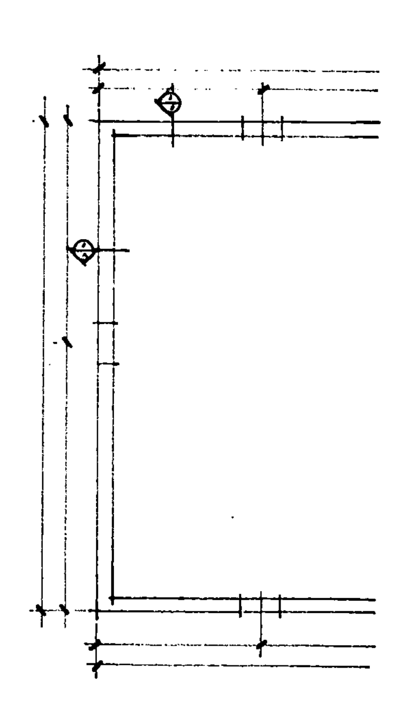

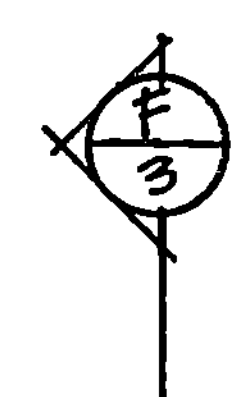

SECTION NUMBER
PAGE SECTION IS ON

THIS SYSTEM IS GOOD IF THERE ARE A LOT OF DRAWINGS. IT MAKES IT EASY TO LOCATE A SPECIFIC DETAIL OR SECTION.

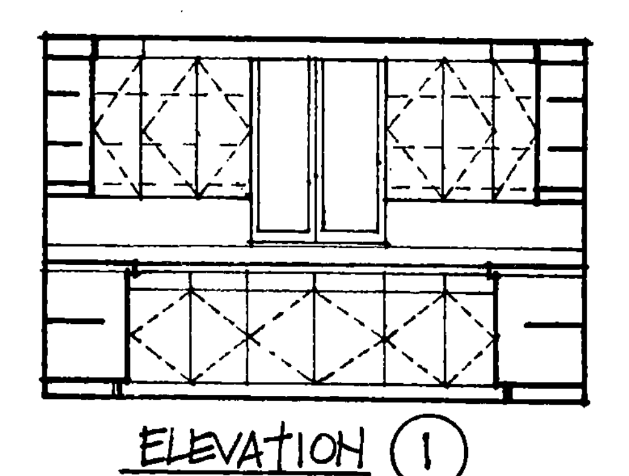

ELEVATION (1)

INTERIOR ROOM ELEVATIONS ARE USEFUL, PARTICULARLY IN THE KITCHEN AND FIREPLACE WALL. YOU CAN SHOW EXACTLY WHAT IS TO BE BUILT THERE. 1/4" SCALE IS USUAL, BUT A LARGER SIZE IS USED WHEN MORE DETAIL IS REQUIRED.

THE SYSTEM FOR SHOWING THE LOCATION ON THE PLAN IS A CIRCLE ARROW COMBINATION. THE TOP NUMBER IS THE ELEVATION NUMBER AND THE BOTTOM ONE IS THE SHEET THE ELEVATION IS ON.

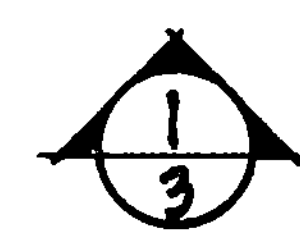

ELEVATION NUMBER
PAGE ELEVATION IS ON

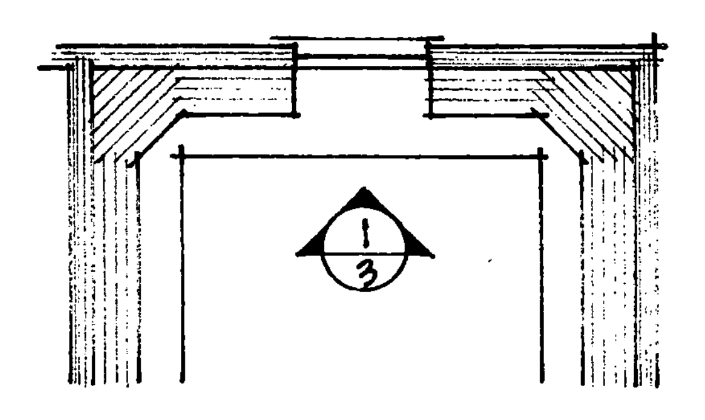

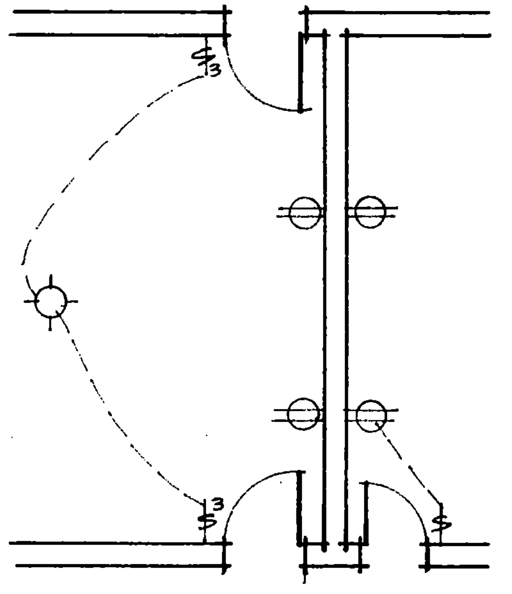

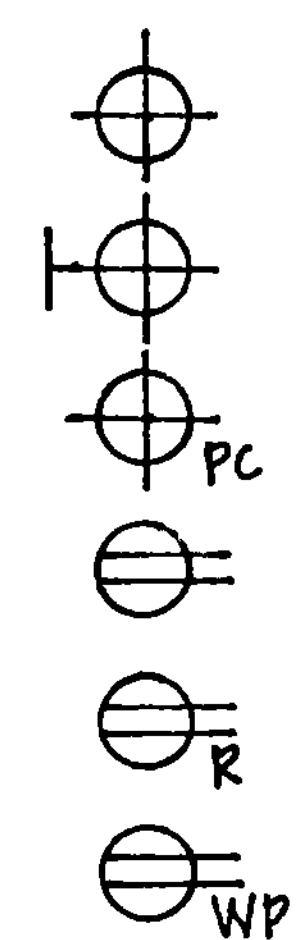

LIGHT - CEILING

LIGHT - WALL

LIGHT - PULL CHAIN

CONVENIENCE OUTLET

RANGE OUTLET

WATER PROOF OUTLET

SWITCH

3 WAY SWITCH

THE ELECTRIC PLAN IS SIMPLE AND USUALLY PUT ON THE FLOOR PLANS. IF THE PLANS ARE VERY "BUSY", A SEPERATE FLOOR PLAN CAN BE TRACED AND THE ELECTICAL LAYOUT DRAWN ON IT.

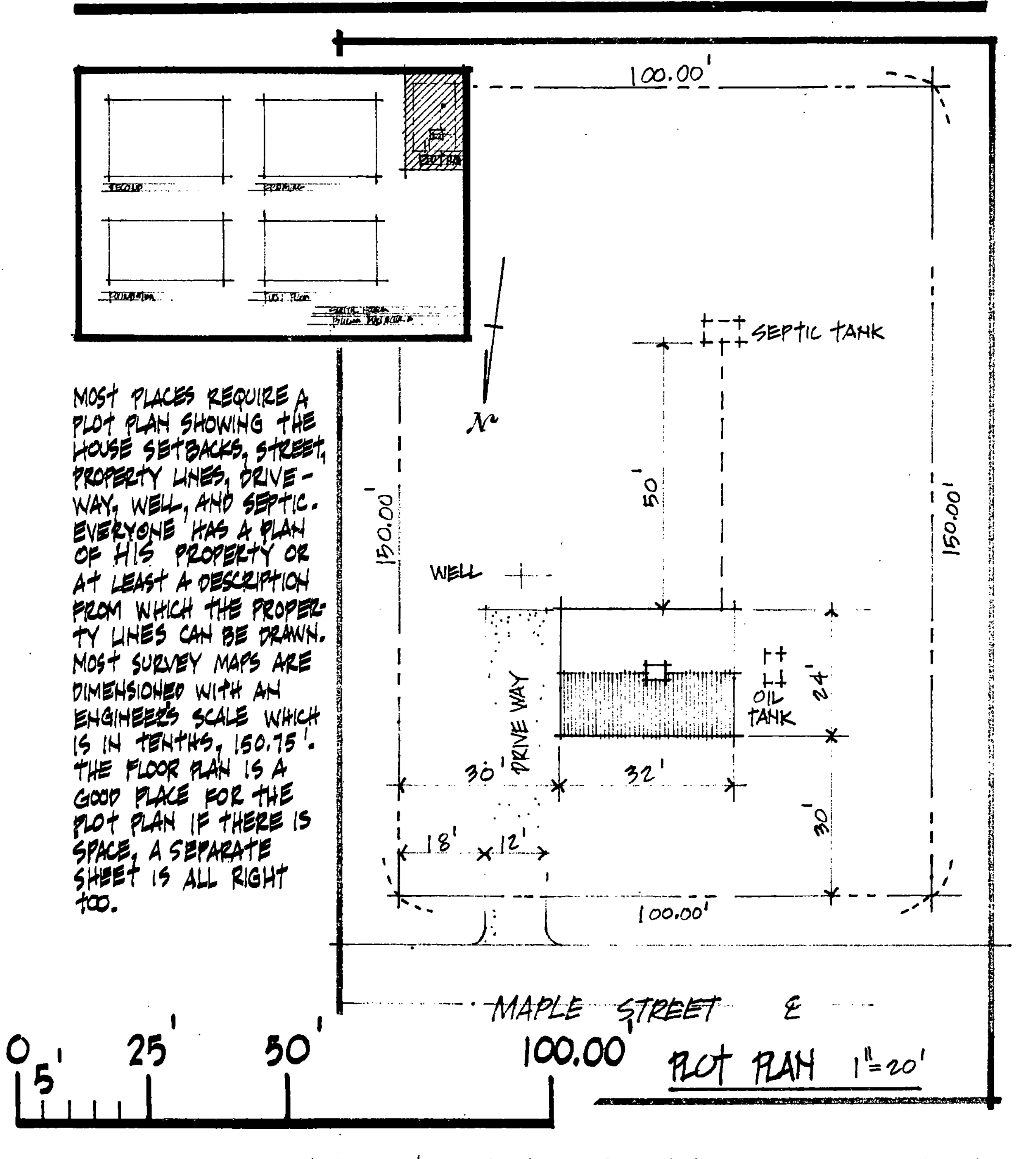

MOST PLACES REQUIRE A PLOT PLAN SHOWING THE HOUSE SETBACKS, STREET, PROPERTY LINES, DRIVEWAY, WELL, AND SEPTIC. EVERYONE HAS A PLAN OF HIS PROPERTY OR AT LEAST A DESCRIPTION FROM WHICH THE PROPERTY LINES CAN BE DRAWN. MOST SURVEY MAPS ARE DIMENSIONED WITH AN ENGINEER'S SCALE WHICH IS IN TENTHS, 150.75'. THE FLOOR PLAN IS A GOOD PLACE FOR THE PLOT PLAN IF THERE IS SPACE, A SEPARATE SHEET IS ALL RIGHT TOO.

IF AN ENGINEERS SCALE IS NOT AVAILABLE, MAKE A SCALE USING YOUR SURVEY PLAN. DIVIDE UP SOME DIMENSION UNTIL YOU GET WHAT YOU NEED. DRAW THE PLOT PLAN AT WHATEVER SCALE IS CONVENIENT.

DOOR SCHEDULE

MARK	SIZE	MATERIAL	PATTERN	REMARKS
(A)	3'-0"x6'-8"x1 3/4"	FIR	108(F-66-2)	
(B)	2'-6"x6'-8"x1 3/4"	FIR	L(F-944)	
(C)	2'-6"x6'-6"x1 3/8"	PINE	M-1051	
(D)	2'-4"x6'-6"x1 3/8"	PINE	M-1051	
(E)				

THE DOOR SCHEDULE LEAVES NO DOUBT AS TO WHAT IS REQUIRED. IT'S PRETTY TOUGH TO FIT ALL THAT INFORMATION ON THE PLAN.

WITH THE DOOR SCHEDULE THE SYMBOL IS SHOWN ON THE PLAN...

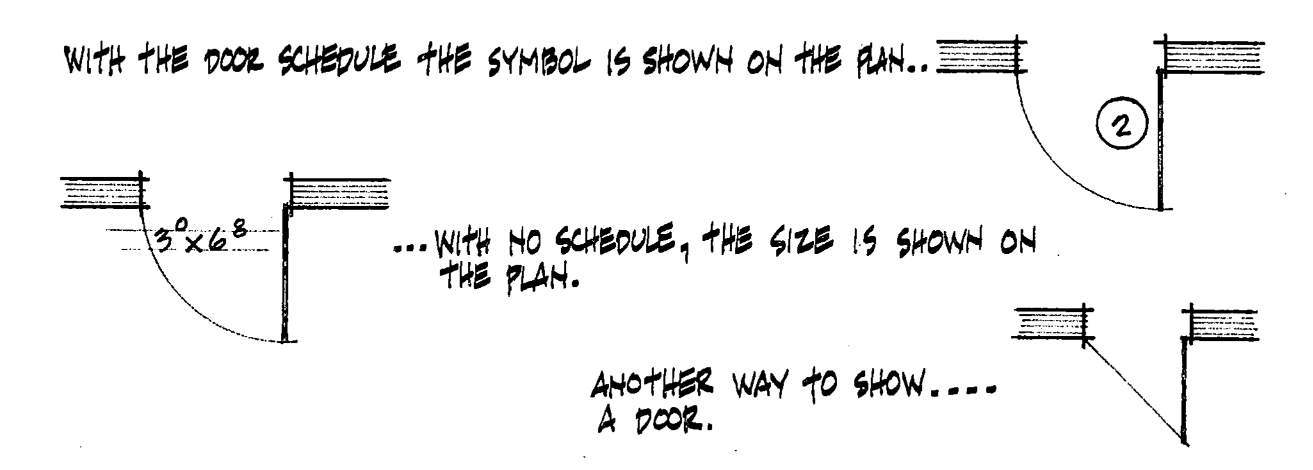

...WITH NO SCHEDULE, THE SIZE IS SHOWN ON THE PLAN.

ANOTHER WAY TO SHOW.... A DOOR.

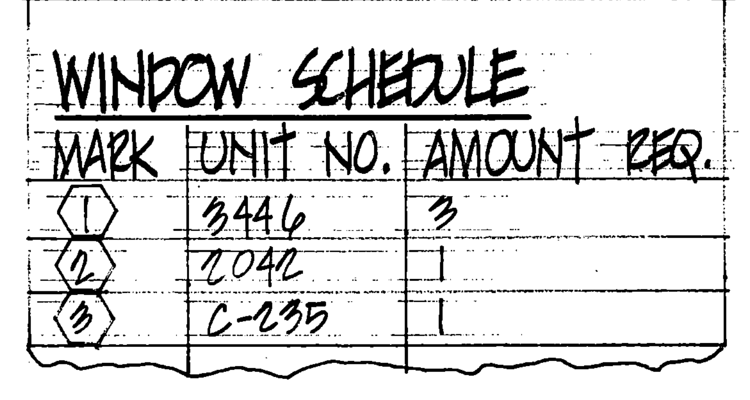

WINDOW SCHEDULE

MARK	UNIT NO.	AMOUNT REQ.
(1)	3446	3
(2)	2042	1
(3)	C-235	1

THE WINDOW SCHEDULE WORKS LIKE THE DOOR SCHEDULE.

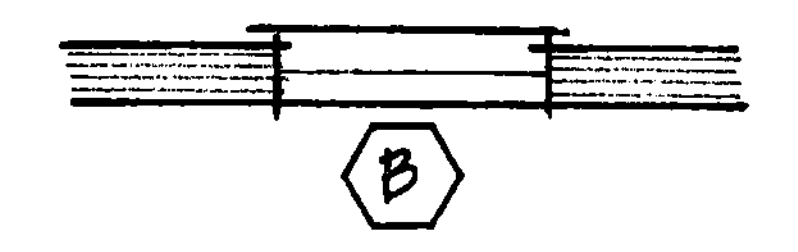

WITH THE SCHEDULE, THE SYMBOL IS SHOWN ON THE PLAN.

WITH NO SCHEDULE, THE UNIT NUMBER IS SHOWN....

ROOM FINISH

ROOM	FLOOR	WALLS	CEILING	BASE	TRIM	REMARKS
ENTRANCE	Vinyl Sheet	½" Sheetrock	½" Sheetrock	Rubber	Pine	
LIVINGROOM	White Oak	½" Sheetrock	½" Sheetrock	1x5	Pine	
DININGROOM	White Oak	½" Sheetrock	½" Sheetrock	1x5	Pine	
KITCHEN	Vinyl Sheet	½" Sheetrock	½" Sheetrock	Rubber	Pine	
BATH #1	Vinyl Sheet	½" Sheetrock	½" Sheetrock	Rubber	Pine	Ceramic tile in tub.

TO AVOID SURPRISING THE BUILDER, A ROOM FINISH SCHEDULE IS A GREAT HELP.

SOME JOBS MIGHT REQUIRE IDENTIFYING EACH WALL IN A ROOM BECAUSE OF MANY DIFFERENT FINISHES. THIS CAN BE DONE WITH THIS SYMBOL ON EVERY FLOOR PLAN, ANY PLACE WILL DO.

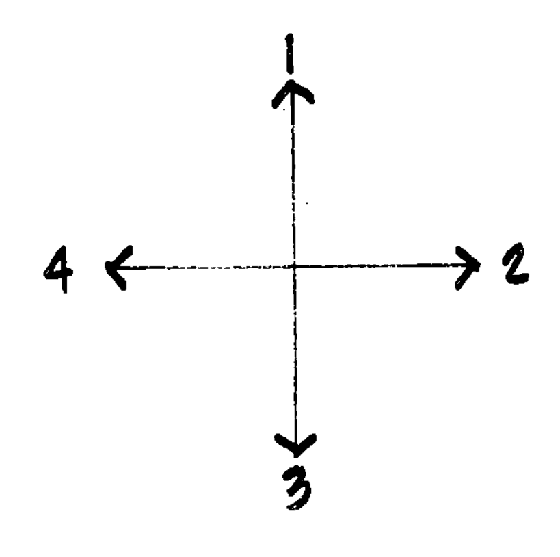

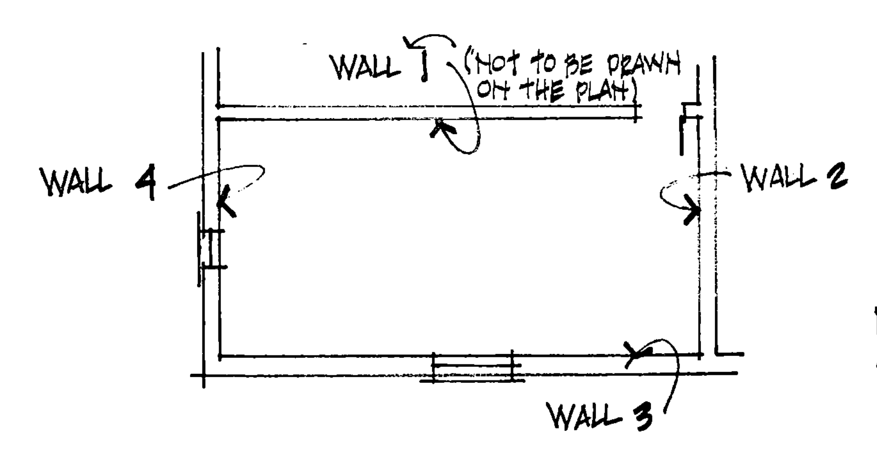

THE ROOM FINISH SCHEDULE WITH THIS SYSTEM WILL SHOW WALL 1, WALL 2, WALL 3, AND WALL 4. THE NUMBER 1 WALL WILL ALWAYS BE THE UPPER, NUMBER 2 THE RIGHT, NUMBER 3 THE BOTTOM, AND NUMBER 4 THE LEFT.

THE LEAD FOR LETTERING SHOULD BE SOFT AND DARK, F OR HB. THE POINT, BEFORE SHARPENING LOOKS LIKE THIS.... NOT UNTIL THE TIP IS SHARPENED BY SANDPAPER OR SHARPENER IS IT READY TO WORK WITH.

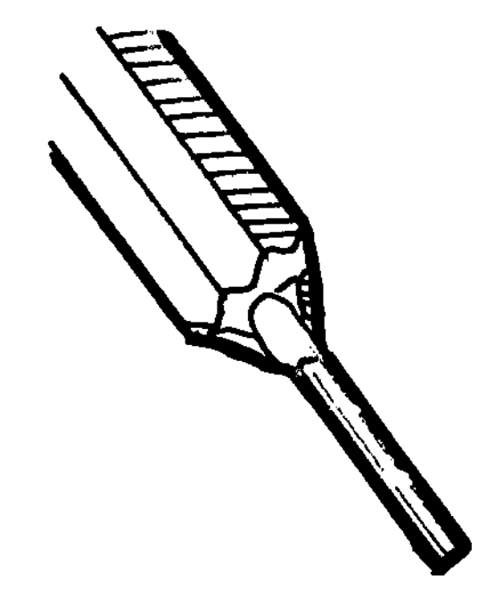

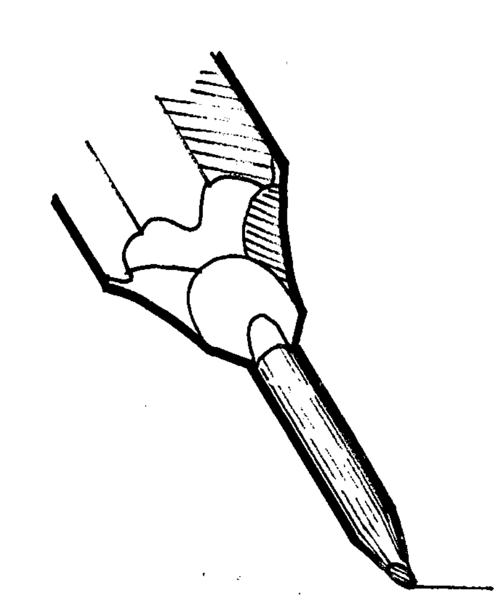

TO GET THIN VERTICAL LINES, ROLL THE PENCIL TO A SHARP CORNER ON THE TIP AND THEN STROKE. FIND THE BROAD PART OF THE TIP FOR THE HORIZONTAL AND CURVED STROKES. THIS COMBINATION MAKES FOR CLEAN CRISP LETTERING.

EXPERIMENT WITH THE TIP; YOU MIGHT LIKE THE CHISEL POINT WHICH IS USED THE SAME WAY, THIN VERTICAL STROKE AND BROAD HORIZONTAL STROKE. MAKE ALL STROKES FROM THE SHOULDER, NOT THE WRIST. USE FIRM SURE VERTICAL STROKES AND QUICK SMOOTH CURVES. THERE IS A LOT OF LETTERING ON A PLAN AND SPEED IS A GREAT HELP IN GETTING A JOB DONE, BUT ABOVE ALL, IT MUST BE CLEAR. IF YOU CAN'T READ IT, THE PRETTY IS WORTHLESS. PRACTICE AND DEVELOPE YOUR OWN STYLE. HOLD THE PENCIL WITH A LIGHT TOUCH AND YOU WILL HAVE BETTER CONTROL.

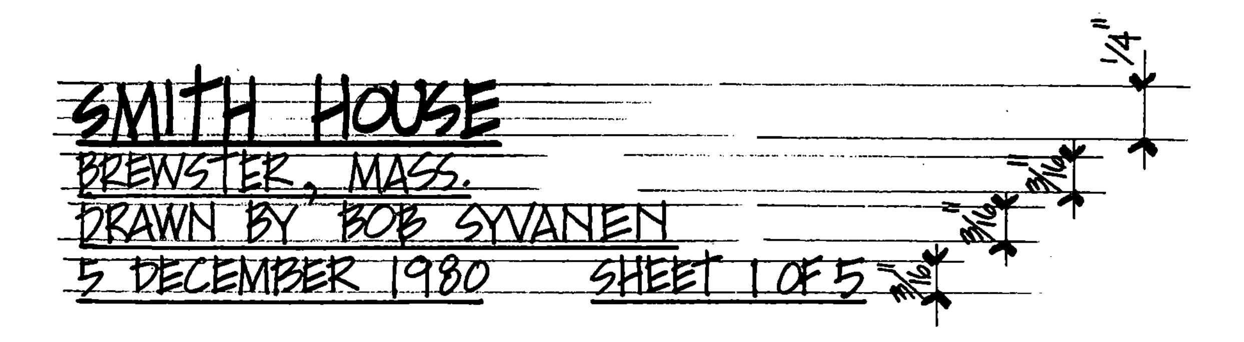

THE TITLES ARE PUT ON ALL THE SHEETS WHEN THE PLANS ARE DONE.

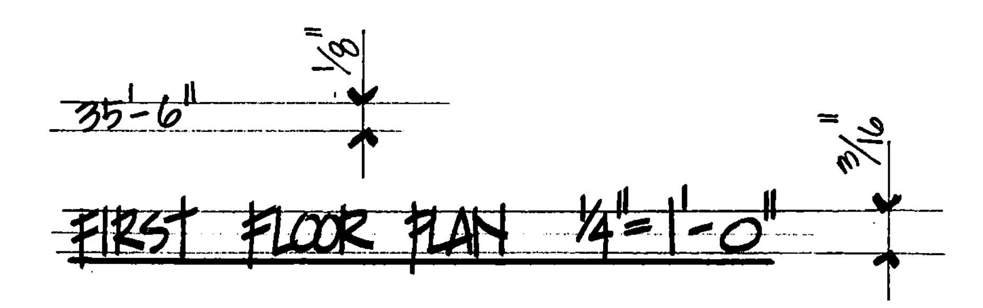

ALL THESE SIZES SUIT ME. SUIT YOURSELF AND THE SPACE AVAILABLE.

BED ROOM 1/8" 35'-4"

ON LETTERS AND NUMBERS, LET VERTICAL LINES EXTEND BEYOND GUIDE LINES. MAKE HORIZONTAL AND CURVES RIDE UP. SOME LETTERS CAN RUN TOGETHER.

BED ROOM 35'-4"

YOU GET A HANG DOG-LOOK WHEN VERTICAL LINES SLANT AND HORIZONTAL LINES DROOP.

| | | QUICK FIRM VERTICAL STROKES.

3 C C D D QUICK SMOOTH CURVES.

LET THE GUIDE LINES SHOW ON THE PRINTS, I THINK THEY HELP THE LOOKS OF THE LETTERING. KEEP THEM THIN SO THEY DON'T CONFLICT WITH DIMENSION LINES.

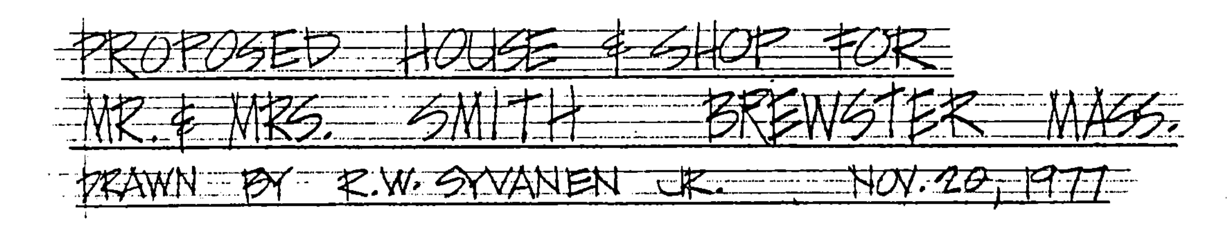

THIS IS MY STYLE FOR FAST LETTERING. I USED THREE LINES.

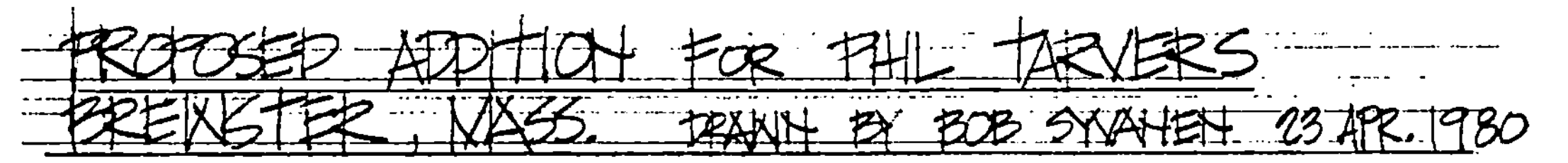

YOU CAN DO THE SAME THING ON TWO LINES.

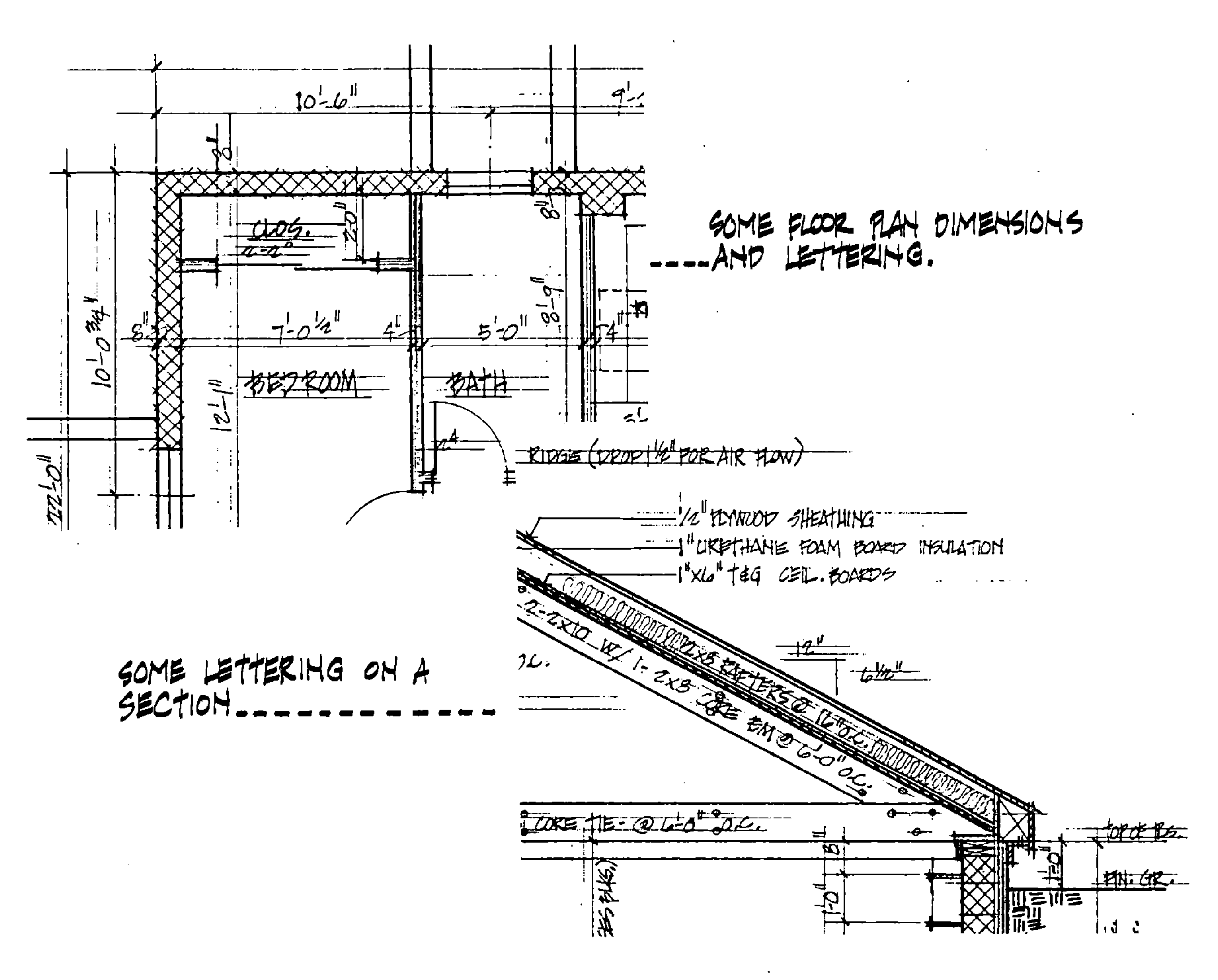

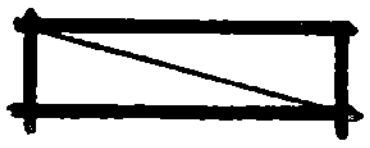

WOOD BLOCKING

WOOD TIMBER

FINISH WOOD OR TRIM

PLYWOOD

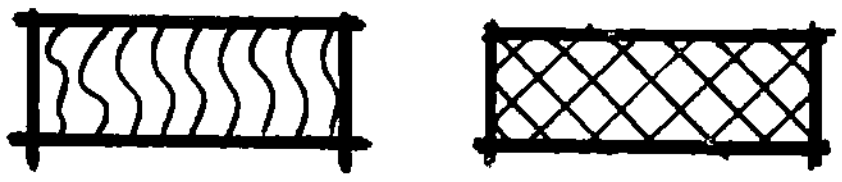

CONCRETE BLOCK

POURED CONCRETE

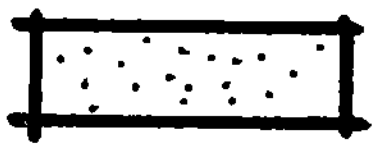

SAND

FIBERGLASS INSULATION

RIGID INSULATION

FOAM BOARD

BRICK (SECTION)

THESE ARE THE MOST COMMON SYMBOLS.

DRAFTING

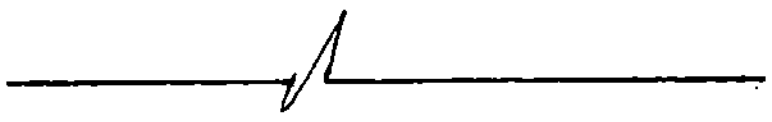

BROKEN LINE OR SECTION

CENTER LINE

BURIED OR HIDDEN LINE

BEAM

GUIDE LINES

DIMENSION LINES

HEAVY OUTLINE

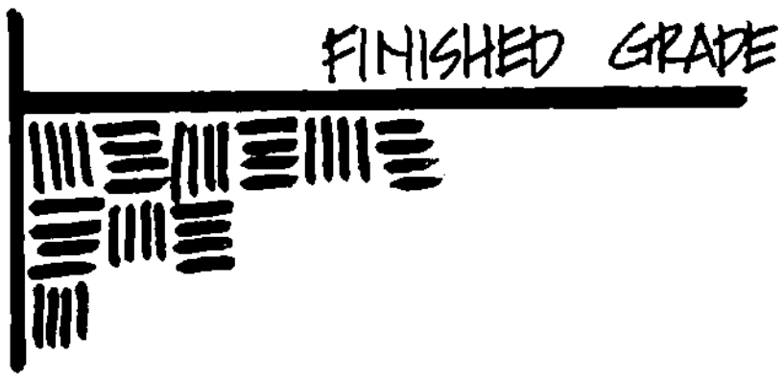

FINISHED GRADE

EARTH

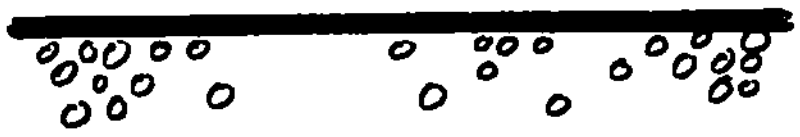

GRAVEL

ABBREVIATIONS THESE ABBREVIATIONS ARE ALWAYS USED. TRY TO SPELL ALL OTHER WORDS FOR CLARITY.

O.C. — ON CENTER — 16" O.C.

& — AND — SAND & GRAVEL

@ — AT — 2x4 STUDS @ 16" O.C.

℄ — CENTER LINE

W/ — WITH — ½" W/PLYWOOD OVER

CONC. — CONCRETE

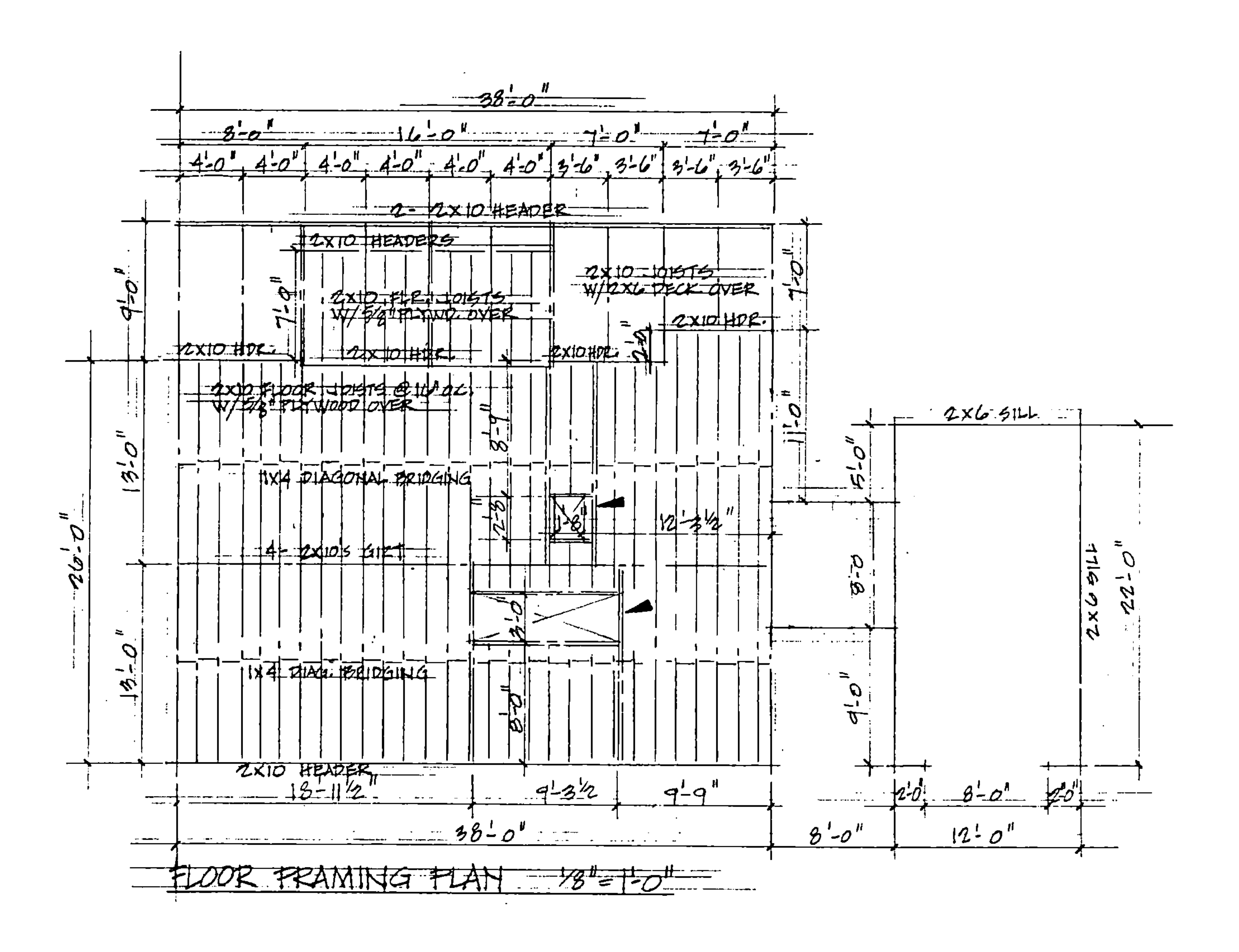

FLOOR FRAMING PLANS CAN BE AT ANY SCALE FROM 1/16"=1'-0" TO 1/4"=1'-0". THIS PLAN WORKED WELL FOR ME AT 1/8"=1'-0"! WHERE CHIMNEY AND STAIRS ARE, X THEM TO SHOW THERE IS NO FRAMING.

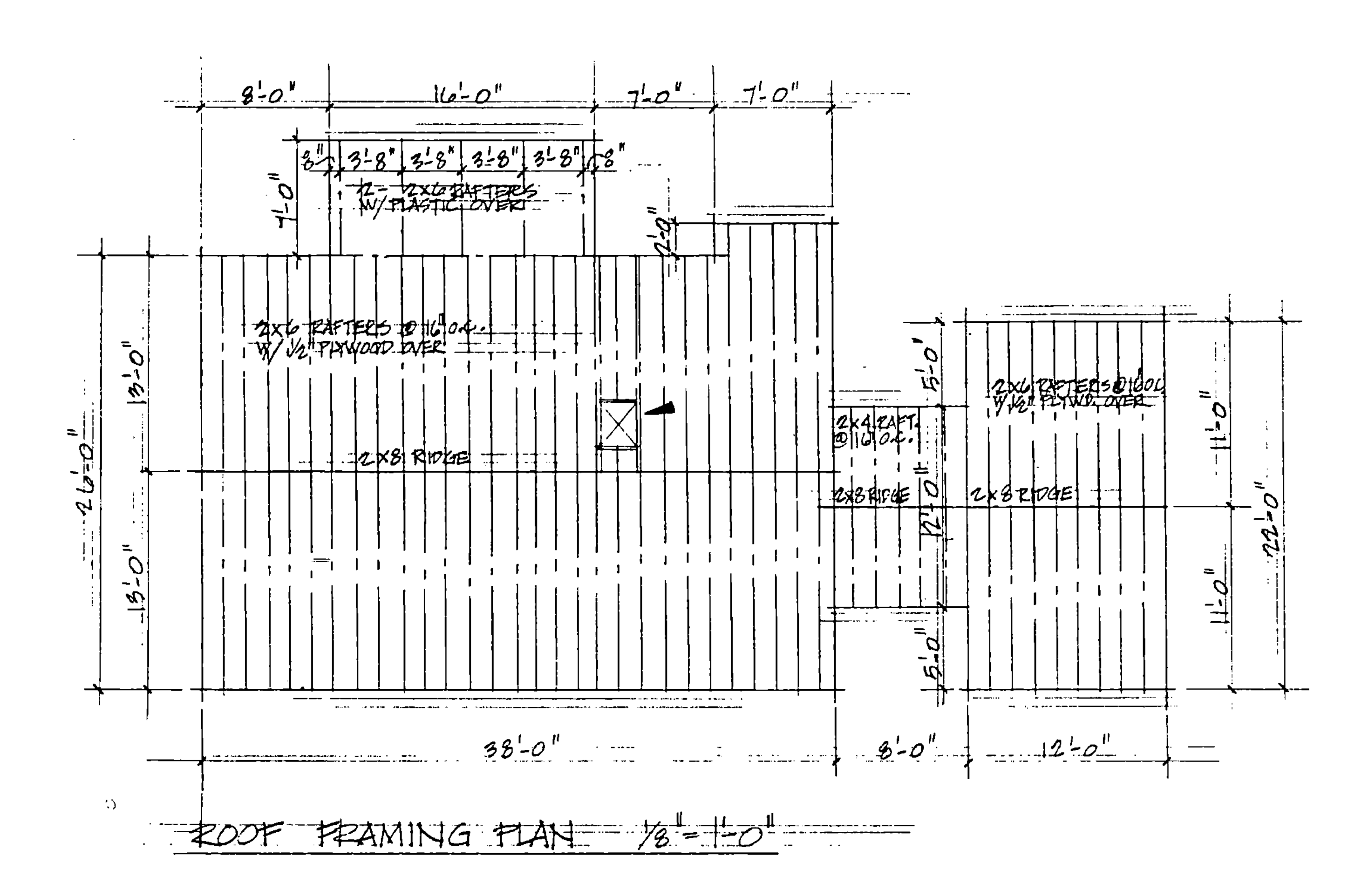

THIS ROOF FRAMING PLAN IS FROM THE SAME SET OF PLANS. AGAIN, X WHERE THERE IS AN OPENING. IN THIS CASE IT IS THE CHIMNEY.

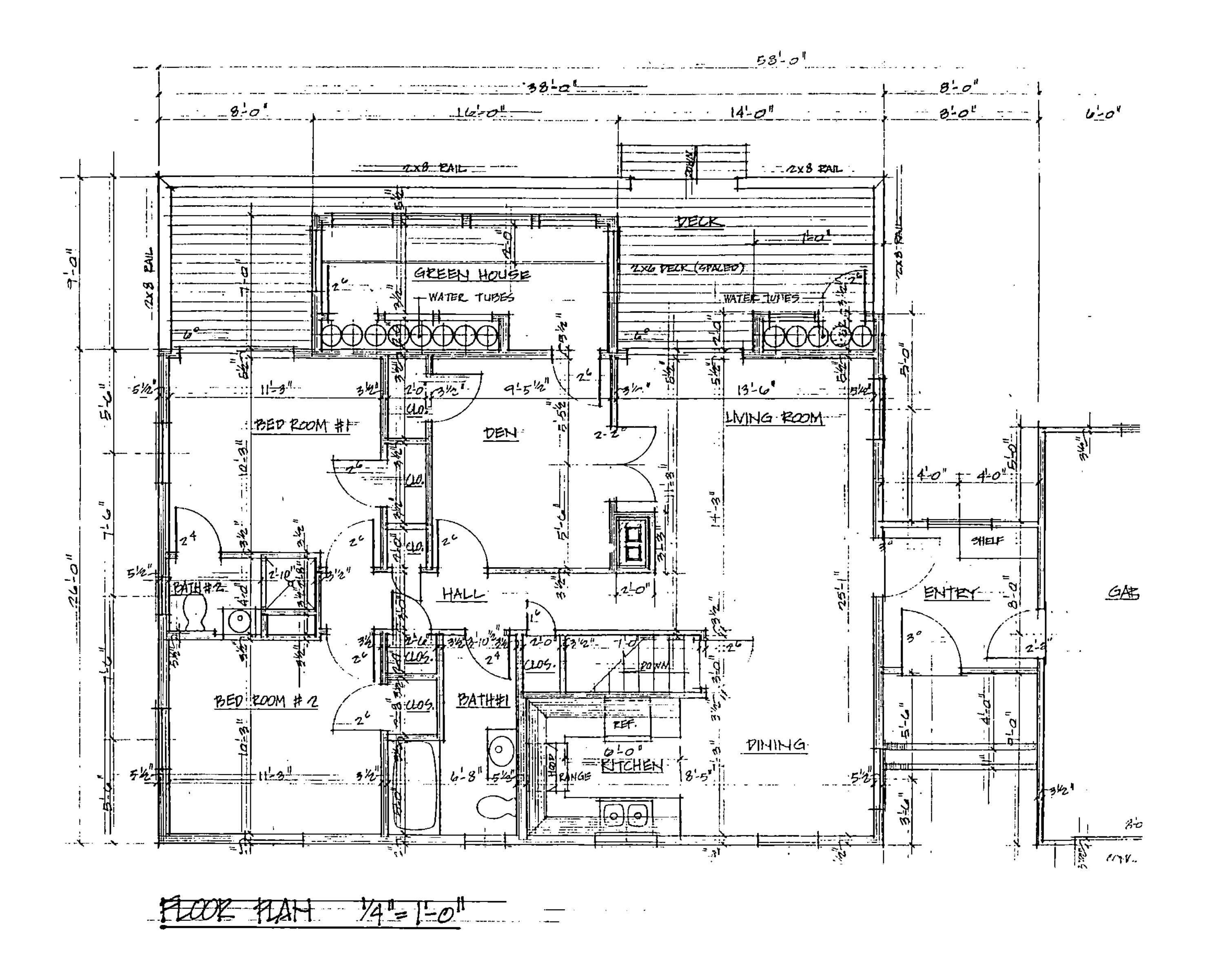

FLOOR PLAN 1/4" = 1'-0"

THIS FLOOR PLAN IS TYPICAL. IT SHOWS ALL THAT IS NECESSARY FOR CONSTRUCTION. THE ROOF AND FLOOR FRAMING PLANS ARE FROM THIS PLAN.

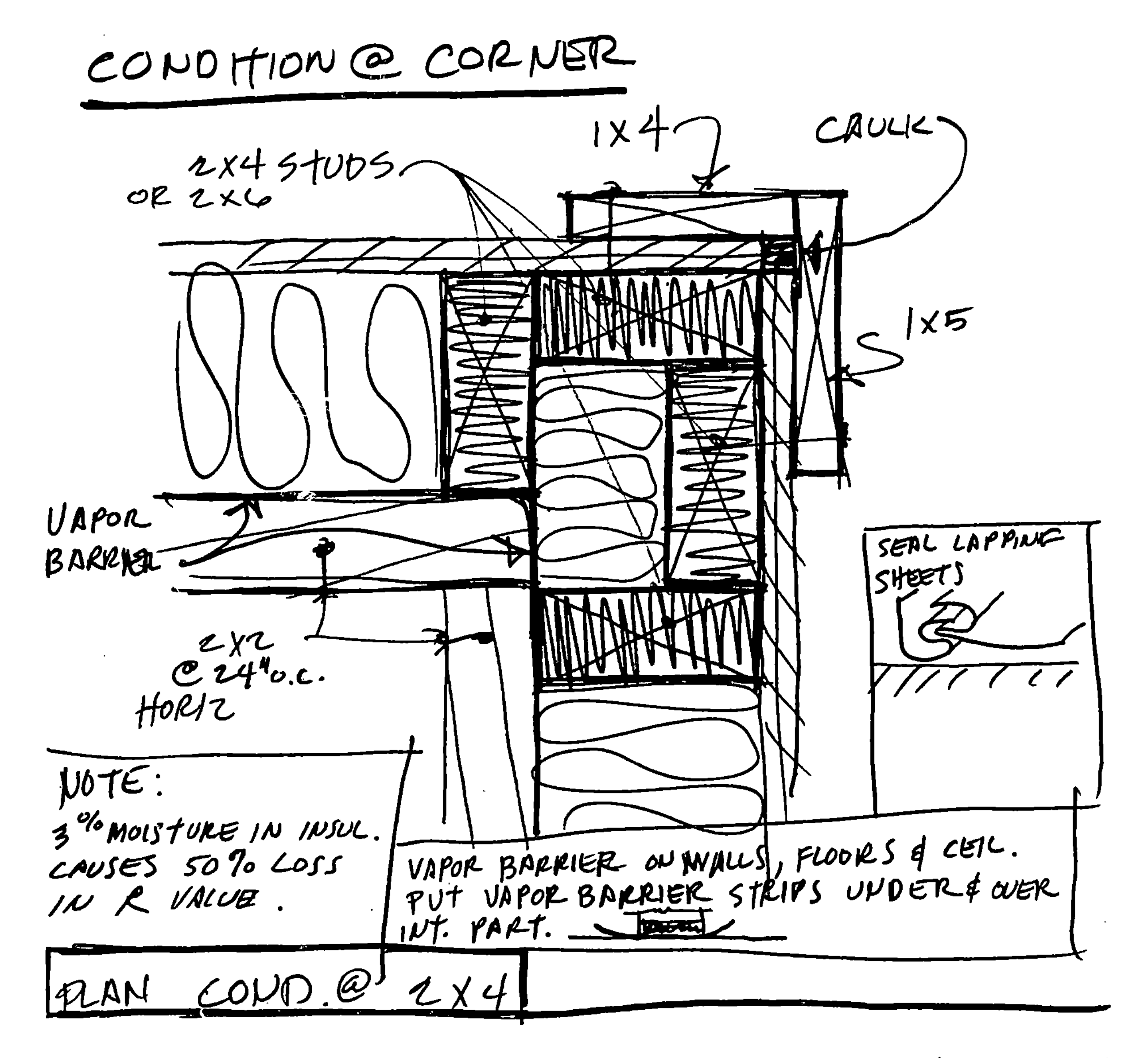

DETAILS CAN BE AS SIMPLE AS THIS TO GET THE JOB DONE. THIS IS WHAT THE CARPENTER DOES ON THE JOB IF THE DETAILS ARE NOT ON THE PLANS. IF YOU WANT THE JOB DONE YOUR WAY IT IS BEST TO DRAW IT FIRST. THIS IS A PLAN OF AN OUTSIDE CORNER FOR A SCANDANAVIAN WALL. THERE IS GOOD NAILING FOR CORNER BOARDS AND ACCESS TO INSULATE THE CORNER FROM INSIDE THE BUILDING. THIS MAKES A NICE TIGHT DRAFT-FREE CORNER.

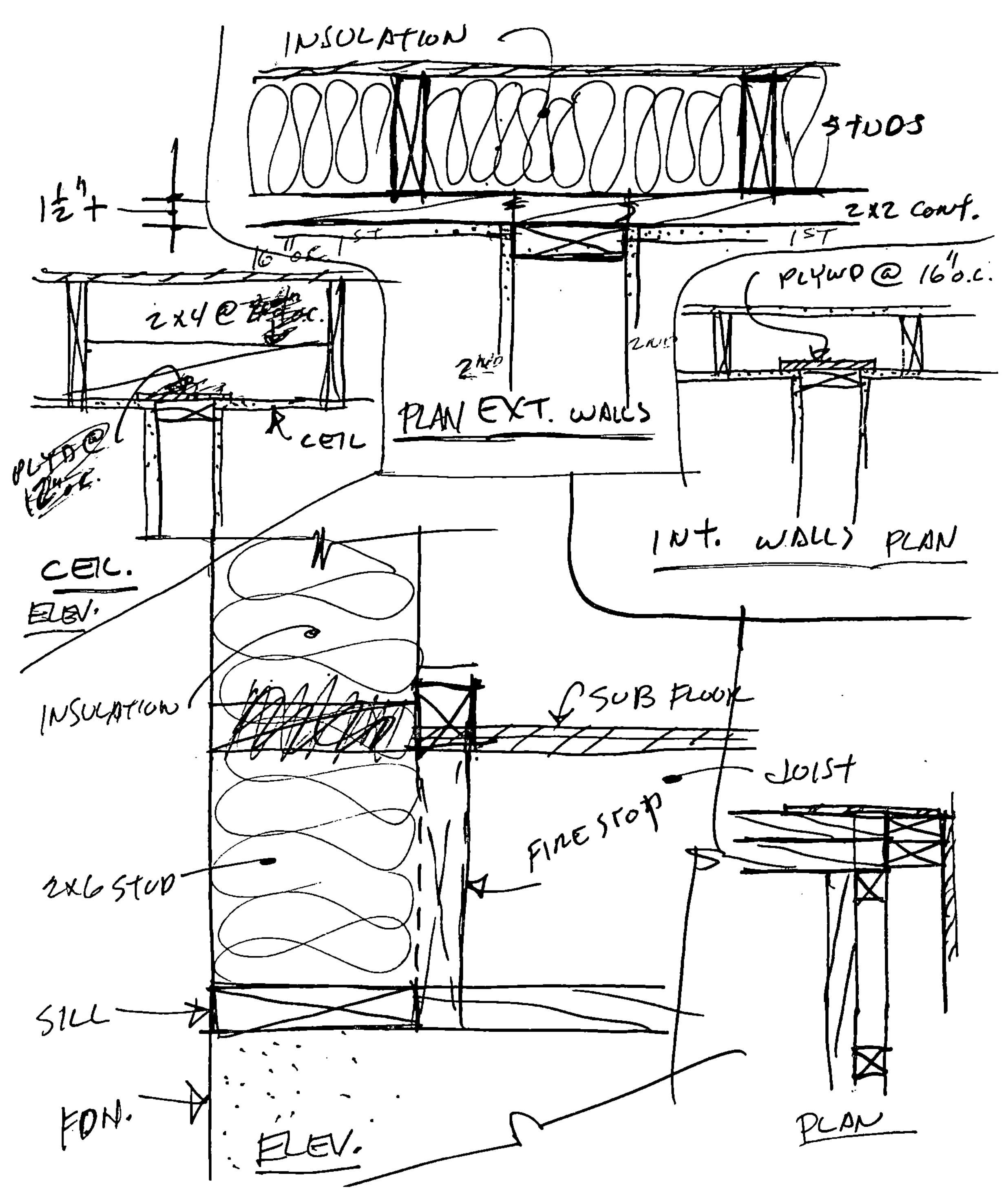

MORE QUICK SKETCH DETAILS THAT WORK, BUT ARE NOT ALWAYS CLEAR TO OTHERS
USE FELT TIP PEN OR SOFT PENCIL.

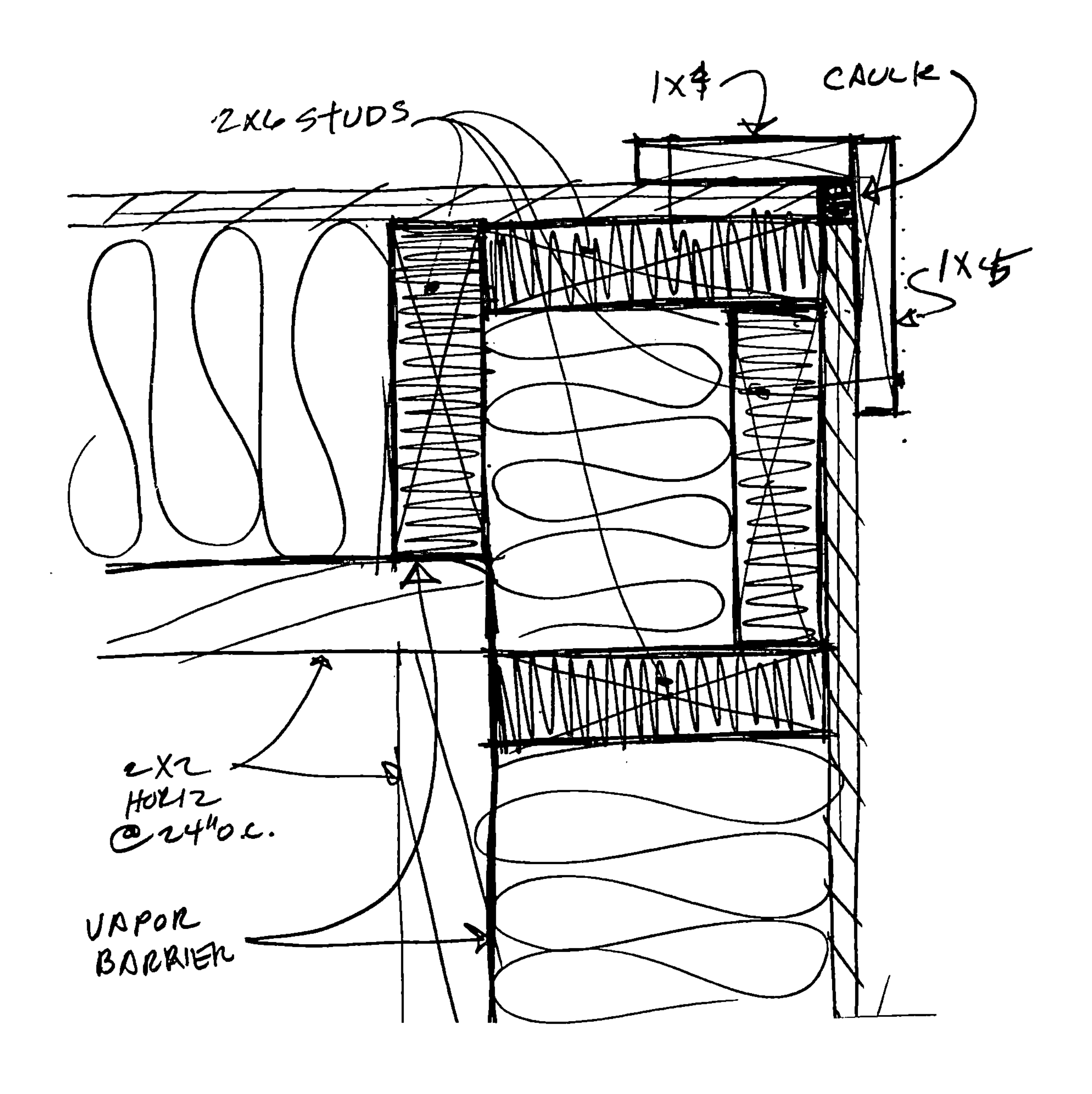

THE SAME SCANDINAVIAN WALL DETAIL WITH 2x6 STUDS. ALL THESE QUICK SKETCH DETAILS CAN BE USED AS IS AND XEROX COPIES MADE TO PASS AROUND.

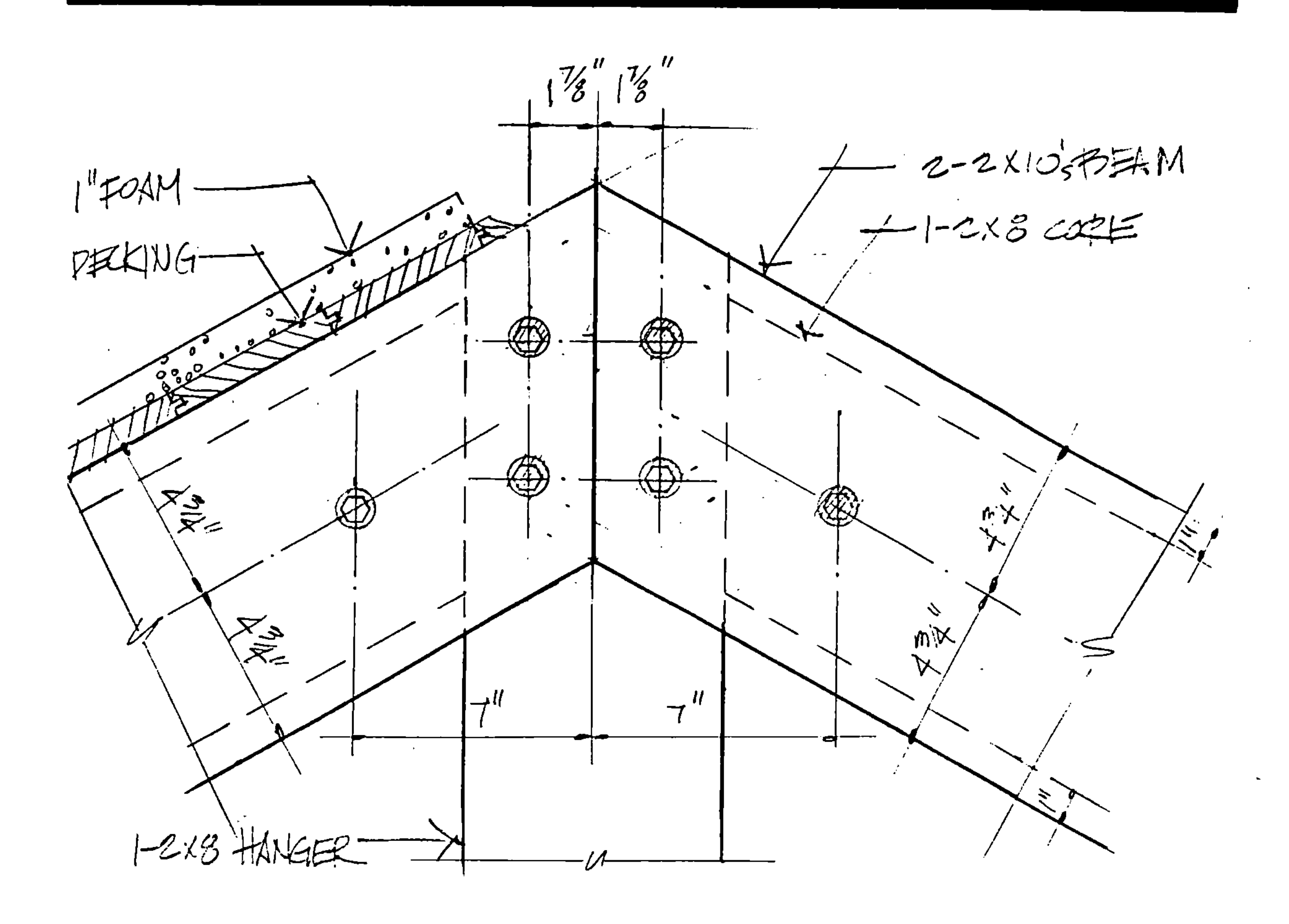

BEAM & HANGER @ RIDGE

3" = 1'-0"

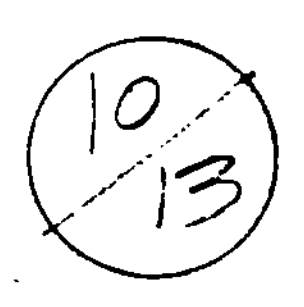

WHEN A LITTLE MORE PRECISION IS REQUIRED, I GO TO THIS SYSTEM. AT THIS SCALE, THINGS CAN BE MEASURED OFF EASILY.

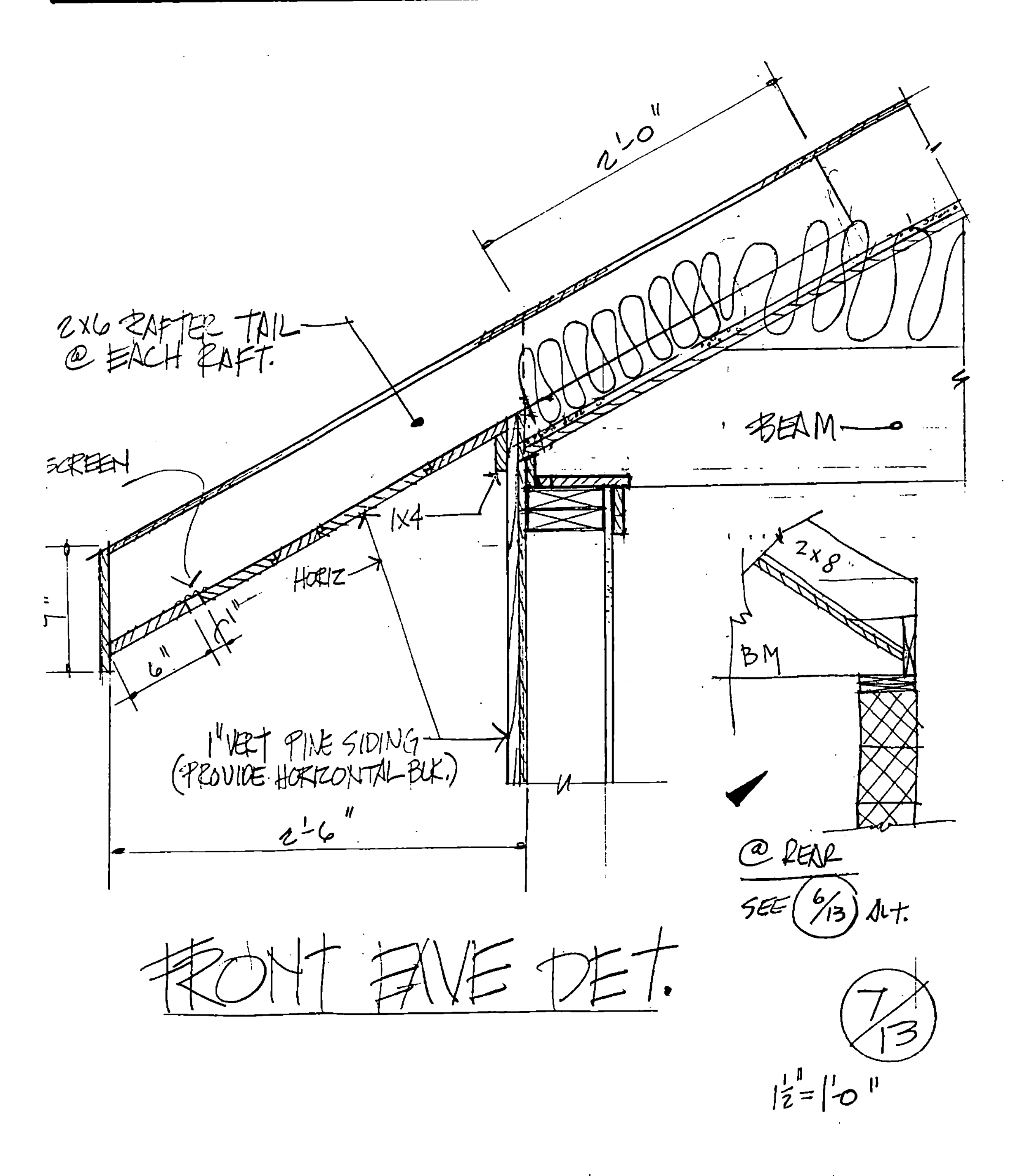

WHAT IS DRAWN DOESN'T ALWAYS WORK BEST AND A REVISION TAKES PLACE IN THE FIELD.

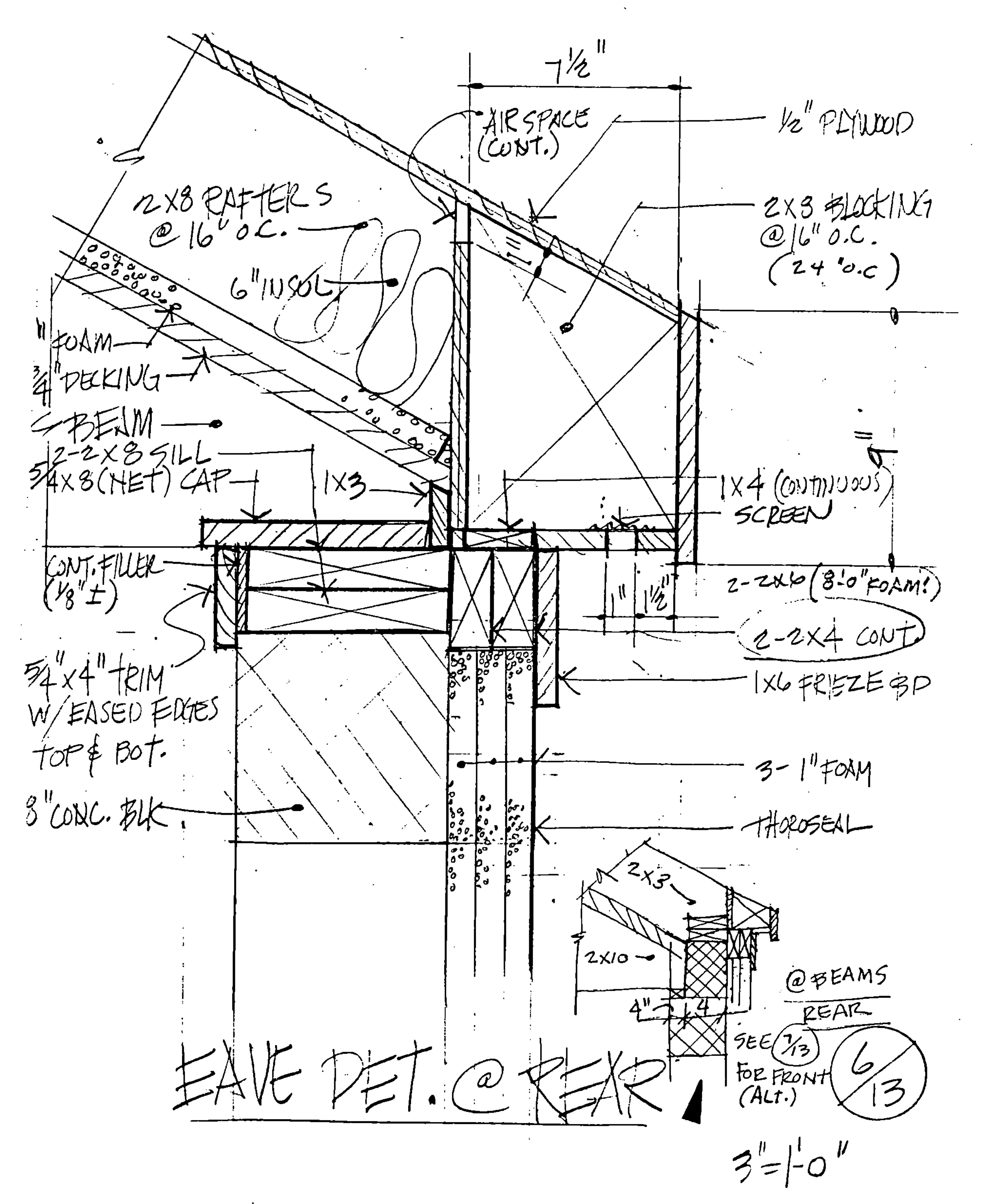

ONCE AGAIN YOU THINK YOU HAVE IT ALL FIGURED, BUT AT LEAST THERE IS A PLACE TO START FROM.

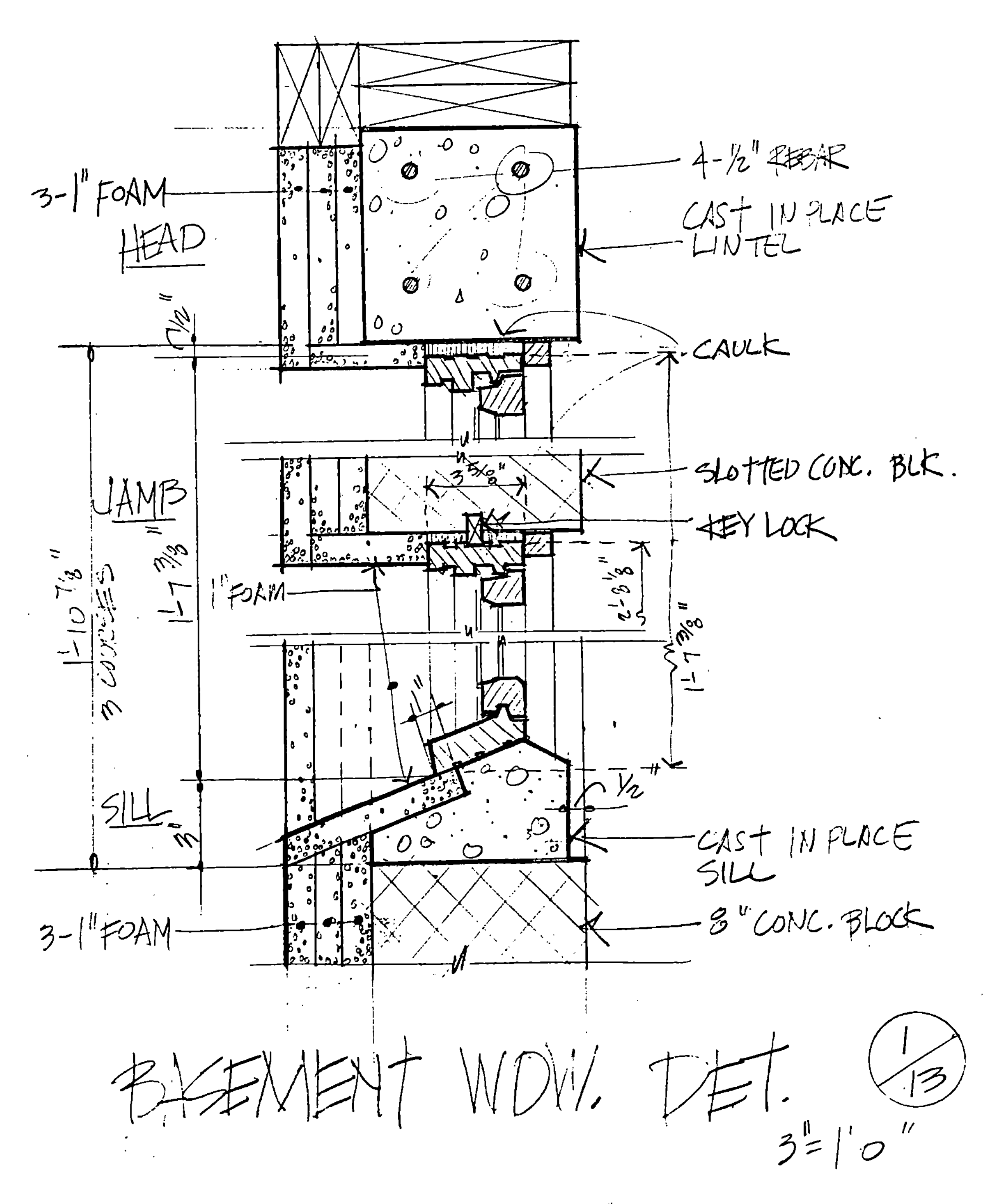

ALL THESE DETAILS WERE DRAWN ON 8½" x 11" SHEETS, IN PENCIL, AND XEROX COPIES MADE FOR USE ON THE JOB.

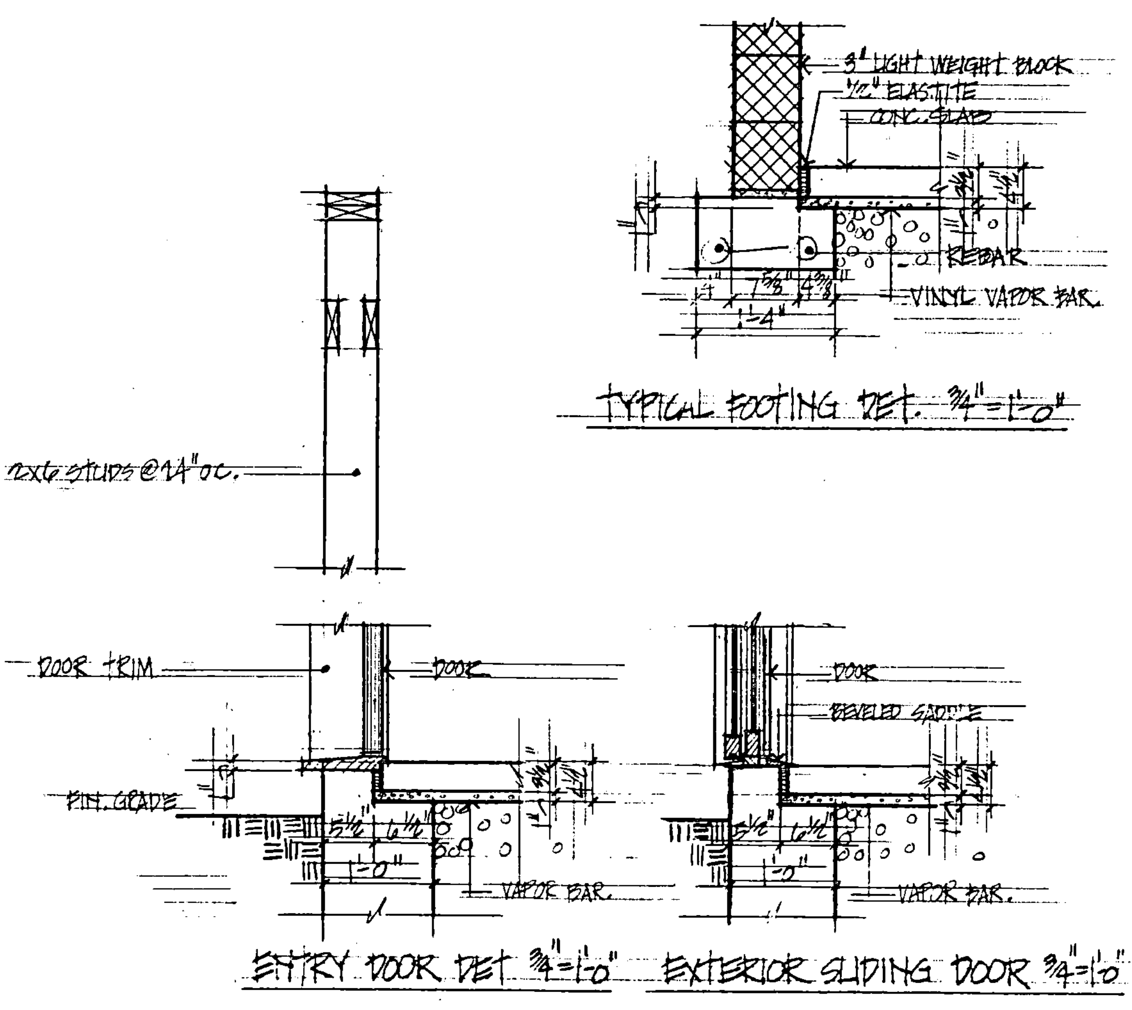

FROM BOTH ROUGH AND REFINED SKETCHES, THESE DETAIL DRAWING WERE MADE ON THE PLANS. PLANS WILL READ BETTER IF ALL RELATED DETAILS LIKE FOUNDATION DETAILS OR WINDOW DETAILS, ARE KEPT TOGETHER.

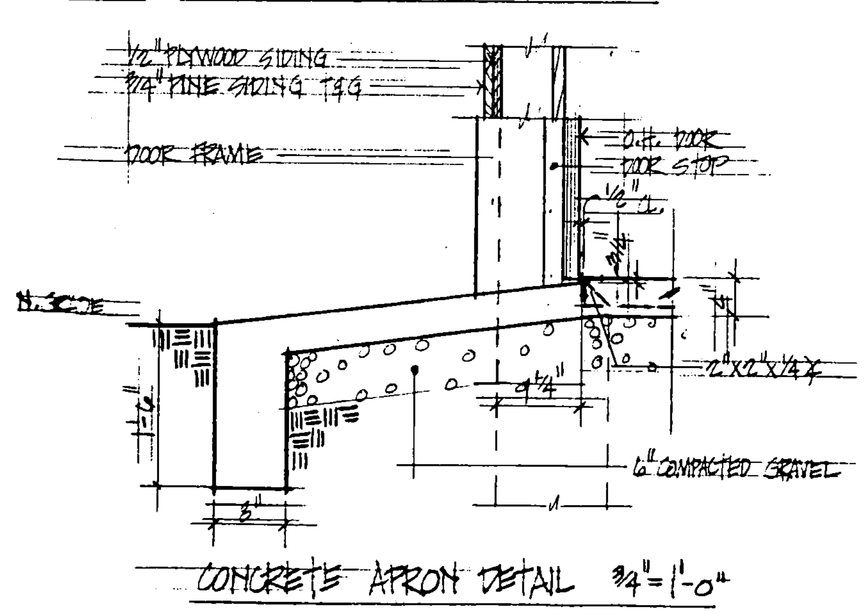

PERSPECTIVES CAN BE A GREAT HELP IN VISUALIZING THE FINISHED PRODUCT. THEY CAN BE PLAIN OR FANCY. THERE IS A LOT WRITTEN ON THIS SUBJECT AND IN FACT MOST ARCHITECTURAL FIRMS FARM THIS WORK OUT.

ELEVATION "A" EXISTING BUILDING — TEXTURE 1-11 PLYWOOD SIDING

YOU MIGHT TRY SIMPLE WORK LIKE THIS THAT I DID FOR A REMODEL JOB.

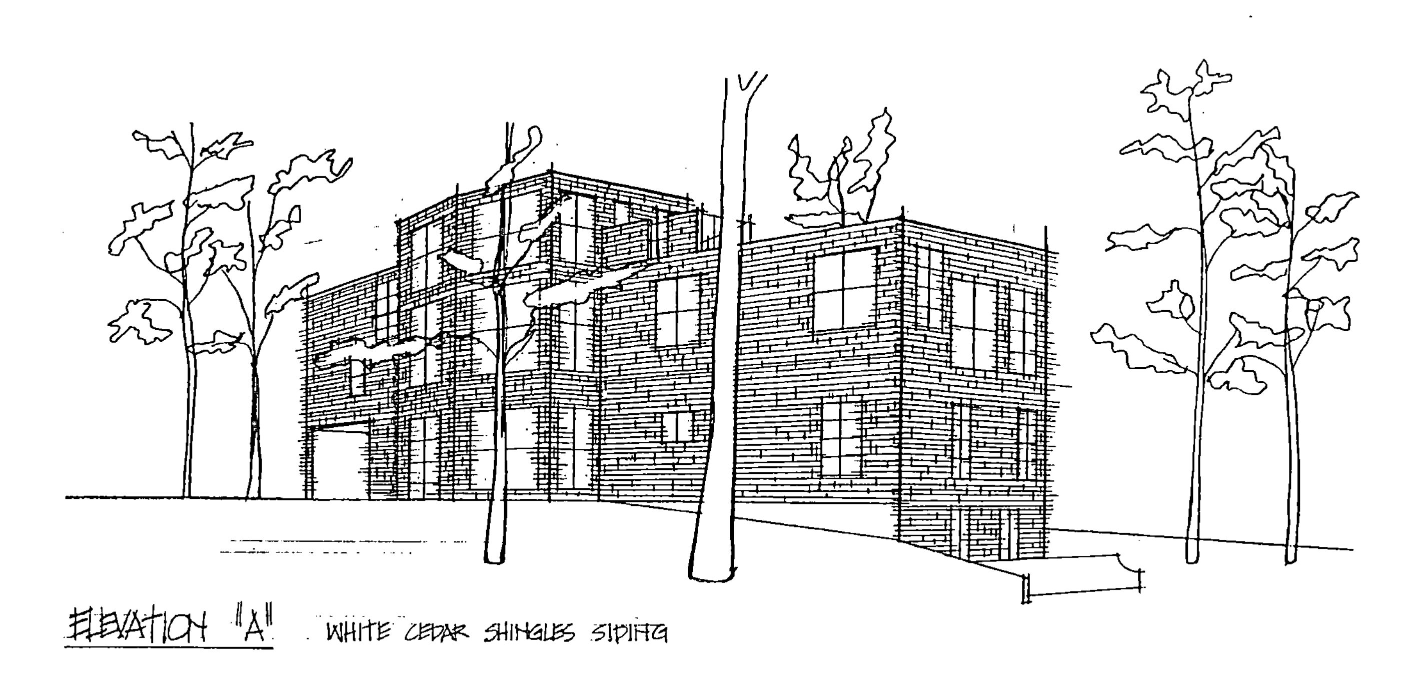

ELEVATION "A" . WHITE CEDAR SHINGLES SIDING

I USED SIMPLE PERSPECTIVE TECHNIQUES, THE OLD "VANISHING POINTS" WE ALL LEARNED ABOUT IN HIGH SCHOOL.

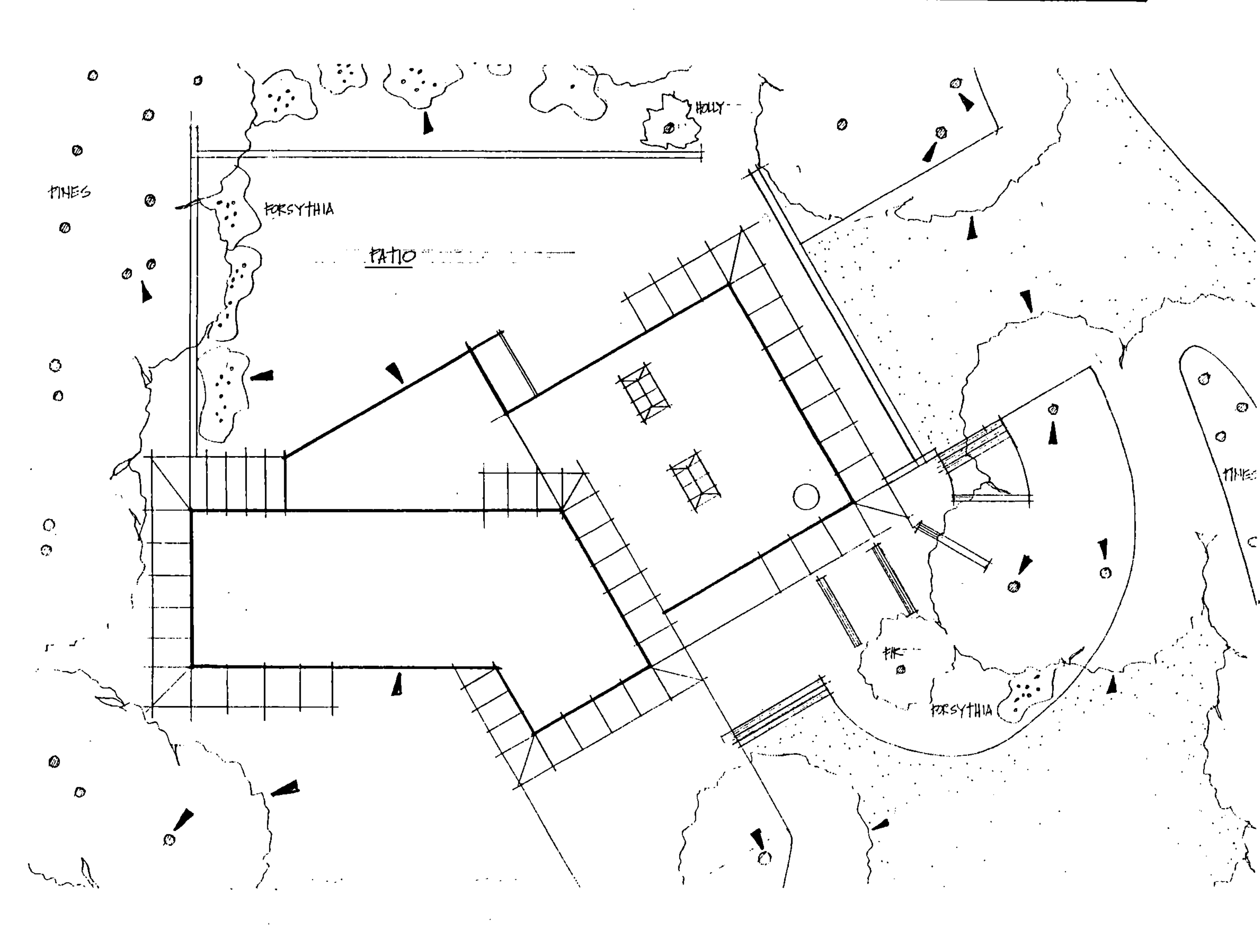

THIS LANDSCAPE PLAN SHOWS THE ROOF AND SURROUNDING FOLIAGE. THE TREE TRUNKS ARE LOCATED AND THEN THE SHAPES ARE SIMULATED AROUND THEM.

A SMALL HOUSE IS USUALLY EASY TO DRAW, BUT NOT ALWAYS EASY TO DESIGN. WITH A LARGE HOUSE THERE ARE ALL TYPES OF SPACE TO WORK WITH- STAIRWAYS, KITCHEN APPLIANCES, TUBS, AND SINKS WILL FIT ANYWHERE. A SMALL HOUSE IS LIMITED AND CLEVER DESIGN IS NECESSARY.

THE FOLLOWING PAGES SHOW A SMALL PASSIVE SOLAR HOUSE THAT WORKED WELL. AS WITH ANY HOUSE, THE OWNERS MADE CHANGES THEY WISH THEY HADN'T, AND THOUGHT OF THINGS THEY SHOULD HAVE DONE. I DON'T KNOW OF ANYONE WHO HAS BUILT THE "PERFECT HOUSE". THERE ARE ALWAYS THINGS THAT SHOULD HAVE BEEN DONE DIFFERENTLY.

SKYLIGHTS ARE GREAT, BUT THEY HAVE A TENDENCY TO LEAK, ARE DIFFICULT TO SHADE IN SUMMER, AND LOOSE MUCH HEAT IN WINTER. A WELL BUILT SKYLIGHT, PROPERLY INSTALLED, SHOULD NOT LEAK. MOTHER NATURE, WITH A TALL LEAFY TREE, OR ANY OTHER WELL DESIGNED SUN SCREEN, WILL TAKE CARE OF THE SUMMER SUN. INSULATING PANELS OR DRAPES (THE SIMPLER THE BETTER) WILL KEEP THE WINTER HEAT IN.

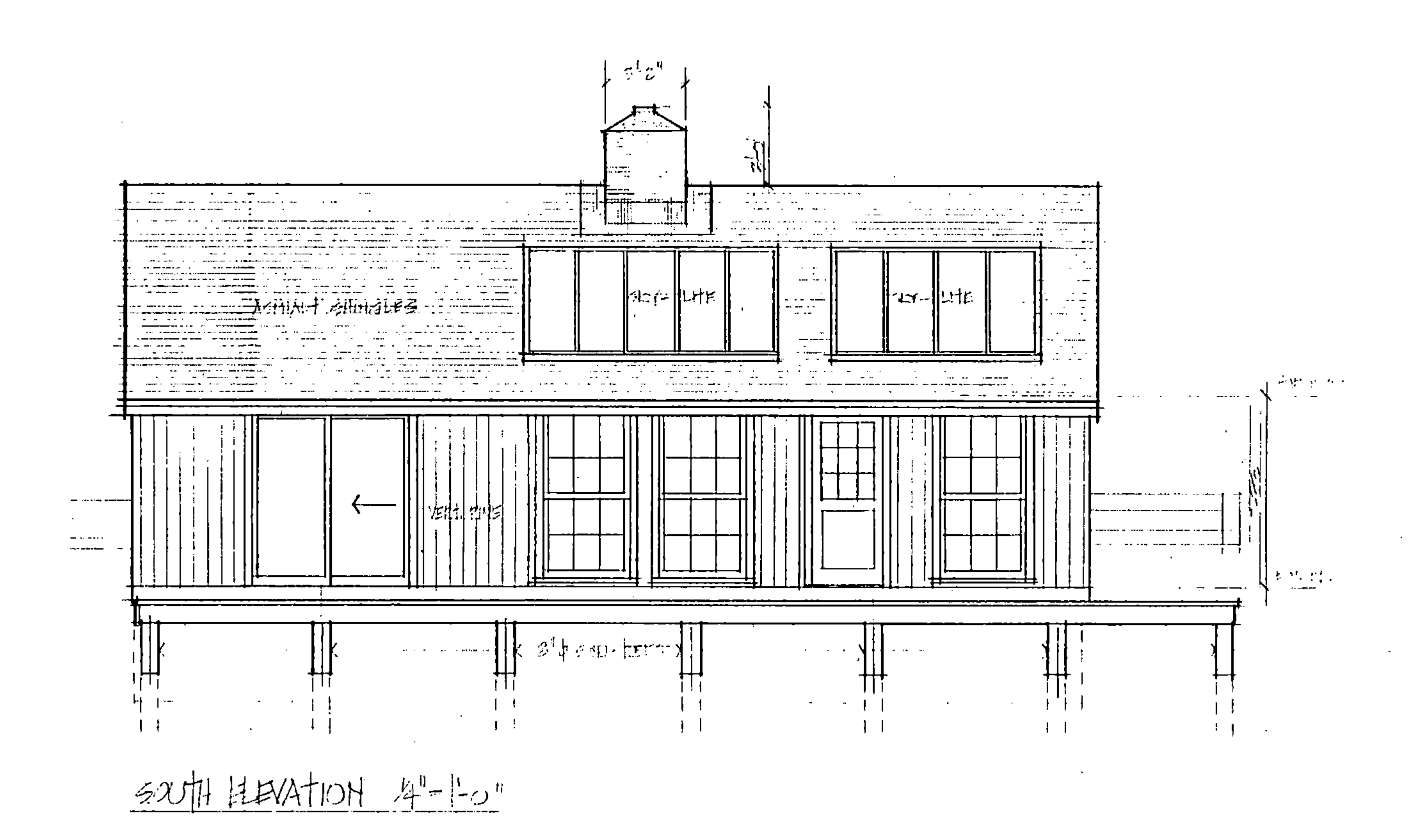

THEY DON'T ALWAYS TURN OUT THE SAME AS THE PLANS.

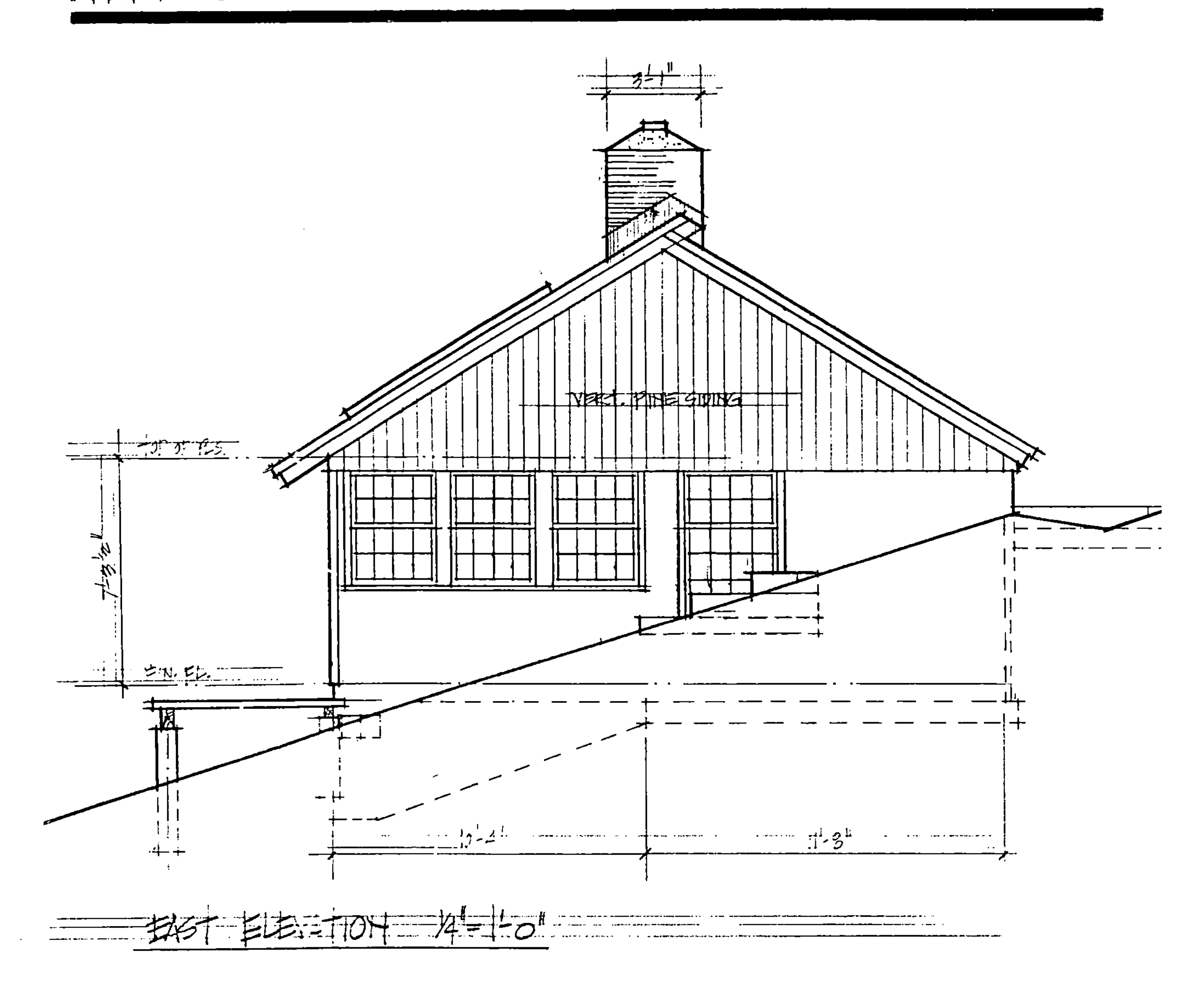

THE HOUSE WASN'T BUILT ON THE ORIGINAL LOCATION AND THAT'S WHY THE DIFFERENCE IN GROUND SLOPE. IT WORKED BETTER THIS WAY.

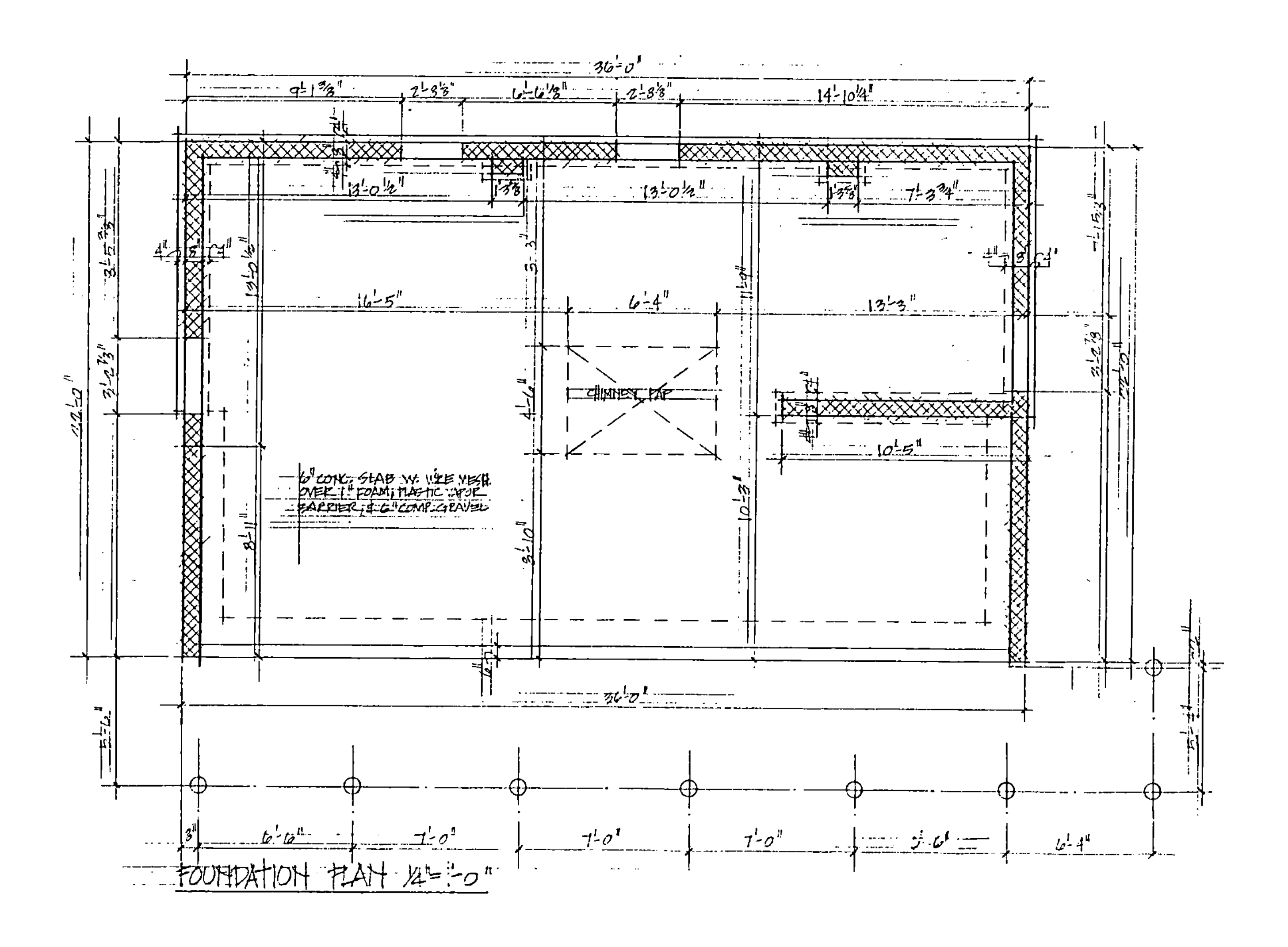

THE FOUNDATION WAS TO BE CONCRETE BLOCK, BUT POURED CONCRETE WAS USED TO SAVE TIME.

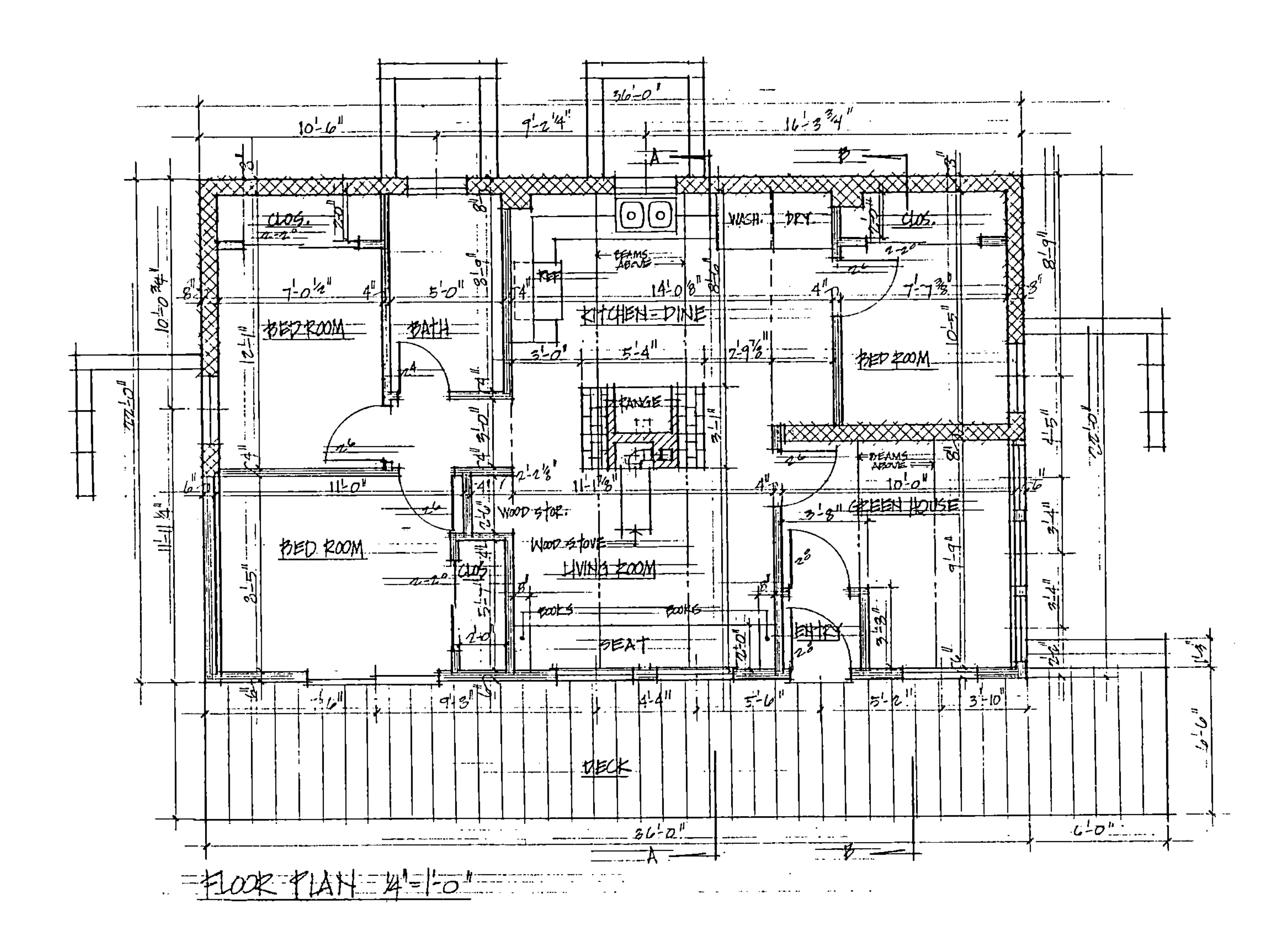

IT WAS A SIMPLE FLOOR PLAN THAT WAS EASY TO DRAW. THE AIR-LOCK ENTRY WAS NOT BUILT; THEY WISH THEY HAD.

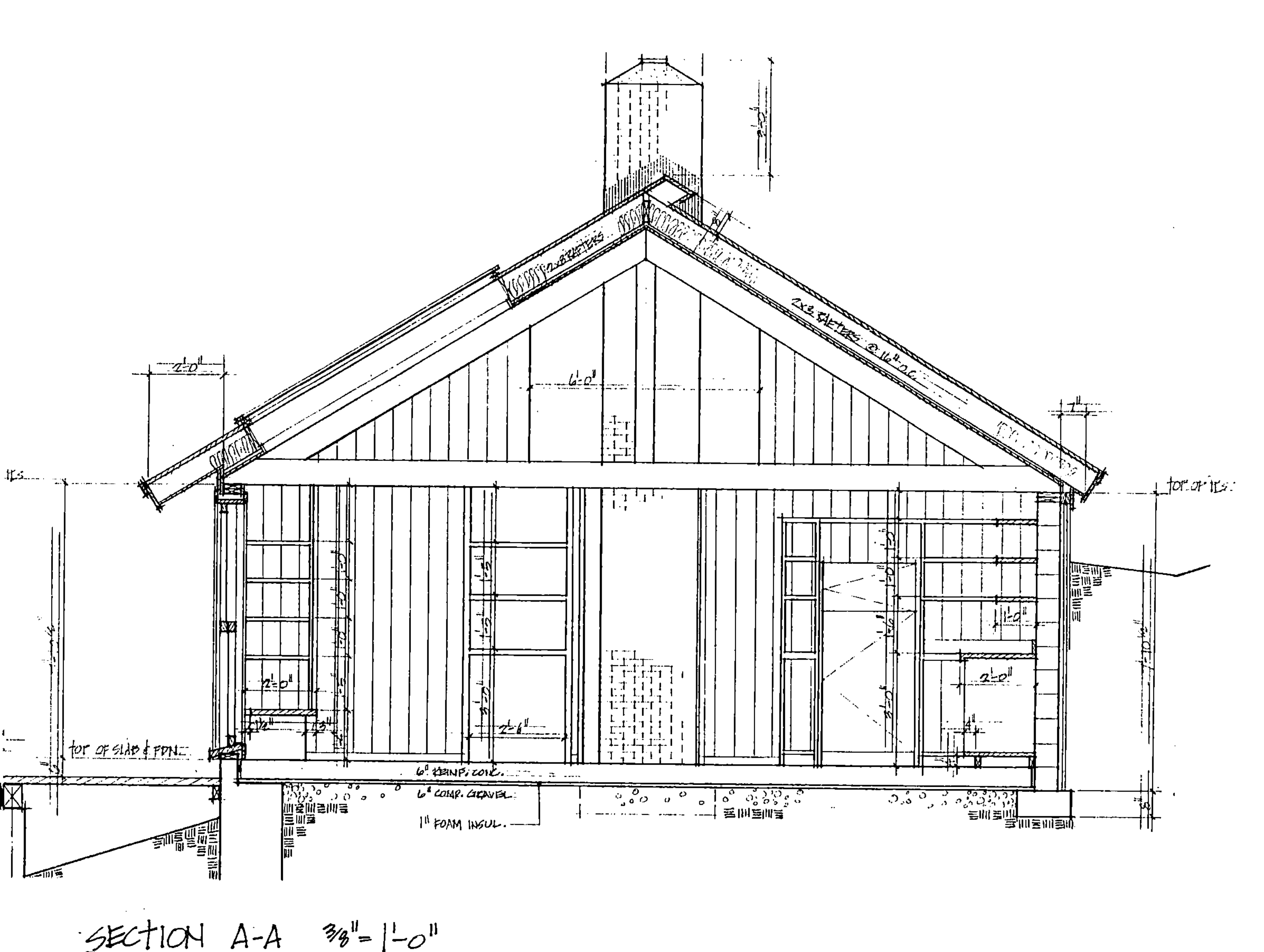

SECTION A-A 3/8" = 1'-0"

A SECTION THROUGH THE BUILDING SHOWING ALL THAT YOU WOULD SEE IF LOOKING IN THAT DIRECTION. THE FLOOR PLAN SHOWS WHERE IT CUTS THROUGH. THE FOAM INSULATION UNDER THE CONCRETE SLAB FLOOR SHOULD BE UNDER THE GRAVEL AND ON TOP OF A VAPOR BARRIER. THE GRAVEL WOULD THEN STORE HEAT WITH THE FLOOR SLAB.

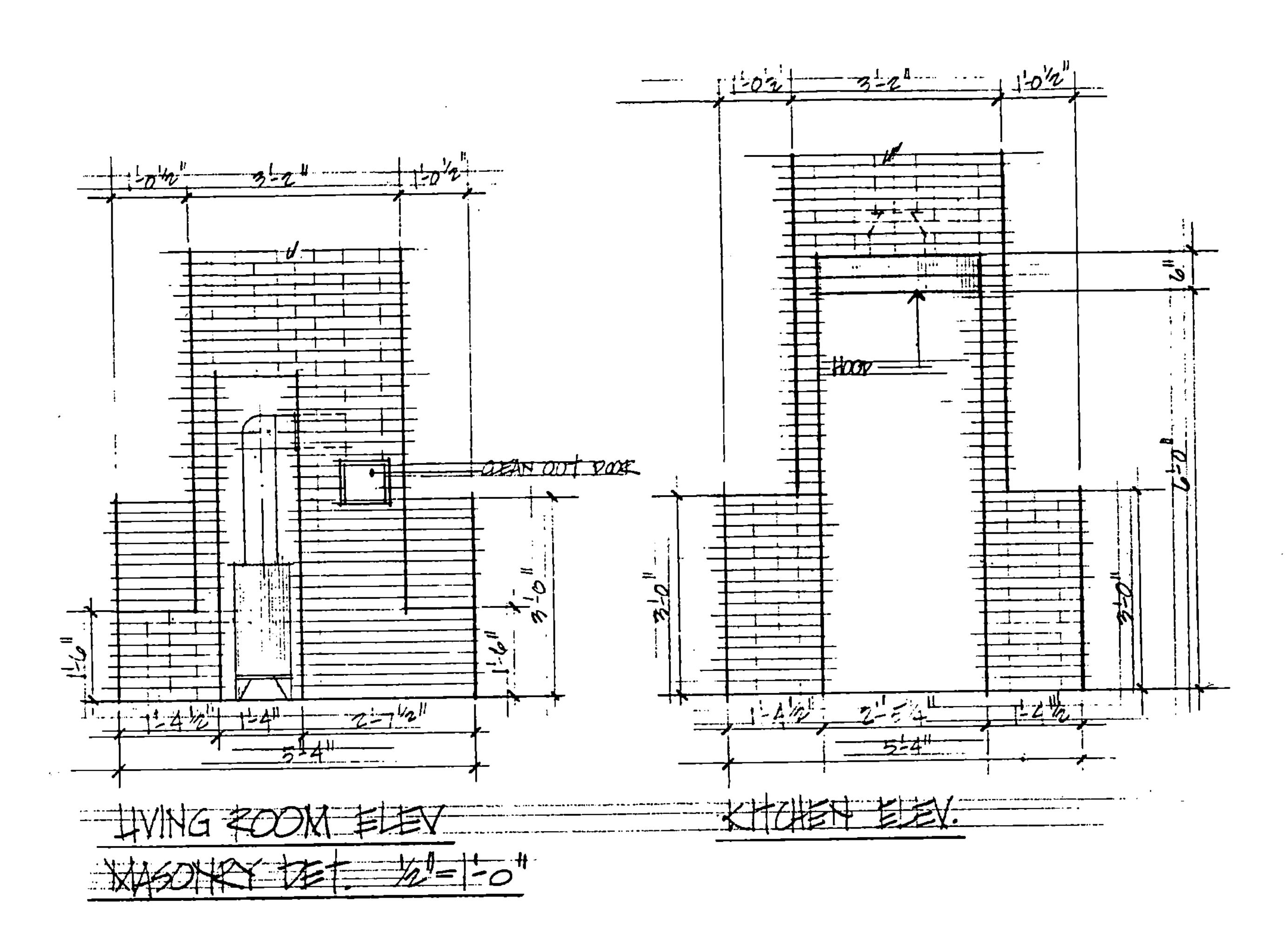

THIS MASONRY UNIT WAS NOT BUILT EITHER. THEY OPTED FOR A FIREPLACE WHICH TURNED OUT VERY NICE. FIREPLACES ARE NOT VERY ENERGY EFFICIENT, BUT THEY CAN BE GOOD HEAT RADIATORS IF THE RUMFORD STYLE IS USED. THE RUMFORD STYLE OPENING IS WIDE AND HIGH (ABOUT AS HIGH AS IT IS WIDE), WITH A SHALLOW HEARTH. THERE IS A GOOD PAPER BACK BOOK ABOUT THIS FIREPLACE.

MOST SMALL HOUSES ARE LOW BUDGET HOUSES WHICH ADD TO THE DESIGN DIFFICULTY. THE FOLLOWING HOUSE-SHOP DESIGN WAS NO EXCEPTION.

THE OWNERS WANTED THREE THINGS, A SHOP WITH AN OLD LOOK, PASSIVE SOLAR LIVING QUARTERS, AND A WARM LOOK IN THE CONCRETE LIVING QUARTERS. THE OLD LOOK IS ACCOMPLISHED WITH UNPAINTED PINE SIDING, LOTS OF GLASS ON THE SOUTH, TWO TROMBE WALLS, AND A GREENHOUSE WHICH GIVE THIS HOUSE TREMENDOUS HEATING POTENTIAL. THE LIVING ROOM, DINING AND KITCHEN AREA HAS 4x6 ADZED BEAMS WITH A WOOD CEILING TO GIVE IT A WARM LOOK.

THE PORCH ON THE SHOP ENTRANCE SIDE IS INVITING. THE SKYLIGHT NEEDS SHADING.

THIS IS AN EXAMPLE OF A SIMPLE, QUICK PERSPECTIVE AND SINCE I BUILT THE HOUSE, THE FINISHED PRODUCT IS PRETTY MUCH THE SAME.

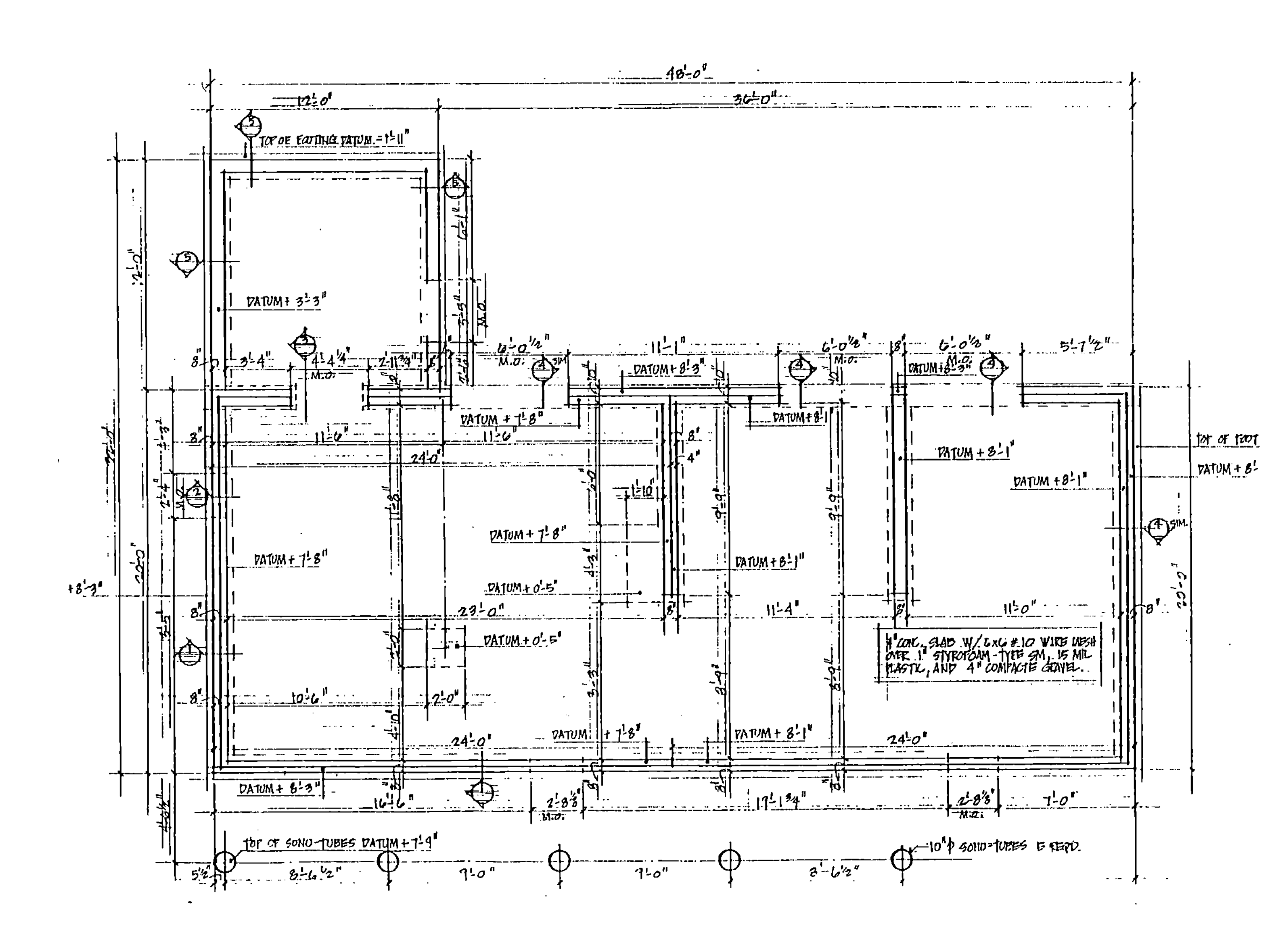

I HAD A LOT OF DIFFERENT CONCRETE HEIGHTS SO I CALLED THEM OUT AS DATUM + 8'-1" OR DATUM + 8'-3". THE CONCRETE CONTRACTOR FOUND IT TOO MUCH TROUBLE SO I MADE SOME CHANGES. THE SIMPLER THE PLANS ARE THE FEWER PROBLEMS YOU WILL ENCOUNTER.

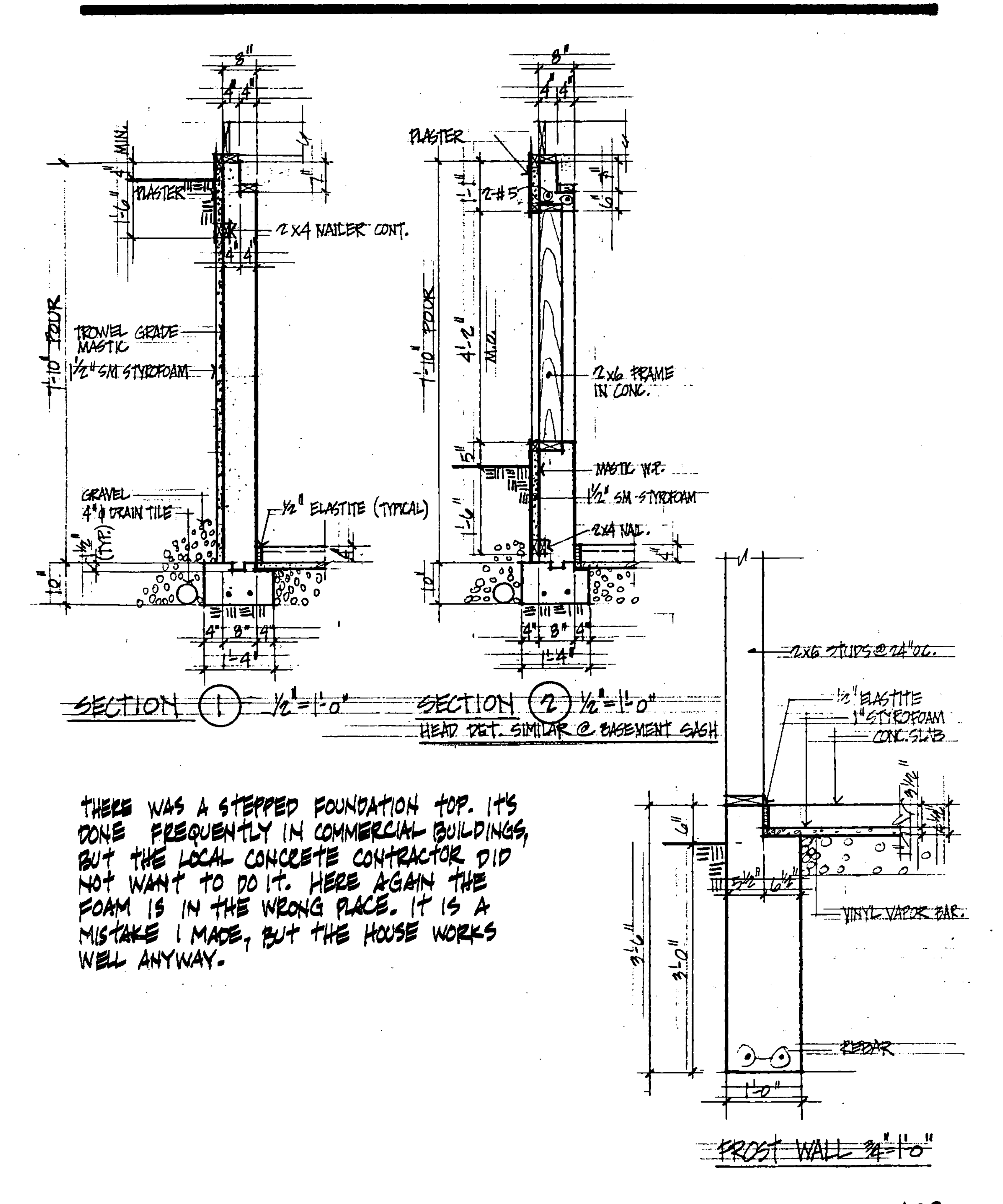

THERE WAS A STEPPED FOUNDATION TOP. IT'S DONE FREQUENTLY IN COMMERCIAL BUILDINGS, BUT THE LOCAL CONCRETE CONTRACTOR DID NOT WANT TO DO IT. HERE AGAIN THE FOAM IS IN THE WRONG PLACE. IT IS A MISTAKE I MADE, BUT THE HOUSE WORKS WELL ANYWAY.

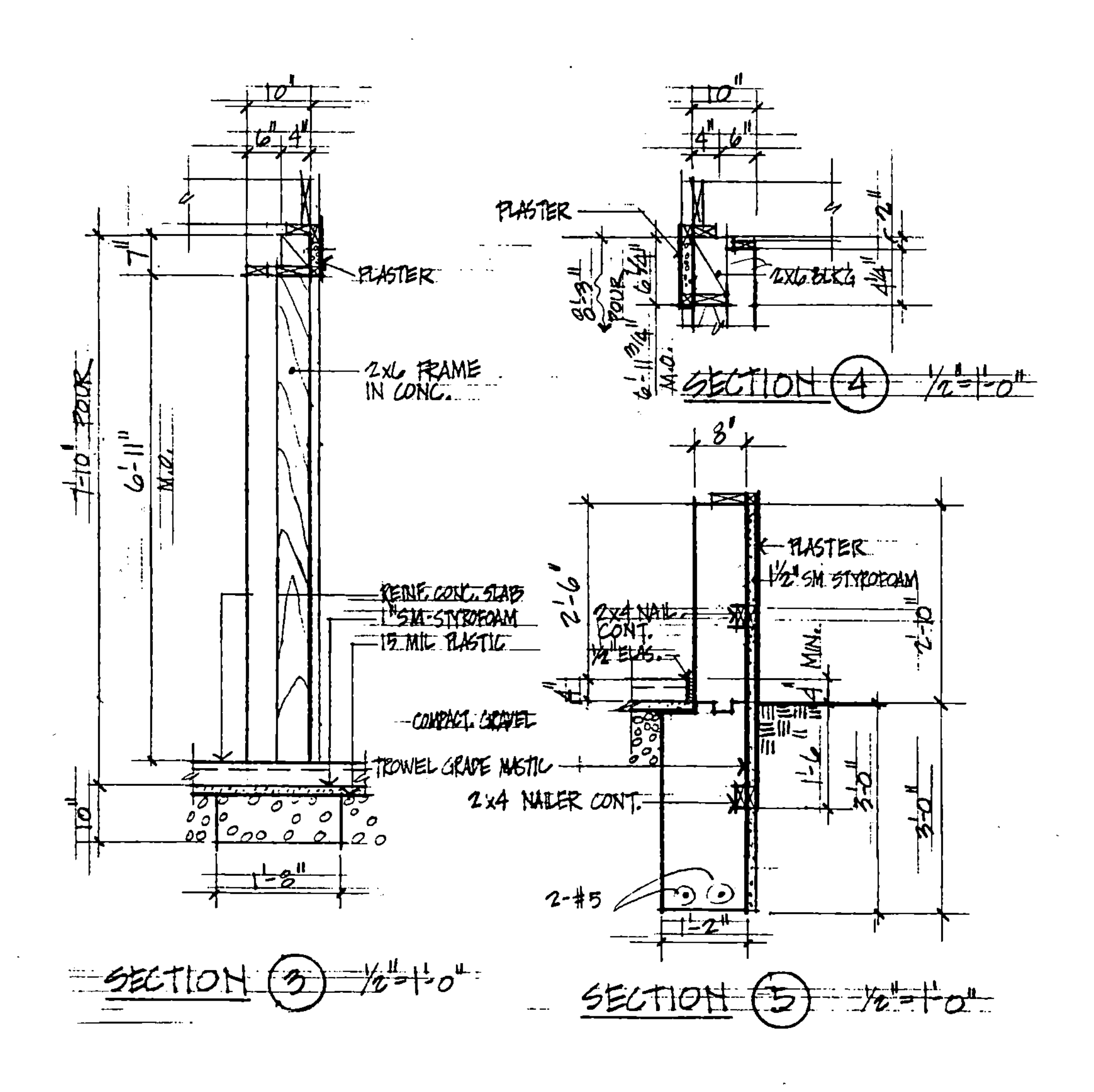

THESE DETAILS WERE DRAWN ON THE SHEET WITH THE FOUNDATION PLANS, IT'S EASIER TO READ THAT WAY. AS YOU LOOK AROUND YOU WILL FIND MANY WAYS TO DO THE SAME JOB AND JUST BECAUSE IT HASN'T BEEN DONE THAT WAY BEFORE DOESN'T MEAN IT'S WRONG.

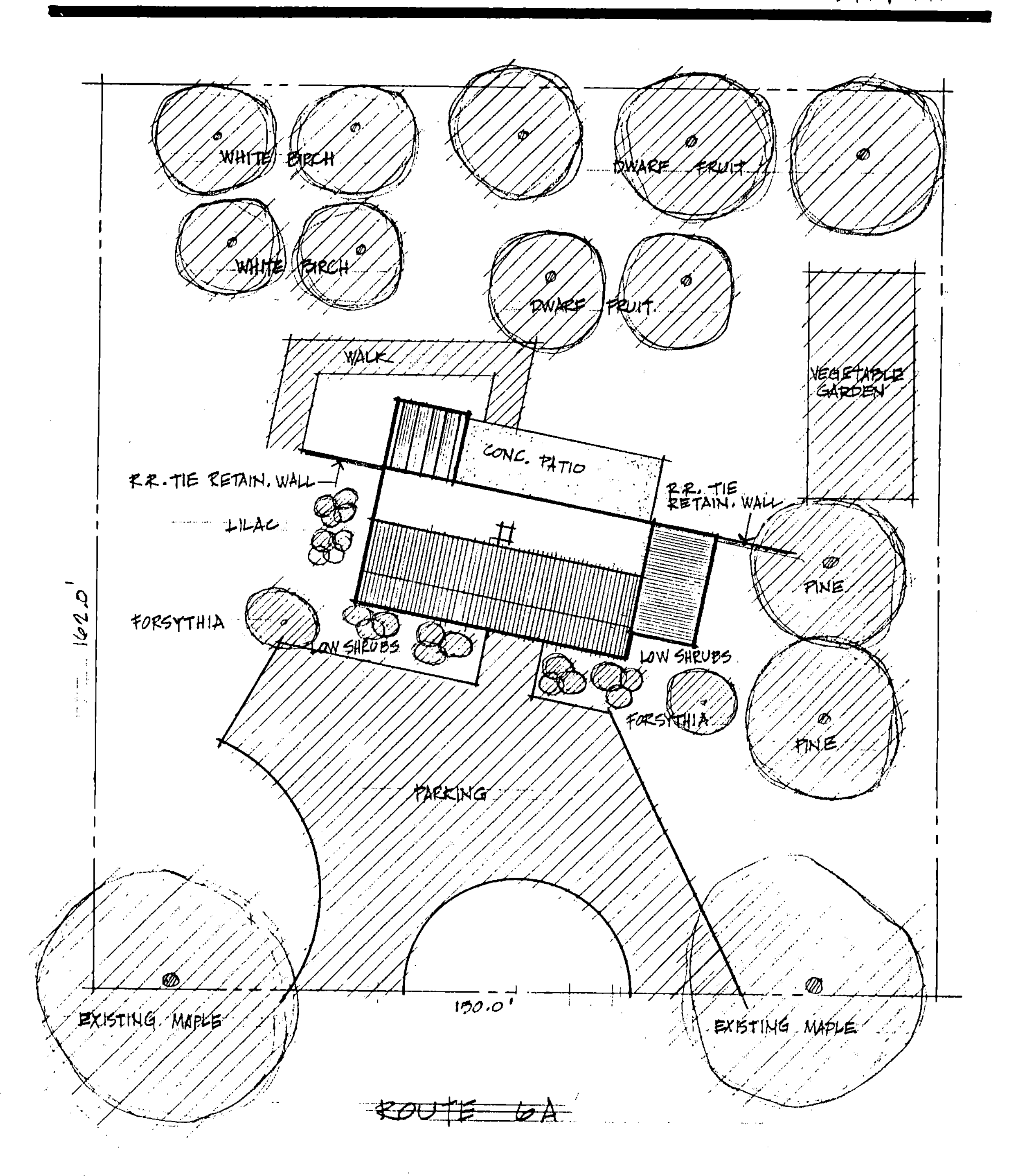

ANOTHER EXAMPLE OF A LANDSCAPE PLOT PLAN. THIS PLAN WAS DONE BECAUSE OF A LOCAL REGULATION. THAT'S WHAT THEY WANTED SO THAT'S WHAT WE GAVE THEM.

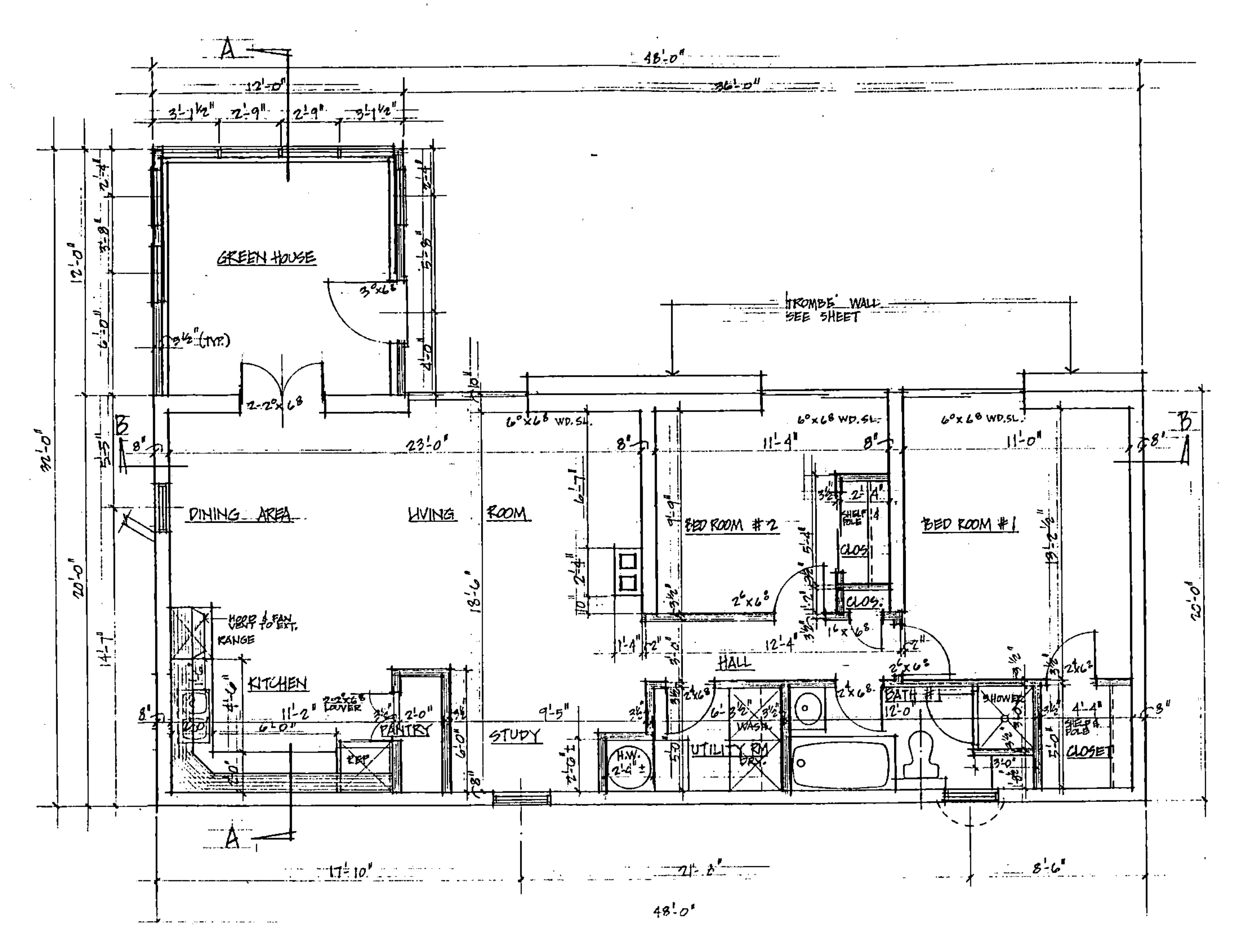

FLOOR PLAN 1/4" = 1'-0"

ANOTHER SIMPLE FLOOR PLAN. I HAVE FOUND THAT THE EASIER IT IS TO DRAW, THE EASIER IT IS TO BUILD.

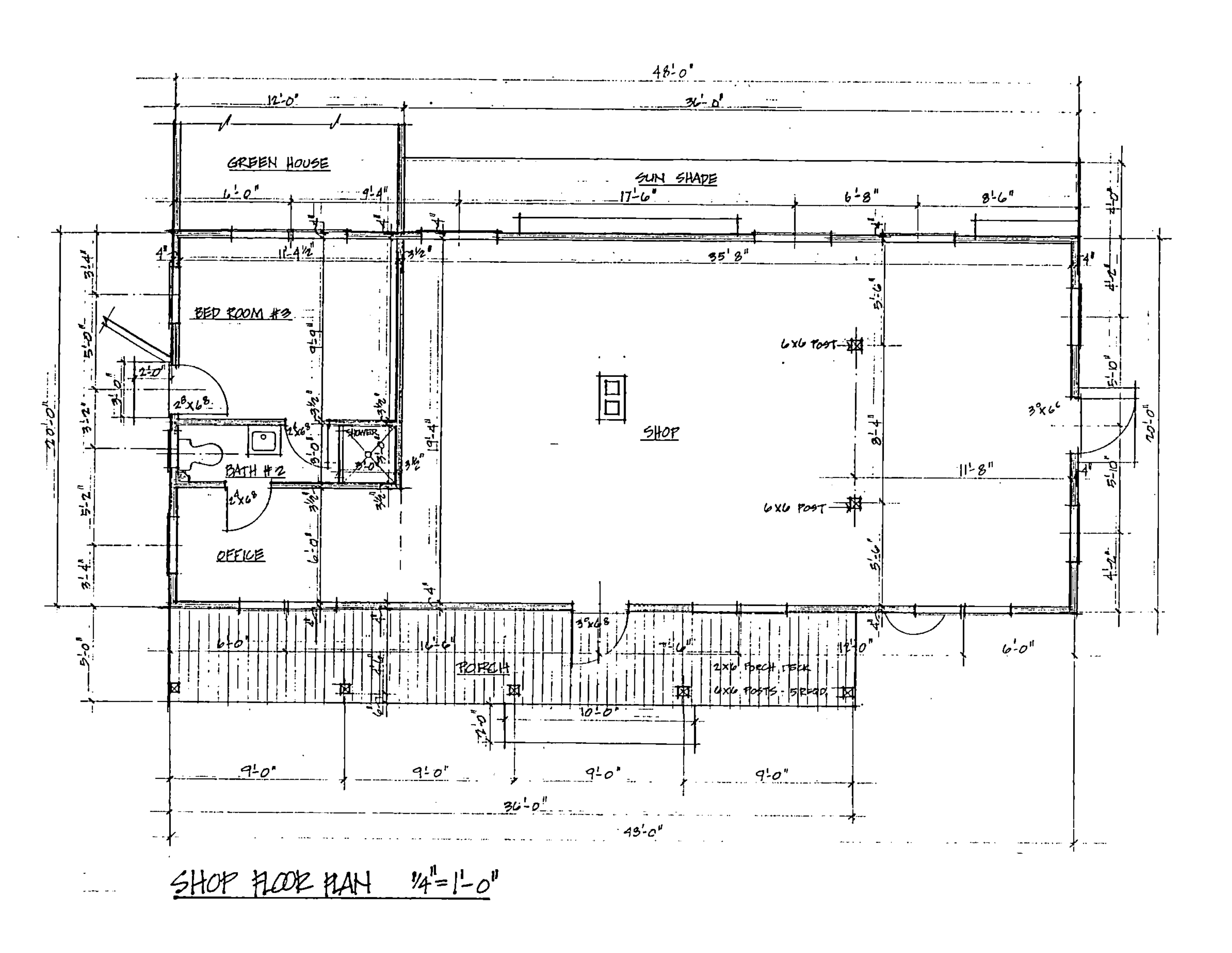

SHOP FLOOR PLAN 1/4" = 1'-0"

I CUT OFF THE GREENHOUSE ON THIS DRAWING BECAUSE I NEEDED THE SPACE.

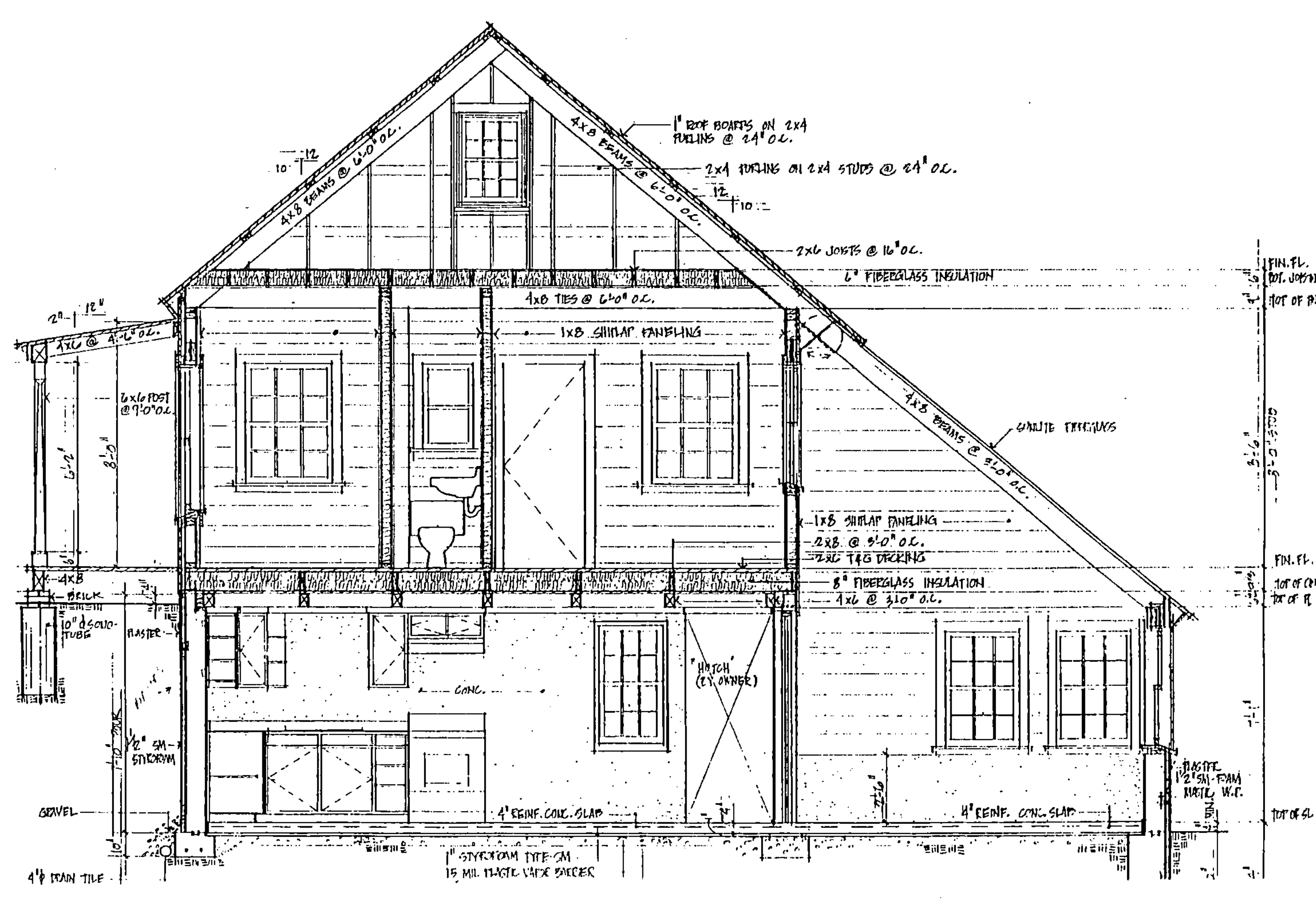

SECTION A-A 3/8" = 1'-0"

A SECTION THROUGH THE MAIN HOUSE AND GREEN HOUSE. THE FOAM INSULATION IS WRONG HERE TOO.

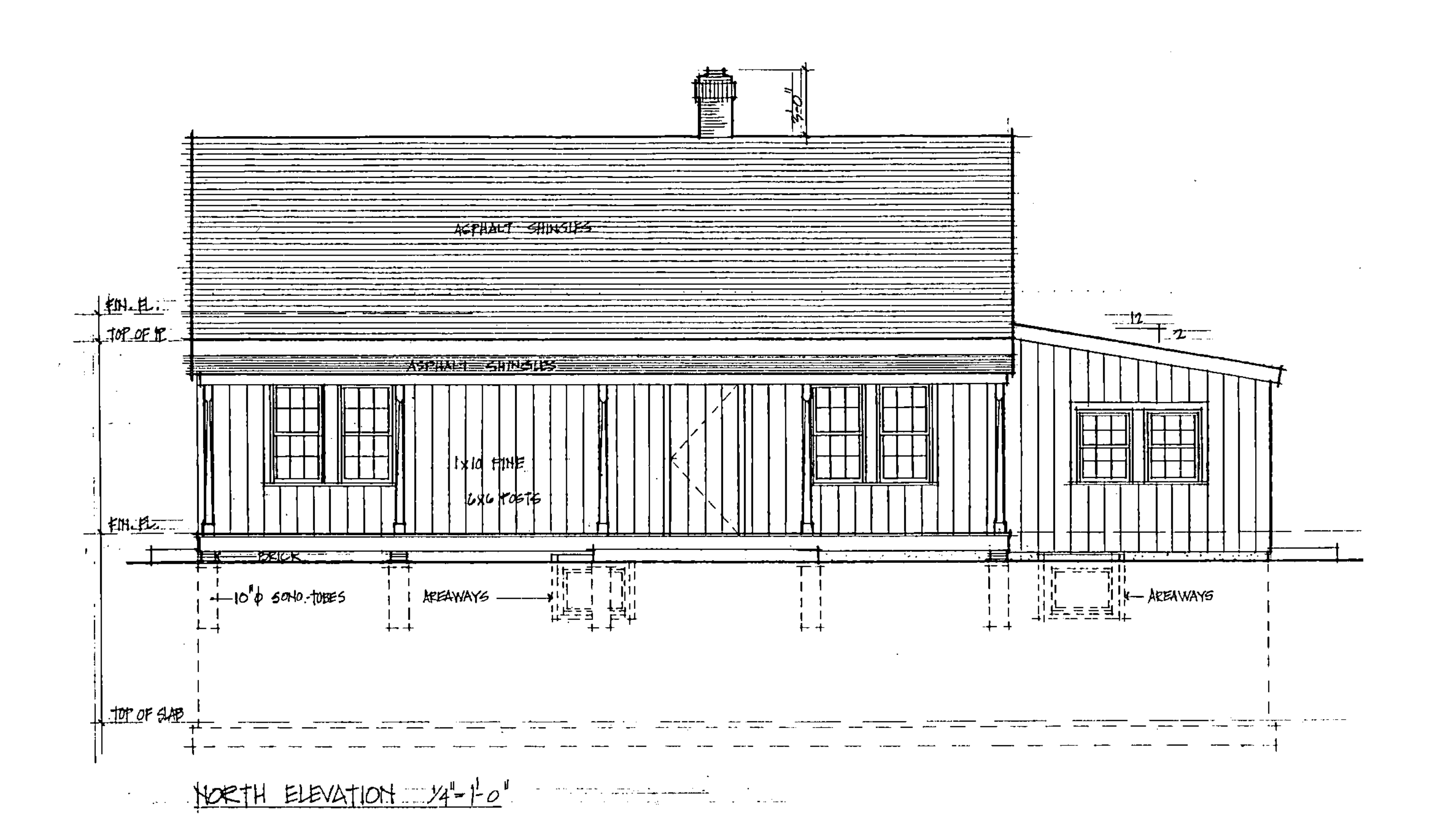

A VERY SIMPLE BUILDING WITH VERY SIMPLE ELEVATIONS.

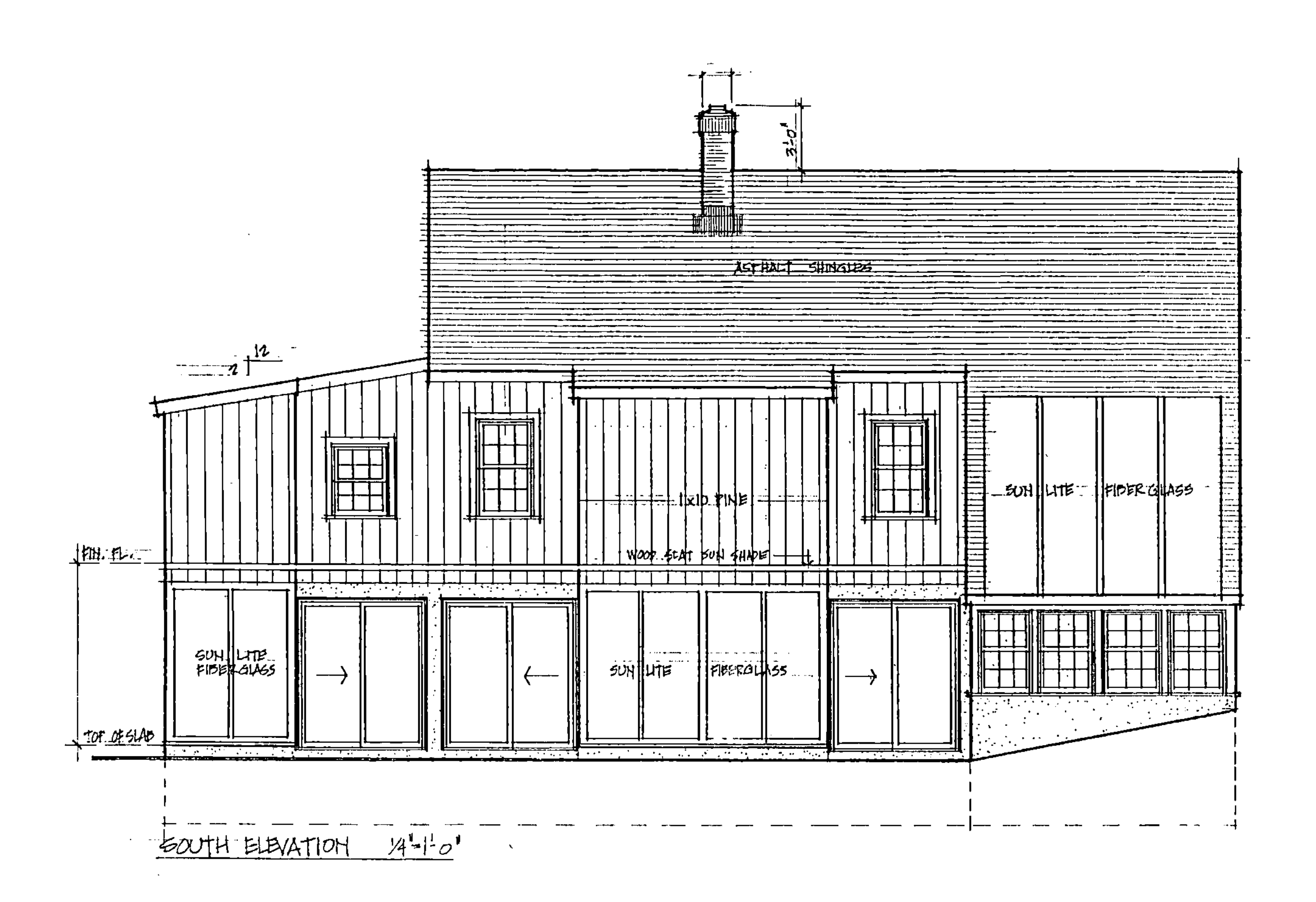

A LITTLE MORE COMPLEX ON THE SOUTH SIDE.

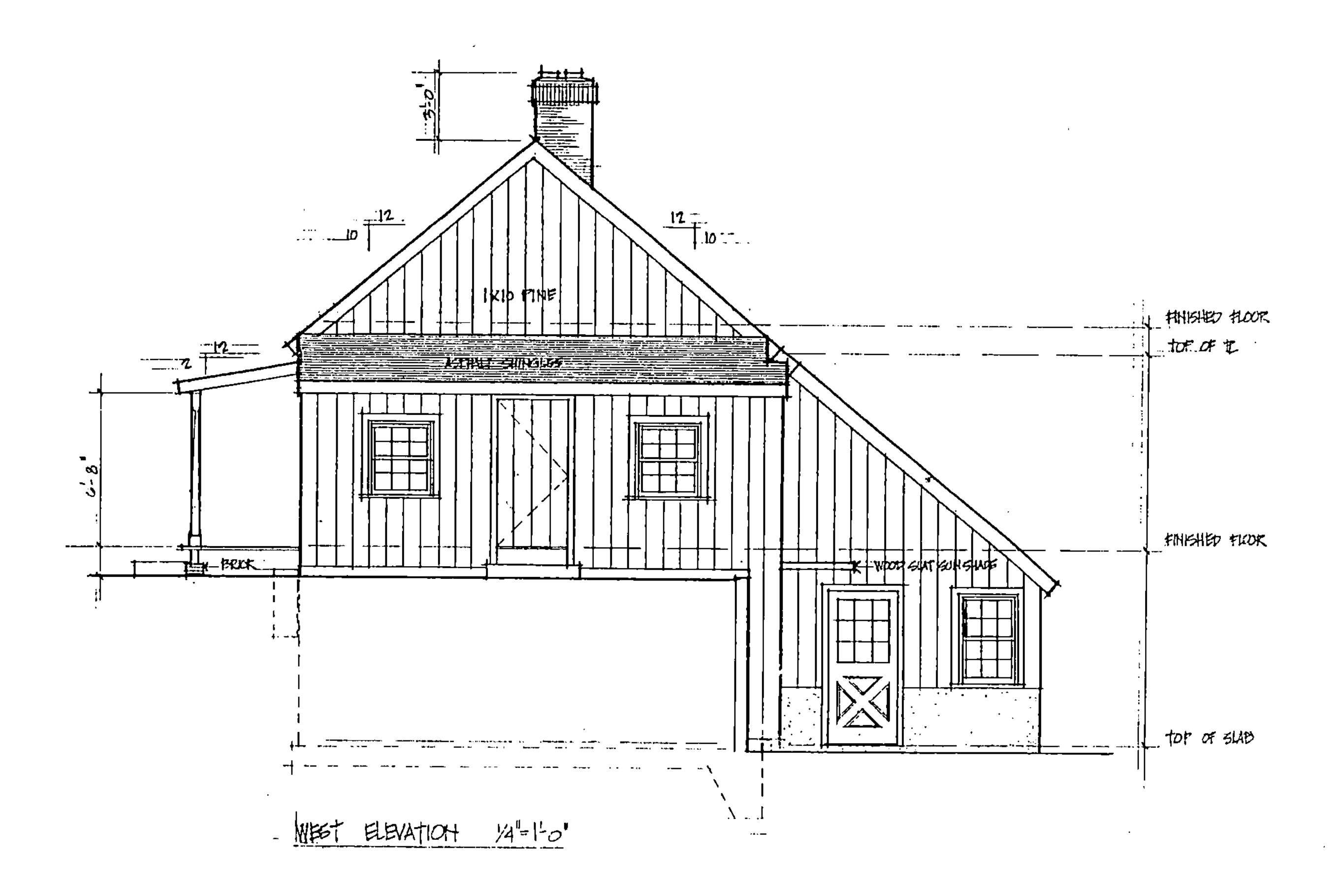

THIS ELEVATION SHOWS UP THE DIFFERENT GROUND LEVELS AS WELL AS THE FOUNDATION CONDITIONS.

East Woods Press Books

Backcountry Cooking
Berkshire Trails for Walking & Ski Touring
Campfire Chillers
Canoeing the Jersey Pine Barrens
Carolina Seashells
Carpentry: Some Tricks of the Trade from an Old-Style Carpenter
Complete Guide to Backpacking in Canada
Drafting: Tips and Tricks on Drawing and Designing House Plans
Exploring Nova Scotia
Florida by Paddle and Pack
Free Attractions, USA
Free Campgrounds, USA
Fructose Cookbook, The
Grand Strand: An Uncommon Guide to Myrtle Beach, The
Healthy Trail Food Book, The
Hiking Cape Cod
Hiking from Inn to Inn
Hiking Virginia's National Forests
Honky Tonkin': Travel Guide to American Music
Hosteling USA, Revised Edition
Inside Outward Bound
Just Folks: Visitin' with Carolina People
Kays Gary, Columnist
Living Land: An Outdoor Guide to North Carolina, The
Maine Coast: A Nature Lover's Guide, The
New England Guest House Book, The
New England: Off the Beaten Path
Parent Power!
A Common-Sense Approach to Raising Your Children In The Eighties
Rocky Mountain National Park Hiking Trails
Sea Islands of the South
Southern Guest House Book, The
Southern Rock: A Climber's Guide to the South
Steppin' Out: A Guide to Live Music in Manhattan
Sweets Without Guilt
Tennessee Trails
Train Trips: Exploring America by Rail
Trout Fishing the Southern Appalachians
Vacationer's Guide to Orlando and Central Florida, A
Walks in the Catskills
Walks in the Great Smokies
Walks with Nature in Rocky Mountain National Park
Whitewater Rafting in Eastern America
Wild Places of the South
Woman's Journey, A
You Can't Live on Radishes

Order from:

The East Woods Press
820 East Blvd.
Charlotte, NC 28203

It's Worth Celebrating—a Cookbook With 506 Taste of Home Recipes!

THE 15TH EDITION in our cookbook series, this *2008 Taste of Home Annual Recipes* marks a truly "tasty" milestone. And to celebrate, we did what we always do—we packed each chapter with plenty of scrumptious, family-pleasing, unforgettable recipes from America's #1 food magazine!

In fact, this convenient collection gives you every recipe from the 2007 issues of *Taste of Home* magazine, plus 18 "bonus" recipes put together into six memorable meals. It all adds up to 506 delicious dishes in one beautiful, full-color cookbook, one you're sure to enjoy for years to come.

Whether you're fixing a weekday dinner for your family, serving a holiday meal to a crowd or simply whipping up something for yourself, you'll have plenty of terrific recipes from which to choose. To get you started, here's a quick glimpse at the winners of *Taste of Home's* six national recipe contests held last year:

• **Go Nuts.** Cooks from around the country shelled out recipes for nut-topped entrees, snacks, pies and more for this crunchy contest. Rustic Nut Bars (p. 99) garnered the Grand Prize, while Mocha Nut Torte (p. 111) took second place.

• **Soup's On!** Our judges were bowled over by everything from meaty broths to elegant bisques. From the pot of entries came first-place winner Colorful Chicken 'n' Squash Soup (p. 36) and runner-up Danish Turkey Dumpling Soup (p. 41).

• **Cupcake Challenge.** What a treat! The *Taste of Home* kitchen staff whipped up and taste-tested hundreds of creative confections. Special Mocha Cupcakes (p. 108) topped the lineup, followed by Lemon Curd Cupcakes (p. 108).

• **Bountiful Harvest.** This bumper crop of mouth-watering recipes grew from fresh fruits and veggies. In the end, our panel picked Spaghetti Squash with Red Sauce (p. 67) as their favorite, with Apple-Brie Spinach Salad (p. 29) next on the list.

• **Potluck Pleasers.** If you're big on cooking for church suppers or other crowd-size events, you'll want to check out the Grand-Prize winner, Duo Tater Bake (p. 148). And turn to page 140 for the runner-up, Special Sesame Chicken Salad.

• **Holiday Baking Bonanza.** What's sweeter at holiday time than oven-fresh desserts, breads and other baked goods? First-place finisher Elegant Chocolate Torte (p. 107) and second-place Special Banana Nut Bread (p. 85) make impressive treats for memorable occasions.

With 506 recipes in this big, keepsake-quality cookbook, you can enjoy a delicious new dish every day for a full year—and beyond. So trust *2008 Taste of Home Annual Recipes* to give you and your family taste-tempting dishes on weekdays, weekends, holidays...every day!

SPECIAL SPOONFULS. Colorful Chicken 'n' Squash Soup (p. 36) won the Grand Prize and Danish Turkey Dumpling Soup (p. 41) took second place in our national "Soup's On!" recipe contest.

2008 Taste of Home Annual Recipes

Editor Michelle Bretl
Art Director Gretchen Trautman
Vice President, Executive Editor/Books
Heidi Reuter Lloyd
Senior Editor/Books Mark Hagen
Associate Editor Jean Steiner
Layout Designer Emma Acevedo
Content Production Supervisor Julie Wagner
Proofreaders Linne Bruskewitz,
Victoria Soukup Jensen
Editorial Assistant Barb Czysz

Taste of Home

Editor Ann Kaiser
Managing Editor Barbara Schuetz
Senior Art Director Sandra L. Ploy
Food Director Diane Werner RD
Food Editor Patricia Schmeling
Senior Recipe Editor Sue A. Jurack
Recipe Editors Mary King, Christine Rukavena
Assistant Editor Melissa Phaneuf
Copy Editor S.K. Enk
Editorial Assistants Jane Stasik,
Mary Ann Koebernik
Graphic Art Associate Ellen Lloyd
Test Kitchen Manager Karen Scales
Test Kitchen Home Economists
Peggy Woodward RD, Tina Johnson, Marie Parker,
Annie Rose, Wendy Stenman, Amy Welk-Thieding RD;
Contributing: Dot Vartan
Test Kitchen Assistants Rita Krajcir, Kris Lehman,
Sue Megonigle, Megan Taylor
Recipe Asset Systems Manager Coleen Martin
Photographers Rob Hagen (Senior), Dan Roberts,
Jim Wieland, Lori Foy
Senior Food Stylist Sarah Thompson
Set Stylists Jenny Bradley Vent (Senior),
Dolores Schaefer
Assistant Food Stylists Kaitlyn Besasie,
Alynna Malson
Photo Studio Coordinator Kathy Swaney

President and Chief Executive Officer
Mary G. Berner
President, Food & Entertaining
Suzanne M. Grimes
Senior Vice President, Editor in Chief
Catherine Cassidy
Creative Director Ardyth Cope

Taste of Home Books

5400 S. 60th St., Greendale WI 53129

International Standard Book Number (10):
0-89821-651-6
International Standard Book Number (13):
978-0-89821-651-6
International Standard Serial Number: 1094-3463

PICTURED AT RIGHT: Clockwise from upper left: Chicken Taco Cups (p. 171), Garlic Pepper Corn (p. 179), Fiesta Rib Eye Steaks (p. 179), Fruit 'n' Cake Kabobs (p. 179), Celebration Braid (p. 85), Barnyard Cupcakes (p. 116), Crusty Roast Leg of Lamb (p. 236), Cracked Pepper Salad Dressing (p. 237), Banana Citrus Sorbet (p. 237), Strawberry-Banana Angel Torte (p. 237) and Springtime Asparagus Medley (p. 236).

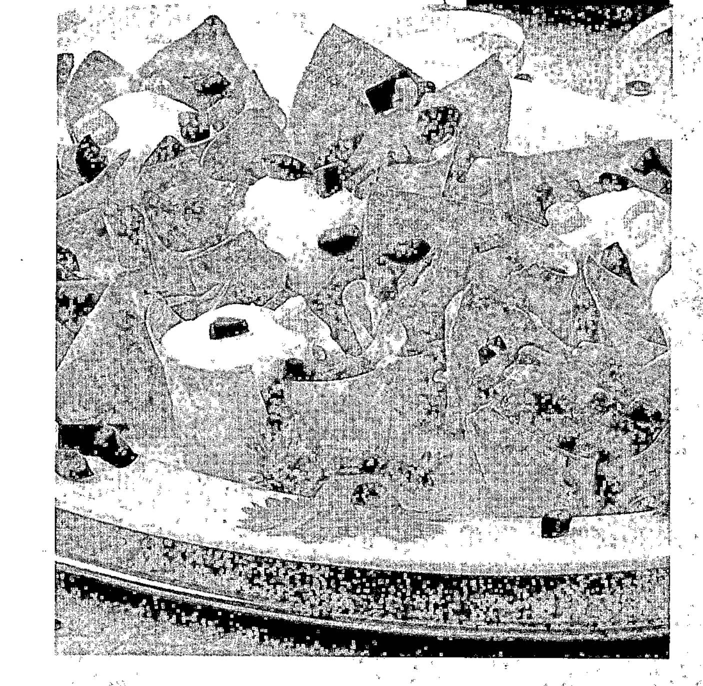
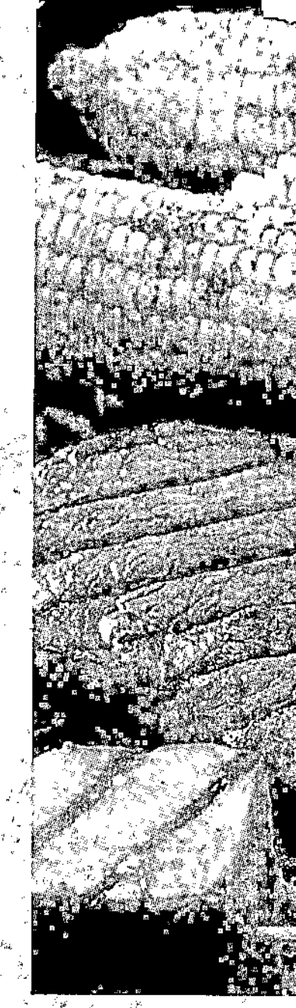

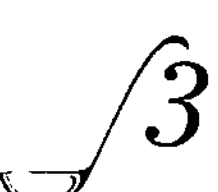

Taste of Home Annual Recipes 2008

PICTURED ON FRONT COVER. Clockwise from upper left: Parker House Dinner Rolls (p. 219), Dijon Green Beans (p. 49), Rolled-Up Turkey (p. 206) and Tart Cherry Lattice Pie (p. 215).

PICTURED ON BACK COVER. Clockwise from upper right: Blueberry Cheesecake (p. 133), Summer Celebration Ice Cream Cake (p. 249) and Peaches 'n' Cream Tart (p. 126).

Front cover photo by Dan Roberts. Food styled by Diane Armstrong and Kaitlyn Besasie. Set styled by Stephanie Marchese.

Snacks & Appetizers

Irresistible hors d'oeurves for dinner parties...yummy munchies for the kids after school...goodies to satisfy those late-night cravings...they're all right here in one convenient chapter.

MOUTH-WATERING MUNCHIES: Clockwise from upper left: Crab Bruschetta (p. 15), BLT Bruschetta (p. 15), Zucchini Feta Bruschetta (p. 14), Crispy Shrimp Poppers (p. 13), Dilly Zucchini Dip (p. 11), Almond Crunch (p. 9) and Georgia Peanut Salsa (p. 7).

Baked Brie with Roasted Garlic

(Pictured below)

Prep: 35 min. + cooling **Bake:** 45 min.

The garlic is mellow and sweet in this recipe. I never fail to get compliments when this brie is my first course.
—Lara Pennell, Mauldin, South Carolina

1 whole garlic bulb
1-1/2 teaspoons plus 1 tablespoon olive oil, *divided*
1 tablespoon minced fresh rosemary *or* 1 teaspoon dried rosemary, crushed
1 round loaf (1 pound) sourdough bread
1 round (8 ounces) Brie *or* Camembert cheese
1 loaf (10-1/2 ounces) French bread baguette, sliced and toasted
Red and green grapes

Remove papery outer skin from garlic (do not peel or separate cloves). Cut top off bulb. Brush with 1-1/2 teaspoons oil; sprinkle with rosemary. Wrap in heavy-duty foil. Bake at 425° for 30-35 minutes or until softened.

Meanwhile, cut the top fourth off the loaf of sourdough bread; carefully hollow out enough of the bottom of the bread so cheese will fit. Cube removed bread; set aside. Place cheese in bread.

Cool garlic for 10-15 minutes. Reduce heat to 375°. Squeeze softened garlic into a bowl and mash with a fork; spread over cheese. Replace bread top; brush outside of bread with remaining oil. Wrap in heavy-duty foil.

Bake for 45-50 minutes or until cheese is melted. Serve with toasted baguette, grapes and reserved bread cubes. **Yield:** 8 servings.

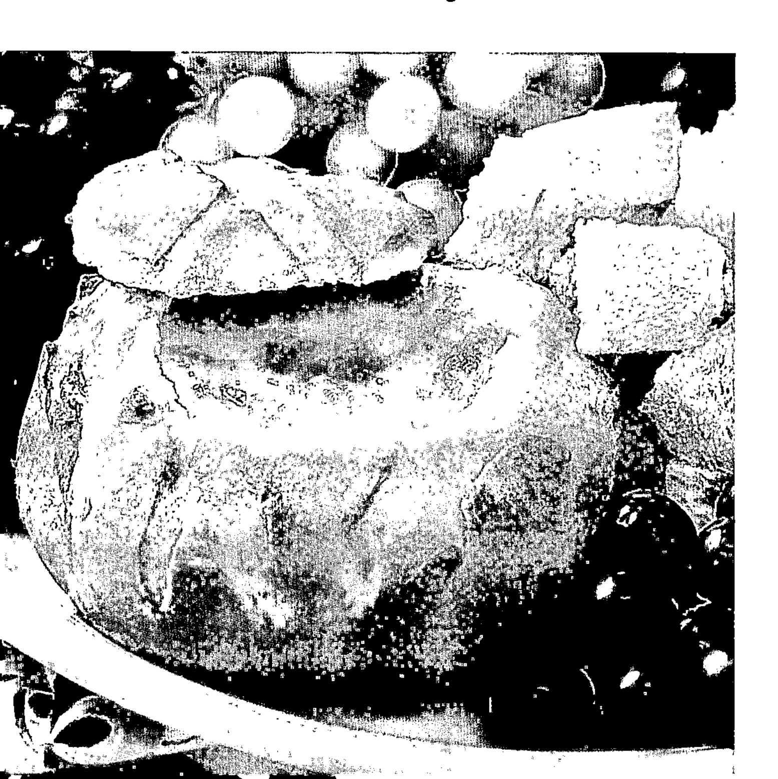

Orange-Glazed Smokies

Prep/Total Time: 15 min.

These mouth-watering sausages are jazzed up with just five ingredients...and can be prepared in a matter of minutes for a hungry group. I always hear rave reviews when I serve them, no matter what the occasion.
—Judy Wilson, Sun City West, Arizona

1 cup packed brown sugar
1 tablespoon all-purpose flour
1/4 cup orange juice concentrate
2 tablespoons prepared mustard
1 tablespoon cider vinegar
1 package (16 ounces) miniature smoked sausages

In a large microwave-safe bowl, combine the first five ingredients. Add sausages; stir to coat. Cover and microwave on high for 3-4 minutes or until bubbly, stirring three times. **Yield:** about 4 dozen.

Editor's Note: This recipe was tested in a 1,100-watt microwave.

Baked Rosemary-Rhubarb Spread

Prep: 20 min. **Bake:** 25 min.

This zippy and unusual spread is sprinkled with toasted pine nuts. The rhubarb adds a tangy accent that's unexpected but always crowd-pleasing.
—Joan Ranzini, Waynesboro, Virginia

2 cups chopped fresh *or* frozen rhubarb
4 ounces cream cheese, softened
1 egg
2 tablespoons all-purpose flour
1 tablespoon minced fresh rosemary *or* 1 teaspoon dried rosemary, crushed
1 garlic clove, minced
1/8 to 1/4 teaspoon crushed red pepper flakes
2 tablespoons pine nuts
Assorted crackers

Place the rhubarb in a large saucepan; cover with water. Bring to a boil. Reduce the heat; cook, uncovered, for 3-4 minutes or until crisp-tender. Drain and rinse in cold water.

In a food processor, combine the cream cheese, egg, flour, rosemary, garlic and pepper flakes; cover and process until blended. Add rhubarb; cover and process until blended. Transfer to a greased 9-in. pie plate. Sprinkle with pine nuts.

Bake, uncovered, at 350° for 25-30 minutes or until golden brown and center is set. Serve with crackers. **Yield:** 1-1/2 cups.

Chili con Queso Artichokes

(Pictured above)

Prep: 10 min. **Cook:** 30 min. + standing

Here at our Ocean Mist Farms, we love finding new recipes for the harvest from our artichoke fields. This one is terrific! —*Kori Tuggle, Castroville, California*

2 large artichokes
1 package (8 ounces) shredded Mexican cheese blend
1 can (5 ounces) evaporated milk
1 can (4 ounces) chopped green chilies, drained
1/2 cup minced fresh cilantro

Using a sharp knife, level the bottom of each artichoke and cut 1 in. from the tops. Using kitchen scissors, snip off tips of outer leaves. Place artichokes in a large saucepan; add 1 in. of water. Bring to a boil. Reduce heat; cover and simmer for 30-35 minutes or until leaves near the center pull out easily.

Invert artichokes to drain for 10 minutes. With a grapefruit spoon, carefully remove the fuzzy centers and discard.

In a small microwave-safe dish, combine the cheese, evaporated milk and chilies. Cover and microwave on high for 1-2 minutes or until cheese is melted, stirring twice. Stir in cilantro. Gently spread artichoke leaves apart; fill with cheese mixture. Serve immediately. **Yield:** 8 servings (2 cups sauce).

Editor's Note: This recipe was tested in a 1,100-watt microwave.

Georgia Peanut Salsa

(Pictured below and on page 4)

Prep: 25 min. + chilling

Former President Jimmy Carter gave first place to this salsa at the Plains Peanut Festival in his Georgia hometown. My daughter, Elizabeth, and I created the recipe just days before the contest. Although we weren't in the judging room, we later saw a tape of President Carter tasting our salsa and saying, "Mmmmmm...that's good." Elizabeth was 9 at the time, but it's a day she'll never forget.
—*Lane McCloud, Siloam Springs, Arkansas*

3 plum tomatoes, seeded and chopped
1 jar (8 ounces) picante sauce
1 can (11 ounces) white *or* shoepeg corn, drained
1/3 cup Italian salad dressing
1 medium green pepper, chopped
1 medium sweet red pepper, chopped
4 green onions, thinly sliced
1/2 cup minced fresh cilantro
2 garlic cloves, minced
2-1/2 cups salted roasted peanuts *or* boiled peanuts
Hot pepper sauce, optional
Tortilla chips

In a large bowl, combine the first nine ingredients. Cover and refrigerate for at least 8 hours.

Just before serving, stir in peanuts and pepper sauce if desired. Serve with tortilla chips. **Yield:** about 6-1/2 cups.

Editor's Note: This recipe was tested with salted peanuts, but the original recipe used boiled peanuts, which are often available in the South.

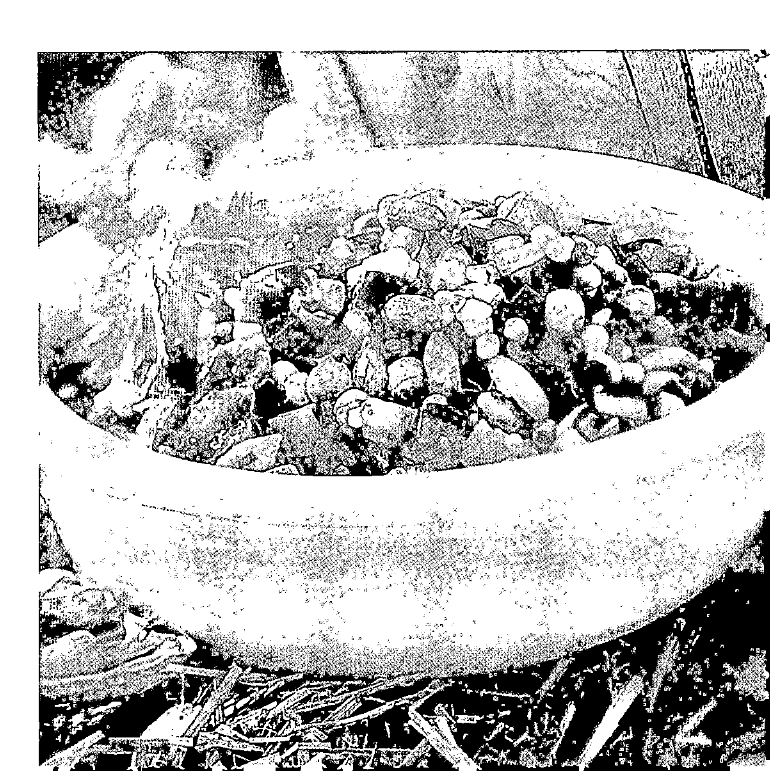

Pinecone-Shaped Blue Cheese Spread

(Pictured below)

Prep: 40 min. + chilling

This pretty "pinecone" consists of crunchy toasted almonds over a flavorful cheese spread. It's a fun, festive appetizer.
—Kathy Johnson, Council Bluffs, Iowa

2 packages (8 ounces *each*) cream cheese, softened
1-1/4 cups process cheese sauce
1 cup (4 ounces) crumbled blue cheese
1/4 cup chopped green onions
1 tablespoon diced pimientos
1/2 teaspoon Worcestershire sauce
1-1/2 cups unblanched almonds, toasted
Fresh rosemary sprigs, optional
Assorted crackers *or* fresh vegetables

In a large mixing bowl, beat the cream cheese, cheese sauce and blue cheese until smooth. Stir in the onions, pimientos and Worcestershire sauce. Cover and refrigerate until firm.

On a serving platter, form cheese spread into a pinecone shape. Beginning at a narrow end, arrange almonds in rows. Garnish with rosemary if desired. Serve with crackers. **Yield:** 3-1/2 cups.

Pastrami Asparagus Roll-Ups

Prep/Total Time: 25 min.

Celebrate spring's bounty with this refreshing asparagus wrap. Once people take that first bite, they're hooked!
—Sharon Waller, Aromas, California

24 fresh asparagus spears (about 1 pound), trimmed
1/2 cup prepared pesto
24 thin slices provolone cheese (about 1 pound)
24 thin slices deli pastrami (about 3/4 pound)

In a large skillet, bring 1/2 in. of water to a boil. Add asparagus; cover and boil for 3 minutes. Drain and immediately place asparagus in ice water. Drain and pat dry.

Spread 1 teaspoon of pesto over each slice of cheese. Top each with an asparagus spear; roll up tightly. Place each on a slice of pastrami; roll up tightly. Refrigerate until serving. **Yield:** 2 dozen.

Chocolate Wheat Cereal Snacks

Prep: 10 min. + cooling

This popular mix really satisfies a sweet tooth. With its chocolate-peanut butter combination, it's great for a snack or party. *—Tracy Golder, Bloomsburg, Pennsylvania*

6 cups frosted bite-size Shredded Wheat
1 cup milk chocolate chips
1/4 cup creamy peanut butter
1 cup confectioners' sugar

Place cereal in a large bowl; set aside. In a small microwave-safe bowl, melt chocolate chips and peanut butter; stir until smooth. Pour over cereal and stir gently to coat. Let stand for 10 minutes.

Sprinkle with sugar and toss to coat. Cool completely. Store in an airtight container. **Yield:** 6 cups.

Toasting Tips

Toasting nuts before using them in a recipe, such as the pinecone-shaped spread above left, intensifies their flavor.

It's easy to toast nuts in your oven. Simply spread the nuts on a baking sheet and bake them at 350° for 5 to 10 minutes or until they are lightly toasted. Be sure to watch them carefully during baking so they don't burn.

Almond Crunch

(Pictured above and on page 4)

Prep: 20 min. **Bake:** 15 min. + cooling

When you start eating this taste-tempting Passover treat, you might not be able to stop! Matzo crackers are topped with buttery caramel, then baked to perfection and spread with melted chocolate and slivered almonds.
—Sharalyn Zander, Jacksonville, Alabama

4 to 6 unsalted saltine matzo crackers
1 cup butter, cubed
1 cup packed brown sugar
3/4 cup semisweet chocolate chips
1 teaspoon shortening
1 cup slivered almonds, toasted

Line a 15-in. x 10-in. x 1-in. baking pan with foil; line the foil with parchment paper. Arrange crackers in the pan; set aside.

In a large heavy saucepan over medium heat, melt butter. Stir in brown sugar. Bring to a boil; cook and stir for 3-4 minutes or until sugar is dissolved. Spread evenly over crackers.

Bake at 350° for 15-17 minutes (cover loosely with foil if top browns too quickly). Cool on a wire rack for 5 minutes. Meanwhile, melt chocolate chips and shortening; stir until smooth. Stir in almonds; spread over top. Cool for 1 hour.

Break into pieces. Cover and refrigerate for at least 2 hours or until set. Store in an airtight container. **Yield:** 1 pound.

Rosemary Veal Meatballs

(Pictured below)

Prep: 25 min. **Cook:** 20 min.

These savory appetizer meatballs, seasoned with rosemary and garlic, get a touch of sweetness from chopped golden raisins. They'll be the hit of your next gathering.
—Rhonda Maiani, Chapel Hill, North Carolina

1 cup (8 ounces) plain yogurt
1 jar (4-1/2 ounces) marinated artichoke hearts, drained and chopped
2 tablespoons prepared Italian salad dressing
1 garlic clove, minced

MEATBALLS:
2 eggs, lightly beaten
3/4 cup soft bread crumbs
1/2 cup golden raisins, finely chopped
3 garlic cloves, minced
4 teaspoons dried rosemary, crushed
1-1/2 teaspoons salt
1 teaspoon pepper
1 pound ground veal
1/4 cup vegetable oil

In a small bowl, combine the yogurt, artichokes, Italian salad dressing and garlic; cover and refrigerate until serving.

In a large bowl, combine the eggs, bread crumbs, raisins, garlic, rosemary, salt and pepper. Crumble veal over mixture and mix well. Shape into 1-in. balls.

In a large skillet, brown meatballs in oil in small batches. Remove with a slotted spoon and keep warm. Serve with yogurt sauce. **Yield:** 3-1/2 dozen (1-1/2 cups sauce).

Cascading Fruit Centerpiece

(Pictured above)

Prep: 3 hours

It's exciting and fun to build this pretty centerpiece, which I've made for special occasions for years. The pineapple dip is wonderful. —*Ellen Brown, Aledo, Texas*

PINEAPPLE GINGER DIP:
- **2 packages (8 ounces *each*) cream cheese, softened**
- **5 tablespoons unsweetened pineapple juice**
- **2 tablespoons confectioners' sugar**
- **1 tablespoon lemon juice**
- **1 tablespoon grated orange peel**
- **1-1/2 to 2 teaspoons ground ginger**
- **1/4 cup flaked coconut, toasted**

CENTERPIECE:
- **1 fresh pineapple**
- **1 large grapefruit**
- **1 large navel orange**
- **1 medium lemon**
- **8 Styrofoam rounds (four 10 inches x 1 inch, four 6 inches x 1 inch)**
- **2 wooden dowels (one 8 inches x 1/4 inch, one 4-1/2 inches x 1/4 inch)**

Floral metal greening pins
- **3 to 5 pounds seedless green grapes**
- **3 to 5 pounds seedless red grapes**
- **1 pound fresh strawberries**

Silk flowers with leaves
- **3 cups yogurt of your choice**

In a small mixing bowl, combine the first six ingredients. Stir in coconut. Chill until serving.

Cut off pineapple top with a fourth of the pineapple attached; set aside. For dip bowl, cut a third from bottom of pineapple; remove fruit, leaving a 1/2-in. shell. Remove peel from center section of pineapple; core and cut pineapple into chunks.

Cut the grapefruit, orange and lemon in half widthwise. Remove pulp from one half of each; set aside. (Save the remaining grapefruit, orange and lemon halves for another use.)

Stack two 10-in. Styrofoam rounds and two 6-in. rounds; cover each stack with heavy-duty foil. Repeat with remaining rounds. On a 19-in. x 15-in. platter, pile the stacks on top of each other, staggering them and anchoring with dowels.

Place pineapple top on the top circle; position lemon, orange and grapefruit cups on the other circles. Place pineapple bowl on platter.

With greening pins, attach clusters of grapes onto circles. Randomly add strawberries, pineapple chunks and more grapes to cover the foil and platter. Decorate with silk flowers. Just before serving, fill pineapple bowl with pineapple ginger dip. Fill lemon, orange and grapefruit cups with yogurt. **Yield:** 1 fruit centerpiece.

How to Create the Cascading Fruit Centerpiece

1. On a large platter, arrange the foil-covered Styrofoam circles in a staggered stack; secure with dowels. Add the pineapple top and fruit bowls for dip.

2. In a cascading line, secure large clusters of green and red grapes using greening pins.

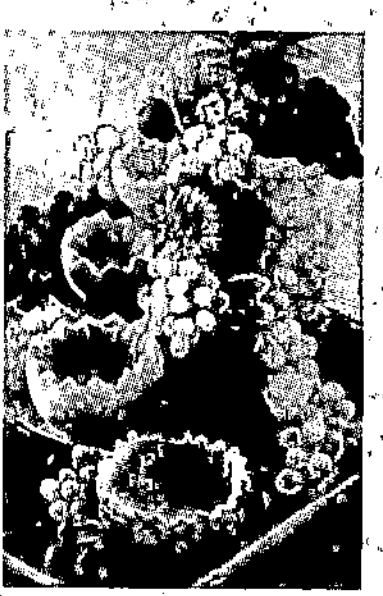

3. Completely fill in the centerpiece with fresh strawberries, pineapple chunks, silk flowers and more grapes.

Dilly Zucchini Dip

(Pictured below and on page 4)

Prep: 15 min. + chilling

This thick and creamy snack will keep your guests coming back for more. When paired with veggies, the colorful dip is pleasing to the eye as well as to the palate.
—Edna Hoffman, Hebron, Indiana

1 cup finely shredded zucchini, squeezed dry
1 cup (4 ounces) shredded sharp cheddar cheese
3/4 cup mayonnaise
1/2 cup chopped walnuts
1 teaspoon lemon juice
1/2 teaspoon dill weed
1/4 teaspoon pepper
Assorted fresh vegetables

In a bowl, combine the first seven ingredients. Cover and refrigerate for 1 hour or until chilled. Serve with vegetables. **Yield:** 2 cups.

Warm Ham 'n' Cheese Spread

(Pictured above)

Prep: 30 min. **Bake:** 15 min.

I'm always looking for creative yet family-pleasing ways to stretch my tight grocery budget. I usually make this delectable spread with ham "ends," available inexpensively at the deli counter.
—Pattie Prescott
Manchester, New Hampshire

4 pita breads (6 inches), split
1/4 cup olive oil
4 cups ground fully cooked ham
1 cup (4 ounces *each*) shredded Swiss, American and cheddar cheeses
1/4 cup mayonnaise
1/2 teaspoon ground mustard
2 tablespoons minced fresh parsley
Additional shredded Swiss cheese, optional

Cut each pita half into eight wedges; brush rough sides with oil. Place on ungreased baking sheets. Bake at 350° for 10-12 minutes or until golden brown, turning once. Remove to wire racks.

In a large bowl, combine the ham, cheeses, mayonnaise and mustard. Transfer to a shallow 1-qt. baking dish. Bake, uncovered, at 350° for 15-20 minutes or until edges are bubbly. Sprinkle with parsley and additional Swiss cheese if desired; serve with pita wedges. **Yield:** 3-1/2 cups.

Editor's Note: Reduced-fat or fat-free mayonnaise is not recommended for this recipe.

On a work surface, carefully remove one sheet of phyllo dough and fold in thirds lengthwise (keep remaining phyllo covered with plastic wrap and a damp towel to prevent it from drying out). Lightly brush phyllo strip with butter. Place a rounded tablespoonful of filling in lower right corner of strip. Fold dough over filling, forming a triangle. Fold triangle up, then fold over, forming another triangle. Continue folding, like a flag, until you come to the end of the strip.

Repeat with remaining phyllo sheets and filling. Place triangles on ungreased baking sheets. Brush with butter. Bake at 375° for 20-25 minutes or until golden brown.

Meanwhile, in a small saucepan, melt butter over medium heat. Whisk in flour until smooth; add milk and salt. Bring to a boil; cook and stir for 2 minutes or until thickened. Remove from heat; stir in Swiss cheese until melted. Sprinkle with parsley if desired. Serve with triangles. **Yield:** 40 appetizers.

Ham 'n' Broccoli Triangles

(Pictured above)

Prep: 40 min. **Bake:** 20 min.

I often make these light and flaky pastry triangles for my family. Leftover spiral ham is perfect for the centers, and the creamy white sauce is sure to please.

—Rebekah Soued, Conifer, Colorado

2-1/2 cups diced fully cooked ham
2-1/2 cups chopped fresh *or* frozen broccoli, thawed
1 cup (4 ounces) shredded part-skim mozzarella cheese
2 eggs
1/2 cup heavy whipping cream
1/2 teaspoon minced fresh basil
1/2 teaspoon pepper
1/4 teaspoon Italian seasoning, optional
Dash cayenne pepper
1 package (16 ounces, 14-inch x 9-inch sheet size) frozen phyllo dough, thawed
3/4 cup butter, melted
SWISS CHEESE SAUCE:
1/4 cup butter
1/4 cup all-purpose flour
2 cups milk
1/2 teaspoon salt
1/2 cup shredded Swiss cheese
Minced fresh parsley, optional

In a large bowl, combine the ham, broccoli and mozzarella cheese. In a small bowl, beat eggs; stir in the cream, basil, pepper, Italian seasoning if desired and cayenne. Stir into ham mixture; set aside.

Smoked Salmon Tomato Pizza

(Pictured below)

Prep/Total Time: 25 min.

This quick and easy appetizer comes in handy when you find yourself in a time crunch. My "honey" likes this pizza so much that he can almost eat one all by himself!

—Natalie Lodes, Plantation, Florida

1 prebaked thin Italian bread shell crust (10 ounces)
1 cup whipped cream cheese

4 ounces smoked salmon (lox), cut into thin strips
1 cup chopped tomato
1/4 cup chopped red onion
2 tablespoons capers, drained
2 tablespoons minced fresh parsley
Pepper to taste

Place the crust on an ungreased 12-in. pizza pan. Bake at 450° for 8-10 minutes or until heated through. Spread with the cream cheese. Sprinkle with salmon, tomato, onion, capers, parsley and pepper. Cut into slices. **Yield:** 8 slices.

Smoky Jalapenos

Prep: 25 min. **Bake:** 20 min.

When I bake these excellent jalapenos, there are no leftovers. They can also be made with mild banana peppers or yellow chili peppers. —Melinda Strable, Ankeny, Iowa

14 jalapeno peppers
4 ounces cream cheese, softened
14 miniature smoked sausages
7 bacon strips

Cut a lengthwise slit in each pepper; remove seeds and membranes. Spread a teaspoonful of cream cheese into each pepper; stuff each with a sausage.

Cut bacon strips in half widthwise; cook in a microwave or skillet until partially cooked. Wrap a bacon piece around each pepper; secure with a toothpick.

Place in an ungreased 13-in. x 9-in. x 2-in. baking dish. Bake, uncovered, at 350° for 20 minutes for spicy flavor, 30 minutes for medium and 40 minutes for mild. **Yield:** 14 appetizers.

Editor's Note: When cutting or seeding hot peppers, use rubber or plastic gloves to protect your hands. Avoid touching your face.

Handling the Heat

Hot peppers have been eaten in many cultures for at least 400 years. The heat-producing component in hot peppers is called capsaicin. It is concentrated in the white membranes and seeds.

Peppers must be handled carefully because capsaicin can irritate the skin and eyes. The heat can linger on the skin for hours. Wear rubber or plastic gloves and avoid touching your face when working with hot peppers. Thoroughly wash your hands, cutting board and knife with hot soapy water afterward.

Crispy Shrimp Poppers

(Pictured above and on page 4)

Prep: 40 min. **Cook:** 5 min. per batch

For this terrific appetizer, a crisp and golden coating surrounds butterflied shrimp stuffed with bacon and cream cheese. —Jacquelynne Stine, Las Vegas, Nevada

20 uncooked medium shrimp, peeled and deveined
4 ounces cream cheese, softened
10 bacon strips
1 cup all-purpose flour
2 eggs, lightly beaten
2 cups panko (Japanese) bread crumbs
Oil for deep-fat frying

Butterfly the shrimp along the outside curves. Spread about 1 teaspoon cream cheese inside each shrimp. Cut bacon strips in half lengthwise; wrap a piece around each shrimp and secure with toothpicks.

In three separate shallow bowls, place the flour, eggs and bread crumbs. Coat the shrimp with flour; dip into eggs, then coat with bread crumbs.

In an electric skillet or deep-fat fryer, heat oil to 375°. Fry shrimp, a few at a time, for 3-4 minutes or until golden brown. Drain on paper towels. Discard toothpicks before serving. **Yield:** 20 appetizers.

A Bite of Bruschetta

CRUNCHY and fresh-tasting, bruschetta starts with grilled bread rubbed with garlic and brushed with olive oil. Add whatever toppings you desire, from artichokes and tomatoes to zucchini. Any way you slice it, this Italian snack is *delizioso*!

Zucchini Feta Bruschetta

(Pictured at far right and on page 4)

Prep: 30 min. + chilling **Cook:** 15 min.

I make these colorful appetizers using tomatoes, zucchini, garlic and basil fresh from the garden. I took them to a family get-together, and everybody said they were great.
—Tootie Ann Webber, East Tawas, Michigan

1 large tomato, seeded and chopped
1 medium zucchini, finely chopped
4 green onions, thinly sliced
2 tablespoons minced fresh basil
4 to 6 garlic cloves, minced
2 tablespoons lemon juice
2 tablespoons olive oil
3/4 teaspoon salt
1/4 teaspoon pepper
1/2 cup crumbled feta cheese
1 loaf (1 pound) unsliced Italian bread
1/4 to 1/3 cup butter, softened

In a large bowl, combine the tomato, zucchini, onions, basil and garlic. In a small bowl, whisk the lemon juice, oil, salt and pepper. Pour over tomato mixture and toss to coat. Stir in feta cheese. Cover and refrigerate for at least 1 hour.

Cut bread into 18 slices; spread butter on both sides. In a large skillet or on a griddle, toast bread on both sides or until lightly browned. Cut each slice in half; use a slotted spoon to top each with tomato mixture. **Yield:** 3 dozen.

Artichoke Bruschetta

(Pictured above far right)

Prep/Total Time: 30 min.

The topping on this bread reminds me of the hot artichoke dips served in restaurants. This recipe also can easily be doubled or tripled. *—Amy Moylan, Omaha, Nebraska*

✓ **Uses less fat, sugar or salt. Includes Nutrition Facts and Diabetic Exchanges.**

1 jar (6-1/2 ounces) marinated artichoke hearts, drained and chopped
1/2 cup grated Romano cheese
1 plum tomato, seeded and chopped
1/3 cup finely chopped red onion
1/3 cup fresh baby spinach, finely chopped
5 tablespoons mayonnaise
1 garlic clove, minced
1 loaf (10-1/2 ounces) French bread baguette

In a large bowl, combine the first seven ingredients. Cut the French bread baguette into 30 slices; top with the artichoke mixture.

Place on ungreased baking sheets. Broil 3-4 in. from the heat for 3-4 minutes or until the edges are lightly browned. **Yield:** 2-1/2 dozen.

Nutrition Facts: 1 piece equals 78 calories, 4 g fat (1 g saturated fat), 3 mg cholesterol, 122 mg sodium, 8 g carbohydrate, 1 g fiber, 2 g protein. **Diabetic Exchanges:** 1/2 starch, 1/2 fat.

Mozzarella Basil Bruschetta

Prep/Total Time: 25 min.

This simple recipe makes a quick, light appetizer for hungry guests or your family. It's like pizza—you can add whatever toppings you're in the mood for each time you make it. *—Cynthia Bent, Newark, Delaware*

Uses less fat, sugar or salt. Includes Nutrition Facts and Diabetic Exchanges.

6 slices Italian bread (1/2 inch thick)
2 tablespoons olive oil
1 large tomato, seeded and chopped
3 tablespoons minced fresh basil
1/4 teaspoon pepper
1/8 teaspoon salt
6 slices part-skim mozzarella cheese, halved

Cut each slice of bread in half; place on an ungreased baking sheet. Brush with oil.

In a small bowl, combine the tomato, basil, pepper and salt. Spoon 1 tablespoonful of the tomato mixture over each piece of bread; top with the pieces of mozzarella cheese and remaining tomato mixture. Bake at 375° for 10-12 minutes or until the cheese is melted. **Yield:** 1 dozen.

Nutrition Facts: 1 piece equals 94 calories, 5 g fat (2 g saturated fat), 8 mg cholesterol, 167 mg sodium, 7 g carbohydrate, 1 g fiber, 5 g protein. **Diabetic Exchanges:** 1 fat, 1/2 starch.

TREAT your family or dinner guests to an Italian tradition by serving any of these fabulous appetizers (clockwise from top left)—Crab Bruschetta, Artichoke Bruschetta, Zucchini Feta Bruschetta or BLT Bruschetta.

BLT Bruschetta

(Pictured above and on page 4)

Prep: 20 min. **Bake:** 10 min. per batch

I dress up my bruschetta with BLT sandwich fixin's. The maple-flavored bacon adds a tasty new dimension. If you like, substitute mesquite-flavored bacon. —Pat Stevens Granbury, Texas

- **5 maple-flavored bacon strips, cooked and crumbled**
- **1/2 cup finely chopped seeded tomato**
- **1/2 cup finely chopped leaf lettuce**
- **1/2 cup prepared pesto, *divided***
- **2 tablespoons minced fresh basil**
- **1/4 teaspoon salt**
- **1/4 teaspoon pepper**
- **1 loaf (10-1/2 ounces) French bread baguette**
- **3 tablespoons olive oil**

In a large bowl, combine the bacon, tomato, lettuce, 2 tablespoons pesto, basil, salt and pepper; set aside.

Cut baguette into 36 slices; place on ungreased baking sheets. Brush with oil. Bake at 400° for 9-11 minutes or until golden brown. Spread with remaining pesto; top each slice with 2 teaspoons bacon mixture. **Yield:** 3 dozen.

Crab Bruschetta

(Pictured above and on page 4)

Prep/Total Time: 20 min.

I usually put only veggies on bruschetta, but I love this crab version. —Mary Petrara, Lancaster, Pennsylvania

- **1/2 cup finely chopped shallots**
- **2 garlic cloves, minced**
- **2 tablespoons plus 1/4 cup olive oil, *divided***
- **2 cans (6 ounces *each*) lump crabmeat, drained**
- **1 cup chopped seeded plum tomatoes**
- **1-1/2 teaspoons minced fresh basil *or* 1/2 teaspoon dried basil**
- **3/4 teaspoon minced fresh oregano *or* 1/4 teaspoon dried oregano**
- **8 slices Italian bread (1/2 inch thick)**

In a large skillet, saute shallots and garlic in 2 tablespoons oil until tender. Add the crab, tomatoes, basil and oregano; cook and stir for 5-6 minutes or until heated through. Remove from the heat.

Brush both sides of each slice of bread with remaining oil. In another large skillet, toast bread for 1-2 minutes on each side. Cut each slice in half; top with crab mixture. **Yield:** 16 appetizers.

Steamed Salmon Kabobs

(Pictured above)

Prep: 45 min. **Cook:** 15 min.

Guests will savor every bite of these tangy and colorful appetizer kabobs. So make sure to have copies of the recipe on hand! —*Kristen Strocchia, Lake Ariel, Pennsylvania*

✓ Uses less fat, sugar or salt. Includes Nutrition Facts and Diabetic Exchanges.

3 tablespoons lemon juice
2 tablespoons olive oil
1 tablespoon minced fresh parsley
1 teaspoon pepper
1/4 teaspoon salt
1 bunch broccoli, cut into 1-inch pieces
1 large red onion, halved and cut into wedges
1 pound salmon fillets, skin removed and cut into 1-inch cubes
1 whole garlic bulb, separated and peeled
1 large yellow summer squash, cut into 1-inch pieces

In a small bowl, combine the lemon juice, oil, parsley, pepper and salt. Thread the broccoli, onion, salmon, garlic cloves and squash onto 24 small wooden skewers; brush with lemon juice mixture.

Place the skewers in a steamer basket; place in a large saucepan over 1 in. of water. Bring to a boil; cover and steam for 12-15 minutes or until the salmon flakes easily with a fork and the vegetables are crisp-tender. **Yield:** 2 dozen.

Nutrition Facts: 1 kabob equals 58 calories, 3 g fat (1 g saturated fat), 11 mg cholesterol, 43 mg sodium, 3 g carbohydrate, 1 g fiber, 5 g protein. **Diabetic Exchanges:** 1/2 lean meat, 1/2 vegetable, 1/2 fat.

Shrimp Toast Cups

(Pictured below)

Prep: 30 min. **Bake:** 15 min. per batch

These fun appetizers always disappear quick as a flash. The pretty toast cups lend themselves to other favorite fillings, too. —*Awynne Thurstenson, Siloam Springs, Arkansas*

24 slices white bread, crusts removed
1 cup butter, melted
2 packages (8 ounces *each*) cream cheese, softened
1/2 cup mayonnaise
3 tablespoons sour cream
3 tablespoons prepared horseradish
3 cans (6 ounces *each*) small shrimp, rinsed and drained
16 green onions, sliced
Fresh dill sprigs, optional

Flatten bread with a rolling pin; cut each slice into four pieces. Place butter in a shallow dish; dip both sides of bread in butter; press into miniature muffin cups. Bake at 325° for 14 minutes or until golden brown. Remove from pans to wire racks to cool.

In a large mixing bowl, beat the cream cheese, mayonnaise, sour cream and horseradish until blended. Just before serving, stir in shrimp and onions; spoon into cups. Garnish with dill if desired. Refrigerate leftovers. **Yield:** 8 dozen.

Zesty Corn Dip

Prep: 10 min. **Bake:** 30 min.

When it comes to crowd-pleasing dips, I recommend this zippy version served in a bread bowl. It goes over well at parties. —Lisa Leaper-Shuck, Worthington, Ohio

1 package (8 ounces) cream cheese, softened
2 tablespoons sour cream
1/2 to 1 teaspoon cayenne pepper
1/2 to 1 teaspoon ground cumin
1 can (15-1/4 ounces) whole kernel corn, drained
1 can (11 ounces) white corn, drained
1 can (10 ounces) diced tomatoes and green chilies, drained
1 cup (4 ounces) shredded cheddar cheese
1 round loaf (1 pound) Italian bread
1 teaspoon minced fresh cilantro

In a large mixing bowl, beat the cream cheese, sour cream, cayenne and cumin until smooth. Stir in the corn, tomatoes and cheese.

Transfer to an ungreased 3-qt. baking dish. Bake, uncovered, at 350° for 30-35 minutes or until bubbly.

Cut a 1-1/2-in. slice off top of bread. Carefully hollow out loaf, leaving a 1/2-in. shell. Cube removed bread and bread top. Cool the dip for 2-3 minutes; spoon into bread shell. Sprinkle with cilantro. Serve warm with bread cubes. **Yield:** 4 cups.

Smoked Salmon Tomato Cups

Prep/Total Time: 30 min.

These easy appetizers pair smoked salmon with fresh cherry tomatoes. —Vicki Raatz, Waterloo, Wisconsin

✓ **Uses less fat, sugar or salt. Includes Nutrition Facts and Diabetic Exchanges.**

24 cherry tomatoes (about 1 pint)
1 package (3 ounces) smoked cooked salmon, finely chopped
1/4 cup fat-free cottage cheese
2 tablespoons finely chopped celery
1 tablespoon finely chopped green onion
1/4 teaspoon dill weed

Cut a thin slice off the top of each tomato. Scoop out and discard pulp; invert tomatoes onto paper towels to drain. In a small bowl, combine the remaining ingredients; stuff into tomatoes. Refrigerate until serving. **Yield:** 2 dozen.

Nutrition Facts: 1 tomato cup equals 9 calories, trace fat (trace saturated fat), 1 mg cholesterol, 81 mg sodium, 1 g carbohydrate, trace fiber, 1 g protein. **Diabetic Exchange:** Free food.

Greek Chicken Wings

(Pictured above)

Prep: 15 min. + marinating **Bake:** 35 min.

Your guests will be delighted with these moist, herbed chicken wings. You may want to make a double batch of the cucumber dipping sauce—it goes quickly! —Lorraine Caland, Thunder Bay, Ontario

3 tablespoons lemon juice
2 tablespoons olive oil
2 tablespoons honey
1 teaspoon dried oregano
1 garlic clove, minced
1/4 teaspoon salt
4 pounds frozen chicken wingettes, thawed

CUCUMBER SAUCE:

1/2 cup plain yogurt
1/2 cup chopped peeled cucumber
1/2 cup crumbled feta cheese
2 tablespoons snipped fresh dill *or* 2 teaspoons dill weed
1 garlic clove, peeled
Dash salt

In a large resealable plastic bag, combine the lemon juice, oil, honey, oregano, garlic and salt. Add chicken wings; seal bag and toss to coat. Refrigerate overnight.

In a blender, combine sauce ingredients. Cover; process until smooth. Transfer to a small bowl. Cover; refrigerate until serving.

Drain and discard marinade. Place wings on a rack in a greased 15-in. x 10-in. x 1-in. baking pan. Bake, uncovered, at 400° for 35-40 minutes or until juices run clear, turning once. Serve with cucumber sauce. **Yield:** about 3 dozen (1 cup sauce).

Buffalo Chicken Dip

Prep/Total Time: 30 min.

This is a great dip for holiday and Super Bowl parties. Everywhere I take it, people ask for the recipe.
—Peggy Foster, Florence, Kentucky

1 package (8 ounces) cream cheese, softened
1 can (10 ounces) chunk white chicken, drained
1/2 cup buffalo sauce
1/2 cup ranch salad dressing
2 cups (8 ounces) shredded Colby-Monterey Jack cheese
Tortilla chips

Spread cream cheese into an ungreased shallow 1-qt. baking dish. Layer with chicken, buffalo sauce and ranch dressing. Sprinkle with cheese.

Bake, uncovered, at 350° for 20-25 minutes or until cheese is melted. Serve warm with tortilla chips. **Yield:** about 2 cups.

Roasted Cumin Cashews

(Pictured below)

Prep: 15 min. **Bake:** 50 min. + cooling

Kick up your get-togethers at Christmas or anytime with these well-seasoned snacks. They're sweet, salty...and so munchable!
—Martha Fehl, Brookville, Indiana

1 egg white
1 tablespoon water
2 cans (9-3/4 ounces *each*) salted whole cashews
1/3 cup sugar
3 teaspoons chili powder
2 teaspoons salt
2 teaspoons ground cumin
1/2 teaspoon cayenne pepper

In a large bowl, whisk egg white and water. Add cashews and toss to coat. Transfer to a colander; drain for 2 minutes. In another bowl, combine the remaining ingredients; add cashews and toss to coat.

Arrange in a single layer in a greased 15-in. x 10-in. x 1-in. baking pan. Bake, uncovered, at 250° for 50-55 minutes, stirring once. Cool on a wire rack. Store in an airtight container. **Yield:** 3-1/2 cups.

Onion Almond Rounds

Prep: 35 min. **Bake:** 10 min.

Onion lovers will rejoice when they taste these cheesy bread rounds topped with sauteed onions and roasted almonds. The robust appetizers can be assembled ahead of time and then popped in the oven when guests arrive.
—Edna Coburn, Tucson, Arizona

2 medium onions, thinly sliced
1 tablespoon butter
1 loaf (24 ounces) sliced sandwich bread
1 package (8 ounces) cream cheese, softened
1/2 cup chopped almonds
2 teaspoons Worcestershire sauce
1/4 to 1/2 teaspoon lemon-herb seasoning
3 tablespoons mayonnaise
1/2 teaspoon paprika
28 roasted salted almonds

In a large skillet, saute onions in butter for 7-10 minutes or until golden brown; set aside. With a 2-in. round biscuit cutter, cut 28 circles from bread slices; place on two baking sheets. Bake at 350° for 5-8 minutes or until lightly toasted.

In a small mixing bowl, combine the cream cheese, chopped almonds, Worcestershire sauce and lemon-herb seasoning. Spread 1 rounded teaspoonful over each bread round.

Top each with onions and about 1/4 teaspoon mayonnaise. Sprinkle with paprika; top each with an almond. Bake at 350° for 5-8 minutes or until heated through. **Yield:** 28 appetizers.

Editor's Note: This recipe was tested with Pepperidge Farm Farmhouse Hearty White Bread.

Chorizo Date Rumaki

Prep: 20 min. **Bake:** 15 min.

My sister-in-law got this recipe from her brother, who's a chef, and shared it with me. It's tasty and disappears quickly. —*Miriam Hershberger, Holmesville, Ohio*

- **1 package (1 pound) sliced bacon**
- **4 ounces uncooked chorizo *or* spicy bulk pork sausage**
- **2 ounces cream cheese, cubed**
- **32 pitted dates**

Cut each bacon strip in half. In a large skillet, cook the bacon in batches over medium heat until partially cooked but not crisp. Remove to paper towels; drain the drippings.

In the same skillet, cook chorizo over medium heat until no longer pink; drain. Stir in cream cheese.

Carefully cut a slit in the center of each date; fill with cream cheese mixture. Wrap a piece of bacon around each stuffed date; secure with toothpicks. Place on ungreased baking sheets. Bake at 350° for 12-15 minutes or until the bacon is crisp. **Yield:** 32 appetizers.

Seaside Prawn Kabobs

Prep: 20 min. + marinating **Cook:** 10 min.

The prawns in this special holiday treat pick up wonderful flavor from a teriyaki marinade, and the pineapple and peppers add color. —*Laura Barrett, Binghamton, New York*

- **1 can (5-1/4 ounces) unsweetened pineapple chunks**
- **3/4 cup roasted garlic teriyaki marinade and sauce**
- **16 uncooked jumbo shrimp, peeled and deveined**
- **1 small sweet red pepper, cut into 1-inch chunks**

Drain the pineapple chunks, reserving 3 tablespoons juice; set the pineapple chunks aside. In a small bowl, combine the garlic teriyaki sauce and reserved pineapple juice; pour 3/4 cup into a large resealable plastic bag. Add the shrimp; seal the bag and turn to coat. Refrigerate for 20 minutes. Set aside remaining marinade for basting.

Drain and discard marinade. On 16 small metal or soaked wooden skewers, alternately thread one shrimp, one pineapple chunk and one red pepper chunk. Broil 3-4 in. from the heat for 4-5 minutes on each side or until shrimp turn pink, basting with reserved marinade. **Yield:** 16 appetizers.

Rainbow Pepper Appetizers

(Pictured above)

Prep/Total Time: 20 min.

Don't be surprised if guests polish off these crisp-tender peppers in no time. —*Marion Karlin, Waterloo, Iowa*

Uses less fat, sugar or salt. Includes Nutrition Facts and Diabetic Exchanges.

- **1/2 *each* medium green, sweet red, yellow and orange peppers**
- **1 cup (4 ounces) shredded Monterey Jack cheese**
- **2 tablespoons chopped ripe olives**
- **1/4 teaspoon crushed red pepper flakes, optional**

Cut each pepper half into nine pieces. Place skin side down in an ungreased ovenproof skillet; sprinkle with cheese, olives and pepper flakes if desired. Broil 3-4 in. from the heat for 5-7 minutes or until peppers are crisp-tender and cheese is melted. **Yield:** 3 dozen.

Nutrition Facts: 3 pieces (prepared with reduced-fat cheese) equals 34 calories, 2 g fat (1 g saturated fat), 7 mg cholesterol, 93 mg sodium, 2 g carbohydrate, trace fiber, 3 g protein. **Diabetic Exchange:** 1 vegetable.

Pepperoni Pinwheels

(Pictured above)

Prep: 20 min. **Bake:** 15 min.

These golden-brown rounds have lots of Italian flavor. They're easy to make...and really good!
—Vikki Rebholz, West Chester, Ohio

- **1/2 cup diced pepperoni**
- **1/2 cup shredded part-skim mozzarella cheese**
- **1/4 teaspoon dried oregano**
- **1 egg, *separated***
- **1 tube (8 ounces) refrigerated crescent rolls**

In a bowl, combine pepperoni, cheese, oregano and egg yolk. In a small bowl, whisk egg white until foamy; set aside. Separate crescent dough into four rectangles; seal perforations. Spread pepperoni mixture over each rectangle to within 1/4 in. of edges. Roll up jelly-roll style, starting with a short side; pinch seams to seal. Cut each into six slices.

Place cut side down on greased baking sheets; brush tops with egg white. Bake at 375° for 12-15 minutes or until golden brown. Serve warm. Refrigerate leftovers. **Yield:** 2 dozen.

Take a Dip

Want to give Pepperoni Pinwheels (above) extra pizzazz? Just serve them with your favorite pizza sauce, marinara sauce or ranch salad dressing on the side for dipping.

Sunflower-Cherry Granola Bars

Prep: 30 min. + cooling

You'll want to try these chewy snack bars. They're loaded with oats and almonds. The dried cherries add just the right amount of tang. *—Laura McDowell, Lake Villa, Illinois*

- **4 cups old-fashioned oats**
- **1 cup sliced almonds**
- **1 cup flaked coconut**
- **1 cup sugar**
- **1 cup light corn syrup**
- **1 cup creamy peanut butter**
- **1/2 cup raisins**
- **1/2 cup dried cherries**
- **1/2 cup sunflower kernels**

Spread oats into an ungreased 15-in. x 10-in. x 1-in. baking pan. Bake at 400° for 15-20 minutes or until lightly browned. Meanwhile, spread almonds and coconut into another ungreased 15-in. x 10-in. x 1-in. baking pan. Bake for 8-10 minutes or until lightly toasted.

In a Dutch oven over medium heat, bring sugar and corn syrup to a boil. Cook and stir for 2-3 minutes or until sugar is dissolved. Remove from the heat; stir in peanut butter until combined. Add the raisins, cherries, sunflower kernels, and toasted oats, almonds and coconut; mix well.

Using a metal spatula, press mixture into an ungreased 15-in. x 10-in. x 1-in. baking pan. Cool to room temperature. Cut into bars. **Yield:** 2-1/2 dozen.

Cinnamon Toasted Almonds

Prep: 15 min. **Bake:** 25 min. + cooling

A handful of these yummy, cinnamon-glazed nuts soon becomes two handfuls, then three, four or more. They taste just like the ones you get at the fair. *—Janice Thompson*
Stacy, Minnesota

- **2 egg whites**
- **6 teaspoons vanilla extract**
- **4 cups unblanched almonds**
- **1/3 cup sugar**
- **1/3 cup packed brown sugar**
- **1 teaspoon salt**
- **1/2 teaspoon ground cinnamon**

In a large mixing bowl, beat egg whites until frothy; beat in vanilla. Add almonds; stir gently to coat. Combine the sugars, salt and cinnamon; add to nut mixture and stir gently to coat.

Spread evenly into two greased 15-in. x 10-in. x 1-in. baking pans. Bake at 300° for 25-30 minutes or until almonds are crisp, stirring once. Cool. Store in an airtight container. **Yield:** about 4 cups.

Smackin' Good Snack Mix

Prep: 15 min. **Bake:** 40 min. + cooling

If you love to munch, this crunchy snack mix is a must! It goes over big with everyone from tailgaters to midnight snackers. —Lucile Cline, Wichita, Kansas

- **6 cups original Bugles**
- **5 cups nacho cheese-flavored Bugles**
- **4 cups miniature cheese crackers**
- **1 package (6 ounces) miniature colored fish-shaped crackers**
- **3 cups miniature pretzels**
- **2 cups Crispix**
- **2 cups lightly salted cashews**
- **3/4 cup butter-flavored popcorn oil**
- **2 envelopes (1 ounce *each*) ranch salad dressing mix**

In a large bowl, combine the first seven ingredients. Combine oil and dressing mix; pour over cracker mixture and toss to coat.

Transfer to three greased 15-in. x 10-in. x 1-in. baking pans. Bake at 250° for 40-45 minutes or until crisp, stirring occasionally. Cool on wire racks. Store in an airtight container. **Yield:** 6 quarts.

Good Granola

Prep: 15 min. **Bake:** 30 min. + cooling

This fun mix can be enjoyed by the handful or as a topping on yogurt, pudding or ice cream. My husband and I have 10 children, and I feel good about them eating this after school or anytime. —Robin MacKenzie Union City, Pennsylvania

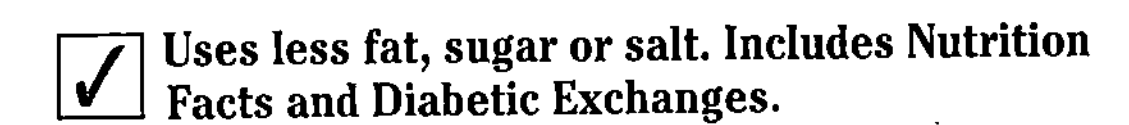

✓ **Uses less fat, sugar or salt. Includes Nutrition Facts and Diabetic Exchanges.**

- **3 cups old-fashioned oats**
- **1 cup toasted wheat germ**
- **1/4 cup slivered almonds**
- **1/4 cup sunflower kernels**
- **1/4 cup wheat *or* oat bran**
- **2 tablespoons sesame seeds**
- **2 tablespoons flaxseed**
- **1 teaspoon ground cinnamon**
- **1/4 teaspoon salt**
- **1/4 cup honey**
- **2 tablespoons canola oil**
- **1-1/2 teaspoons vanilla extract**

In a large bowl, combine the first nine ingredients; set aside. In a small saucepan, combine the honey, oil and vanilla. Cook over low heat for 1-2 minutes or until heated through. Pour over oat mixture; toss to coat.

Spread into a 15-in. x 10-in. x 1-in. baking pan coated with nonstick cooking spray. Bake at 275° for 30-35 minutes or until lightly browned, stirring twice. Cool. Store in an airtight container. **Yield:** 5 cups.

Nutrition Facts: 1/4 cup (calculated without ice cream) equals 119 calories, 5 g fat (1 g saturated fat), 0 cholesterol, 41 mg sodium, 16 g carbohydrate, 3 g fiber, 5 g protein. **Diabetic Exchanges:** 1 starch, 1 fat.

Crab Deviled Eggs

(Pictured below)

Prep/Total Time: 30 min.

Tasty crabmeat and fresh chopped veggies dress up this all-time favorite. My family likes crab salad and deviled eggs, so I tried combining them. What a hit! —Kevon Shuler, Chelsea, Michigan

- **12 hard-cooked eggs**
- **1 can (6 ounces) crabmeat, drained, flaked and cartilage removed**
- **1/2 cup mayonnaise**
- **1 green onion, finely chopped**
- **1 tablespoon finely chopped celery**
- **1 tablespoon finely chopped green pepper**
- **2 teaspoons Dijon mustard**
- **1 teaspoon minced fresh parsley**
- **1/2 teaspoon salt**
- **1/8 teaspoon pepper**
- **3 dashes hot pepper sauce**
- **3 dashes Worcestershire sauce**
- **Additional minced fresh parsley**

Cut eggs in half lengthwise. Remove yolks; set whites aside. In a bowl, mash the yolks. Add the crab, mayonnaise, onion, celery, green pepper, mustard, parsley, salt, pepper, hot pepper sauce and Worcestershire sauce; mix well.

Pipe or spoon into egg whites. Sprinkle with additional parsley. Refrigerate until serving. **Yield:** 2 dozen.

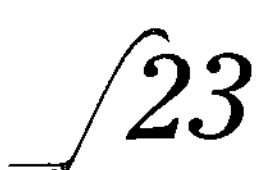

Salads & Dressings

Which of these mouth-watering pasta, vegetable, rice and other medleys will your family enjoy most? It's truly a toss-up!

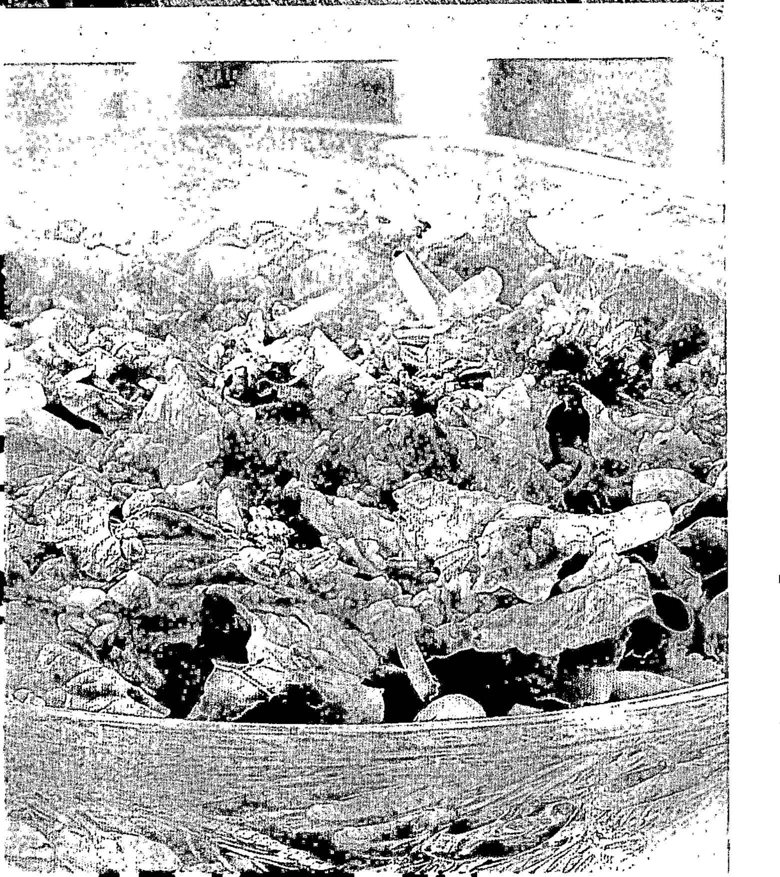

PICK OF THE CROP. Clockwise from upper left: Onion Beet Salad (p. 25), Layered Summertime Salad (p. 27), Sesame-Almond Romaine Salad (p. 29), Peppery Vegetable Salad (p. 26) and Festive Rice Salad (p. 26).

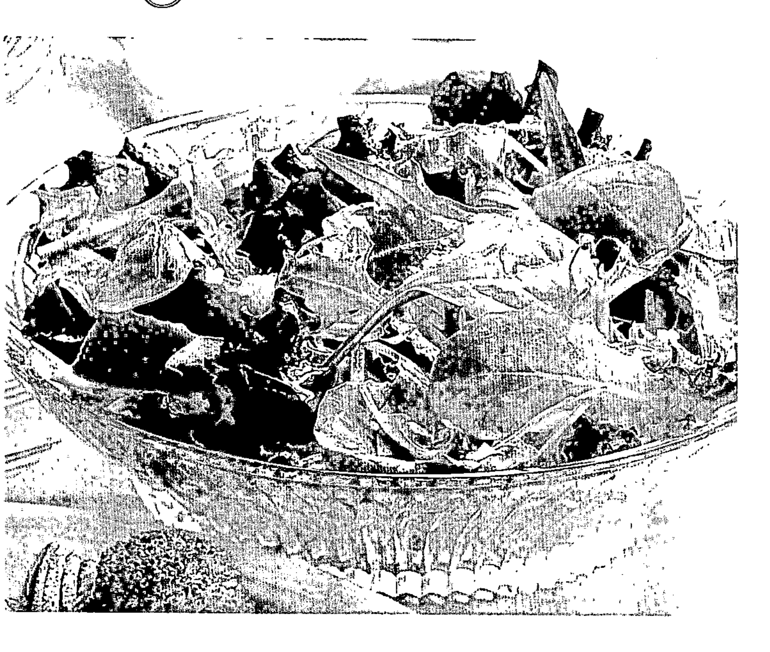

Garlic Vinaigrette

(Pictured above)

Prep/Total Time: 10 min.

I use this flavorful salad dressing year-round. It's especially nice with assorted greens, tomatoes and sweet onions.
—Carol Birkemeier, Nashville, Tennessee

 **Uses less fat, sugar or salt. Includes Nutrition Facts and Diabetic Exchanges.**

- **2 tablespoons vegetable broth**
- **2 tablespoons red wine vinegar**
- **2 tablespoons olive oil**
- **1 tablespoon sugar**
- **2 teaspoons lemon juice**
- **2 teaspoons Worcestershire sauce**
- **1 teaspoon Dijon mustard**
- **2 garlic cloves, minced**
- **1/4 teaspoon salt**
- **1/4 teaspoon minced fresh thyme**
- **1/4 teaspoon pepper**

In a jar with a tight-fitting lid, combine all ingredients; shake well. Serve dressing over salad greens. **Yield:** about 1/2 cup.

Nutrition Facts: 4-1/2 teaspoons equals 65 calories, 6 g fat (1 g saturated fat), 0 cholesterol, 191 mg sodium, 4 g carbohydrate, trace fiber, trace protein. **Diabetic Exchanges:** 1 vegetable, 1 fat.

Veggie Salad in Lettuce Cups

Prep/Total Time: 20 min.

Radishes, carrots and cucumber give this special salad a nice crunch. *—Amy Short, Lesage, West Virginia*

✓ **Uses less fat, sugar or salt. Includes Nutrition Facts and Diabetic Exchanges.**

- **1 large cucumber, seeded and finely chopped**
- **2 small carrots, shredded**
- **12 radishes, coarsely chopped**
- **2 tablespoons olive oil**
- **1 tablespoon white balsamic vinegar**
- **1 tablespoon lemon juice**
- **3 teaspoons minced fresh thyme**
- **1 teaspoon honey mustard**
- **1/2 teaspoon salt**
- **1/4 teaspoon pepper**
- **8 Boston lettuce leaves**

In a small bowl, combine the cucumber, carrots and radishes. In a jar with a tight-fitting lid, combine the oil, vinegar, lemon juice, thyme, mustard, salt and pepper; shake well. Pour over vegetables and toss to coat.

For each serving, shape two lettuce leaves into a cup; fill with vegetable mixture. Serve immediately. **Yield:** 4 servings.

Nutrition Facts: 3/4 cup equals 92 calories, 7 g fat (1 g saturated fat), trace cholesterol, 321 mg sodium, 7 g carbohydrate, 2 g fiber, 1 g protein. **Diabetic Exchanges:** 1-1/2 fat, 1 vegetable.

Mediterranean Lamb And Bean Salad

Prep: 30 min. + chilling

This meaty main dish combines lamb, artichokes, beans and feta cheese with a tangy dressing.
—Lora Winckler, Sunnyside, Washington

✓ **Uses less fat, sugar or salt. Includes Nutrition Facts and Diabetic Exchanges.**

- **1 pound boneless leg of lamb**
- **2 jars (6-1/2 ounces *each*) marinated artichoke hearts, drained**
- **1 can (16 ounces) kidney beans, rinsed and drained**
- **2 cups frozen cut green beans, thawed**
- **1/2 cup julienned sweet red pepper**
- **1/4 cup chopped red onion**
- **1/2 cup fat-free Italian salad dressing**
- **1/4 cup red wine vinegar**
- **1/4 teaspoon pepper**

Crumbled reduced-fat feta cheese, optional

Grill lamb, covered, over medium heat for 10-20 minutes or until meat reaches desired doneness (for medium-rare, a meat thermometer should read 145°; medium, 160°; well-done, 170°). Cut into cubes.

In a large bowl, combine the lamb, artichokes, kidney beans, green beans, red pepper and onion. In a small bowl, combine the salad dressing, vinegar and pepper; drizzle over salad and toss to coat. Cover and refrigerate for at least 4 hours. Serve with feta cheese if desired. **Yield:** 6 servings.

Nutrition Facts: 1 cup (calculated without feta cheese) equals 289 calories, 14 g fat (4 g saturated fat), 43 mg cholesterol, 700 mg sodium, 22 g carbohydrate, 5 g fiber, 19 g protein. **Diabetic Exchanges:** 2-1/2 lean meat, 1 starch, 1 vegetable, 1 fat.

Crowd-Pleasing Taco Salad

Prep/Total Time: 30 min.

While this recipe might seem complicated at first glance, it can be fixed in just 30 minutes. It's always a big hit.
—Ann Cahoon, Bradenton, Florida

- **1 pound ground beef**
- **1/2 cup ketchup**
- **1 teaspoon dried oregano**
- **1 teaspoon chili powder**
- **1/2 teaspoon salt**
- **1/4 teaspoon pepper**
- **1 medium head iceberg lettuce, torn**
- **2 medium tomatoes, diced**
- **1 cup (4 ounces) shredded taco cheese**
- **1 can (2-1/4 ounces) sliced ripe olives, drained**
- **1/2 cup mayonnaise**
- **1/4 cup taco sauce**
- **1 package (10-1/2 ounces) corn chips**

In a large saucepan, cook beef over medium heat until no longer pink; drain. Stir in the ketchup, oregano, chili powder, salt and pepper. Bring to a boil. Reduce heat; cover and simmer for 10 minutes.

In a large bowl, combine the lettuce, tomatoes, cheese, olives and beef mixture. Combine mayonnaise and sauce. Pour over salad; toss to coat. Sprinkle with chips. Serve immediately. **Yield:** 14-16 servings.

The Basics of Beets

Look for beets with smooth, unblemished skin. The greens, if attached, should be brightly colored and not wilted. Before storing beets, remove the greens. Place beets in a plastic bag in the refrigerator for up to 3 weeks.

Just before using beets, gently wash them. Stains from beets are difficult to remove, so protect work surfaces and consider wearing rubber or plastic gloves when handling them.

Onion Beet Salad

(Pictured below and on page 22)

Prep: 30 min. **Bake:** 1 hour + chilling

Everyone loves the homemade dressing that coats these baked beets and onions. I'm often asked to bring this pretty dish to family gatherings.
—Barbara Van Lanen, Salinas, California

- **12 whole fresh beets (about 2-1/2 pounds), peeled and halved**
- **5 tablespoons olive oil, *divided***
- **1 large red onion, chopped**
- **1/2 cup balsamic vinegar**
- **1/3 cup red wine vinegar**
- **1/4 cup sugar**
- **1 teaspoon salt**
- **1 teaspoon dried basil**
- **1/2 teaspoon pepper**

Place beets in a large resealable plastic bag; add 2 tablespoons oil. Seal bag and shake to coat. Place an 18-in. x 12-in. piece of heavy-duty foil in a 15-in. x 10-in. x 1-in. baking pan. Arrange beets on foil; fold foil over beets and seal tightly. Bake at 400° for 1 to 1-1/4 hours or until tender.

Cool to room temperature. Cut beets into cubes; place in a large bowl. Add onion. In a small bowl, whisk the vinegars, sugar, salt, basil, pepper and remaining oil. Pour over beet mixture; gently toss to coat. Cover and refrigerate for at least 1 hour, stirring several times. Serve with a slotted spoon. **Yield:** 9 servings.

Festive Rice Salad

(Pictured above and on page 22)

Prep: 30 min. + chilling

I tasted a salad similar to this one at a friend's house, and I was determined to re-create it at home. After several tries, I came up with this sweet-tart concoction. It's easy to prepare and colorful, too. —*Terri Simpson Palm Harbor, Florida*

✓ **Uses less fat, sugar or salt. Includes Nutrition Facts and Diabetic Exchanges.**

3/4 cup uncooked long grain rice
1 package (10 ounces) frozen peas, thawed
1 small sweet red pepper, chopped
3/4 cup chopped green onions
1/2 cup dried cranberries

DRESSING:

1/2 cup canola oil
1/3 cup white vinegar
3 tablespoons sugar
1/2 teaspoon dill weed
1/4 teaspoon salt
1/4 teaspoon ground mustard
1/8 teaspoon pepper

Cook rice according to package directions; cool. In a large bowl, combine rice, peas, red pepper, onions and cranberries. In a small bowl, whisk dressing ingredients. Drizzle over salad and toss to coat. Refrigerate until serving. **Yield:** 12 servings.

Nutrition Facts: 1/2 cup equals 172 calories, 9 g fat (1 g saturated fat), 0 cholesterol, 77 mg sodium, 21 g carbohydrate, 2 g fiber, 2 g protein. **Diabetic Exchanges:** 1-1/2 fat, 1 starch, 1 vegetable.

Peppery Vegetable Salad

(Pictured below and on page 22)

Prep: 20 min. + chilling

To use up the harvest our large garden produces every year, I've had to find some great recipes, and this one is a favorite. It's loaded with crunch and well-seasoned flavor. —*Andrea Sheatz, Knox, Pennsylvania*

15 poblano *and/or* banana peppers, seeded and coarsely chopped (about 7 cups)
1-1/2 cups fresh cauliflowerets, cut into bite-size pieces
3 small carrots, coarsely chopped
1 large sweet red pepper, coarsely chopped
1 cup pitted ripe olives
3 garlic cloves, minced
1 cup water
1 cup white vinegar
3/4 cup olive oil
4 teaspoons dried oregano
1 teaspoon salt

In a large bowl, combine the peppers, cauliflower, carrots, red pepper, olives and garlic. In a jar with a tight-fitting lid, combine the remaining ingredients; shake well. Pour over vegetable mixture and toss to coat. Cover and refrigerate overnight. Serve with a slotted spoon. **Yield:** 12 servings.

Editor's Note: When cutting or seeding hot peppers, use rubber or plastic gloves to protect your hands. Avoid touching your face.

Corn Bread Layered Salad

Prep: 20 min. **Bake:** 15 min. + cooling

My mom's corn bread salad is so filling, it can be a meal in itself. The recipe has been in our family for years.
—Jody Miller, Oklahoma City, Oklahoma

- **1 package (8-1/2 ounces) corn bread/muffin mix**
- **6 green onions, chopped**
- **1 medium green pepper, chopped**
- **1 can (15-1/4 ounces) whole kernel corn, drained**
- **1 can (15 ounces) pinto beans, rinsed and drained**
- **3/4 cup mayonnaise**
- **3/4 cup sour cream**
- **2 medium tomatoes, seeded and chopped**
- **1/2 cup shredded cheddar cheese**

Prepare and bake corn bread according to package directions. Cool on a wire rack.

Crumble corn bread into a 2-qt. glass serving bowl. Layer with onions, green pepper, corn and beans. In a small bowl, combine mayonnaise and sour cream; spread over vegetables. Sprinkle with tomatoes and cheese. Refrigerate until serving. **Yield:** 6-8 servings.

Creamy Blueberry Gelatin Salad

Prep: 30 min. + chilling

Plump blueberries and a fluffy topping star in this pretty, refreshing treat. My grandchildren often request it.
—Sharon Hoefert, Greendale, Wisconsin

- **2 packages (3 ounces *each*) grape gelatin**
- **2 cups boiling water**
- **1 can (21 ounces) blueberry pie filling**
- **1 can (20 ounces) unsweetened crushed pineapple, undrained**

TOPPING:

- **1 package (8 ounces) cream cheese, softened**
- **1 cup (8 ounces) sour cream**
- **1/2 cup sugar**
- **1 teaspoon vanilla extract**
- **1/2 cup chopped walnuts**

In a large bowl, dissolve gelatin in boiling water. Cool for 10 minutes. Stir in pie filling and pineapple until blended. Transfer to a 13-in. x 9-in. x 2-in. dish. Cover and refrigerate until partially set, about 1 hour.

For topping, in a small mixing bowl, combine the cream cheese, sour cream, sugar and vanilla. Carefully spread over gelatin; sprinkle with walnuts. Cover and refrigerate until firm. **Yield:** 12-15 servings.

Layered Summertime Salad

(Pictured above and on page 23)

Prep/Total Time: 30 min.

Luscious layers of pasta and veggies make up this super summer salad, which can be prepared ahead of time for warm-weather parties. *—Betty Fulks, Onia, Arkansas*

- **2 cups uncooked gemelli *or* spiral pasta**
- **1 cup mayonnaise**
- **2 tablespoons lemon juice**
- **1 teaspoon sugar**
- **1/2 teaspoon garlic powder**
- **1/2 cup sliced green onions**
- **4 bacon strips, cooked and crumbled, *divided***
- **4 cups torn romaine**
- **1 cup fresh snow peas, trimmed and halved**
- **1 cup fresh cauliflowerets**
- **1 cup fresh broccoli florets**
- **1 large sweet red pepper, chopped**
- **1/2 cup shredded Swiss cheese**

Cook pasta according to package directions. Meanwhile, in a small bowl, combine the mayonnaise, lemon juice, sugar and garlic powder; set aside. Drain pasta and rinse in cold water; toss with green onions and half of the bacon.

In a large salad bowl, layer half of the romaine, pasta mixture, peas, cauliflower, broccoli, red pepper, mayonnaise mixture and cheese. Repeat layers. Sprinkle with remaining bacon. Cover and refrigerate until serving. **Yield:** 16 servings.

Watermelon Race Car

(Pictured below)

Prep: 1 hour

The movie Cars inspired my mom to make this fun melon race car for my son's third birthday. When he saw it, he absolutely beamed! —*Camile Hixon, Galena, Ohio*

2 ounces white candy coating
1 teaspoon shortening
2 cream-filled chocolate sandwich cookies
2 miniature cream-filled chocolate sandwich cookies
1 large watermelon
4 lemon slices
4 lime slices
1 medium cantaloupe, cut into balls *or* cubes
1 medium honeydew, cut into balls *or* cubes

In a small microwave-safe bowl, heat candy coating and shortening, uncovered, at 30% power for 3-4 minutes or until melted, stirring every 30 seconds. For eyes, coat the large cookies with candy coating; place a miniature cookie in the center of each for pupil. Place on waxed paper to harden.

With a sharp knife, cut a thin slice from bottom of watermelon so it sits flat. Beginning above center at one end of melon, lightly score a horizontal cutting line (leaving a fourth of the opposite end unmarked). Repeat on other side of melon.

With a long sharp knife, make a vertical cut from top of melon to end of cutting marks. Cut along the cutting mark, making sure to cut all the way through. Gently pull off cut section; cut a thin slice for a spoiler. Trim spoiler piece to fit smoothly against the end of the melon; attach with three toothpicks.

Remove fruit from cut section and inside melon. Cut into balls or cubes; set aside.

Using a sharp razor blade, carefully carve a smile in the front of watermelon; carve a number in the side. Add other designs if desired. For wheels, attach lemon and lime slices with toothpicks. Attach cookie eyes with toothpicks.

In a large bowl, combine the cantaloupe, honeydew and reserved watermelon. Spoon into melon car. **Yield:** 14 servings.

Editor's Note: This recipe was tested in a 1,100-watt microwave.

Beef Pilaf Salad

Prep: 45 min. + chilling

Spicy, sweet, crunchy, chewy...this main-dish salad has it all! I've enjoyed this recipe for more than 20 years. —*Debbie Shivers, Lake Ozark, Missouri*

 **Uses less fat, sugar or salt. Includes Nutrition Facts and Diabetic Exchanges.**

2 cups thinly sliced cooked beef sirloin steak (about 3/4 pound)
1 can (16 ounces) kidney beans, rinsed and drained
1 cup cooked brown rice
1 cup canned garbanzo beans *or* chickpeas, rinsed and drained
1 cup chopped apple
1/2 cup sliced celery
1 small green pepper, chopped
1/4 cup sliced ripe olives
1/4 cup sliced green onions
1/4 cup minced fresh parsley

DRESSING:

1/3 cup red wine vinegar
3 tablespoons olive oil
2 tablespoons water
2 teaspoons sugar
2 garlic cloves, minced
1 teaspoon ground mustard
1 teaspoon lemon juice
1/4 teaspoon salt
1/4 teaspoon paprika
Dash cayenne pepper

In a large bowl, combine the first 10 ingredients. In a jar with a tight-fitting lid, combine dressing ingredients; shake well. Pour over salad; toss to coat. Cover and refrigerate for at least 1 hour. **Yield:** 6 servings.

Nutrition Facts: 1 cup equals 351 calories, 13 g fat (3 g saturated fat), 50 mg cholesterol, 375 mg sodium, 34 g carbohydrate, 7 g fiber, 25 g protein. **Diabetic Exchanges:** 3 lean meat, 2 starch, 1 vegetable.

Green Bean and Mozzarella Salad

Prep: 20 min. + chilling

Your family will love this bright, tasty salad. It's easy to make and take to a potluck, too. —*Stasha Wampler Clinchport, Virginia*

- **2 cups cut fresh green beans (2-inch pieces)**
- **6 plum tomatoes, sliced**
- **1 block (8 ounces) mozzarella cheese, cubed**
- **1/2 cup Italian salad dressing**
- **1/3 cup minced fresh basil**
- **1/4 teaspoon salt**
- **1/8 teaspoon pepper**

Place beans in a small saucepan and cover with water. Bring to a boil; cook, uncovered, for 6-8 minutes or until crisp-tender. Drain and rinse in cold water.

Place beans in a large salad bowl. Add the remaining ingredients; gently toss to coat. Cover and refrigerate for 1 hour before serving. **Yield:** 6-8 servings.

Sesame-Almond Romaine Salad

(Pictured on page 23)

Prep/Total Time: 25 min.

With ramen noodles and a homemade dressing, this crisp, tangy salad is one of my favorites. We especially enjoy this recipe when the lettuce is fresh from the garden. —*Diane Alger, Hebron, Connecticut*

- **1/2 cup butter, cubed**
- **1 tablespoon sugar**
- **2 packages (3 ounces *each*) ramen noodles, crushed**
- **1/3 cup sesame seeds**
- **1/4 cup slivered almonds**
- **2 bunches romaine, torn**
- **4 green onions, thinly sliced**

DRESSING:

- **3/4 cup olive oil**
- **1/3 cup sugar**
- **1/4 cup red wine vinegar**
- **1 tablespoon soy sauce**

In a large skillet, melt butter and sugar over medium heat. Add crushed noodles, sesame seeds and almonds. (Discard the seasoning packets from noodles or save for another use.) Cook and stir for 6-8 minutes or until browned; set aside. In a large salad bowl, toss the romaine and green onions.

In a jar with a tight-fitting lid, combine the dressing ingredients; shake well. Just before serving, drizzle over romaine mixture; top with noodle mixture. Toss to coat. **Yield:** 12 servings.

Apple-Brie Spinach Salad

(Pictured above)

Prep/Total Time: 30 min.

In the summer, I don't like to prepare or eat large meals, so I often fix salads. I'm always on the lookout for new and interesting recipes. This one's a winner that I frequently make for company. —*Rhonda Crowe Victoria, British Columbia*

- **4 large apples, cut into 1/2-inch wedges**
- **4 tablespoons maple syrup, *divided***
- **8 cups fresh baby spinach**
- **1 round (8 ounces) Brie *or* Camembert cheese, cubed**
- **1/2 cup pecan halves, toasted**

DRESSING:

- **1/4 cup apple cider *or* juice**
- **1/4 cup vegetable oil**
- **3 tablespoons cider vinegar**
- **1 teaspoon Dijon mustard**
- **1 garlic clove, minced**

Place apples on an ungreased baking sheet; brush with 2 tablespoons syrup. Broil 3-4 in. from the heat for 3 minutes. Turn; brush with remaining syrup. Broil 3-5 minutes longer or until crisp-tender.

In a large salad bowl, combine the spinach, cheese cubes, pecans and apples. In a small saucepan, combine the dressing ingredients; bring to a boil. Pour over the salad and toss to coat. Serve immediately. **Yield:** 10 servings.

Soups & Sandwiches

When it comes to casual comfort food, nothing hits the spot like a simmering pot of soup or a piled-high sandwich. You're sure to discover new favorites in this chock-full chapter.

HEARTWARMING HELPINGS. Clockwise from upper left: Colorful Chicken 'n' Squash Soup (p. 36), Blue Cheese Clubs (p. 33), Cheeseburger Paradise Soup (p. 34), Zesty Vegetarian Wraps (p. 45) and Asparagus Brunch Pockets (p. 38).

Minestrone with Italian Sausage

(Pictured above)

Prep: 25 min. **Cook:** 1 hour

I make this satisfying soup all the time, and it's my dad's absolute favorite. The recipe freezes well and tastes just as good reheated. —*Linda Reis, Salem, Oregon*

1 pound bulk Italian sausage
1 large onion, chopped
2 large carrots, chopped
2 celery ribs, chopped
1 medium leek (white portion only), chopped
3 garlic cloves, minced
1 medium zucchini, cut into 1/2-inch pieces
1/4 pound fresh green beans, trimmed and cut into 1/2-inch pieces
6 cups beef broth
2 cans (14-1/2 ounces *each*) diced tomatoes with basil, oregano and garlic
3 cups shredded cabbage
1 teaspoon dried basil
1 teaspoon dried oregano
1/4 teaspoon pepper
1 can (15 ounces) garbanzo beans *or* chickpeas, rinsed and drained
1/2 cup uncooked small pasta shells
3 tablespoons minced fresh parsley
1/3 cup grated Parmesan cheese

In a soup kettle, cook sausage and onion over medium heat until meat is no longer pink; drain. Stir in the carrots, celery, leek and garlic; cook for 3 minutes. Add zucchini and green beans; cook 2 minutes longer.

Stir in broth, tomatoes, cabbage, basil, oregano and pepper. Bring to a boil. Reduce heat. Cover; simmer for 45 minutes.

Return to a boil. Stir in garbanzo beans, pasta and parsley. Cook for 6-9 minutes or until the pasta is tender. Serve with Parmesan cheese. **Yield:** 11 servings (about 3 quarts).

Southwestern Backyard Burgers

Prep/Total Time: 30 min.

Whether you're in a stadium parking lot or on your patio, these burgers are great on the grill. Sometimes I shape six patties rather than eight with this mixture because I like my burgers big! —*Robert Hodges, San Diego, California*

1 can (4 ounces) chopped green chilies
1/4 cup Worcestershire sauce
1/2 teaspoon hickory Liquid Smoke, optional
1/2 cup crushed butter-flavored crackers (about 12 crackers)
4-1/2 teaspoons chili powder
3 teaspoons ground cumin
1/2 teaspoon salt
1/2 teaspoon pepper
2 pounds lean ground beef
1/2 pound bulk pork sausage
8 slices pepper Jack cheese
8 sesame seed hamburger buns, split
Lettuce leaves, optional

In a bowl, combine the first eight ingredients. Crumble beef and sausage over mixture and mix well. Shape into eight patties.

Grill, covered, over medium heat for 5-7 minutes on each side or until no longer pink. Top with cheese. Grill 1 minute longer or until cheese is melted.

Grill buns, cut side down, for 1-2 minutes or until toasted. Serve burgers on buns with lettuce if desired. **Yield:** 8 servings.

Beef 'n' Bean Braid

Prep: 30 min. + rising **Bake:** 20 min.

I hear "oohs" and "aahs" when I bring this beautiful loaf to a potluck. With a hearty filling, each slice is a robust sandwich. —*Val O'Connell, Hotchkiss, Colorado*

1 package (1/4 ounce) active dry yeast
3/4 cup warm water (110° to 115°)
2 tablespoons butter, melted
1 tablespoon sugar
1/2 teaspoon salt
1 egg
2 cups all-purpose flour

FILLING:

1 pound ground beef
1 medium onion, chopped
1 teaspoon garlic salt
1 can (16 ounces) kidney beans, rinsed and drained
2/3 cup water
2 tablespoons chili powder
1/4 teaspoon ground cumin
1/4 teaspoon cayenne pepper
1-1/2 cups (6 ounces) shredded cheddar cheese
1 egg, lightly beaten

In a large mixing bowl, dissolve yeast in warm water. Add butter, sugar, salt, egg and 1-1/2 cups flour; beat until smooth. Stir in enough remaining flour to form a soft dough. Turn onto a floured surface; knead until smooth and elastic, about 6-8 minutes. Place in a greased bowl, turning once to grease top. Cover; let rise in a warm place until doubled, about 1 hour.

Meanwhile, in a large skillet, cook the beef, onion and garlic salt over medium heat until meat is no longer pink; drain. Stir in the beans, water and seasonings; bring to a boil. Reduce heat; simmer, uncovered, for 5-10 minutes or until thickened.

Punch dough down. Turn onto a lightly floured surface; roll into a 16-in. x 11-in. rectangle. Transfer to a greased 15-in. x 10-in. baking sheet. Spread beef mixture down center of dough; sprinkle with cheese.

On each long side, cut 1-in.-wide strips about 1/2 in. from filling. Starting at one end, fold alternating strips at an angle across filling; seal ends. Cover and let rise until doubled, about 30 minutes.

Brush egg over dough. Bake at 350° for 20-25 minutes or until golden brown. Remove to a wire rack. Let stand for 5 minutes before slicing. **Yield:** 6-8 servings.

Savory Ham Wraps

Prep/Total Time: 20 min.

The flavorful dressing is what makes these tender roll-ups so special. —*Ruth Peterson, Jenison, Michigan*

1/4 cup mayonnaise
1 tablespoon milk
3/4 teaspoon sugar
1/4 teaspoon prepared mustard
1/8 teaspoon celery seed
Dash salt
2 flour tortillas (10 inches), warmed
1/4 pound thinly sliced deli ham
1/3 cup shredded Swiss cheese
2/3 cup shredded lettuce
1 medium tomato, seeded and chopped
1 green onion, chopped

In a small bowl, whisk the first six ingredients; spread evenly over each tortilla. Layer with ham and cheese. Top with lettuce, tomato and onion. Roll up tightly; secure with toothpicks if desired. **Yield:** 2 servings.

Blue Cheese Clubs

(Pictured below and on page 30)

Prep/Total Time: 25 min.

The secret to this recipe is the mild blue cheese spread that complements the turkey. Cut into triangles, these sandwiches look elegant and are easy to fix.
—*Nancy Jo Leffler, Depauw, Indiana*

1 package (3 ounces) cream cheese, softened
1/2 cup crumbled blue cheese
4 tablespoons mayonnaise, *divided*
1 teaspoon dried minced onion
Dash salt and pepper
Dash Worcestershire sauce
8 slices white bread, toasted
8 slices tomato
8 slices deli turkey
4 slices Swiss cheese
4 slices whole wheat bread, toasted
8 bacon strips, cooked
4 lettuce leaves

In a small mixing bowl, beat cream cheese. Add blue cheese, 1 tablespoon mayonnaise, onion, salt, pepper and Worcestershire sauce; beat until combined.

Spread over four slices of white bread; layer with tomato, turkey, Swiss cheese, wheat bread, bacon and lettuce. Spread remaining mayonnaise over remaining white bread; place over lettuce. Secure with toothpicks; cut into triangles. **Yield:** 4 servings.

Cheeseburger Paradise Soup

(Pictured above and on page 30)

Prep: 30 min. **Cook:** 25 min.

I've never met a person who didn't enjoy this creamy soup, which is hearty enough to have as a main course along with bread or rolls. —*Nadina Iadimarco, Burton, Ohio*

6 medium potatoes, peeled and cubed
1 small carrot, grated
1 small onion, chopped
1/2 cup chopped green pepper
2 tablespoons chopped seeded jalapeno pepper
3 cups water
2 tablespoons plus 2 teaspoons beef bouillon granules
2 garlic cloves, minced
1/8 teaspoon pepper
2 pounds ground beef
1/2 pound sliced fresh mushrooms
2 tablespoons butter
5 cups milk, *divided*
6 tablespoons all-purpose flour
1 package (16 ounces) process cheese (Velveeta), cubed
Crumbled cooked bacon

In a soup kettle, combine the first nine ingredients; bring to a boil. Reduce heat; cover and simmer for 15-20 minutes or until potatoes are tender.

Meanwhile, in a large skillet, cook beef and mushrooms in butter over medium heat until meat is no longer pink; drain. Add to soup. Stir in 4 cups milk; heat through.

In a small bowl, combine flour and remaining milk until smooth; gradually stir into soup. Bring to a boil; cook and stir for 2 minutes or until thickened. Reduce heat; stir in cheese until melted. Garnish with bacon. **Yield:** 14 servings (about 3-1/2 quarts).

Editor's Note: When cutting or seeding hot peppers, use rubber or plastic gloves to protect your hands. Avoid touching your face.

Turkey Burgers with Caramelized Onions

Prep: 2 hours **Grill:** 10 min.

Worcestershire and hot pepper sauces add extra flavor to these moist turkey burgers...but it's the sweet-savory onion topping that makes them something special.
—*Phyl Broich-Wessling, Garner, Iowa*

✓ **Uses less fat, sugar or salt. Includes Nutrition Facts and Diabetic Exchanges.**

2 large sweet onions, cut into 1/8-inch slices
2 teaspoons olive oil
1 teaspoon butter
1 teaspoon sugar
1/2 teaspoon salt
1/4 teaspoon pepper
2 tablespoons minced fresh thyme *or* 2 teaspoons dried thyme
2 green onions, chopped
2 tablespoons Worcestershire sauce
1 teaspoon hot pepper sauce
1-1/4 pounds lean ground turkey
6 hamburger buns, split

In a large nonstick skillet over low heat, cook onions in oil and butter for 1 hour or until tender, stirring occasionally. Sprinkle with sugar, salt and pepper; cook over medium-low heat for 45-60 minutes or until golden brown, stirring occasionally. Sprinkle with thyme; keep warm.

In a large bowl, combine the green onions, Worcestershire sauce and pepper sauce; crumble turkey over mixture and mix well. Shape into six patties.

Coat grill rack with nonstick cooking spray before starting the grill. Grill burgers, covered, over medium-hot heat for 5-7 minutes on each side or until a meat thermometer reads 165° and juices run clear. Serve on buns with caramelized onions. **Yield:** 6 servings.

Nutrition Facts: 1 burger with about 1/3 cup onions equals 311 calories, 12 g fat (3 g saturated fat), 76 mg cholesterol, 596 mg sodium, 28 g carbohydrate, 2 g fiber, 21 g protein. **Diabetic Exchanges:** 3 lean meat, 2 starch.

Chunky Tomato-Basil Bisque

Prep: 20 min. **Cook:** 50 min.

My husband, Patrick Maas, whips up this wonderful bisque. Sweet red pepper enhances the tomato and basil.
—Veronique Deblois, Mine Hill, New Jersey

6 celery ribs, chopped
1 large onion, chopped
1 medium sweet red pepper, chopped
1/4 cup butter, cubed
3 cans (14-1/2 ounces *each*) diced tomatoes, undrained
1 tablespoon tomato paste
3/4 cup loosely packed basil leaves, coarsely chopped
3 teaspoons sugar
2 teaspoons salt
1/2 teaspoon pepper
1-1/2 cups heavy whipping cream

In a large saucepan, saute the celery, onion and red pepper in butter for 5-6 minutes or until tender. Add tomatoes and tomato paste. Bring to a boil. Reduce heat; cover and simmer for 40 minutes.

Remove from the heat. Stir in the basil, sugar, salt and pepper; cool slightly. Transfer half of the soup mixture to a blender. While processing, gradually add cream; process until pureed. Return to the pan; heat through (do not boil). **Yield:** 5 servings.

Marinated Beef on Buns

Prep: 15 min. + marinating
Bake: 2 hours + standing

The fresh garlic really comes through in these robust sandwiches. I served them at our family reunion, and everyone wanted the recipe.—Nancy Yarlett, Thornhill, Ontario

3-1/2 cups ketchup
2 medium onions, finely chopped
1-1/2 cups packed brown sugar
1 cup soy sauce
1/2 cup white vinegar
1/2 cup vegetable oil
4 garlic cloves, minced
1/4 teaspoon ground ginger
1 beef eye round roast (3 pounds)
3/4 cup water
12 sandwich buns, split

In a large bowl, combine the first eight ingredients. Pour half of the marinade into a large resealable plastic bag; add roast. Seal bag and turn to coat; refrigerate overnight. Cover and refrigerate remaining marinade.

Drain and discard marinade. Place the roast in a large roasting pan. Combine water and reserved marinade; pour over roast. Cover and bake at 350° for 2 hours or until meat is tender. Let stand for 15 minutes before slicing. Serve on buns. Skim fat from pan juices; serve with sandwiches. **Yield:** 12 servings.

Buffalo Wing Hoagies

(Pictured below)

Prep/Total Time: 30 min.

The first time I had buffalo wings, I was hooked! Since then, I've used that idea for other foods, including these zippy sandwiches. —Kelly Williams, La Porte, Indiana

1 package (9 ounces) ready-to-use Southwestern chicken strips
1 cup butter, softened, *divided*
2 tablespoons Louisiana-style hot sauce
1 tablespoon barbecue sauce
1/2 teaspoon chili powder
6 garlic cloves, minced
1 tablespoon minced fresh parsley
4 hoagie buns, split
4 cups shredded lettuce
2 tablespoons finely chopped celery
Dash salt and pepper
4 plum tomatoes, sliced
2 tablespoons blue cheese dressing

In a large skillet, combine the chicken, 1/4 cup butter, hot sauce, barbecue sauce and chili powder. Cook, uncovered, over low heat for 5 minutes or until heated through.

Meanwhile, in a small bowl, combine the garlic, parsley and remaining butter; spread over cut sides of buns. Place on a baking sheet; broil 8 in. from the heat for 3-5 minutes or until lightly browned.

Spoon chicken mixture onto bun bottoms; top with lettuce, celery, salt, pepper, tomatoes and salad dressing. Replace bun tops. **Yield:** 4 servings.

Colorful Chicken 'n' Squash Soup

(Pictured below and on page 30)

Prep: 25 min. **Cook:** 1-1/2 hours

When I turned 40, I decided to live a healthier lifestyle, which included cooking healthier for my family. This soup is loaded with nutritious squash, kale and carrots. I make it every week, and my family loves it. —*Trina Bigha Fairhaven, Massachusetts*

- **1 broiler/fryer chicken (4 pounds), cut up**
- **13 cups water**
- **5 pounds butternut squash, peeled and cubed (about 10 cups)**
- **1-1/4 pounds fresh kale, chopped**
- **6 medium carrots, chopped**
- **2 large onions, chopped**
- **3 teaspoons salt**

Place chicken and water in a soup kettle. Bring to a boil. Reduce heat; cover and simmer for 1 hour or until chicken is tender.

Remove chicken from the broth. Strain the broth and skim the fat. Return broth to the pan; add the squash, kale, carrots and onions. Bring to a boil. Reduce heat; cover and simmer for 25-30 minutes or until the vegetables are tender.

When chicken is cool enough to handle, remove meat from bones and cut into bite-size pieces. Discard bones and skin. Add chicken and salt to soup; heat through. **Yield:** 14 servings (5-1/2 quarts).

Shrimp Salad Croissants

Prep: 15 min. + chilling

The scrumptious salad filling in these sandwiches is packed with shrimp. I often pair them with potato sticks on the side. You could also serve the salad on a bed of iceberg lettuce. —*Molly Seidel, Edgewood, New Mexico*

- **1 pound cooked small shrimp**
- **2 celery ribs, diced**
- **2 small carrots, shredded**
- **1 cup mayonnaise**
- **1/3 cup finely chopped onion**
- **Dash salt and pepper**
- **2 packages (2-1/4 ounces *each*) sliced almonds**
- **8 croissants, split**

In a large bowl, combine the shrimp, celery, carrots, mayonnaise, onion, salt and pepper. Cover and refrigerate for at least 2 hours. Just before serving, stir in almonds. Serve on croissants. **Yield:** 8 servings.

Roasted Yellow Pepper Soup

Prep: 25 min. **Cook:** 40 min.

We got this recipe from a good friend and merchant marine in New Hampshire. My husband and our two children liked it so much that I started raising yellow peppers. —*Amy Spurrier, Wellsburg, West Virginia*

- **6 large sweet yellow peppers**
- **1 large onion, chopped**
- **1 cup chopped leeks (white portion only)**
- **1/4 cup butter, cubed**
- **3 small potatoes, peeled and cubed**
- **5 cups chicken broth**
- **1/2 teaspoon salt**
- **1/2 teaspoon pepper**
- **Shredded Parmesan cheese, optional**

Halve peppers; remove and discard tops and seeds. Broil peppers 4 in. from the heat until skins blister, about 4 minutes. Immediately place peppers in a bowl; cover and let stand for 15-20 minutes.

Meanwhile, in a large saucepan, saute onion and leeks in butter until tender. Add the potatoes, broth, salt and pepper. Bring to a boil. Reduce heat; cover and simmer for 30 minutes or until potatoes are tender.

Peel off and discard charred skin from peppers. Finely chop peppers; add to potato mixture. Cool slightly.

In a blender, cover and process soup in batches until smooth. Return to the pan; heat through (do not boil). Serve with shredded Parmesan cheese if desired. **Yield:** 8 cups (2 quarts).

Cuban Pork Sandwiches

(Pictured above)

Prep: 30 min. + marinating
Bake: 25 min. + standing

Seasoned pork, a homemade relish and slices of Swiss come together deliciously in this mouth-watering sandwich. The tangy flavor is terrific. —*Connie Zangla Annandale, Minnesota*

1 small red onion, thinly sliced
1 cup water
1 jar (7-1/4 ounces) roasted sweet red peppers, drained and chopped
1/3 cup cider vinegar
2 garlic cloves, peeled and halved
1/2 teaspoon dried oregano
1/4 teaspoon salt
1/4 teaspoon pepper
1/4 teaspoon ground cumin

SANDWICH:

1 garlic clove, minced
1 teaspoon ground cumin
1/2 teaspoon salt
1 pork tenderloin (about 1 pound)
1 teaspoon olive oil
8 slices sourdough bread
8 slices Swiss cheese

For relish, in a small saucepan over medium heat, bring onion and water to a boil. Cook and stir for 1 minute; drain. Transfer to a bowl; add roasted peppers, vinegar, halved garlic cloves, oregano, salt, pepper and cumin. Let stand at room temperature for 1 hour. Discard garlic. (Relish can be made ahead and stored in the refrigerator for up to 1 week.)

In a small bowl, mash the minced garlic, cumin and salt. Rub over tenderloin; place in a shallow baking pan. Bake, uncovered, at 425° for 25-30 minutes or until a meat thermometer reads 160°. Let stand for 10 minutes; thinly slice pork.

Heat oil in a large skillet over medium heat. Top four slices of bread with pork, desired amount of relish and two slices of cheese; top with remaining bread. Cook sandwiches for 2-4 minutes on each side or until golden brown. **Yield:** 4 servings (2 cups relish).

Hearty Beef Vegetable Soup

Prep: 20 min. **Cook:** 2 hours

Chock-full of tasty ingredients, this stew-like soup is actually easy to make. I serve it with fresh-baked bread or breadsticks. If you prefer, substitute chuck roast for the stew meat. —*Sherman Snowball, Salt Lake City, Utah*

3 tablespoons all-purpose flour
1/2 teaspoon salt
1/4 teaspoon pepper
1 pound beef stew meat, cut into 1/2-inch cubes
2 tablespoons olive oil
1 can (14-1/2 ounces) Italian diced tomatoes
1 can (8 ounces) tomato sauce
2 tablespoons red wine vinegar
2 tablespoons Worcestershire sauce
3 garlic cloves, minced
1 teaspoon dried oregano
3 cups hot water
4 medium potatoes, peeled and cubed
6 medium carrots, sliced
2 medium turnips, peeled and cubed
1 medium zucchini, halved lengthwise and sliced
1 medium green pepper, julienned
1 cup sliced fresh mushrooms
1 medium onion, chopped
1 can (4 ounces) chopped green chilies
2 tablespoons sugar

In a large resealable plastic bag, combine the flour, salt and pepper. Add beef, a few pieces at a time, and shake to coat.

In a soup kettle or Dutch oven, brown beef in oil. Stir in tomatoes, tomato sauce, vinegar, Worcestershire sauce, garlic and oregano. Bring to a boil. Reduce heat; cover and simmer for 1 hour.

Stir in the remaining ingredients. Bring to a boil. Reduce heat; cover and simmer for 1 hour or until meat and vegetables are tender. **Yield:** 8 servings (about 2-1/2 quarts).

Asparagus Brunch Pockets

(Pictured above and on page 30)

Prep: 20 min. **Bake:** 15 min.

I always receive compliments on these yummy bundles. They're great not only for brunch but also as a hot lunch or side dish. —Cynthia Linthicum, Towson, Maryland

- **1 pound fresh asparagus, trimmed and cut into 1-inch pieces**
- **4 ounces cream cheese, softened**
- **1 tablespoon milk**
- **1 tablespoon mayonnaise**
- **1 tablespoon finely chopped onion**
- **1 tablespoon diced pimientos**
- **1/8 teaspoon salt**
- **Pinch pepper**
- **1 tube (8 ounces) refrigerated crescent rolls**
- **2 teaspoons butter, melted**
- **1 tablespoon seasoned bread crumbs**

In a large saucepan, bring 1/2 in. of water to a boil. Add asparagus; cover and boil for 3 minutes. Drain and set aside.

In a small mixing bowl, beat the cream cheese, milk and mayonnaise until smooth. Stir in onion, pimientos, salt and pepper.

Unroll crescent dough and separate into triangles; place on an ungreased baking sheet. Spoon 1 teaspoon of cream cheese mixture into the center of each triangle; top with asparagus. Top each with another teaspoonful of cream cheese mixture. Bring three corners of dough together and twist; pinch edges to seal.

Brush with butter; sprinkle with seasoned bread crumbs. Bake at 375° for 15-18 minutes or until golden brown. **Yield:** 8 servings.

Bavarian Meatball Stew

Prep: 35 min. **Cook:** 35 min.

This beefy stew became a favorite when my daughter gave me the recipe years ago. It really warms you up on chilly evenings. —Janice Mitchell, Aurora, Colorado

- **1 egg, lightly beaten**
- **1/2 cup soft bread crumbs**
- **3 tablespoons dried parsley flakes**
- **1/4 teaspoon ground allspice**
- **1/4 teaspoon ground nutmeg**
- **1/4 teaspoon pepper**
- **1-1/2 pounds ground beef**
- **2 cans (14-1/2 ounces *each*) beef broth**
- **1 can (14-1/2 ounces) diced tomatoes, undrained**
- **1 can (14 ounces) Bavarian sauerkraut, rinsed and well drained**
- **2 medium potatoes, peeled and cubed**
- **2 medium carrots, sliced**
- **2 celery ribs, sliced**
- **1 envelope onion soup mix**
- **1 tablespoon sugar**
- **1/2 teaspoon pepper**
- **1 bay leaf**

In a large bowl, combine the first six ingredients. Crumble beef over mixture and mix well. Shape into 1-in. balls. Place on a greased rack in a shallow baking pan. Bake at 400° for 15 minutes.

Meanwhile, in a large saucepan, combine the remaining ingredients. Add meatballs. Bring to a boil. Reduce heat; cover and simmer for 30-35 minutes or until the vegetables are tender. Discard bay leaf before serving. **Yield:** 8 servings.

Creamy Clam Chowder

Prep: 15 min. **Cook:** 30 min.

This is one of the easiest recipes for clam chowder I've ever made. It's especially good with sourdough bread. —Lori Kimble, McDonald, Pennsylvania

- **1 large onion, chopped**
- **3 medium carrots, chopped**
- **2 celery ribs, sliced**
- **3/4 cup butter, cubed**
- **2 cans (10-3/4 ounces *each*) condensed cream of potato soup, undiluted**
- **3 cans (6-1/2 ounces *each*) minced clams**
- **3 tablespoons cornstarch**
- **1 quart half-and-half cream**

In a large saucepan, saute the onion, carrots and celery in butter until tender. Stir in the potato soup and two cans of undrained clams. Drain and discard the juice from remaining can of clams; add clams to soup.

Combine cornstarch and a small amount of cream until smooth; stir into soup. Add the remaining cream. Bring to a boil; cook and stir for 2 minutes or until thickened. **Yield:** 9 servings (about 2 quarts).

Zippy Chicken Mushroom Soup

(Pictured below)

Prep: 15 min. **Cook:** 25 min.

When my sister-in-law telephoned me looking for a good cream of mushroom soup recipe, I gave her this hearty one. A splash of hot pepper sauce gives it a boost.

—Julia Thornely, Layton, Utah

1/2 pound fresh mushrooms, chopped
1/4 cup *each* chopped onion, celery and carrot
1/4 cup butter, cubed
1/2 cup all-purpose flour
5-1/2 cups chicken broth
1 teaspoon pepper
1/2 teaspoon white pepper
1/4 teaspoon dried thyme
Pinch dried tarragon
1/2 teaspoon hot pepper sauce
3 cups half-and-half cream
2-1/2 cups cubed cooked chicken
1 tablespoon minced fresh parsley
1-1/2 teaspoons lemon juice
1/2 teaspoon salt

In a Dutch oven, saute the mushrooms, onion, celery and carrot in butter until tender. Stir in the flour until blended. Add the chicken broth and seasonings; mix well. Bring to a boil. Reduce heat; simmer, uncovered, for 10 minutes.

Stir in the half-and-half cream, chicken, parsley, lemon juice and salt; heat through (do not boil). **Yield:** 11 servings (2-3/4 quarts).

All-Day Soup

Prep: 25 min. **Cook:** 8 hours

Don't let the title fool you. It's the slow cooker that works all day, not you! I just start this meaty soup in the morning, and my part's done. —Cathy Logan, Sparks, Nevada

✓ **Uses less fat, sugar or salt. Includes Nutrition Facts and Diabetic Exchanges.**

1 beef flank steak (1-1/2 pounds), cut into 1/2-inch cubes
1 medium onion, chopped
1 tablespoon olive oil
5 medium carrots, thinly sliced
4 cups shredded cabbage
4 medium red potatoes, diced
2 celery ribs, diced
2 cans (14-1/2 ounces *each*) diced tomatoes, undrained
2 cans (14-1/2 ounces *each*) beef broth
1 can (10-3/4 ounces) condensed tomato soup, undiluted
1 tablespoon sugar
2 teaspoons Italian seasoning
1 teaspoon dried parsley flakes

In a large skillet, brown steak and onion in oil; drain. Transfer to a 5-qt. slow cooker. Stir in the remaining ingredients. Cover and cook on low for 8-10 hours or until meat is tender. **Yield:** 8 servings.

Nutrition Facts: 1-3/4 cups (prepared with reduced-sodium broth and reduced-fat reduced-sodium tomato soup) equals 279 calories, 9 g fat (3 g saturated fat), 37 mg cholesterol, 501 mg sodium, 31 g carbohydrate, 6 g fiber, 18 g protein. **Diabetic Exchanges:** 2 starch, 2 lean meat, 1 vegetable.

Freezer Convenience

Most soups freeze well (for up to 3 months), so they're great as leftovers or to make ahead of time. Keep these tips in mind:

- Cool the soup before freezing it. To cool it quickly, put the kettle in a sink full of ice water. When cool, put the soup in an airtight freezer-safe container, leaving 1/4 inch of headspace for expansion.

- Pasta may get mushy in the freezer, so add it to the soup just before serving.

- Soups made with cream or potatoes are better eaten fresh. They can taste grainy if frozen, thawed and reheated.

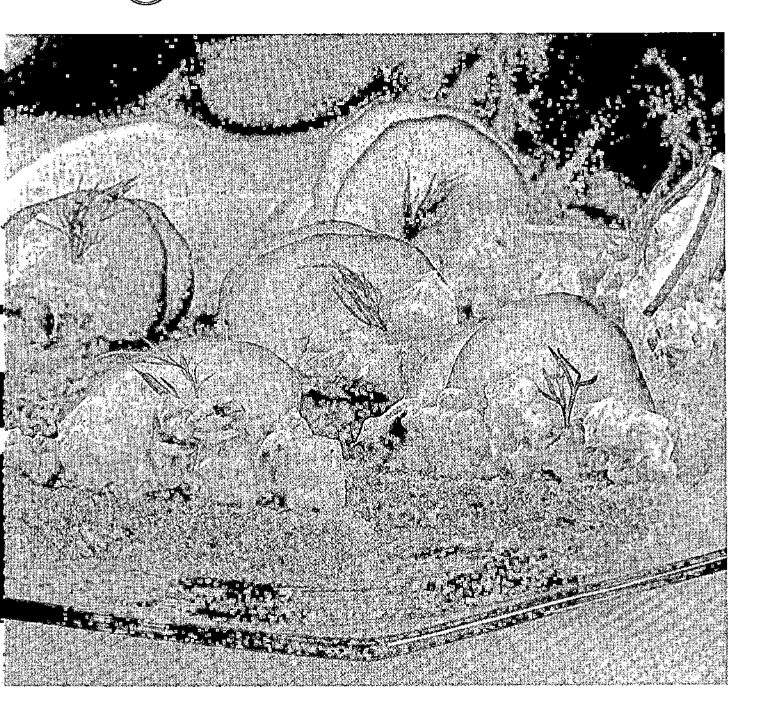

Cucumber Tea Sandwiches

(Pictured above)

Prep: 25 min. + chilling

My children wanted to plant a garden, and we ended up with buckets of cucumbers. When I got tired of making pickles, I came up with these pretty sandwiches. We made 200 of them for a family gathering, and everyone asked for the recipe. —*Kimberly Smith, Brighton, Tennessee*

1 package (8 ounces) cream cheese, softened
1/4 cup mayonnaise
1 tablespoon snipped fresh dill
1 tablespoon lemon juice
1/2 teaspoon Worcestershire sauce
1/4 teaspoon salt
1/8 teaspoon cayenne pepper
1/8 teaspoon pepper
2 large cucumbers, seeded and chopped
1/2 cup chopped sweet red pepper
1/4 cup chopped onion
1/4 cup pimiento-stuffed olives, chopped
1/4 cup minced fresh parsley
12 slices whole wheat bread
Cucumber slices and fresh dill sprigs, optional

In a small mixing bowl, combine the first eight ingredients; beat until blended. Stir in the cucumbers, red pepper, onion, olives and parsley. Cover and refrigerate for up to 2 hours.

Remove the crusts from bread; cut each slice into four triangles. Spread with cream cheese mixture. Garnish with cucumber slices and dill sprigs if desired. **Yield:** 4 dozen.

Lemony Chicken Soup

Prep: 5 min. **Cook:** 30 min.

While living in California, I enjoyed a delicious chicken-lemon soup at a local restaurant. When I returned to Texas, I wanted to make my own but never found a recipe. I experimented with many versions before creating this one. —*Brenda Tollett, San Antonio, Texas*

1/3 cup butter, cubed
3/4 cup all-purpose flour
6 cups chicken broth, *divided*
1 cup milk
1 cup half-and-half cream
1-1/2 cups cubed cooked chicken
1 tablespoon lemon juice
1/2 teaspoon salt
1/8 teaspoon pepper
Dash nutmeg
8 lemon slices

In a soup kettle or large saucepan, melt butter. Stir in flour until smooth; gradually add 2 cups broth, milk and cream. Bring to a boil; cook and stir for 2 minutes or until thickened.

Stir in the chicken, lemon juice, salt, pepper, nutmeg and remaining broth. Cook over medium heat until heated through, stirring occasionally. Garnish each serving with a lemon slice. **Yield:** 8 servings (2 quarts).

Stroganoff Sandwiches

Prep: 10 min. **Cook:** 30 min.

This recipe is great when you're watching the big game on TV. I often make the meat mixture ahead of time and add the sour cream when I reheat it just before serving. —*Susan Graham, Cherokee, Iowa*

1-1/2 pounds ground beef
1 medium onion, chopped
6 to 8 bacon strips, cooked and crumbled
2 garlic cloves, minced
2 tablespoons all-purpose flour
1/2 teaspoon salt
1/2 teaspoon paprika
1/8 teaspoon ground nutmeg
1 can (10-3/4 ounces) condensed cream of mushroom soup, undiluted
1/2 cup sliced fresh mushrooms
1 cup (8 ounces) sour cream
8 hamburger buns, split

In a large skillet, cook beef and onion over medium heat until meat is no longer pink; drain. Add bacon and garlic. Combine the flour, salt, paprika and nutmeg; gradually stir into beef mixture until blended.

Stir in soup and mushrooms (mixture will be thick). Bring to a boil. Reduce heat; simmer, uncovered, for 4-5 minutes or until heated through. Stir in sour cream. Cook 3-4 minutes longer or until heated through, stirring occasionally (do not boil). Serve on buns. **Yield:** 8 servings.

Cheeseburger Meat Loaf Hoagies

Prep: 20 min. **Bake:** 30 min.

I worked as a school cook for 15 years and still love to try new recipes. My family likes meat loaf sandwiches, and this one is something special. —*Connie Boucher*
Dixon, Missouri

- **1 egg, lightly beaten**
- **1 can (8 ounces) tomato sauce**
- **1 cup quick-cooking oats**
- **1/4 cup chopped onion**
- **1/2 teaspoon salt**
- **1/2 teaspoon pepper**
- **1-1/2 pounds ground beef**
- **1/3 cup mayonnaise**
- **2 tablespoons ketchup**
- **2 medium tomatoes, sliced**
- **8 slices cheddar cheese**
- **8 bacon strips, cooked and halved**
- **8 hoagie buns, split and toasted**

In a large bowl, combine the egg, tomato sauce, oats, onion, salt and pepper. Crumble beef over mixture and mix well. Press evenly into an ungreased 13-in. x 9-in. x 2-in. baking dish.

Bake, uncovered, at 350° for 30-35 minutes or until no pink remains in the meat and a meat thermometer reads 160°; drain.

Combine mayonnaise and ketchup; spread over meat loaf. Cut into eight rectangles; top each with tomatoes, cheese and bacon. Place on a baking sheet; broil 3-4 in. from the heat for 2-3 minutes or until cheese is melted. Serve on buns. **Yield:** 8 sandwiches.

Danish Turkey Dumpling Soup

(Pictured at right)

Prep: 35 min. **Cook:** 2-1/2 hours

This recipe was handed down from my grandmother, who was a Danish caterer. My 100 percent Italian husband has come to expect this on chilly winter evenings. It not only warms the body, it also warms your heart!
—*Karen Sue Garback-Pristera, Albany, New York*

- **1 leftover turkey carcass (from a 12- to 14-pound turkey)**
- **9 cups water**
- **3 teaspoons chicken bouillon granules**
- **1 bay leaf**
- **1 can (14-1/2 ounces) stewed tomatoes, cut up**
- **1 medium turnip, peeled and diced**
- **2 celery ribs, chopped**
- **1 medium onion, chopped**
- **1 medium carrot, chopped**
- **1/4 cup minced fresh parsley**
- **1 teaspoon salt**

DUMPLINGS:

- **1/2 cup water**
- **1/4 cup butter, cubed**
- **1/2 cup all-purpose flour**
- **1 teaspoon baking powder**
- **1/8 teaspoon salt**
- **2 eggs**
- **1 tablespoon minced fresh parsley**

Place carcass, water, bouillon and bay leaf in a soup kettle. Bring to a boil. Reduce heat; cover and simmer for 1-1/2 hours.

Remove carcass. Strain broth and skim fat; discard bay leaf. Return broth to pan. Add vegetables, parsley and salt. Remove turkey from bones and cut into bite-size pieces; add to soup. Discard bones. Bring to a boil. Reduce heat; cover and simmer for 25-30 minutes or until vegetables are crisp-tender.

For dumplings, in a large saucepan, bring water and butter to a boil. Combine flour, baking powder and salt; add all at once to pan and stir until a smooth ball forms. Remove from heat; let stand 5 minutes. Add eggs, one at a time, beating well after each addition. Continue beating until smooth and shiny. Stir in parsley.

Drop batter in 12 mounds onto simmering soup. Cover and simmer for 20 minutes or until a toothpick inserted in a dumpling comes out clean (do not lift cover while simmering). **Yield:** 6 servings (about 2 quarts).

Roasted Vegetable Chili

(Pictured below)

Prep: 35 min. **Cook:** 30 min.

I serve this delicious and satisfying chili with corn chips, cheese, sour cream and a small salad. To save time, purchase vegetables that have already been diced.
—Hannah Barringer, Loudon, Tennessee

✓ **Uses less fat, sugar or salt. Includes Nutrition Facts and Diabetic Exchanges.**

- **1 medium butternut squash, peeled and cut into 1-inch pieces**
- **3 large carrots, sliced**
- **2 medium zucchini, cut into 1-inch pieces**
- **2 tablespoons olive oil, *divided***
- **1-1/2 teaspoons ground cumin**
- **2 medium green peppers, diced**
- **1 large onion, chopped**
- **3 cans (14-1/2 ounces *each*) reduced-sodium chicken broth**
- **3 cans (14-1/2 ounces *each*) diced tomatoes, undrained**
- **2 cans (15 ounces *each*) cannellini *or* white kidney beans, rinsed and drained**
- **1 cup water**
- **1 cup salsa**
- **3 teaspoons chili powder**
- **6 garlic cloves, minced**

Place the squash, carrots and zucchini in a 15-in. x 10-in. x 1-in. baking pan. Combine 1 tablespoon oil and cumin; drizzle over vegetables and toss to coat. Bake, uncovered, at 450° for 25-30 minutes or until tender, stirring once.

Meanwhile, in a soup kettle, saute green peppers and onion in remaining oil for 3-4 minutes or until tender. Stir in the broth, tomatoes, beans, water, salsa, chili powder and garlic. Bring to a boil. Reduce heat; simmer, uncovered, for 10 minutes.

Stir in roasted vegetables. Return to a boil. Reduce heat; simmer, uncovered, for 5-10 minutes or until heated through. **Yield:** 13 servings (5 quarts).

Nutrition Facts: 1-1/3 cups equals 156 calories, 3 g fat (trace saturated fat), 0 cholesterol, 559 mg sodium, 28 g carbohydrate, 9 g fiber, 6 g protein. **Diabetic Exchanges:** 2 vegetable, 1 starch, 1/2 fat.

Creamy Bacon Mushroom Soup

Prep/Total Time: 30 min.

This rich, flavorful soup is always a hit. If you like, garnish each bowl with chopped green onion or shredded Swiss.
—Nathan Mercer, Inman, South Carolina

- **10 bacon strips, diced**
- **1 pound sliced fresh mushrooms**
- **1 medium onion, chopped**
- **3 garlic cloves, minced**
- **1 quart heavy whipping cream**
- **1 can (14-1/2 ounces) chicken broth**
- **1 package (5 ounces) shredded Swiss cheese**
- **3 tablespoons cornstarch**
- **1/2 teaspoon salt**
- **1/2 teaspoon pepper**
- **3 tablespoons water**

In a soup kettle or large saucepan, cook bacon over medium heat until crisp. Using a slotted spoon, remove to paper towels; drain, reserving 2 tablespoons drippings. In the drippings, saute mushrooms, onion and garlic. Stir in cream and broth. Gradually stir in cheese until melted.

In a small bowl, combine cornstarch, salt, pepper and water until smooth. Stir into soup. Bring to a boil; cook and stir for 2 minutes or until thickened. Garnish with bacon. **Yield:** 8 servings (2 quarts).

Land of Enchantment Posole

Prep: 30 min. **Cook:** 1 hour

My family named this spicy soup after the moniker of our home state, New Mexico: "Land of Enchantment." We usually make a big batch of it around Christmas, when we have lots of family over...and we never have extras.
—Suzanne Caldwell, Artesia, New Mexico

- **1-1/2 pounds pork stew meat, cut into 3/4-inch cubes**
- **1 large onion, chopped**
- **2 garlic cloves, minced**
- **2 tablespoons vegetable oil**
- **3 cups beef broth**

2 cans (15-1/2 ounces *each*) hominy, rinsed and drained
2 cans (4 ounces *each*) chopped green chilies
1 to 2 jalapeno peppers, seeded and chopped, optional
1/2 teaspoon salt
1/2 teaspoon ground cumin
1/2 teaspoon dried oregano
1/4 teaspoon pepper
1/4 teaspoon cayenne pepper
1/2 cup minced fresh cilantro
Tortilla strips, optional

In a soup kettle or Dutch oven, cook the pork, onion and garlic in oil over medium heat until meat is no longer pink; drain. Stir in the broth, hominy, chilies, jalapeno if desired, salt, cumin, oregano, pepper and cayenne.

Bring to a boil. Reduce heat; cover and simmer for 45-60 minutes or until meat is tender. Stir in cilantro. Serve with tortilla strips if desired. **Yield:** 5 servings.

Editor's Note: When cutting or seeding hot peppers, use rubber or plastic gloves to protect your hands. Avoid touching your face.

Christmas Clam Chowder

Prep: 15 min. **Cook:** 30 min.

I whip up this cheesy chowder every year on Christmas or New Year's Eve. Sometimes I substitute broccoli for the potatoes and clams. —Joy Schuster, Glentana, Montana

4 cups cubed red potatoes
3 cups water
1 medium carrot, grated
1 small onion, chopped
2 teaspoons chicken bouillon granules
1 teaspoon dried parsley flakes
1/2 teaspoon pepper
2 tablespoons all-purpose flour
1/2 cup cold water
3 cans (6-1/2 ounces *each*) chopped clams, drained
2/3 cup cubed process cheese (Velveeta)
1 can (12 ounces) evaporated milk

In a large saucepan, combine the first seven ingredients. Bring to a boil. Reduce heat; cover and simmer for 20 minutes or until potatoes are tender.

In a small bowl, combine flour and cold water until smooth. Stir into potato mixture. Bring to a boil; cook and stir for 2 minutes or until thickened. Reduce heat. Add clams and cheese; cook and stir until cheese is melted. Stir in milk; heat through. **Yield:** 9 servings (about 2 quarts).

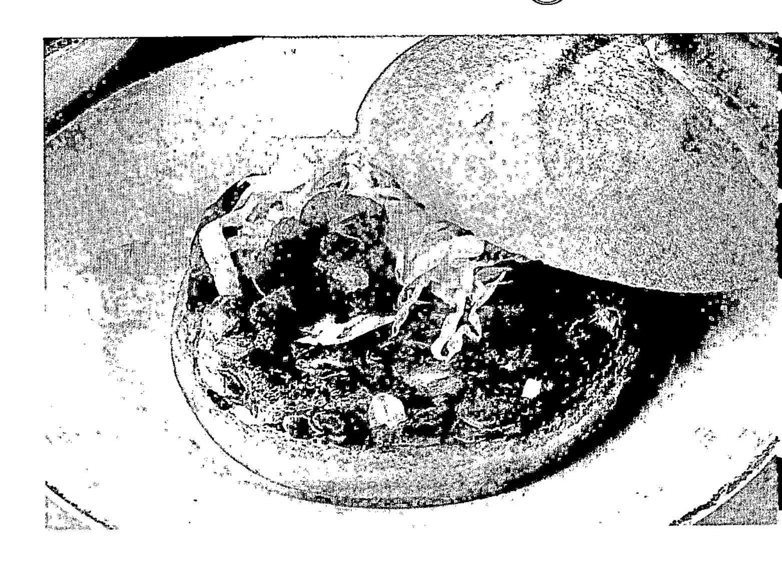

Black Bean Burgers

(Pictured above)

Prep/Total Time: 25 min.

My son encouraged me to come up with a good veggie burger for him, and he gave this recipe a triple A+! —Clara Honeyager, North Prairie, Wisconsin

1 cup frozen mixed vegetables, thawed
1 small onion, chopped
1/2 cup chopped sweet red pepper
1 can (15 ounces) black beans, rinsed and drained, *divided*
1 tablespoon cornstarch
2 tablespoons cold water
1 cup mashed potato flakes
1/4 cup quick-cooking oats
3 tablespoons whole wheat flour
2 tablespoons nonfat dry milk powder
1 egg, lightly beaten
1/2 teaspoon salt
1/4 teaspoon pepper
4 teaspoons vegetable oil
6 kaiser rolls, split
2 cups shredded lettuce
3/4 cup salsa

In a large microwave-safe bowl, combine the mixed vegetables, onion and red pepper. Cover and microwave on high for 2 minutes.

Coarsely mash 3/4 cup black beans. In a bowl, combine the cornstarch and water until smooth; stir in the mashed beans, potato flakes, oats, flour, milk powder, egg, salt and pepper. Stir in vegetable mixture and remaining black beans. Shape into six 5/8-in.-thick patties.

In a large nonstick skillet, cook patties in oil for 4-5 minutes on each side or until lightly browned. Serve on rolls with lettuce and salsa. **Yield:** 6 servings.

Editor's Note: This recipe was tested in a 1,100-watt microwave.

Golden Seafood Chowder

(Pictured above)

Prep: 25 min. **Cook:** 25 min.

Loaded with crab, shrimp and cheddar cheese, this chowder is so good that I make it weekly. Sometimes I use chicken or ham instead of the seafood, leaving out the Clamato juice. Either way, this eye-catching soup is a winner.
—Ami Paton, Waconia, Minnesota

1/2 cup finely chopped onion
1/4 cup butter, cubed
1 can (14-1/2 ounces) chicken broth
1 cup cubed peeled potatoes
2 celery ribs, chopped
2 medium carrots, chopped
1/4 cup Clamato juice
1/4 teaspoon lemon-pepper seasoning
1/4 cup all-purpose flour
2 cups milk
2 cups (8 ounces) shredded sharp cheddar cheese
1 can (6 ounces) crabmeat, drained, flaked and cartilage removed
1 cup cooked medium shrimp, peeled and deveined

In a large saucepan, saute onion in butter until tender. Stir in the broth, potato, celery, carrots, Clamato juice and lemon-pepper. Bring to a boil. Reduce heat; cover and simmer for 15-20 minutes or until vegetables are tender.

In a small bowl, whisk flour and milk until smooth; add to soup. Bring to a boil; cook and stir for 2 minutes or until thickened. Reduce heat. Add the cheese, crab and shrimp; cook and stir until cheese is melted. **Yield:** 4 servings.

Mushroom Tomato Bisque

Prep: 30 min. **Cook:** 10 min.

After tasting a similar soup in a restaurant, I tinkered with a few recipes at home, and this was the result. It might seem complicated at first glance, but it's really not.
—Connie Stevens, Schaefferstown, Pennsylvania

1-1/2 pounds plum tomatoes, halved lengthwise
5 tablespoons olive oil, *divided*
2 garlic cloves, minced
1/2 teaspoon salt
1/2 teaspoon dried basil
1/2 teaspoon dried oregano
1/2 teaspoon pepper
1/2 pound sliced fresh mushrooms
1/2 cup finely chopped sweet onion
1-1/4 cups chicken broth
1/3 to 1/2 cup tomato paste
Pinch sugar, optional
3/4 cup heavy whipping cream
2 tablespoons grated Parmesan cheese

Place tomatoes cut side down in a greased 15-in. x 10-in. x 1-in. baking pan. Brush with 3 tablespoons oil. Combine garlic, salt, basil, oregano and pepper; sprinkle over tomatoes. Bake, uncovered, at 450° for 20-25 minutes or until edges are well browned.

Cool slightly. Place tomatoes and pan drippings in a blender. Cover and process until blended; process 1 minute longer.

In a large saucepan, saute mushrooms and onion in remaining oil for 5-8 minutes or until tender. Stir in broth, tomato paste, sugar if desired and tomato puree. Bring to a boil. Remove from the heat; stir in cream. Garnish with Parmesan cheese. **Yield:** 4 servings.

Gingered Butternut Squash Soup

Prep: 35 min. **Cook:** 40 min.

Roasting the squash adds a wonderful flavor to this delightful pureed soup. If you want a true vegetarian dish, substitute vegetable stock for the chicken broth.
—Kim Pettipas, Oromocto, New Brunswick

4 pounds butternut squash, peeled and cubed (about 8 cups)
6 teaspoons olive oil, *divided*
1 large onion, chopped
2 tablespoons butter
1 tablespoon minced fresh gingerroot
2-1/2 teaspoons curry powder
3/4 teaspoon salt
1/4 teaspoon pepper
3 large potatoes, peeled and cubed

6 cups chicken broth
1-1/2 cups milk
Sour cream, optional

Place squash in a greased 15-in. x 10-in. x 1-in. baking pan. Drizzle with 4-1/2 teaspoons oil; toss to coat. Bake, uncovered, at 450° for 30 minutes, stirring every 15 minutes. Bake 5-10 minutes longer or until tender. Set aside.

In a soup kettle, saute onion in butter and remaining oil for 5 minutes or until tender. Stir in ginger, curry, salt and pepper; cook for 2 minutes. Stir in potatoes; cook 2 minutes longer. Stir in broth. Bring to a boil. Reduce heat; cover and simmer for 15-20 minutes or until potatoes are tender. Cool slightly.

Stir in the reserved squash. In a blender, puree the soup in batches until smooth. Return to the pan. Stir in the milk; heat through. Garnish with sour cream if desired. **Yield:** 9 servings (about 3 quarts).

Chicken Soup with Potato Dumplings

Prep: 25 min. **Cook:** 40 min.

Our family calls this comforting, old-fashioned soup our "Sunday dinner soup" because it's almost a complete dinner in a bowl. We love the dumplings on top.
—Marie McConnell, Las Cruces, New Mexico

1/4 cup chopped onion
2 garlic cloves, minced
1 tablespoon vegetable oil
6 cups chicken broth
2 cups cubed cooked chicken
2 celery ribs, chopped
2 medium carrots, sliced
1/4 teaspoon dried sage leaves
DUMPLINGS:
1-1/2 cups biscuit/baking mix
1 cup cold mashed potatoes (with added milk)
1/4 cup milk
1 tablespoon chopped green onion
1/8 teaspoon pepper

In a large saucepan, saute onion and garlic in oil for 3-4 minutes or until onion is tender. Stir in the broth, chicken, celery, carrots and sage. Bring to a boil. Reduce heat; cover and simmer for 10-15 minutes or until the vegetables are tender.

In a small bowl, combine the dumpling ingredients. Drop heaping tablespoonfuls of batter onto simmering soup. Cover and simmer for 20 minutes or until a toothpick inserted in a dumpling comes out clean (do not lift cover while simmering). **Yield:** 5 servings.

Zesty Vegetarian Wraps

(Pictured below and on page 30)

Prep/Total Time: 10 min.

The pretty tortilla in this satisfying vegetarian roll-up holds crisp veggies, spicy cheese slices and a Southwest-style dressing. —Cori Lehman, South Milwaukee, Wisconsin

Uses less fat, sugar or salt. Includes Nutrition Facts.

2 tablespoons mayonnaise
1 teaspoon lime juice
2 to 4 drops Louisiana-style hot sauce
2 spinach tortillas *or* flour tortillas of your choice (8 inches)
2 lettuce leaves
1/2 medium green pepper, julienned
2 slices pepper Jack cheese

In a small bowl, combine the mayonnaise, lime juice and hot sauce. Spread over the tortillas. Top with the lettuce, green pepper and cheese slices; roll up tightly. **Yield:** 2 servings.

Nutrition Facts: 1 wrap (prepared with fat-free mayonnaise) equals 242 calories, 11 g fat (5 g saturated fat), 32 mg cholesterol, 427 mg sodium, 28 g carbohydrate, 2 g fiber, 10 g protein.

Side Dishes & Condiments

Round out a memorable menu by serving any of these outstanding accompaniments—from delicious veggie dishes to special spreads.

STANDOUT SIDES. Clockwise from upper left: Corn Bread Vegetable Cobbler (p. 53), Maple-Ginger Root Vegetables (p. 48), Herbed Tomatoes 'n' Green Beans (p. 52), Company Corn (p. 51) and Woolly Butter Lamb (p. 51).

Maple-Ginger Root Vegetables

(Pictured above and on page 47)

Prep: 35 min. **Bake:** 45 min.

This recipe is a favorite because it brings out the different flavors of the vegetables. Even my children enjoy it... they like the drizzle of maple syrup! It's a tasty way to introduce kids to turnips, rutabaga and parsnips, too. —*Kelli Ritz Innisfail, Alberta*

✓ **Uses less fat, sugar or salt. Includes Nutrition Facts and Diabetic Exchanges.**

- **5 medium parsnips, peeled and sliced**
- **5 small carrots, sliced**
- **3 medium turnips, peeled and cubed**
- **2 small sweet potatoes, peeled and cubed**
- **1 small rutabaga, peeled and cubed**
- **1 large sweet onion, cut into wedges**
- **1 small red onion, cut into wedges**
- **2 tablespoons olive oil**
- **1 tablespoon minced fresh gingerroot**
- **1 teaspoon salt**
- **1/2 teaspoon pepper**
- **1 cup maple syrup**

Place the first seven ingredients in a large resealable plastic bag; add the oil, ginger, salt and pepper. Seal bag and shake to coat. Arrange vegetables in a single layer in two 15-in. x 10-in. x 1-in. baking pans coated with nonstick cooking spray.

Bake, uncovered, at 425° for 25 minutes, stirring once. Drizzle with the syrup. Bake 20-25 minutes longer or until the vegetables are tender, stirring once. **Yield:** 24 servings.

Nutrition Facts: 3/4 cup equals 92 calories, 1 g fat (trace saturated fat), 0 cholesterol, 119 mg sodium, 20 g carbohydrate, 2 g fiber, 1 g protein. **Diabetic Exchange:** 1 starch.

Gingered Orange Beets

(Pictured below)

Prep: 10 min. **Bake:** 70 min.

My husband was pleasantly surprised when he tried my new twist on beets. The orange and ginger are a nice complement and make this vegetable dish special. —*Marion Tipton, Phoenix, Arizona*

- **1-1/2 pounds whole fresh beets (about 4 medium), trimmed and cleaned**
- **6 tablespoons olive oil, *divided***
- **1/4 teaspoon salt**
- **1/4 teaspoon white pepper**
- **1 tablespoon rice wine vinegar**
- **1 tablespoon orange juice concentrate**
- **1-1/2 teaspoons grated orange peel, *divided***
- **1/2 teaspoon minced fresh gingerroot**
- **1 medium navel orange, peeled, sectioned and chopped**
- **1/3 cup pecan halves, toasted**

Brush beets with 4 tablespoons oil; sprinkle with salt and pepper. Wrap loosely in foil; place on a baking sheet. Bake at 425° for 70-75 minutes or until fork-tender. Cool slightly.

In a small bowl, whisk vinegar, orange juice concentrate, 1 teaspoon orange peel, ginger and remaining oil; set aside.

Peel beets and cut into wedges; place in a serving bowl. Add orange sections and pecans. Drizzle with orange sauce and toss to coat. Sprinkle with remaining orange peel. **Yield:** 4 servings.

Broccoli-Stuffed Potatoes

(Pictured above)

Prep: 10 min. **Bake:** 70 min.

These hearty broccoli-and-cheese potatoes get extra flavor from sour cream and fresh dill. Serve them as a side... or even alone as a light lunch. —*Fran Scott Birmingham, Michigan*

Uses less fat, sugar or salt. Includes Nutrition Facts.

- **4 medium baking potatoes (8 ounces *each*)**
- **2 cups fresh broccoli florets**
- **1/2 cup chopped onion**
- **2 tablespoons reduced-fat butter**
- **1/3 cup fat-free milk**
- **1/3 cup reduced-fat sour cream**
- **2 tablespoons snipped fresh dill**
- **1/2 teaspoon salt**
- **1/4 teaspoon pepper**
- **1/2 cup shredded reduced-fat cheddar cheese**

Scrub and pierce potatoes. Bake at 400° for 1 hour or until tender. Cut a thin slice off the top of each potato and discard. Carefully scoop out pulp, leaving thin shells.

In a small skillet, saute the broccoli and onion in butter for 5 minutes or until tender. In a large bowl, mash the potato pulp with milk, sour cream, dill, salt and pepper until smooth. Fold in broccoli mixture.

Stuff into potato shells; sprinkle with cheese. Place on a baking sheet. Bake for 10-15 minutes or until heated through. **Yield:** 4 servings.

Nutrition Facts: 1 potato equals 296 calories, 8 g fat (5 g saturated fat), 27 mg cholesterol, 468 mg sodium, 47 g carbohydrate, 5 g fiber, 12 g protein.

Dijon Green Beans

(Pictured below and on front cover)

Prep/Total Time: 20 min.

I love this recipe because it combines the freshness of garden green beans with a tangy dressing. It's a wonderful side dish. —*Jannine Fisk, Malden, Massachusetts*

✓ Uses less fat, sugar or salt. Includes Nutrition Facts and Diabetic Exchanges.

- **1-1/2 pounds fresh green beans, trimmed**
- **2 tablespoons red wine vinegar**
- **2 tablespoons olive oil**
- **2 teaspoons Dijon mustard**
- **1/2 teaspoon salt**
- **1/4 teaspoon pepper**
- **1 cup grape tomatoes, halved**
- **1/2 small red onion, sliced**
- **2 tablespoons grated Parmesan cheese**

Place beans in a large saucepan; cover with water. Bring to a boil. Cook, uncovered, for 8-10 minutes or until crisp-tender.

Meanwhile, for dressing, whisk the vinegar, oil, mustard, salt and pepper in a small bowl. Drain beans; place in a large bowl. Add tomatoes and onion. Drizzle with dressing and toss to coat. Sprinkle with Parmesan cheese. **Yield:** 10 servings.

Nutrition Facts: 3/4 cup equals 54 calories, 3 g fat (1 g saturated fat), 1 mg cholesterol, 167 mg sodium, 6 g carbohydrate, 2 g fiber, 2 g protein. **Diabetic Exchanges:** 1 vegetable, 1/2 fat.

Hearty Calico Bean Bake

(Pictured above)

Prep: 10 min. **Bake:** 1 hour

For years, my mother made this savory-sweet bean dish. I was always thrilled when it was on the menu, and now I serve it often to my own family. You can vary the types of beans if you wish. —*Heather Biedler, Laura, Illinois*

- **1 can (16 ounces) pork and beans, undrained**
- **1 can (16 ounces) kidney beans, rinsed and drained**
- **1 can (15-1/2 ounces) great northern beans, rinsed and drained**
- **1 can (15-1/2 ounces) chili beans, undrained**
- **1 can (14-1/2 ounces) cut wax beans, drained**
- **1-1/2 cups packed brown sugar**
- **1-1/2 cups cubed fully cooked ham**
- **1-1/2 cups cubed cheddar cheese**
- **1/2 cup ketchup**
- **1 small onion, chopped**
- **2 tablespoons Worcestershire sauce**

In a large bowl, combine all ingredients. Transfer to a greased shallow 3-qt. baking dish. Bake, uncovered, at 350° for 1 hour or until bubbly and heated through. **Yield:** 10 servings.

Rhubarb Barbecue Sauce

Prep: 20 min. **Cook:** 20 min.

Full of zippy flavor, this barbecue sauce is wonderful on poultry and pork. Your taste buds get an extra boost from fresh garlic, pepper sauce and other seasonings.
—*Carol Anderson, Coaldale, Alberta*

✓ **Uses less fat, sugar or salt. Includes Nutrition Facts.**

- **1 cup chopped fresh *or* frozen rhubarb**
- **2/3 cup water**
- **1 medium onion, finely chopped**
- **1 teaspoon canola oil**
- **1 garlic clove, minced**
- **1 cup ketchup**
- **2/3 cup packed brown sugar**
- **1/2 cup dark corn syrup**
- **2 tablespoons cider vinegar**
- **2 tablespoons Worcestershire sauce**
- **1 tablespoon Dijon mustard**
- **1-1/2 teaspoons hot pepper sauce**
- **1/4 teaspoon salt**

In a small saucepan, bring rhubarb and water to a boil. Reduce heat; simmer, uncovered, for 5-6 minutes or until tender. Remove from the heat; cool slightly.

Place the rhubarb in a blender or food processor. Cover; process until smooth. Set aside.

In the same saucepan, saute onion in oil until tender. Add garlic; saute 1 minute longer. Add the remaining ingredients.

Whisk in rhubarb puree until blended. Bring to a boil. Reduce heat; simmer, uncovered, for 5 minutes. Use as a basting sauce for grilled meats. Store in the refrigerator. **Yield:** 2-1/3 cups.

Nutrition Facts: 2 tablespoons equals 80 calories, trace fat (trace saturated fat), 0 cholesterol, 251 mg sodium, 20 g carbohydrate, trace fiber, trace protein.

Relishing Rhubarb

The tart flavor of rhubarb lends itself well to sugar-enhanced pies, desserts, relishes, jams and sauces, including Rhubarb Barbecue Sauce (above). One pound of rhubarb equals 3 cups chopped raw or 2 cups cooked.

The stalks of rhubarb vary in color from pale pink to cherry red. If you are purchasing fresh rhubarb, select stalks that are firm and crisp, not limp.

Always trim off and discard any rhubarb leaves, which contain oxalic acid and are toxic. Thick stalks can be peeled with a vegetable peeler to remove the fibrous strings.

Woolly Butter Lamb

(Pictured below and on page 46)

Prep: 2 hours

Traditionally placed in Polish Easter baskets, a butter lamb symbolizes the goodness and richness of Christ. I began learning how to make them when I was 2 years old, and now my daughter helps me. —Marya LaRoche Cheshire, Massachusetts

2 sticks cold butter (1/2 pound)
1 tablespoon butter, softened
2 whole cloves
1/8 teaspoon ground cinnamon
Fresh parsley sprigs and edible pansies, optional

Cut a third from the end of one stick of butter. Place larger piece on a serving dish for lamb's body. Spread some of the softened butter on cut side of smaller piece; position vertically on left side of larger piece for neck and head. Trim edges.

Cut a 1/4-in. slice from the second stick of butter; cut diagonally in half. Spread softened butter on cut edge of one triangle; secure to front of head/neck piece for nose. Set remaining triangle aside.

Cut a diagonal slice from each end of the second butter stick. Spread softened butter on cut long sides; secure to back of head for ears.

Cut remaining butter and reserved triangle to fit into a garlic press. Squeeze butter through press in batches. Use toothpicks to curl pieces. Beginning at the top and working down, place curls on body. (If butter softens while assembling, place in refrigerator for 10 minutes or until firm.)

Insert cloves for eyes; add two dots of cinnamon on nose for nostrils. Refrigerate until serving. Garnish plate with parsley and edible pansies if desired. **Yield:** 1 butter lamb.

Company Corn

(Pictured above and on page 46)

Prep/Total Time: 25 min.

This simple Thanksgiving dish is a cherished family favorite. It's been passed around for so long, no one can remember who it came from first! —Shannon Schirm Green River, Wyoming

2 packages (10 ounces *each*) frozen corn
1 medium onion, chopped
1/4 cup chopped celery
1/3 cup butter, cubed
2 tablespoons minced fresh parsley
1 teaspoon salt
1/2 teaspoon dried savory
1/2 teaspoon white pepper
3/4 cup sour cream
1 teaspoon lemon juice

Cook the corn according to the package directions. Meanwhile, in a large saucepan, saute the onion and celery in butter until tender. Stir in the parsley, salt, savory and pepper.

Drain the corn; add to the onion mixture. Stir in the sour cream and lemon juice. Serve immediately. **Yield:** 6-8 servings.

Garden-Grown Goodness

WHEN backyard gardens and farmers markets overflow with fruits and veggies, put them to good use with the fresh ideas here. Each flavorful recipe bursts with a bounty of wholesome produce.

Whether you choose savory Corn Bread Vegetable Cobbler, Seasoned Red Potatoes, Herbed Tomatoes 'n' Green Beans or sweet Cherry Rhubarb Jam, you'll enjoy the pick of the crop!

Seasoned Red Potatoes

(Pictured below)

Prep: 10 min. **Bake:** 50 min.

These simple but tasty potaoes are on the menu at my SideTrack Cafe in Metamora, Indiana. I've collected recipes for years, and this is a favorite. —*Nancy Johnson Connersville, Indiana*

- **12 to 14 small red potatoes**
- **1/4 cup olive oil**
- **1/4 cup butter, melted**
- **1 teaspoon salt**
- **1 teaspoon garlic powder**
- **1 teaspoon dried basil**
- **1/2 teaspoon dried thyme**
- **1/2 teaspoon pepper**

Peel a strip from around each potato. Place potatoes in an ungreased 3-qt. baking dish. In a bowl, combine the oil, butter and seasonings; drizzle over potatoes. Bake, uncovered, at 350° for 50-55 minutes or until tender, stirring every 15 minutes. **Yield:** 6 servings.

Herbed Tomatoes 'n' Green Beans

(Pictured above and on page 47)

Prep/Total Time: 30 min.

Looking for new ways to dress up fresh-picked green beans? With just the right amount of oregano and parsley, this colorful side dish is seasoned to please. —*Maryalice Wood, Langley, British Columbia*

✓ **Uses less fat, sugar or salt. Includes Nutrition Facts and Diabetic Exchanges.**

- **3 green onions, coarsely chopped**
- **2 garlic cloves, minced**
- **2 teaspoons olive oil**
- **1/2 pound fresh green beans, trimmed**
- **1/4 cup chicken broth**
- **2 medium tomatoes, diced**
- **1 tablespoon minced fresh oregano**
- **1 tablespoon minced fresh parsley**
- **1/8 teaspoon salt**
- **1/8 teaspoon pepper**

In a small skillet, saute the onions and garlic in oil until tender. Add the beans and broth. Bring to a boil. Reduce the heat; cover and simmer for 6-9 minutes or until crisp-tender.

Stir in the tomatoes and seasonings; heat through. **Yield:** 4 servings.

Nutrition Facts: 2/3 cup equals 59 calories, 3 g fat (trace saturated fat), 0 cholesterol, 144 mg sodium, 9 g carbohydrate, 3 g fiber, 2 g protein. **Diabetic Exchanges:** 2 vegetable, 1/2 fat.

Corn Bread Vegetable Cobbler

(Pictured below and on page 46)

Prep: 25 min. **Bake:** 1-1/4 hours

For a change of pace, try this savory cobbler. It's a medley of tender, nourishing vegetables topped with generous dollops of golden-brown corn bread. —*Edna Hoffman*
Hebron, Indiana

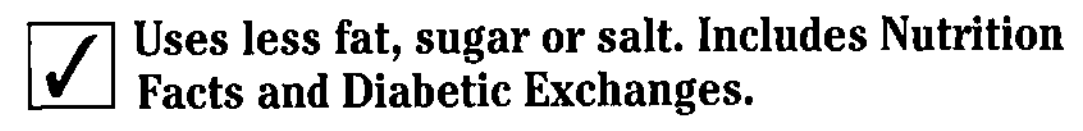

Uses less fat, sugar or salt. Includes Nutrition Facts and Diabetic Exchanges.

- **1 butternut squash (2 pounds), peeled and cut into 1/2-inch pieces**
- **1 pound red potatoes, cut into 1/2-inch wedges**
- **3 medium parsnips, peeled and cut into 1/2-inch pieces**
- **1 medium red onion, cut into 1/2-inch wedges**
- **3 tablespoons olive oil**
- **1 teaspoon salt**
- **1 teaspoon dried tarragon**
- **1 can (14-1/2 ounces) vegetable *or* chicken broth**
- **2 cups fresh broccoli florets**
- **1/2 teaspoon grated lemon peel**
- **4 teaspoons cornstarch**
- **1-1/2 cups milk, *divided***
- **1-3/4 cups biscuit/baking mix**
- **1/2 cup yellow cornmeal**
- **Dash cayenne pepper**

Place the squash, potatoes, parsnips and onion in a shallow 3-qt. baking dish. Combine the oil, salt and tarragon; drizzle over vegetables and toss to coat. Bake, uncovered, at 375° for 1 hour or until tender, stirring once.

Meanwhile, in a large saucepan, bring broth to a boil. Add broccoli and lemon peel. Reduce heat; cover and cook for 2 minutes or until the broccoli is crisp-tender.

In a small bowl, combine cornstarch and 1/2 cup milk until smooth. Add to broccoli. Bring to a boil; cook and stir for 2 minutes or until thickened. Add to roasted vegetables; stir to combine.

In a bowl, combine the baking mix, cornmeal, cayenne and remaining milk until smooth. Drop batter in 12 mounds over hot vegetables. Bake, uncovered, for 15-20 minutes or until topping is browned. **Yield:** 12 servings.

Nutrition Facts: 1 cup (prepared with reduced-sodium chicken broth and reduced-fat baking mix) equals 205 calories, 6 g fat (1 g saturated fat), 4 mg cholesterol, 514 mg sodium, 34 g carbohydrate, 4 g fiber, 5 g protein. **Diabetic Exchanges:** 2 starch, 1 fat.

Cherry Rhubarb Jam

(Pictured above)

Prep: 10 min. + standing **Cook:** 15 min. + cooling

This yummy spread is "jam-packed" with lots of cherry and rhubarb flavor. My mother gives jars of it away during rhubarb season. —*Faye Sampson, Radcliffe, Iowa*

- **4 cups diced fresh *or* frozen rhubarb**
- **1-1/2 cups sugar**
- **1 package (3 ounces) cherry gelatin**
- **1 can (21 ounces) cherry pie filling**
- **1/8 teaspoon almond extract, optional**

In a large saucepan, combine the rhubarb and sugar; let stand for 1-1/2 hours, stirring occasionally.

Bring to a boil; cook, uncovered, for 10 minutes or until rhubarb is tender. Remove from the heat; stir in gelatin until dissolved. Stir in pie filling and extract if desired. Transfer to jars; cool. Cover and store in the refrigerator for up to 3 weeks. **Yield:** 5 cups.

Biscuit Mushroom Bake

(Pictured above)

Prep: 20 min. **Bake:** 15 min.

Mushroom lovers will appreciate this home-style, biscuit-topped dish. The recipe was passed down from my aunt, and it's a hit with everyone who tries it. —*Dawn Esterly Meadville, Pennsylvania*

1 pound sliced fresh mushrooms
2 tablespoons butter
3 tablespoons all-purpose flour
1 cup chicken broth
1/2 cup milk
1 tablespoon lemon juice
1 teaspoon onion powder
1 teaspoon garlic powder
1/4 teaspoon salt
1/4 teaspoon pepper
1/4 teaspoon paprika
1 tube (12 ounces) refrigerated biscuits

In a large skillet, saute mushrooms in butter. Stir in flour until blended. Gradually add broth and milk. Bring to a boil; cook and stir for 2 minutes or until thickened and bubbly. Remove from the heat. Stir in the lemon juice, onion powder, garlic powder, salt, pepper and paprika.

Pour into a greased 11-in. x 7-in. x 2-in. baking dish. Arrange biscuits over the top. Bake, uncovered, at 375° for 15-20 minutes or until biscuits are golden brown. Let stand for 5 minutes before serving. **Yield:** 5 servings.

Pierogi Pasta Shells

(Pictured below)

Prep: 30 min. **Bake:** 30 min.

My family loves pierogies, so I decided to create my own version. I took them to a Christmas party, and they won rave reviews. I left with an empty pan—not a crumb to be found! —*Kim Wallace, Dover, Ohio*

51 uncooked jumbo pasta shells
2 packages (32 ounces *each*) refrigerated mashed potatoes
2 tablespoons dried minced onion
1/2 teaspoon onion powder
1/2 teaspoon garlic powder
4 cups (16 ounces) shredded cheddar cheese, *divided*
1/2 cup chopped green onions

Cook pasta shells according to package directions; drain and rinse in cold water. Place mashed potatoes in a large microwave-safe bowl. Cover and microwave on high for 4 minutes, stirring once. Add the minced onion, onion powder and garlic powder. Stir in 2 cups of the cheese until blended.

Stuff into shells. Place in two greased 13-in. x 9-in. x 2-in. baking dishes. Sprinkle with the green onions and remaining cheese. Cover and bake at 350° for 20 minutes. Uncover; bake 10 minutes longer or until heated through. **Yield:** 17 servings.

Cherry Tomato Mozzarella Saute

(Pictured above)

Prep/Total Time: 25 min.

Fast to fix and full of flavor, this warm, wholesome side dish makes the most of cherry tomatoes. They're jazzed up with fresh mozzarella cubes and fresh thyme.
—Summer Jones, Pleasant Grove, Utah

1/4 cup chopped shallots
1 garlic clove, minced
1 teaspoon minced fresh thyme
2 teaspoons olive oil
2-1/2 cups cherry tomatoes, halved
1/4 teaspoon salt
1/4 teaspoon pepper
4 ounces fresh mozzarella cheese, cut into 1/2-inch cubes

In a large skillet, saute the shallots, garlic and thyme in oil until tender. Add the tomatoes, salt and pepper; heat through. Remove from the heat; stir in cheese. **Yield:** 4 servings.

Versatile Chili-Cheese Sauce

Prep/Total Time: 30 min.

This sauce is so good, there's no limit to the ways it can be enjoyed. We use it as a fondue...as a dip for tortilla chips or vegetables...or poured over grilled chicken, baked potatoes and more. *—Darlene Brenden, Salem, Oregon*

1/4 cup butter, cubed
1/4 cup all-purpose flour
2 cups heavy whipping cream
1 cup (8 ounces) sour cream
1 can (4 ounces) chopped green chilies, undrained
3 teaspoons chicken bouillon granules
1 cup (4 ounces) shredded cheddar cheese
1/2 cup shredded Monterey Jack cheese
1/4 teaspoon pepper

In a large saucepan, melt butter. Stir in flour until smooth; gradually add cream. Bring to a boil; cook and stir for 2 minutes or until thickened.

Stir in the sour cream, chilies and bouillon. Reduce heat to medium; cook and stir for 3-4 minutes or until heated through. Add cheeses and pepper. Cook until bubbly and cheese is melted, stirring occasionally. **Yield:** 4 cups.

Homemade Almond Paste

Prep/Total Time: 10 min.

When a recipe I wanted to try called for almond paste, I came up with my own. It saves the expense of the store-bought kind, and I've found it results in a lighter baked product. *—Anne Keenan, Nevada City, California*

1-1/2 cups blanched almonds
1-1/2 cups confectioners' sugar
1 egg white
1-1/2 teaspoons almond extract
1/4 teaspoon salt

Place the blanched almonds in a food processor; cover and process until smooth. Add the confectioners' sugar, egg white, almond extract and salt; cover and process until smooth.

Divide almond paste into 1/2-cup portions; place in airtight containers. Refrigerate for up to 1 month or freeze for up to 3 months. **Yield:** 1-1/2 cups.

About Almond Paste

Almond paste is used to add great almond flavor to such recipes as breads, pastries, cookies, cakes and other desserts.

Almond paste is similar to marzipan, but they are not interchangeable in recipes. Marzipan is a sweeter, more pliable confection that is often tinted with food coloring and molded to make decorative shapes...or rolled into thin sheets and used to cover cakes.

Raspberry Rhubarb Sauce

(Pictured above)

Prep/Total Time: 20 min.

I serve this tart, ruby-red sauce over vanilla frozen yogurt, ice cream or cake for a colorful and refreshing dessert after a big dinner. The topping is also a great way to perk up a stack of pancakes for breakfast or brunch.
—Inge Schermerhorn, Kingston, New Hampshire

✓ **Uses less fat, sugar or salt. Includes Nutrition Facts and Diabetic Exchanges.**

4-1/2 cups sliced fresh *or* frozen rhubarb
1 cup warm water
1 package (.3 ounce) sugar-free raspberry gelatin
6 cups fat-free vanilla frozen yogurt

Place rhubarb in a large saucepan. In a small bowl, combine water and gelatin; pour over rhubarb. Bring to a boil; reduce heat and simmer, uncovered, for 5 minutes or until rhubarb is tender.

Serve warm or chilled over frozen yogurt. Refrigerate leftovers. **Yield:** 2-1/2 cups.

Nutrition Facts: 3 tablespoons sauce with 1/2 cup frozen yogurt equals 108 calories, trace fat (trace saturated fat), 2 mg cholesterol, 82 mg sodium, 21 g carbohydrate, 1 g fiber, 6 g protein. **Diabetic Exchanges:** 1 fat-free milk, 1/2 fruit.

Walnut Cranberry Butter

(Pictured below)

Prep: 20 min. + chilling

This pink butter is heavenly on warm bread or biscuits. Be sure to use real butter—low-fat substitutes won't give the same results. *—Corky Huffsmith, Indio, California*

3/4 cup butter, softened
2 tablespoons brown sugar
2 tablespoons honey
1 cup chopped fresh cranberries
2 tablespoons chopped walnuts, toasted

In a small mixing bowl, beat butter, brown sugar and honey until fluffy, about 5 minutes. Add cranberries and nuts; beat 5 minutes longer or until butter turns pink.

Transfer to a sheet of plastic wrap; roll into a log. Refrigerate until chilled. Unwrap and slice or place on a butter dish. **Yield:** 1-1/3 cups.

Saucy Sprouts and Oranges

Prep/Total Time: 30 min.

With this mouth-watering recipe, you'll win over folks who say they don't care for brussels sprouts. I serve it with turkey or ham. *—Carolyn Hannay, Antioch, Tennessee*

✓ **Uses less fat, sugar or salt. Includes Nutrition Facts and Diabetic Exchanges.**

3 medium navel oranges
1 pound fresh brussels sprouts, trimmed and halved

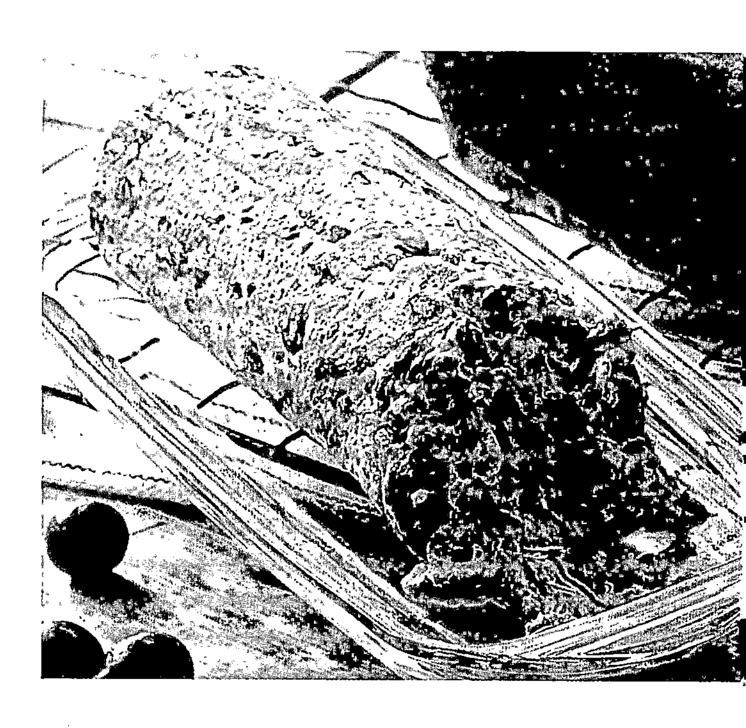

- 1 tablespoon butter
- 2 teaspoons cornstarch
- 2 tablespoons honey mustard
- 1/4 teaspoon Chinese five-spice powder
- 2 tablespoons slivered almonds, toasted

Finely grate peel of one orange; set peel aside. Cut that orange in half; squeeze juice into a 1-cup measuring cup. Add enough water to measure 1/2 cup; set aside. Peel and discard white membranes from remaining oranges; section them and set aside.

In a large saucepan, bring 1 in. of water and brussels sprouts to a boil. Cover and cook for 8-10 minutes or until crisp-tender.

Meanwhile, in a small saucepan, melt butter. Whisk cornstarch and reserved orange juice mixture until smooth; add to the butter. Stir in mustard and five-spice powder. Bring to a boil over medium heat; cook and stir for 1-2 minutes or until thickened and bubbly.

Drain sprouts; gently stir in orange sections. Transfer to a serving bowl; drizzle with sauce. Sprinkle with almonds and grated orange peel. **Yield:** 6 servings.

Nutrition Facts: 3/4 cup equals 97 calories, 4 g fat (1 g saturated fat), 5 mg cholesterol, 81 mg sodium, 15 g carbohydrate, 3 g fiber, 4 g protein. **Diabetic Exchanges:** 1 vegetable, 1 fat, 1/2 fruit.

Rosemary-Onion Green Beans

Prep/Total Time: 25 min.

I came across this recipe in a cookbook from France. People always want to know what gives the green beans their unique flavor. It's the rosemary, of course!

—Lucy Banks, Jackson, Mississippi

Uses less fat, sugar or salt. Includes Nutrition Facts and Diabetic Exchanges.

- 2 small onions, thinly sliced
- 1 fresh rosemary sprig
- 2 teaspoons butter
- 1-1/4 pounds fresh green beans, trimmed
- 1/4 cup water
- 1/4 teaspoon salt
- 1/4 teaspoon pepper

In a large nonstick skillet, saute onions and rosemary in butter for 3-5 minutes or until onions are tender.

Add the beans, water, salt and pepper. Bring to a boil. Reduce heat; cover and cook for 7-9 minutes or just until beans are tender. Discard rosemary. **Yield:** 4 servings.

Nutrition Facts: 1 cup equals 70 calories, 2 g fat (1 g saturated fat), 5 mg cholesterol, 175 mg sodium, 12 g carbohydrate, 5 g fiber, 3 g protein. **Diabetic Exchanges:** 2 vegetable, 1/2 fat.

Mishmash Applesauce

(Pictured above)

Prep: 30 min. **Cook:** 30 min. + standing

This fun recipe is so easy because you don't need to peel the apples. Berries and rhubarb give it a tongue-tingling twist. Try it hot or cold, over ice cream or just as is.

—Beverly Rice, Elm Grove, Wisconsin

☑ **Uses less fat, sugar or salt. Includes Nutrition Facts and Diabetic Exchanges.**

- 3 pounds tart apples, chopped
- 2 cups chopped fresh *or* frozen rhubarb
- 1 cup chopped fresh *or* frozen strawberries
- 1 cup fresh *or* frozen blueberries
- 1 cup fresh *or* frozen cranberries
- 1 cup orange juice
- 2 packages (.3 ounce *each*) sugar-free strawberry gelatin

In a Dutch oven, combine the fruit and orange juice. Bring to a boil over medium heat, stirring frequently. Sprinkle gelatin over fruit mixture; mix well. Reduce heat; cover and simmer for 15-20 minutes or until apples are tender.

Remove from the heat; mash fruit. Let stand for 15 minutes. Serve warm or chilled. **Yield:** 8 cups.

Nutrition Facts: 1/2 cup equals 75 calories, trace fat (trace saturated fat), 0 cholesterol, 24 mg sodium, 18 g carbohydrate, 3 g fiber, 1 g protein. **Diabetic Exchange:** 1 fruit.

Main Dishes

Whether you want an extra-special entree for your holiday dinner or an easy-but-pleasing main dish for a busy weeknight, you'll find a wide variety of meaty—and meatless—choices right here.

MAIN ATTRACTIONS. Clockwise from upper left: Herbed Lamb Kabobs (p. 61), Spaghetti Squash with Red Sauce (p. 67), Roasted Pheasants with Oyster Stuffing (p. 68), Rice-Stuffed Acorn Squash (p. 63) and Seafood Casserole (p. 64).

French-Style Chicken

(Pictured above)

Prep/Total Time: 25 min.

When I have friends over, I make this classy, light recipe and serve it with a green salad and crisp French bread.
—Catherine Johnston, Stafford, New York

✓ **Uses less fat, sugar or salt. Includes Nutrition Facts and Diabetic Exchanges.**

6 boneless skinless chicken breast halves (4 ounces *each*)
3/4 teaspoon salt-free lemon-pepper seasoning
1-1/3 cups reduced-sodium chicken broth
3 medium unpeeled apples, cut into wedges
1 medium onion, thinly sliced
4 tablespoons apple cider *or* juice, *divided*
1/4 teaspoon ground cinnamon
1/8 teaspoon ground nutmeg
1 tablespoon cornstarch
Minced fresh parsley

Sprinkle chicken with lemon-pepper. In a large nonstick skillet coated with nonstick cooking spray, cook chicken for 5-6 minutes on each side or until juices run clear. Remove and keep warm.

In the same skillet, combine broth, apples, onion, 3 tablespoons cider, cinnamon and nutmeg. Bring to a boil. Combine cornstarch and remaining cider until smooth; stir into apple mixture. Bring to a boil; cook and stir for 1-2 minutes or until thickened. Top with chicken; sprinkle with parsley. **Yield:** 6 servings.

Nutrition Facts: 1 serving equals 186 calories, 3 g fat (1 g saturated fat), 63 mg cholesterol, 194 mg sodium, 16 g carbohydrate, 2 g fiber, 24 g protein. **Diabetic Exchanges:** 3 lean meat, 1 fruit.

Thai Chicken Fettuccine

Prep: 25 min. **Cook:** 15 min. + cooling

Try this chicken and pasta dish, and it'll bowl you over with its unique salsa-peanut dressing and bold taste.
—Michelle Van Loon, Round Lake, Illinois

1 cup salsa
1/4 cup creamy peanut butter
2 tablespoons orange juice
2 tablespoons honey
1 teaspoon soy sauce
8 ounces uncooked fettuccine
3/4 pound boneless skinless chicken breasts, cut into strips
1 tablespoon vegetable oil
1 medium sweet red pepper, julienned
1/4 cup minced fresh cilantro

For sauce, in a microwave-safe bowl, combine the first five ingredients. Cover and microwave on high for 1 minute; stir. Set aside.

Cook the fettuccine according to the package directions. Meanwhile, in a large skillet, cook the chicken strips in oil over medium heat for 3-5 minutes or until browned. Add the sweet red pepper; cook and stir until crisp-tender.

Drain fettuccine; top with chicken mixture and sauce. Sprinkle with cilantro. **Yield:** 4 servings.

Editor's Note: This recipe was tested in a 1,100-watt microwave.

Ham 'n' Cheese Squares

Prep: 15 min. **Bake:** 20 min.

This delicious egg bake is loaded with ham, Swiss cheese and caraway. The dish cuts nicely into squares, making it an ideal addition to a weekend brunch buffet.
—Sue Ross, Casa Grande, Arizona

1-1/2 cups cubed fully cooked ham
1 carton (6 ounces) plain yogurt
1/4 cup crushed saltines (about 6)
1/4 cup shredded Swiss cheese
2 tablespoons butter, melted
2 teaspoons caraway seeds
6 eggs

In a large bowl, combine the first six ingredients. In a small mixing bowl, beat eggs until thickened and lemon-colored; fold into ham mixture. Transfer to a greased 8-in. square baking dish.

Bake at 375° for 20-25 minutes or until a knife inserted near the center comes out clean. Let stand for 5 minutes before cutting. **Yield:** 9 servings.

Asparagus Beef Stir-Fry

Prep/Total Time: 30 min.

I love filet mignon—but not its price! For an alternative, I came up with this unusual stir-fry recipe. Now I cook it once a week. —*Linda Flynn, Ellicott City, Maryland*

1 pound beef tenderloin, cubed
1 green onion, sliced
2 garlic cloves, minced
1/2 teaspoon salt
1/4 teaspoon pepper
1 tablespoon vegetable oil
1 pound fresh asparagus, trimmed and cut into 2-inch pieces
1/2 pound sliced fresh mushrooms
1/4 cup butter, cubed
1 tablespoon soy sauce
1-1/2 teaspoons lemon juice
Hot cooked rice

In a wok or large skillet, stir-fry the cubed beef, onion, garlic, salt and pepper in oil for 3-5 minutes; remove and keep warm.

In the same pan, stir-fry the asparagus and mushrooms in butter until asparagus is tender. Return beef mixture to the pan. Stir in soy sauce and lemon juice; heat through. Serve with rice. **Yield:** 4 servings.

Grilled Ham Steaks

Prep: 10 min. + marinating **Grill:** 10 min.

These mouth-watering ham slices are marinated in a molasses-spice sauce and then grilled to perfection. —*Sandy Morris, Lititz, Pennsylvania*

1-1/2 cups orange juice
1/2 cup cider vinegar
1/4 cup packed brown sugar
1 to 2 tablespoons ground cloves
1 tablespoon molasses
2 to 3 teaspoons ground mustard
2 to 3 teaspoons ground ginger
3 bone-in fully cooked ham steaks (1 pound *each* and 3/8 inch thick)

In a small bowl, combine the first seven ingredients. Pour 1-1/2 cups into a large resealable plastic bag; add ham. Seal bag; turn to coat. Refrigerate for 1-2 hours. Cover and refrigerate remaining marinade for basting.

Coat grill rack with nonstick cooking spray before starting the grill. Drain and discard marinade. Grill ham steaks, covered, over medium heat for 10-15 minutes or until heated through, basting frequently with reserved marinade. **Yield:** 12 servings.

Herbed Lamb Kabobs

(Pictured below and on page 58)

Prep: 15 min. + marinating **Grill:** 20 min.

It's hard to resist these attractive kabobs. The herbed marinade not only tenderizes the meat, but also adds delightful flavor to the lamb and veggies. —*Janet Dingler, Cedartown, Georgia*

1 cup vegetable oil
1 medium onion, chopped
1/2 cup lemon juice
1/2 cup minced fresh parsley
3 to 4 garlic cloves, minced
2 teaspoons salt
2 teaspoons dried marjoram
2 teaspoons dried thyme
1/2 teaspoon pepper
2 pounds boneless lamb, cut into 1-inch cubes
1 medium red onion, cut into wedges
1 large green pepper, cut into 1-inch pieces
1 large sweet red pepper, cut into 1-inch pieces

In a bowl, combine the first nine ingredients. Pour 1 cup into a large resealable plastic bag; add lamb. Seal bag and turn to coat; refrigerate for 6-8 hours. Cover and refrigerate remaining marinade for basting.

Drain and discard marinade. On eight metal or soaked wooden skewers, alternately thread the lamb, onion and peppers. Grill, uncovered, over medium-hot heat for 8-10 minutes on each side or until meat reaches desired doneness, basting frequently with reserved marinade. **Yield:** 8 servings.

Pressure-Cooker Pork Ribs

(Pictured below)

Prep/Total Time: 25 min.

You'll enjoy melt-in-your-mouth ribs when you try this pressure-cooker recipe. When I was younger, my mom made these saucy spareribs for special weekend dinners.
—Paula Zsiray, Logan, Utah

2 pounds boneless country-style pork ribs, cut into 2-inch chunks
1 teaspoon onion salt
1 teaspoon pepper
1 teaspoon paprika
1 tablespoon vegetable oil
1 cup water
3 tablespoons ketchup
4-1/2 teaspoons white vinegar
1 teaspoon Worcestershire sauce
1 teaspoon prepared mustard
1/8 teaspoon celery seed

Sprinkle the ribs with onion salt, pepper and paprika. In a pressure cooker, brown ribs in oil on all sides. Remove from the pressure cooker and drain. Return meat to the pressure cooker. Combine the remaining ingredients; pour over meat.

Close cover securely; place pressure regulator on vent pipe. Bring cooker to full pressure over high heat. Reduce heat to medium-high and cook for 15 minutes. (Pressure regulator should maintain a slow steady rocking motion; adjust heat if needed.)

Remove from the heat; allow pressure to drop on its own. Skim fat from sauce if necessary and serve with ribs if desired. **Yield:** 4 servings.

Editor's Note: This recipe was tested at 13 pounds of pressure (psi).

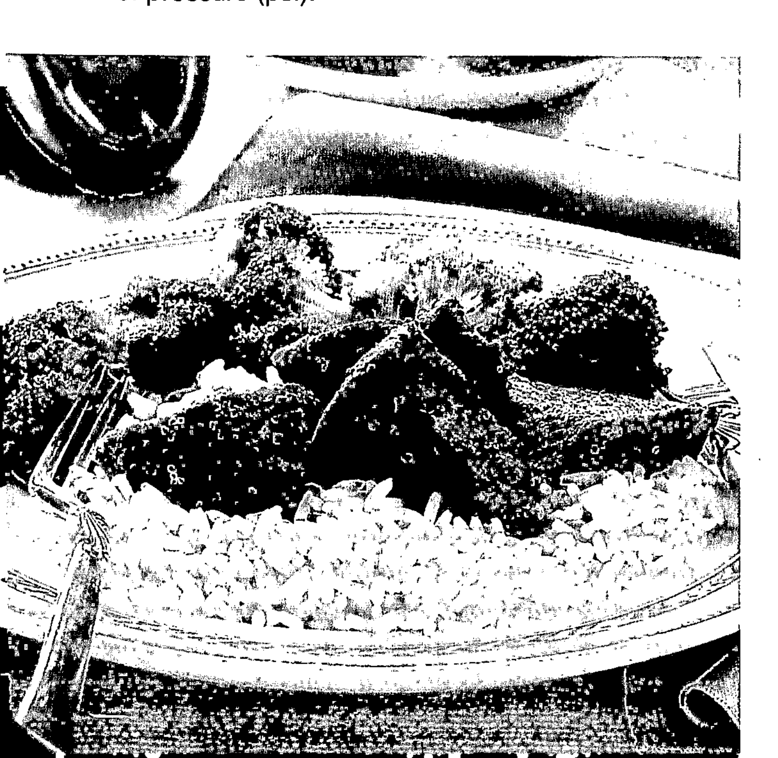

Roasted Sea Scallops

Prep: 10 min. **Bake:** 25 min.

We like scallops because of their availability here year-round. These are delicately seasoned with roasted tomatoes and can be served over either rice or toasted garlic bread.
—Marguerite Shaeffer, Sewell, New Jersey

✓ **Uses less fat, sugar or salt. Includes Nutrition Facts and Diabetic Exchanges.**

1 large tomato, chopped
1 medium onion, chopped
1 tablespoon minced fresh parsley
1 tablespoon olive oil
1-1/2 teaspoons paprika
1/2 teaspoon salt
1/4 teaspoon pepper
12 sea scallops (2 ounces *each*)
Hot cooked rice, optional

In an ungreased 3-qt. baking dish, combine the first seven ingredients. Bake, uncovered, at 400° for 10 minutes or until bubbly.

Stir in scallops. Bake 15 minutes longer or until scallops are firm and opaque. Serve with rice if desired. **Yield:** 4 servings.

Nutrition Facts: 3 scallops (calculated without rice) equals 209 calories, 5 g fat (1 g saturated fat), 56 mg cholesterol, 575 mg sodium, 11 g carbohydrate, 2 g fiber, 30 g protein. **Diabetic Exchanges:** 4 very lean meat, 1 vegetable, 1/2 starch.

Beef Rib Roast

Prep: 15 min. **Bake:** 2 hours + standing

My mother topped beef roast with bacon and onion. Whenever I prepare this delicious entree, I can't help but reminisce about her wonderful cooking.
—Betty Abel Jellencich, Utica, New York

1 bone-in beef rib roast (4 to 5 pounds)
1 garlic clove, minced
1 teaspoon salt
1/2 teaspoon pepper
1 small onion, sliced
6 to 8 bacon strips

Place roast, fat side up, on a rack in a shallow roasting pan. Rub with garlic, salt and pepper; top with onion and bacon.

Bake, uncovered, at 325° for 2-3 hours or until meat reaches desired doneness (for medium-rare, a meat thermometer should read 145°; medium, 160°; well-done, 170°). Transfer to warm serving platter. Let stand for 10-15 minutes before slicing. **Yield:** 10-12 servings.

Ham-Potato Phyllo Bake

(Pictured above)

Prep: 30 min. **Bake:** 20 min.

This one-dish wonder features a luscious ham and potato filling layered between phyllo dough pastry.I get many requests for it. —Tracy Hartsuff, Lansing, Michigan

- **3 pounds red potatoes, peeled and thinly sliced**
- **1 medium onion, chopped**
- **8 tablespoons butter, *divided***
- **20 sheets phyllo dough (14 inches x 9 inches)**
- **2 cups (16 ounces) sour cream**
- **2 cups cubed fully cooked ham**
- **2 cups (8 ounces) shredded cheddar cheese**
- **7 teaspoons dill weed, *divided***
- **2 teaspoons garlic powder**
- **1 teaspoon salt**
- **1/2 teaspoon pepper**
- **1 egg, beaten**
- **2 tablespoons half-and-half cream**

Place potatoes in a Dutch oven and cover with water. Bring to a boil; reduce heat. Cover and cook for 10-15 minutes or until tender; drain. In a small skillet, saute onion in 1 tablespoon butter until tender; set aside.

Melt remaining butter. Brush a 13-in. x 9-in. x 2-in. baking dish with some of the butter. Unroll phyllo sheets; trim to fit into dish. (Keep dough covered with plastic wrap and a damp cloth while assembling.) Place one phyllo sheet in prepared dish; brush with butter. Repeat twice.

Top with half of the sour cream, potatoes, onion, ham and cheese. Combine 6 teaspoons dill, garlic powder, salt and pepper; sprinkle half over cheese. Layer with three phyllo sheets, brushing each with butter. Top with remaining sour cream, potatoes, onion, ham, cheese and seasoning mixture.

Layer with remaining phyllo dough, brushing each sheet with butter. Combine egg and cream; brush over top. Sprinkle with remaining dill.

Bake, uncovered, at 350° for 20-25 minutes or until heated through. Let stand for 5 minutes. Cut into squares. **Yield:** 12-15 servings.

Rice-Stuffed Acorn Squash

(Pictured below and on page 58)

Prep: 45 min. **Bake:** 20 min.

We often make this meatless dish in fall, after harvesting fresh squash from our garden.A hint of Oriental flavor offers a different, unexpected accent. I especially enjoy the pleasant combination of mozzarella and ginger. —Lydia Garcia, Gouldsboro, Maine

- **2 large acorn squash**
- **3/4 cup uncooked long grain rice**
- **1-1/2 cups water**
- **2 tablespoons soy sauce**
- **1 medium onion, chopped**
- **1/4 cup butter, cubed**
- **2 medium tart apples, peeled and chopped**
- **1 cup (4 ounces) shredded part-skim mozzarella cheese**
- **1/2 cup chopped walnuts**
- **1/2 cup half-and-half cream**
- **1/4 cup balsamic vinegar**
- **3 tablespoons honey**
- **3 teaspoons minced fresh gingerroot**
- **1 teaspoon curry powder**

Cut squash in half; remove seeds. Place cut side down in a greased 13-in. x 9-in. x 2-in. baking dish. Cover and bake at 350° for 40-45 minutes or until tender.

Meanwhile, in a large saucepan, bring the rice, water and soy sauce to a boil. Reduce heat; cover and simmer for 15-18 minutes or until liquid is absorbed and rice is tender.

In a large skillet, saute onion in butter until almost tender. Add apples; saute for 3 minutes. Remove from the heat; stir in the rice, cheese, walnuts, cream, vinegar, honey, ginger and curry.

Turn squash over; stuff with rice mixture. Bake, uncovered, for 20-25 minutes or until heated through. **Yield:** 4 servings.

Comforting Casseroles

SOMETIMES you need more than a hug—you need comfort food! Bubbling, golden casseroles have a special way of warming hearts. Each tasty bake here is sure to make your family feel cozy and content.

Seafood Casserole

(Pictured below and on page 58)

Prep: 20 min. **Bake:** 40 min.

This easy rice casserole is teeming with succulent shrimp and crabmeat. If you like, substitute scallops or other seafood. —Nancy Billups, Princeton, Iowa

- **1 package (6 ounces) long grain and wild rice**
- **1 pound frozen crabmeat, thawed *or* 2-1/2 cups canned lump crabmeat, drained**
- **1 pound cooked medium shrimp, peeled, deveined and cut into 1/2-inch pieces**
- **2 celery ribs, chopped**
- **1 medium onion, finely chopped**
- **1/2 cup finely chopped green pepper**
- **1 can (4 ounces) mushroom stems and pieces, drained**
- **1 jar (2 ounces) diced pimientos, drained**
- **1 cup mayonnaise**
- **1 cup milk**
- **1/2 teaspoon pepper**
- **Dash Worcestershire sauce**
- **1/4 cup dry bread crumbs**

Cook rice according to package directions. Meanwhile, in a large bowl, combine crab, shrimp, celery, onion, green pepper, mushrooms and pimientos. In a small bowl, whisk mayonnaise, milk, pepper and Worcestershire sauce; stir into seafood mixture. Stir in rice.

Transfer mixture to a greased 13-in. x 9-in. x 2-in. baking dish. Sprinkle with the bread crumbs. Bake, uncovered, at 375° for 40-50 minutes or until bubbly. **Yield:** 6 servings.

Editor's Note: Reduced-fat or fat-free mayonnaise is not recommended for this recipe.

Hominy Sausage Bake

Prep: 40 min. **Bake:** 35 min.

My cousin made this spicy corn dish for a special event, and it was an instant hit. The Cajun flavor is fantastic! —Frances Bowman, Lilesville, North Carolina

- **1 pound smoked sausage, cut into 1/4-inch slices**
- **1 teaspoon olive oil**
- **2 cups cubed fully cooked ham**
- **2 packages (8 ounces *each*) red beans and rice mix**
- **6 cups water**
- **2 tablespoons butter**
- **1/4 teaspoon cayenne pepper**
- **1 can (29 ounces) hominy, rinsed and drained**
- **1 jar (12 ounces) pickled jalapeno peppers, drained and chopped**
- **1 can (15-1/4 ounces) whole kernel corn, drained**
- **1 cup (4 ounces) shredded cheddar cheese**
- **1 cup corn chips, crushed**

In a Dutch oven, brown sausage in oil; drain and set aside. In same pan, brown ham. Stir in beans and rice mix, water, butter and cayenne. Bring to a boil. Reduce heat; cover and simmer for 25 minutes or until beans and rice are tender, stirring occasionally.

Transfer to a greased 3-qt. baking dish. Layer with hominy and sausage; top with peppers and corn. Bake, uncovered, at 350° for 30-35 minutes or until heated through. Sprinkle with cheese; bake 5 minutes longer or until the cheese is melted. Sprinkle with corn chips. **Yield:** 8 servings.

Editor's Note: When chopping hot peppers, use rubber or plastic gloves to protect your hands. Avoid

touching your face. This recipe was tested with Zatarain's New Orleans-style red beans and rice.

Pizza Casserole

(Pictured above)

Prep: 20 min. **Bake:** 30 min.

Friends and family love this spin on pizza. It's a tomatoey pasta casserole packed with mozzarella and beef.
—Nancy Foust, Stoneboro, Pennsylvania

3 cups uncooked spiral pasta
2 pounds ground beef
1 medium onion, chopped
2 cans (8 ounces *each*) mushroom stems and pieces, drained
1 can (15 ounces) tomato sauce
1 jar (14 ounces) pizza sauce
1 can (6 ounces) tomato paste
1/2 teaspoon sugar
1/2 teaspoon garlic powder
1/2 teaspoon onion powder
1/2 teaspoon dried oregano
4 cups (16 ounces) shredded part-skim mozzarella cheese, *divided*
1 package (3-1/2 ounces) sliced pepperoni
1/2 cup grated Parmesan cheese

Cook pasta according to package directions. Meanwhile, in a Dutch oven, cook beef and onion over medium heat until meat is no longer pink; drain. Stir in the mushrooms, sauces, tomato paste, sugar and seasonings. Drain pasta; stir into meat sauce.

Divide half of the mixture between two greased 11-in. x 7-in. x 2-in. baking dishes; sprinkle each with 1 cup of mozzarella cheese. Repeat layers. Top with pepperoni and Parmesan cheese. Cover and bake at 350° for 20 minutes. Uncover; bake 10-15 minutes longer or until heated through. **Yield:** 2 casseroles (8 servings each).

Spanish Rice Turkey Casserole

(Pictured below)

Prep: 30 min. **Bake:** 20 min.

Green chilies and tender turkey star in this cheesy entree. Even my grandparents, who aren't fans of spicy foods, enjoy this.
—Ann Herren, Pulaski, Tennessee

2 packages (6.8 ounces *each*) Spanish rice and vermicelli mix
1/4 cup butter, cubed
4 cups water
1 can (14-1/2 ounces) diced tomatoes, undrained
1 can (10 ounces) diced tomatoes and green chilies, undrained
3 cups cubed cooked turkey *or* chicken
1 can (11 ounces) whole kernel corn, drained
1/2 cup sour cream
1 cup (4 ounces) shredded Mexican cheese blend, *divided*

In a large skillet, saute rice and vermicelli in butter until golden brown. Gradually stir in the water, tomatoes and contents of rice seasoning packets. Bring to a boil. Reduce heat; cover and simmer for 15-20 minutes or until rice is tender.

Meanwhile, in a large bowl, combine the turkey, corn, sour cream and 1/2 cup cheese. Stir in rice mixture. Transfer to a greased 3-qt. baking dish. Sprinkle with remaining cheese (dish will be full). Bake, uncovered, at 375° for 20-25 minutes or until heated through. **Yield:** 8 servings.

Macadamia-Crusted Tilapia

(Pictured above)

Prep: 20 min. **Bake:** 15 min.

With a refreshing pineapple salsa, these crispy golden fillets are special enough to serve company. I like to garnish each one with whole macadamia nuts.
—Jennifer Fisher, Austin, Texas

- **2 eggs**
- **1/8 teaspoon cayenne pepper**
- **1 cup all-purpose flour**
- **1-3/4 cups macadamia nuts, finely chopped**
- **4 tilapia fillets (6 ounces *each*)**
- **1 tablespoon butter, melted**

PINEAPPLE SALSA:

- **1 cup cubed fresh pineapple**
- **1/4 cup chopped sweet red pepper**
- **3 tablespoons thinly sliced green onions**
- **2 tablespoons sugar**
- **1 jalapeno pepper, seeded and chopped**
- **1 tablespoon lime juice**
- **1/2 teaspoon minced fresh gingerroot**
- **2 tablespoons minced fresh cilantro**

In a shallow bowl, whisk eggs and cayenne. Place flour and macadamia nuts in separate shallow bowls. Coat tilapia with flour, then dip in egg mixture and coat with nuts. Place on a greased baking sheet; drizzle with butter. Bake at 375° for 15-20 minutes or until fish flakes easily with a fork.

Meanwhile, in a small serving bowl, combine the pineapple, red pepper, onions, sugar, jalapeno, lime juice and ginger; sprinkle with cilantro. Serve with fish. **Yield:** 4 servings (1-1/2 cups salsa).

Editor's Note: When cutting or seeding hot peppers, use rubber or plastic gloves to protect your hands. Avoid touching your face.

Creamy Celery Beef Stroganoff

Prep: 20 min. **Cook:** 8 hours

A creamy sauce coats the tender beef cubes in this satisfying classic. It's easy to make but doesn't taste that way.
—Kim Wallace, Dover, Ohio

- **2 pounds beef stew meat, cut into 1-inch cubes**
- **1 can (10-3/4 ounces) condensed cream of celery soup, undiluted**
- **1 can (10-3/4 ounces) condensed cream of mushroom soup, undiluted**
- **1 medium onion, chopped**
- **1 jar (6 ounces) sliced mushrooms, drained**
- **1 envelope onion soup mix**
- **1/2 teaspoon pepper**
- **1 cup (8 ounces) sour cream**

Hot cooked noodles

In a 3-qt. slow cooker, combine the first seven ingredients. Cover and cook on low for 8 hours or until the beef is tender. Stir in sour cream. Serve with noodles. **Yield:** 6 servings.

Seasoned Pork Loin Roast

Prep: 20 min. **Bake:** 1-1/2 hours + standing

This guaranteed crowd-pleaser can be served all year long. In summer I grill it, and in winter I roast it in the oven. Delicious! *—Elaine Seip, Medicine Hat, Alberta*

- **2 teaspoons garlic salt**
- **2 teaspoons garlic-pepper blend**
- **2 teaspoons lemon-pepper seasoning**
- **1 boneless rolled pork loin roast (about 5 pounds)**

BASTING SAUCE:

- **3 cups water**
- **2 tablespoons lemon juice**
- **1-1/2 teaspoons dried minced onion**
- **1/2 teaspoon garlic salt**
- **1/2 teaspoon garlic-pepper blend**
- **1/2 teaspoon lemon-pepper seasoning**
- **1/2 teaspoon crushed red pepper flakes**
- **1/2 teaspoon grated lemon peel**

Combine the garlic salt, garlic-pepper and lemon-pepper; rub over roast. Place on a rack in a shallow roasting pan. Bake, uncovered, at 325° for 1-1/2 to 2 hours or until a meat thermometer reads 160°.

Meanwhile, in a large saucepan, combine the basting sauce ingredients. Bring to a boil; reduce heat. Simmer, uncovered, for 10 minutes. Brush over roast occasionally while baking. Let the roast stand for 10 minutes before slicing. **Yield:** 15-18 servings.

Three-Cheese Sausage Lasagna

Prep: 30 min. **Bake:** 30 min. + standing

Hearty appetites will enjoy this non-traditional lasagna. It has a simple white sauce and plenty of savory Italian sausage and cheese. —*Lesley Cormier Pepperell, Massachusetts*

1-1/2 pounds bulk Italian sausage
6 tablespoons butter
6 tablespoons all-purpose flour
1 teaspoon salt
1 teaspoon pepper
3 cups milk
9 lasagna noodles, cooked and drained
6 ounces sliced part-skim mozzarella cheese
4 ounces sliced provolone cheese
1/3 cup grated Romano cheese

In a large skillet, cook sausage over medium heat until no longer pink; drain and set aside. In a small saucepan, melt butter. Stir in flour, salt and pepper until smooth. Gradually stir in milk. Bring to a boil over medium heat; cook and stir for 2 minutes or until thickened. Remove from the heat.

In a greased 13-in. x 9-in. x 2-in. baking dish, layer a fourth of the sauce, three noodles, half of the sausage and a third of the mozzarella and provolone slices. Repeat layers once. Spoon half of the remaining sauce over the top. Layer with remaining noodles, sauce and sliced cheeses; sprinkle with Romano cheese.

Bake, uncovered, at 350° for 30-35 minutes or until heated through. Let stand for 15 minutes before cutting. **Yield:** 12 servings.

Spaghetti Squash with Red Sauce

(Pictured at right and on page 58)

Prep: 15 min. **Cook:** 25 min.

This fabulous meatless main dish is a great way to get the kids to eat lots of vegetables...and a great way to use some of the fresh harvest from your garden. —*Kathryn Pehl, Prescott, Arizona*

✓ Uses less fat, sugar or salt. Includes Nutrition Facts and Diabetic Exchanges.

1 medium spaghetti squash (about 4 pounds)
2 cups chopped fresh tomatoes
1 cup sliced fresh mushrooms
1 cup diced green pepper
1/2 cup shredded carrot
1/4 cup diced red onion
2 garlic cloves, minced
2 teaspoons Italian seasoning
1/8 teaspoon pepper
1 tablespoon olive oil
1 can (15 ounces) tomato sauce
Grated Parmesan cheese, optional

Cut squash in half lengthwise; discard seeds. Place squash, cut side down, on a microwave-safe plate. Microwave, uncovered, on high for 14-16 minutes or until tender.

Meanwhile, in a large skillet, saute the tomatoes, mushrooms, green pepper, carrot, onion, garlic, Italian seasoning and pepper in oil for 6-8 minutes or until tender. Add tomato sauce; heat through.

When squash is cool enough to handle, use a fork to separate strands. Place squash on a serving platter; top with sauce. Sprinkle with Parmesan cheese if desired. **Yield:** 6 servings.

Nutrition Facts: 3/4 cup squash with 1/3 cup sauce (calculated without Parmesan cheese) equals 135 calories, 4 g fat (1 g saturated fat), 0 cholesterol, 372 mg sodium, 25 g carbohydrate, 6 g fiber, 4 g protein. **Diabetic Exchanges:** 2 vegetable, 1 starch, 1/2 fat.

Editor's Note: This recipe was tested in a 1,100-watt microwave.

Roasted Pheasants with Oyster Stuffing

(Pictured below and on page 58)

Prep: 20 min. **Bake:** 1-1/4 hours + standing

This hunter's holiday dinner features bacon-topped pheasants with savory oyster stuffing. If you don't have hunters in the family, ask your meat department to order pheasant, or substitute two broiler-fryer chickens.
—Gloria Warczak, Cedarburg, Wisconsin

1 can (8 ounces) whole oysters
2 cups herb stuffing mix
2 cups corn bread stuffing mix
1 can (14-1/2 ounces) chicken broth, *divided*
1 medium onion, chopped
1/2 cup chopped celery
1/4 cup egg substitute
2 pheasants (2 to 3 pounds *each*)
2 tablespoons Worcestershire sauce
1 teaspoon poultry seasoning
6 bacon strips

Drain oysters, reserving liquid; coarsely chop oysters. In a large bowl, combine the oysters and liquid, stuffing mixes, 3/4 cup broth, onion, celery and egg substitute. Loosely stuff into pheasants. Skewer or fasten openings. Tie drumsticks together. Place breast side up on a rack in a roasting pan.

Combine Worcestershire sauce and remaining broth; spoon over pheasants. Sprinkle with poultry seasoning. Place three bacon strips over each pheasant.

Bake, uncovered, at 350° for 1-1/4 to 1-1/2 hours or until a meat thermometer reads 180° for poultry and 165° for stuffing. Cover and let stand for 10 minutes before removing stuffing and slicing. **Yield:** 6-8 servings.

Pesto Pork Roast

Prep: 30 min. **Bake:** 1-1/2 hours + standing

I came up with this pork and pasta meal when I had abundant tomatoes from the garden and leftover pesto.
—Jennifer Magrey, Sterling, Connecticut

1/4 cup plus 2 tablespoons olive oil, *divided*
2 cups loosely packed basil leaves
1/2 cup grated Parmesan cheese
4 garlic cloves, peeled
12 plum tomatoes
1-1/2 teaspoons pepper, *divided*
1 teaspoon kosher salt, *divided*
1 bone-in pork loin roast (4 to 5 pounds)
1 package (16 ounces) egg noodles

For pesto, in a blender, combine 1/4 cup oil, basil, Parmesan cheese and garlic. Cover and process until blended. Remove 2 tablespoons pesto to a small bowl; stir in remaining oil and set aside.

Cut each tomato into four slices; place in a greased shallow roasting pan. Sprinkle with 1 teaspoon pepper and 1/2 teaspoon salt. Sprinkle remaining pepper and salt over roast; place over tomatoes. Spread remaining pesto over roast. Bake, uncovered, at 350° for 1-1/2 to 2 hours or until a meat thermometer reads 160°.

Remove roast and keep warm. Let stand for 10 minutes before slicing. Cook noodles according to package directions; drain and place in a large bowl.

With a slotted spoon, add tomatoes to the noodles; add the reserved pesto and toss to coat. Serve with the pork. **Yield:** 6-8 servings.

Chicken Manicotti

Prep: 30 min. **Bake:** 35 min.

Rich and creamy, this unusual manicotti is wonderful for special occasions. *—Liz Lorch, Spirit Lake, Iowa*

2 packages (3 ounces *each*) cream cheese, softened
1 cup (8 ounces) sour cream
1/4 cup minced fresh parsley
1/2 teaspoon salt
1/4 teaspoon pepper
4 cups cubed cooked chicken
1 medium onion, finely chopped
1 can (8 ounces) mushroom stems and pieces, drained
1 tablespoon butter
14 manicotti shells, cooked and drained

SAUCE:

6 tablespoons butter
6 tablespoons all-purpose flour
1/4 teaspoon salt

3-1/2 cups milk
3 cups (12 ounces) shredded Monterey Jack *or* cheddar cheese
4 tablespoons shredded Parmesan cheese, *divided*

In a large mixing bowl, beat cream cheese until fluffy. Beat in the sour cream, parsley, salt and pepper. Stir in chicken. In a small skillet, cook onion and mushrooms in butter until tender; add to chicken mixture. Stuff into manicotti shells.

In a large saucepan, melt the butter. Stir in flour and salt until smooth. Gradually whisk in milk. Bring to a boil; cook and stir for 2 minutes or until thickened. Stir in 2-1/2 cups Monterey Jack cheese and 2 tablespoons Parmesan cheese just until melted.

Spread about 1/2 cup cheese sauce in each of two greased 11-in. x 7-in. x 2-in. dishes. Top with stuffed shells and remaining sauce. Cover and bake at 350° for 25 minutes. Uncover; sprinkle with remaining cheeses. Bake 10-15 minutes longer or until bubbly and cheese is melted. **Yield:** 7 servings.

Cheese Ravioli with Zucchini

Prep: 15 min. **Cook:** 20 min.

Whipping cream lends delectable flavor to the sauce for this colorful medley, which starts with packaged ravioli.
—Maria Regakis, Somerville, Massachusetts

1 cup heavy whipping cream
1/2 cup chicken broth
1 package (9 ounces) refrigerated cheese ravioli
1 small onion, finely chopped
1 tablespoon butter
1 medium sweet red pepper, julienned
3 cups julienned zucchini
1/2 teaspoon salt
1/4 teaspoon garlic powder
3/4 cup grated Parmesan cheese, *divided*
1 to 2 tablespoons minced fresh basil
1 tablespoon minced fresh parsley

In a large saucepan, bring cream and broth to a boil. Reduce heat; simmer, uncovered, for 10-15 minutes or until reduced to 1 cup. Meanwhile, cook ravioli according to package directions.

In a large skillet, saute onion in butter for 2 minutes. Add red pepper; cook 2 minutes longer. Stir in the zucchini, salt and garlic powder; cook for 1-2 minutes or until vegetables are crisp-tender. Keep warm.

Stir 1/2 cup Parmesan cheese, basil and parsley into cream sauce; cook for 1 minute. Drain ravioli; add to skillet with cream sauce. Toss to coat. Sprinkle with remaining Parmesan cheese. **Yield:** 4 servings.

Almond Chicken Casserole

(Pictured above)

Prep: 15 min. **Bake:** 25 min.

Crunchy cornflakes and almonds dress up this wholesome dish. It's delicious and so easy to assemble.
—Michelle Krzmarzick, Redondo Beach, California

2 cups cubed cooked chicken
1 can (10-3/4 ounces) condensed cream of chicken soup, undiluted
1 cup (8 ounces) sour cream
3/4 cup mayonnaise
2 celery ribs, chopped
3 hard-cooked eggs, chopped
1 can (8 ounces) water chestnuts, drained and chopped
1 can (4 ounces) mushroom stems and pieces, drained
1 tablespoon finely chopped onion
2 teaspoons lemon juice
1/2 teaspoon salt
1/4 teaspoon pepper
1 cup (4 ounces) shredded cheddar cheese
1/2 cup crushed cornflakes
2 tablespoons butter, melted
1/4 cup sliced almonds

In a large bowl, combine the first 12 ingredients. Transfer to a greased 13-in. x 9-in. x 2-in. baking dish; sprinkle with cheese.

Toss cornflakes with butter; sprinkle over cheese. Top with almonds. Bake, uncovered, at 350° for 25-30 minutes or until heated through. **Yield:** 6-8 servings.

Spinach Lasagna

(Pictured above)

Prep: 25 min. **Bake:** 45 min. + standing

I love Italian food, and this meatless lasagna is the best. Plus, it's a little lighter, with just 8 grams of fat per serving.
—E. Marie Goetz, Morgantown, Kentucky

✓ **Uses less fat, sugar or salt. Includes Nutrition Facts and Diabetic Exchanges.**

12 uncooked lasagna noodles
2 packages (10 ounces *each*) frozen chopped spinach, thawed and squeezed dry
2 cartons (15 ounces *each*) fat-free ricotta cheese
1 egg
1 egg white
1 cup grated Parmesan cheese, *divided*
1 garlic clove, minced
1 teaspoon dried basil
1/2 teaspoon pepper
1/4 teaspoon ground nutmeg
3 cups (12 ounces) shredded part-skim mozzarella cheese
2 cans (15 ounces *each*) tomato sauce

Cook noodles according to package directions. Rinse in cold water; drain well. In a large bowl, combine the spinach, ricotta cheese, egg, egg white, 1/4 cup Parmesan cheese, garlic, basil, pepper and nutmeg. Combine mozzarella cheese and remaining Parmesan cheese.

In a 13-in. x 9-in. x 2-in. baking dish coated with nonstick cooking spray, layer three noodles, 3/4 cup tomato sauce, 1 cup spinach mixture and 3/4 cup cheese mixture. Repeat layers three times.

Bake, uncovered, at 375° for 45-50 minutes or until bubbly. Let stand for 10-15 minutes before cutting. **Yield:** 12 servings.

Nutrition Facts: 1 piece equals 300 calories, 8 g fat (5 g saturated fat), 53 mg cholesterol, 753 mg sodium, 31 g carbohydrate, 3 g fiber, 23 g protein. **Diabetic Exchanges:** 2 lean meat, 1-1/2 starch, 1 vegetable, 1/2 fat.

Homemade Spaghetti Sauce

Prep: 40 min. **Cook:** 50 min.

Turkey sausage links bring extra spice to this hearty sauce. It transforms pasta into a satisfying main dish.
—Laurinda Johnston, Belchertown, Massachusetts

✓ **Uses less fat, sugar or salt. Includes Nutrition Facts and Diabetic Exchanges.**

4 celery ribs, chopped
1 large onion, chopped
1 large green pepper, chopped
2-1/2 cups water, *divided*
3 Italian turkey sausage links (4 ounces *each*), casings removed
1 can (29 ounces) tomato sauce
1 can (28 ounces) diced tomatoes, undrained
3 cans (6 ounces *each*) tomato paste
1/2 cup minced fresh parsley
5 to 6 garlic cloves, minced
3 teaspoons Italian seasoning
1 teaspoon sugar
1/8 teaspoon salt
1/8 teaspoon pepper
6 cups hot cooked spaghetti
Shredded Parmesan cheese, optional

In a Dutch oven, combine the celery, onion, green pepper and 1 cup water. Bring to a boil. Reduce heat to medium; cook, uncovered, for 10-12 minutes or until vegetables are tender and water is reduced.

Crumble sausage over vegetable mixture; cook until meat is no longer pink. Stir in the tomato sauce, tomatoes, tomato paste, parsley, garlic, Italian seasoning, sugar, salt, pepper and remaining water.

Bring to a boil. Reduce heat; simmer, uncovered, for 45-50 minutes, stirring occasionally. Serve over spaghetti. Garnish with Parmesan cheese if desired. **Yield:** 12 servings.

Nutrition Facts: 1 cup sauce with 1/2 cup spaghetti (calculated without Parmesan cheese) equals 230 calories, 3 g fat (1 g saturated fat), 15 mg cholesterol, 645 mg sodium, 40 g carbohydrate, 6 g fiber, 12 g protein. **Diabetic Exchanges:** 2 starch, 2 vegetable, 1 lean meat.

Diner Meat Loaf

(Pictured below)

Prep: 20 min. **Bake:** 1 hour + standing

Homemade tomato sauce dresses up this traditional meat loaf. To bake it, I line my broiler pan with foil that has holes poked into it so the fat drains through. The meat browns beautifully. —*Joan Airey, Rivers, Manitoba*

1 large onion, finely chopped
1 celery rib, chopped
2 garlic cloves, minced
1 tablespoon olive oil
2 eggs, lightly beaten
1/3 cup ketchup
3/4 cup soft bread crumbs
1 tablespoon minced fresh parsley
1 teaspoon salt
1/2 teaspoon pepper
1/4 teaspoon ground nutmeg
2 pounds ground beef

BAKED TOMATO SAUCE:
1 can (28 ounces) diced tomatoes, undrained
1 tablespoon olive oil
2 garlic cloves, minced
1/4 teaspoon salt
1/4 teaspoon crushed red pepper flakes

In a large skillet, saute onion, celery and garlic in oil until tender; cool. In a large bowl, combine eggs, ketchup, crumbs, parsley, salt, pepper, nutmeg and onion mixture. Crumble beef over mixture; mix well.

Shape into a loaf. Place in a greased 11-in. x 7-in. x 2-in. baking dish. Bake, uncovered, at 350° for 1 to 1-1/4 hours or until no pink remains and a meat thermometer reads 160°.

Meanwhile, combine sauce ingredients; pour into an ungreased 8-in. square baking dish. Bake, uncovered, at 350° for 1 hour or until thickened, stirring occasionally. Let meat loaf stand for 10 minutes before slicing; serve with sauce. **Yield:** 8 servings.

Romano Chicken Supreme

(Pictured above)

Prep: 20 min. **Bake:** 20 min.

Plenty of Romano cheese and golden-brown bread crumbs add flavor and crunch to this tender chicken and mushroom dinner. —*Anna Minegar, Zolfo Springs, Florida*

 **Uses less fat, sugar or salt. Includes Nutrition Facts and Diabetic Exchanges.**

6 boneless skinless chicken breast halves (5 ounces *each*)
1/4 teaspoon salt
1 pound fresh mushrooms, chopped
1 tablespoon lemon juice
2 garlic cloves, minced
1 teaspoon dried basil
3 tablespoons butter
1/2 cup reduced-sodium chicken broth
2 tablespoons orange juice
1 cup soft bread crumbs
1/3 cup grated Romano cheese

In a large skillet coated with nonstick cooking spray, brown chicken on both sides over medium heat. Transfer to a 13-in. x 9-in. x 2-in. baking dish coated with nonstick cooking spray; sprinkle with salt.

In the same skillet, saute mushrooms, lemon juice, garlic and basil in butter. Stir in broth and orange juice; bring to a boil. Reduce heat; simmer, uncovered, for 2-3 minutes or until heated through. Spoon over chicken; sprinkle with crumbs and cheese.

Bake, uncovered, at 400° for 20-25 minutes or until lightly browned and the chicken juices run clear. **Yield:** 6 servings.

Nutrition Facts: 1 chicken breast half with about 1/4 cup mushroom mixture equals 274 calories, 11 g fat (6 g saturated fat), 100 mg cholesterol, 413 mg sodium, 8 g carbohydrate, 1 g fiber, 34 g protein. **Diabetic Exchanges:** 4-1/2 very lean meat, 2 fat, 1/2 starch.

Eggplant Parmigiana

(Pictured below)

Prep: 2 hours **Bake:** 35 min.

This delicious eggplant casserole from my mom makes a wonderful meatless meal. The homemade marinara sauce can't be beat. —*Valerie Belley, St. Louis, Missouri*

- **2 medium eggplant, peeled and cut into 1/2-inch slices**
- **2 teaspoons salt**
- **2 large onions, chopped**
- **2 tablespoons minced fresh basil *or* 2 teaspoons dried basil**
- **2 bay leaves**
- **1 tablespoon minced fresh oregano *or* 1 teaspoon dried oregano**
- **1 tablespoon minced fresh thyme *or* 1 teaspoon dried thyme**
- **3 tablespoons olive oil**
- **1 can (14-1/2 ounces) diced tomatoes, undrained**
- **1 can (12 ounces) tomato paste**
- **1 tablespoon honey**
- **1-1/2 teaspoons lemon-pepper seasoning**
- **4 garlic cloves, minced**
- **2 eggs, beaten**
- **1/2 teaspoon pepper**
- **1-1/2 cups dry bread crumbs**
- **1/4 cup butter, *divided***
- **2 pounds (32 ounces) shredded part-skim mozzarella cheese**
- **1 cup grated Parmesan cheese**

Place eggplant in a colander; sprinkle with salt. Let stand for 30 minutes. Meanwhile, in a large skillet, saute onions, basil, bay leaves, oregano and thyme in oil until onions are tender.

Add the tomatoes, tomato paste, honey and lemon-pepper. Bring to a boil. Reduce heat; cover and simmer for 30 minutes. Add garlic; simmer 10 minutes longer. Discard bay leaves.

Rinse the eggplant slices; pat dry with paper towels. In a shallow bowl, combine the eggs and pepper; place the bread crumbs in another shallow bowl. Dip the eggplant into the egg, then coat with the bread crumbs. Let stand for 5 minutes.

In a large skillet, cook half of the eggplant in 2 tablespoons butter for 3 minutes on each side or until lightly browned. Repeat with the remaining half of eggplant and butter.

In each of two greased 11-in. x 7-in. x 2-in. baking dishes, layer half of the eggplant, half of the tomato sauce and half of the mozzarella cheese. Repeat the layers. Sprinkle with the Parmesan cheese. Bake, uncovered, at 375° for 35 minutes or until bubbly. **Yield:** 10-12 servings.

Pecan Chicken with Chutney

Prep: 15 min. **Cook:** 20 min.

Hot pepper sauce adds zip to these tender pecan-crusted chicken breasts, and easy-to-fix peach chutney adds a sweet-sour accent. I serve this dish with rice on the side. —*Carisa Bravoco, Furlong, Pennsylvania*

- **3/4 cup all-purpose flour**
- **1/8 teaspoon salt**
- **1/8 teaspoon pepper**
- **2 eggs**
- **1/3 cup buttermilk**
- **1/8 teaspoon hot pepper sauce**
- **1 cup finely chopped pecans**
- **3/4 cup dry bread crumbs**
- **6 boneless skinless chicken breast halves (6 ounces *each*)**
- **2 tablespoons butter**
- **2 tablespoons vegetable oil**

PEACH MANGO CHUTNEY:

- **2 cups sliced peeled fresh *or* frozen peaches, thawed**
- **1 cup mango chutney**

In a shallow bowl, combine the flour, salt and pepper. In another shallow bowl, whisk the eggs, buttermilk and hot pepper sauce. In a third bowl, combine the pecans and bread crumbs. Flatten the chicken to 1/4-in. thickness. Coat the chicken with flour mixture, then dip in egg mixture and coat with pecan mixture.

In a large skillet over medium heat, cook chicken in butter and oil for 8-10 minutes on each side or until juices run clear.

Meanwhile, in a small saucepan, combine the peaches and mango chutney. Bring to a boil. Reduce the heat; simmer, uncovered, for 15-20 minutes or until heated through. Serve the chutney with the chicken. **Yield:** 6 servings (1-3/4 cups chutney).

Rosemary-Skewered Artichoke Chicken

Prep: 20 min. + marinating **Grill:** 20 min.

These attractive kabobs have a lovely, fresh herb flavor whether you choose to use the rosemary stems as skewers or not. —Lisa White, San Diego, California

Uses less fat, sugar or salt. Includes Nutrition Facts and Diabetic Exchanges.

- **1/3 cup olive oil**
- **2 tablespoons snipped fresh dill**
- **1 tablespoon minced fresh oregano**
- **2 teaspoons grated lemon peel**
- **2 garlic cloves, minced**
- **1/2 teaspoon salt**
- **1/4 teaspoon pepper**
- **1-1/2 pounds boneless skinless chicken breasts, cut into 1-inch cubes**
- **6 fresh rosemary stems (18 inches)**
- **1 can (14 ounces) water-packed artichoke hearts, rinsed, drained and halved**
- **2 medium yellow summer squash, cut into 1-inch slices**
- **6 cherry tomatoes**

In a large resealable plastic bag, combine the oil, dill, oregano, lemon peel, garlic, salt and pepper; add chicken. Seal bag and turn to coat; refrigerate for at least 2 hours.

Using a vegetable peeler, peel bark from the bottom half of each rosemary stem and make a point at each end; soak in water until ready to use.

Drain and discard marinade. On soaked rosemary stems, alternately thread the chicken, artichokes, squash and tomatoes. Position the leaf parts of the rosemary stems so they are outside of the grill cover. Grill, covered, over medium heat for 10-15 minutes on each side or until chicken juices run clear and vegetables are tender. **Yield:** 6 servings.

Nutrition Facts: 1 skewer equals 215 calories, 9 g fat (2 g saturated fat), 63 mg cholesterol, 321 mg sodium, 8 g carbohydrate, 1 g fiber, 25 g protein. **Diabetic Exchanges:** 3 very lean meat, 1 vegetable, 1 fat.

Sweet Pepper Venison Stir-Fry

(Pictured above)

Prep: 15 min. + marinating **Cook:** 10 min.

Each year, our friends have a "wild game feed" where everyone brings a different dish. This one knocked their socks off—they practically licked the wok clean! —Kathy Gasser, Waukesha, Wisconsin

- **1/4 cup cornstarch**
- **2 teaspoons sugar**
- **6 tablespoons soy sauce**
- **1/4 cup white wine vinegar**
- **1/2 teaspoon pepper**
- **1 venison tenderloin (about 1 pound), cut into 2-inch strips**
- **1 medium green pepper, julienned**
- **1 medium sweet red pepper, julienned**
- **3 tablespoons vegetable oil**
- **Hot cooked rice**

In a small bowl, combine the cornstarch, sugar, soy sauce, vinegar and pepper; stir until smooth. Pour half into a large resealable plastic bag; add venison. Seal bag and turn to coat; refrigerate for 1-2 hours. Cover and refrigerate remaining marinade.

Drain and discard marinade. In a large skillet or wok, stir-fry venison and peppers in oil for 4-6 minutes or until meat is no longer pink and peppers are crisp-tender. Stir reserved marinade; add to the pan. Bring to a boil; cook and stir for 1-2 minutes or until thickened. Serve with rice. **Yield:** 2 servings.

Lemon-Dill Chicken Popover

(Pictured above)

Prep: 30 min. **Bake:** 25 min.

Don't be surprised if this one-dish meal disappears quickly. The crisp, golden popover has a delectable filling of chicken, carrots, broccoli and more. —*Patricia Tjugum*
Tomahawk, Wisconsin

- **3 tablespoons butter, melted, *divided***
- **1 cup all-purpose flour**
- **1/2 teaspoon salt**
- **6 eggs**
- **1 cup milk**

FILLING:

- **2 medium carrots, chopped**
- **1 cup fresh broccoli florets**
- **1 medium onion, chopped**
- **1 tablespoon butter**
- **2 cups cubed cooked chicken**
- **1 can (10-3/4 ounces) condensed cream of chicken soup, undiluted**
- **1 medium sweet red pepper, diced**
- **1 tablespoon lemon juice**
- **1 teaspoon dill weed**
- **1/2 cup shredded cheddar cheese**

Brush the bottom of a deep-dish pie plate with 1 tablespoon butter; set aside. In a small mixing bowl, combine flour and salt. Beat in the eggs, milk and remaining butter until smooth. Pour into prepared pie plate.

Bake at 400° for 20 minutes. Reduce heat to 350°; bake 5-10 minutes longer or until golden brown and center is set.

In a microwave-safe bowl, combine carrots, broccoli, onion and butter. Cover and microwave on high for 5-10 minutes or until vegetables are crisp-tender; set aside. In another microwave-safe bowl, combine chicken, soup, red pepper, lemon juice and dill. Cover and microwave on high for 3-4 minutes or until red pepper is tender, stirring once. Stir in cheese and half of the vegetable mixture.

Spoon chicken mixture into center of popover. Surround with remaining vegetable mixture. Cut into wedges; serve immediately. **Yield:** 6 servings.

Editor's Note: This recipe was tested in a 1,100-watt microwave.

Stuffed Flank Steak

Prep: 15 min. **Bake:** 50 min.

Guests always admire the pretty spiral slices and flavor of this succulent entree. A moist bread stuffing dresses up the tender flank steak. —*Adelaide Muldoon*
Springfield, Virginia

✓ **Uses less fat, sugar or salt. Includes Nutrition Facts and Diabetic Exchanges.**

- **1 small onion, finely chopped**
- **1 tablespoon butter**
- **3 cups soft bread crumbs**
- **1/2 to 3/4 teaspoon poultry seasoning**
- **1/2 teaspoon salt**
- **1/4 teaspoon pepper, *divided***
- **1 beef flank steak (1-1/2 pounds)**
- **2 teaspoons all-purpose flour**
- **1 cup reduced-sodium beef broth**

In a small nonstick skillet, saute the onion in the butter until tender. In a large bowl, combine the soft bread crumbs, poultry seasoning, salt, 1/8 teaspoon pepper and sauteed onion.

Flatten steak to 1/2-in. thickness; spread with stuffing to within 1 in. of edges. Roll up jelly-roll style, starting with a long side; tie with kitchen string. Rub with remaining pepper.

Place in an 11-in. x 7-in. x 2-in. baking dish coated with nonstick cooking spray. Bake, uncovered, at 350° for 50-55 minutes or until meat is tender.

Remove the meat and discard the string. Cut into slices and keep warm. Skim the fat from the pan juices; pour into a small saucepan. Combine the flour and broth until smooth; stir into the juices. Bring to a boil; cook and stir for 1-2 minutes or until thickened. Serve with the steak. **Yield:** 6 servings.

Nutrition Facts: 1 serving equals 249 calories, 12 g fat (5 g saturated fat), 54 mg cholesterol, 460 mg sodium, 13 g carbohydrate, 1 g fiber, 21 g protein. **Diabetic Exchanges:** 3 lean meat, 1 starch.

Garden Vegetable Quiche

Prep: 20 min. **Bake:** 40 min. + standing

Make your next weekend or holiday brunch extra special with this deep-dish quiche. Fresh rosemary enhances the delightful egg bake, which is chock-full of wholesome garden ingredients. It cuts nicely, too. —Kristina Ledford Indianapolis, Indiana

- **1 unbaked deep-dish pastry shell (9 inches)**
- **1 small red onion, sliced**
- **1/2 cup sliced fresh mushrooms**
- **1/4 cup diced yellow summer squash**
- **3 garlic cloves, minced**
- **1 tablespoon butter**
- **1/2 cup fresh baby spinach**
- **1 cup (4 ounces) shredded Swiss cheese**
- **4 eggs**
- **1-2/3 cups heavy whipping cream**
- **1/2 teaspoon salt**
- **1/2 teaspoon minced fresh rosemary**
- **1/4 teaspoon pepper**

Line unpricked pastry shell with a double thickness of heavy-duty foil. Bake at 450° for 8 minutes. Remove foil; bake 5 minutes longer. Cool on a wire rack. Reduce heat to 350°.

In a large skillet, saute the onion, mushrooms, squash and garlic in butter until tender. Add the spinach; saute 1 minute longer. Spoon into the baked crust; top with the Swiss cheese. In a bowl, whisk the eggs, whipping cream, salt, rosemary and pepper until blended; pour over cheese.

Cover edges of crust loosely with foil. Bake for 40-45 minutes or until a knife inserted near the center comes out clean. Let stand for 10 minutes before cutting. **Yield:** 6-8 servings.

Very Rosemary

Long before rosemary became popular in the kitchen, this pungent herb was prized for its "health-related" powers, such as improving memory and relieving headaches. It was also believed that rosemary could protect against evil spirits.

Cooks today turn to rosemary for its powerful, pine-like flavor that complements vegetables, breads, eggs, meats, soups, cream sauces and dishes such as Garden Vegetable Quiche (above).

Use rosemary fresh when possible. Sprigs will keep for a week or longer when the stems are immersed in water. The fresh leaves can also be stored in the freezer.

Seafood Au Gratin

(Pictured below)

Prep: 30 min. **Bake:** 15 min.

A seafood casserole is a "must" for my buffets. My father was a fisherman, so we ate fish almost every day. Over the years, I've tried many seafood dishes, but none better than this one. —Hazel McMullin, Amherst, Nova Scotia

- **4 tablespoons butter, *divided***
- **2 tablespoons all-purpose flour**
- **1/8 teaspoon pepper**
- **1 cup chicken broth**
- **1/2 cup milk**
- **1/2 cup grated Parmesan cheese, *divided***
- **1/2 pound sea scallops**
- **1 pound haddock *or* cod fillets, cut into six pieces**
- **1-1/2 cups sliced fresh mushrooms**
- **1/2 cup shredded part-skim mozzarella cheese**
- **1/2 cup shredded cheddar cheese**

In a large saucepan, melt 2 tablespoons butter. Stir in flour and pepper until smooth; gradually add broth and milk. Bring to a boil; cook and stir for 2 minutes or until thickened. Stir in 1/4 cup Parmesan cheese; set aside.

Place scallops in another saucepan; cover with water. Simmer, uncovered, for 4-5 minutes or until firm and opaque. Meanwhile, place fillets in a shallow 2-qt. microwave-safe dish. Cover and microwave on high for 2-4 minutes or until fish flakes easily with a fork. Drain scallops. Arrange fish and scallops in a greased 11-in. x 7-in. x 2-in. baking dish.

In a small skillet, saute mushrooms in remaining butter until tender; stir into cheese sauce. Spoon over seafood. Sprinkle with mozzarella, cheddar and remaining Parmesan cheese. Cover and bake at 350° for 15-20 minutes or until bubbly and cheese is melted. **Yield:** 6 servings.

Meatball Calzones

(Pictured above)

Prep: 1-1/2 hours + standing **Bake:** 25 min.

My family can't get enough of this savory entree—we have to have it at least once a month! Leftovers freeze well for a quick meal later. —*Cori Cooper, Flagstaff, Arizona*

3 eggs, lightly beaten
1 cup seasoned bread crumbs
1 cup grated Parmesan cheese
3 teaspoons Italian seasoning
2 pounds ground beef
3 loaves (1 pound *each*) frozen bread dough, thawed
3 cups (12 ounces) shredded part-skim mozzarella cheese
1 egg white, lightly beaten
Additional Italian seasoning
1 jar (14 ounces) spaghetti sauce, warmed

In a large bowl, combine the eggs, bread crumbs, Parmesan cheese and Italian seasoning. Crumble beef over mixture and mix well. Shape into 1-in. balls.

Place the meatballs on a rack in a shallow baking pan. Bake, uncovered, at 400° for 10-15 minutes or until no longer pink. Drain on paper towels. Reduce the heat to 350°.

On a floured surface, roll each portion of dough into an 18-in. x 12-in. rectangle. Spoon a third of the meatballs and mozzarella cheese down the center of each rectangle. Fold dough over filling; press edges firmly to seal.

Place seam side down on greased baking sheets. Brush tops with egg white; sprinkle with Italian seasoning. Let stand for 15-30 minutes. Bake for 25-30 minutes or until golden brown. Serve with spaghetti sauce. **Yield:** 3 calzones (4 servings each).

Seasoned Crab Cakes

(Pictured below)

Prep: 20 min. + chilling **Cook:** 10 min.

At the National Hard Crab Derby in Chrisfield, Maryland, these scrumptious, golden patties won first place. I entered them on a whim after trying many crab cake recipes for my family. We think these are unbeatable.
—*Betsy Hedeman, Timonium, Maryland*

3 cans (6 ounces *each*) crabmeat, drained, flaked and cartilage removed
1 cup cubed bread
2 eggs
3 tablespoons mayonnaise
3 tablespoons half-and-half cream
1 tablespoon lemon juice
1 tablespoon butter, melted
1-1/2 teaspoons seafood seasoning
1 teaspoon Worcestershire sauce
1 teaspoon salt
1/2 cup dry bread crumbs
1/2 cup vegetable oil

In a large bowl, combine crab and bread cubes. In another bowl, whisk the eggs, mayonnaise, cream, lemon juice, butter, seafood seasoning, Worcestershire sauce and salt. Add to crab mixture and mix gently (mixture will be moist).

Place bread crumbs in a shallow dish. Drop crab mixture by 1/3 cupfuls into crumbs; shape each into a 3/4-in.-thick patty. Carefully turn to coat. Cover and refrigerate for at least 2 hours.

In a large skillet, cook crab cakes in oil for 4-5 minutes on each side or until golden brown and crispy. **Yield:** 8 crab cakes.

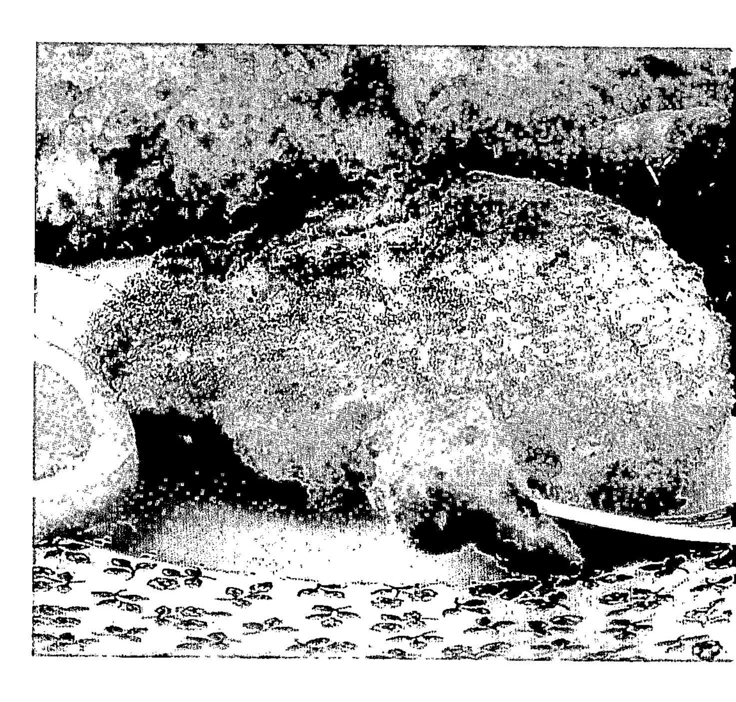

Beef and Spinach Lasagna

(Pictured above)

Prep: 10 min. **Bake:** 40 min. + standing

Using no-cook noodles gives you a jump start on assembling this popular main dish. It cuts nicely after standing a few minutes, revealing luscious layers.
—Carolyn Schmeling, Brookfield, Wisconsin

✓ **Uses less fat, sugar or salt. Includes Nutrition Facts.**

- **1 pound ground beef**
- **1 medium onion, chopped**
- **2 jars (26 ounces *each*) meatless spaghetti sauce**
- **4 garlic cloves, minced**
- **1 teaspoon dried basil**
- **1 teaspoon dried oregano**
- **1 package (10 ounces) frozen chopped spinach, thawed and squeezed dry**
- **2 cups ricotta cheese**
- **2 cups (8 ounces) shredded part-skim mozzarella cheese, *divided***
- **9 no-cook lasagna noodles**

In a large skillet, cook beef and onion over medium heat until meat is no longer pink; drain. Stir in the spaghetti sauce, garlic, basil and oregano. Bring to a boil. Reduce heat; cover and simmer for 10 minutes. In a bowl, combine the spinach, ricotta and 1 cup mozzarella until combined.

Spread 1-1/2 cups meat sauce into a greased 13-in. x 9-in. x 2-in. baking dish. Top with three noodles. Spread 1-1/2 cups sauce to edges of noodles. Top with half of the spinach mixture. Repeat layers. Top with remaining noodles, sauce and mozzarella.

Cover and bake at 375° for 30 minutes. Uncover; bake 10-15 minutes longer or until bubbly. Let stand for 10 minutes before cutting. **Yield:** 12 servings.

Nutrition Facts: 1 piece (prepared with lean ground beef) equals 281 calories, 11 g fat (6 g saturated fat), 50 mg cholesterol, 702 mg sodium, 26 g carbohydrate, 3 g fiber, 20 g protein.

Curried Chicken with Peaches

Prep: 15 min. **Cook:** 3-1/4 hours

This chicken is slow-cooked in a curry-ginger sauce, and peaches round out the amazing blend of flavors.
—Heidi Martinez, Colorado Springs, Colorado

- **1 broiler/fryer chicken (3 pounds), cut up**
- **1/8 teaspoon salt**
- **1/8 teaspoon pepper**
- **1 can (29 ounces) sliced peaches**
- **1/2 cup chicken broth**
- **2 tablespoons butter, melted**
- **1 tablespoon dried minced onion**
- **2 teaspoons curry powder**
- **2 garlic cloves, minced**
- **1/4 teaspoon ground ginger**
- **3 tablespoons cornstarch**
- **3 tablespoons cold water**
- **1/4 cup raisins**
- **Toasted flaked coconut, optional**

Place chicken in a 5-qt. slow cooker; sprinkle with salt and pepper. Drain peaches, reserving 1/2 cup juice; set peaches aside. In a small bowl, combine the broth, butter, onion, curry, garlic, ginger and reserved juice; pour over chicken. Cover and cook on low for 3-4 hours or until chicken juices run clear.

Remove chicken and keep warm. In a small bowl, combine cornstarch and water until smooth; stir into the cooking juices. Add raisins. Cover and cook on high for 10-15 minutes or until thickened. Stir in peaches; heat through. Serve over chicken. Garnish with coconut if desired. **Yield:** 4 servings.

Cheese Choice

Like cottage cheese, ricotta cheese (used in the Beef and Spinach Lasagna recipe at left) is considered a "fresh" or unripened cheese. Both cheeses are soft, mild in flavor and usually sold in a round plastic container.

Ricotta has a fine, moist and grainy texture, while cottage cheese—whether the curds are small or large—is "lumpier" and has more liquid. Although the two cheeses can be used interchangeably in some recipes, the recipe often dictates which is the better choice.

Herbed Turkey Breasts

(Pictured below)

Prep: 15 min. + marinating **Cook:** 3-1/2 hours

These slow-cooked turkey breasts simmer for several hours in a well-seasoned marinade, resulting in remarkable flavor. —Laurie Mace, Los Osos, California

2 cans (14-1/2 ounces *each*) chicken broth
1 cup lemon juice
1/2 cup packed brown sugar
1/2 cup fresh sage
1/2 cup minced fresh thyme
1/2 cup lime juice
1/2 cup cider vinegar
1/2 cup olive oil
2 envelopes onion soup mix
1/4 cup Dijon mustard
2 tablespoons minced fresh marjoram
1 tablespoon paprika
2 teaspoons garlic powder
2 teaspoons pepper
1 teaspoon salt
2 boneless turkey breasts (2 pounds *each*)

In a blender, process the first 15 ingredients in batches until blended. Pour 3-1/2 cups marinade into a large resealable plastic bag; add the turkey. Seal bag and turn to coat; refrigerate for 8 hours or overnight. Cover and refrigerate remaining marinade.

Drain and discard the marinade. Place the turkey in a 5-qt. slow cooker; add the reserved marinade. Cover and cook on high for 3-1/2 to 4-1/2 hours or until the juices run clear and a meat thermometer reads 170°. **Yield:** 14-16 servings.

Slow-Cooked Beef Brisket

(Pictured above)

Prep: 10 min. **Cook:** 8-1/2 hours

When my husband and I were working full-time, we loved coming home to this mushroom-topped brisket. —Anna Stodolak, Volant, Pennsylvania

1 large onion, sliced
1 fresh beef brisket (3 to 4 pounds), cut in half
1/4 teaspoon pepper
1 jar (4-1/2 ounces) sliced mushrooms, drained
3/4 cup beef broth
1/2 cup chili sauce
1/4 cup packed brown sugar
2 garlic cloves, minced
1/4 cup all-purpose flour
1/4 cup cold water

Place the onion in a 5-qt. slow cooker. Rub the brisket with the pepper; place over the onion. Top with the mushrooms. In a small bowl, combine the beef broth, chili sauce, brown sugar and garlic; pour over the brisket. Cover and cook on low for 8-9 hours or until the meat is tender.

Remove the brisket and keep warm. In a small bowl, combine the flour and water until smooth; stir into the cooking juices. Cover and cook on high for 30 minutes or until thickened. Slice brisket; serve with gravy. **Yield:** 6-8 servings.

Editor's Note: This is a fresh beef brisket, not corned beef. The meat comes from the first cut of the brisket.

Marinated Flank Steak

Prep: 10 min. + marinating **Grill:** 15 min.

An Oriental-style marinade accents this grilled flank steak. It's an ideal summer entree, but you can enjoy it year-round...just broil it indoors instead. —*Isabel Fowler*
Fairbanks, Alaska

Uses less fat, sugar or salt. Includes Nutrition Facts and Diabetic Exchanges.

- **1 beef flank steak (2 pounds)**
- **1/2 cup canola oil**
- **1/2 cup reduced-sodium soy sauce**
- **1/4 cup red wine vinegar**
- **2 tablespoons water**
- **2 teaspoons brown sugar**
- **2 teaspoons minced fresh gingerroot**
- **2 garlic cloves, minced**
- **1/4 teaspoon pepper**

Score surface of the steak, making diamond shapes 1/2 in. deep. In a large resealable plastic bag, combine the remaining ingredients; add the steak. Seal the bag and turn to coat; refrigerate for at least 2 hours, turning occasionally.

Drain and discard marinade. Grill steak, covered, over medium heat for 6-8 minutes on each side or until meat reaches desired doneness (for medium-rare, a meat thermometer should read 145°; medium, 160°; well-done, 170°). To serve, thinly slice across the grain. **Yield:** 8 servings.

Nutrition Facts: 3 ounces cooked beef equals 204 calories, 12 g fat (4 g saturated fat), 54 mg cholesterol, 219 mg sodium, 1 g carbohydrate, trace fiber, 22 g protein. **Diabetic Exchanges:** 3 lean meat, 1 fat.

New Orleans-Style Chicken

(Pictured at right)

Prep: 30 min. **Cook:** 50 min.

This hearty, one-dish meal is loaded with tender chunks of chicken, colorful veggies, beans and rice. It's a nutritious entree that's a favorite of mine. —*Jason Bagley*
Kennewick, Washington

Uses less fat, sugar or salt. Includes Nutrition Facts.

- **1-1/4 pounds boneless skinless chicken breasts, cubed**
- **3 teaspoons canola oil, *divided***
- **2 medium carrots, chopped**
- **1 large onion, chopped**
- **1 medium sweet red pepper, chopped**
- **1 medium green pepper, chopped**
- **2 portobello mushrooms (3 ounces *each*), chopped**
- **3 garlic cloves, minced**
- **2-3/4 cups hot water**
- **1 can (15 ounces) black beans, rinsed and drained**
- **1 package (8 ounces) red beans and rice mix**
- **1 can (14-1/2 ounces) diced tomatoes, drained**
- **1/3 cup shredded Asiago cheese**

In a large nonstick skillet, brown the chicken in 1 teaspoon oil over medium-high heat; remove chicken and set aside. In the same skillet, saute the carrots, onion and peppers in the remaining oil for 10 minutes. Add the mushrooms and garlic; saute 1-2 minutes longer or until the vegetables are tender.

Stir in the water, black beans and red beans and rice mix. Return chicken to the pan; bring to a boil. Reduce heat; cover and simmer for 30-35 minutes or until liquid is absorbed and rice is tender. Stir in tomatoes; heat through. Just before serving, sprinkle with cheese. **Yield:** 6 servings.

Nutrition Facts: 1-1/3 cups equals 381 calories, 7 g fat (2 g saturated fat), 58 mg cholesterol, 891 mg sodium, 50 g carbohydrate, 9 g fiber, 30 g protein.

Editor's Note: This recipe was prepared with Zatarain's New Orleans-style red beans and rice.

Breads, Rolls & Muffins

You can practically smell the aroma of fresh-baked braids, coffee cakes, buns and other treats when you page through this delightful chapter.

GOLDEN GOODIES. Clockwise from upper left: Cream Cheese Carrot Muffins (p. 83), White Chocolate Cranberry Bread (p. 86), Cream Cheese Cranberry Muffins (p. 87), Celebration Braid (p. 85), Maple Walnut Rolls (p. 84) and Rich Coffee Cake (p. 83).

Christmas Morning Croissants

(Pictured below)

Prep: 50 min. + chilling **Bake:** 20 min.

Growing up in France, we often enjoyed buttery croissants for breakfast with steaming cups of hot chocolate. I tried to re-create the experience for my family with this recipe, and now it's a Christmas tradition. —*Tish Stevenson Grand Rapids, Michigan*

2 packages (1/4 ounce *each*) active dry yeast
1 cup warm water (110° to 115°)
1-1/4 cups cold butter, *divided*
5 cups all-purpose flour
1/3 cup sugar
1-1/2 teaspoons salt
3/4 cup evaporated milk
2 eggs
1 tablespoon water

In a large mixing bowl, dissolve yeast in warm water; let stand for 5 minutes. Melt 1/4 cup butter; set aside. Combine 1 cup flour, sugar and salt; add to yeast mixture. Add the milk, 1 egg and melted butter; beat until smooth.

Place remaining flour in a large bowl; cut in remaining butter until crumbly. Add yeast mixture; mix well. Do not knead. Cover and refrigerate overnight.

Punch dough down. Turn onto a lightly floured surface; knead about six times. Divide dough into four pieces. Roll each piece into a 16-in. circle; cut each circle into eight wedges.

Roll up wedges from the wide end and place point side down 3 in. apart on ungreased baking sheets. Curve ends to form a crescent shape. Cover and let rise in a warm place for 1 hour.

Beat water and remaining egg; brush over rolls. Bake at 325° for 20-25 minutes or until lightly browned. Serve warm. **Yield:** 32 rolls.

Featherlight Rolls

Prep: 25 min. + rising **Bake:** 15 min.

Aptly named for their featherlight texture, these delectable treats from my mother were a favorite of family and friends alike. The rolls are just as tasty the next day with a slice of cheese on each section of the cloverleaf. —*Terri Duhon, Bryan, Texas*

2 packages (1/4 ounce *each*) active dry yeast
1/2 cup warm water (110° to 115°)
1 tablespoon plus 1/3 cup sugar, *divided*
1 cup warm milk (110° to 115°)
1/3 cup shortening
2 eggs
1-1/2 teaspoons salt
4 to 5 cups all-purpose flour

In a large mixing bowl, dissolve yeast in warm water. Stir in 1 tablespoon sugar; let stand for 5 minutes. Add the milk, shortening, eggs, salt, 3 cups flour and remaining sugar. Beat on medium speed for 2 minutes.

Stir in enough remaining flour to form a soft dough (mixture will be sticky). Do not knead. Cover and refrigerate overnight.

Punch dough down. Turn onto a lightly floured surface; divide into 24 portions. Divide each portion into three pieces; shape each into a ball. Place three balls in each greased muffin cup. Cover and let rise in a warm place until doubled, about 1-3/4 hours.

Bake at 350° for 13-15 minutes or until golden brown. Remove from pans to wire racks. Serve warm. **Yield:** 2 dozen.

Storing Yeast

Store unopened packages of dry yeast in a cool, dark and dry place. You should use dry yeast by the "best if used by" date that is printed on the package.

Opened packages of bulk dry yeast should be stored in an airtight container in the refrigerator for about 6 weeks...or in the freezer for up to 6 months.

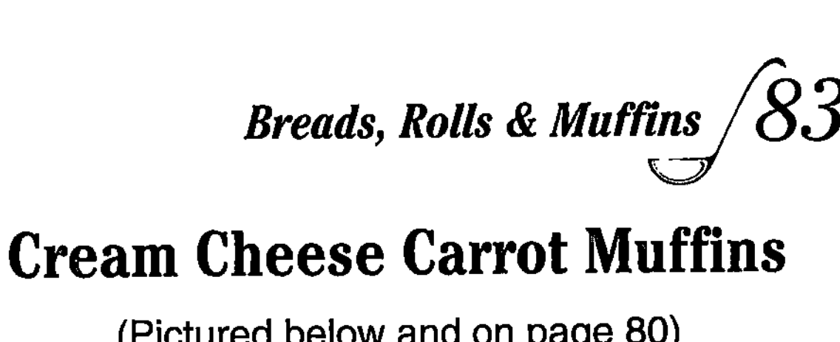

Rich Coffee Cake

(Pictured above and on page 80)

Prep: 30 min. **Bake:** 1 hour + cooling

Cinnamon and pecans are swirled throughout this tender, golden cake. Drizzled with chocolate, it's sure to please sweet tooths. —*Gaytha Holloway, Marion, Indiana*

1 cup butter, softened
2 cups sugar
2 eggs
2 cups all-purpose flour
1-1/2 teaspoons baking powder
1/2 teaspoon salt
1 cup (8 ounces) sour cream
1/2 teaspoon vanilla extract

TOPPING:

1 cup chopped pecans
2 tablespoons sugar
1 teaspoon ground cinnamon

CHOCOLATE GLAZE:

1/2 cup semisweet chocolate chips
1/4 cup butter, cubed

In a large mixing bowl, cream butter and sugar. Add eggs, one at a time, beating well after each addition. Combine the flour, baking powder and salt. Combine sour cream and vanilla; add to creamed mixture alternately with dry ingredients.

Combine topping ingredients; sprinkle 2 tablespoons into a greased and floured 10-in. tube pan. For glaze, in a microwave-safe bowl, melt chocolate chips and butter; stir until smooth. Spoon half of the batter over topping; sprinkle with half of the remaining topping. Drizzle with half of the glaze. Top with remaining batter; sprinkle with remaining topping.

Bake at 350° for 60-70 minutes or until toothpick inserted near the center comes out clean. Cool for 10 minutes before removing from pan to a wire rack.

Warm remaining glaze; drizzle over coffee cake. Serve warm if desired. **Yield:** 12 servings.

Cream Cheese Carrot Muffins

(Pictured below and on page 80)

Prep: 25 min. **Bake:** 20 min.

I revised a pumpkin bread recipe to make these yummy spiced muffins. Pureed canned carrots and a cream cheese filling make them wonderfully moist.
—*Francy Schneidecker, Tillamook, Oregon*

1 can (14-1/2 ounces) sliced carrots, drained
1-3/4 cups all-purpose flour
1 cup sugar
1-1/4 teaspoons baking soda
1/2 teaspoon salt
1/2 teaspoon ground cinnamon
1/8 teaspoon *each* ground allspice, cloves and nutmeg
1 egg
1/3 cup vegetable oil

FILLING:

1 package (8 ounces) cream cheese, softened
1 egg
1/4 cup sugar

Place carrots in a food processor; cover and process until smooth. In a large bowl, combine the flour, sugar, baking soda, salt and spices. In a small bowl, whisk the pureed carrots, egg and oil; stir into the dry ingredients just until moistened.

Fill 12 greased muffin cups one-third full. In a small mixing bowl, beat the filling ingredients until smooth. Drop by tablespoonfuls into the center of each muffin. Top with remaining batter.

Bake at 350° for 20-25 minutes or until a toothpick comes out clean. Cool for 5 minutes before removing from pan to a wire rack. Serve warm. **Yield:** 1 dozen.

Maple Walnut Rolls

(Pictured above and on page 80)

Prep: 45 min. + rising **Bake:** 20 min. + cooling

For old-fashioned taste, you can't beat this attractive nut bread. The recipe makes four rolls, so you'll have enough for family and friends all through the Christmas holiday season. —*Elleen Oberreuter, Danbury, Iowa*

6 to 7 cups all-purpose flour
3 tablespoons sugar
2 packages (1/4 ounce *each*) active dry yeast
1 teaspoon salt
1 cup (8 ounces) sour cream
1 cup butter, softened
1/2 cup water
3 eggs, lightly beaten

FILLING:

3/4 cup butter, melted
1/2 cup sugar
3 tablespoons maple syrup
5 cups ground walnuts

ICING:

2 cups confectioners' sugar
2 to 3 tablespoons milk

In a large mixing bowl, combine 2 cups flour, sugar, yeast and salt. In a small saucepan, heat sour cream, butter and water to 120°-130°; add to dry ingredients. Beat on medium speed for 2 minutes. Add eggs and 1/2 cup flour; beat 2 minutes longer. Stir in enough remaining flour to form a soft dough.

Turn onto a floured surface; knead until smooth and elastic, about 6-8 minutes. Place in a greased bowl; turn once to grease top. Cover and let rise in a warm place until doubled, about 1-1/4 hours. Punch dough down; divide into four portions. Roll each portion into a 14-in. x 12-in. rectangle.

In a large bowl, combine the butter, sugar and syrup; stir in the walnuts. Sprinkle 1 cup over each rectangle. Roll up each, jelly-roll style, starting with a long side; pinch the seams to seal. Place seam side down on greased baking sheets. Cover; let rise in a warm place until doubled, about 45 minutes.

Bake at 350° for 20-25 minutes or until lightly browned. Remove from pans to wire racks to cool. Combine the icing ingredients; drizzle over the rolls. **Yield:** 4 rolls (14 slices each).

Vanilla Cinnamon Rolls

Prep: 30 min. + rising **Bake:** 20 min.

This is the best recipe I've ever found for cinnamon rolls. Nice and tender, they have delightful vanilla flavor and a yummy frosting. Every time I bake them, they disappear in a flash. —*Linda Martin, Warsaw, Indiana*

2 cups cold milk
1 package (3.4 ounces) instant vanilla pudding mix
2 packages (1/4 ounce *each*) active dry yeast
1/2 cup warm water (110° to 115°)
1/2 cup plus 2 tablespoons butter, melted, *divided*
2 eggs
2 tablespoons sugar
1 teaspoon salt
6 cups all-purpose flour
1/2 cup packed brown sugar
1 teaspoon ground cinnamon

FROSTING:

1 cup packed brown sugar
1/2 cup heavy whipping cream
1/2 cup butter, cubed
2 cups confectioners' sugar

In a bowl, whisk milk and pudding mix for 2 minutes; set aside. In a large mixing bowl, dissolve yeast in warm water. Add 1/2 cup butter, eggs, sugar, salt and 2 cups flour. Beat on medium speed for 3 minutes. Add pudding; beat until smooth. Stir in enough remaining flour to form a soft dough (dough will be sticky).

Turn onto a floured surface; knead until smooth and elastic, about 6-8 minutes. Place in a greased bowl, turning once to grease top. Cover and let rise in a warm place until doubled, about 1 hour.

Punch dough down. Turn onto a floured surface; divide in half. Roll each portion into an 18-in. x 11-in. rectangle; brush with remaining butter. Combine brown sugar and cinnamon; sprinkle over dough to within 1/2 in. of edges.

Roll up each rectangle jelly-roll style, starting with a long side; pinch seam to seal. Cut each into 16 slices. Place cut side down in two greased 13-in. x 9-in. x 2-in. baking dishes. Cover and let rise until doubled, about 30 minutes. Bake at 350° for 20-25 minutes or until golden brown.

Meanwhile, in a large saucepan, combine the brown sugar, cream and butter. Bring to a boil; cook and stir for 2 minutes. Remove from the heat. Beat in confectioners' sugar with a hand mixer until creamy. Frost rolls. Serve warm. **Yield:** 32 rolls.

Special Banana Nut Bread

Prep: 25 min. **Bake:** 1 hour + cooling

I like to give this glazed banana bread to my friends and neighbors. The recipe yields two loaves, so I can serve one and freeze the other in case I need a last-minute gift.

—Beverly Sprague, Catonsville, Maryland

- **3/4 cup butter, softened**
- **1 package (8 ounces) cream cheese, softened**
- **2 cups sugar**
- **2 eggs**
- **1-1/2 cups mashed ripe bananas (about 4 medium)**
- **1/2 teaspoon vanilla extract**
- **3 cups all-purpose flour**
- **1/2 teaspoon baking powder**
- **1/2 teaspoon baking soda**
- **1/2 teaspoon salt**
- **2 cups chopped pecans, *divided***

ORANGE GLAZE:

- **1 cup confectioners' sugar**
- **3 tablespoons orange juice**
- **1 teaspoon grated orange peel**

In a large mixing bowl, cream the butter, cream cheese and sugar until light and fluffy. Add eggs, one at a time, beating well after each addition. Add bananas and vanilla; mix well. Combine the flour, baking powder, baking soda and salt; add to creamed mixture. Fold in 1 cup pecans.

Transfer to two greased 8-in. x 4-in. x 2-in. loaf pans. Sprinkle with remaining pecans. Bake at 350° for 1 to 1-1/4 hours or until a toothpick inserted near the center comes out clean.

In a small bowl, whisk the glaze ingredients; drizzle over loaves. Cool for 10 minutes before removing from pans to wire racks. **Yield:** 2 loaves.

Celebration Braid

(Pictured below and on page 81)

Prep: 35 min. + rising **Bake:** 20 min. + cooling

During the holiday season, I sometimes bake a couple of these golden loaves a day. Everyone in our family loves them any time of year. The recipe originated with one for Jewish challah, which I began making years ago.

—Marcia Vermaire, New Era, Michigan

- **2 packages (1/4 ounce *each*) active dry yeast**
- **1 cup warm water (110° to 115°)**
- **1/3 cup butter, softened**
- **1/4 cup sugar**
- **1 teaspoon salt**
- **2 eggs**
- **4-1/2 to 5 cups all-purpose flour**
- **1 egg yolk**
- **1 tablespoon cold water**

In a large mixing bowl, dissolve yeast in warm water. Add the butter, sugar, salt, eggs and 3 cups flour. Beat on medium speed for 3 minutes. Stir in enough remaining flour to form a soft dough.

Turn onto a floured surface; knead until smooth and elastic, about 6-8 minutes. Place in a greased bowl, turning once to grease top. Cover and let rise in a warm place until doubled, about 1 hour.

Punch dough down. Turn onto a lightly floured surface; divide into four pieces. Shape each piece into an 18-in. rope. Place ropes parallel to each other on a greased baking sheet. Beginning from the right side, braid dough by placing the first rope over the second rope, under the third and over the fourth. Repeat three or four times, beginning each time from the right side. Pinch ends to seal and tuck under.

Cover and let rise until doubled, about 45 minutes. Beat egg yolk and water; brush over braid. Bake at 350° for 20-25 minutes or until golden brown. Remove from pan to a wire rack to cool. **Yield:** 1 loaf.

Merry Cranberries

WITH their ruby-red color and burst of tartness, cranberries are perennial favorites at Christmastime. The tangy and festive treats here are sure to brighten up your holiday season.

Cranberry Almond Coffee Cake

(Pictured below)

Prep: 20 min. **Bake:** 45 min. + cooling

The berries add a tongue-tingling accent to this coffee cake. It's a Christmas morning tradition in my family.
—Anne Keenan, Nevada City, California

- **1/2 cup almond paste**
- **6 tablespoons butter, softened**
- **1/2 cup plus 2 tablespoons sugar, *divided***
- **3 eggs**
- **1-1/3 cups all-purpose flour, *divided***
- **1 teaspoon baking powder**
- **1 teaspoon almond extract**
- **1/2 teaspoon vanilla extract**
- **2-1/4 cups fresh *or* frozen cranberries**

In a small mixing bowl, cream almond paste, butter and 1/2 cup sugar until fluffy. Add two eggs, beating well after each addition. Combine 1 cup flour and baking powder; add to creamed mixture. Beat in remaining egg and flour. Stir in extracts. Gently fold in berries.

Spread evenly into a greased 8-in. square baking dish; sprinkle with remaining sugar. Bake at 325° for 45-55 minutes or until a toothpick inserted near center comes out clean. Cool on a wire rack. **Yield:** 9 servings.

White Chocolate Cranberry Bread

(Pictured at right and on page 81)

Prep: 15 min. **Bake:** 55 min. + cooling

This quick bread has yummy vanilla-citrus flavor in addition to the tangy cranberries. The fine texture is similar to a pound cake's. *—Ruth Burrus, Zionsville, Indiana*

- **1/2 cup butter-flavored shortening**
- **1 cup sugar**
- **3 eggs**
- **1/2 cup buttermilk**
- **3 tablespoons orange juice**
- **1 teaspoon grated lemon peel**
- **1 teaspoon vanilla extract**
- **1/2 cup vanilla *or* white chips, melted and cooled**
- **2-1/4 cups all-purpose flour**
- **1/2 teaspoon salt**
- **1/4 teaspoon baking soda**
- **1 cup dried cranberries**

In a large mixing bowl, cream the shortening and sugar. Add the eggs, one at a time, beating well after each addition. Beat in the buttermilk, orange juice, lemon peel and vanilla extract. Stir in the melted and cooled vanilla chips. Combine the flour, salt and baking soda; gradually add to the creamed mixture. Stir in the dried cranberries.

Pour into a greased and floured 9-in. x 5-in. x 3-in. loaf pan. Bake at 350° for 55-60 minutes or until a toothpick inserted near the center comes out clean. Cool for 10 minutes before removing from pan to a wire rack to cool completely. **Yield:** 1 loaf.

Cranberry Creativity

A mixture of sugared fresh cranberries and sugared strips of orange peel can make a pretty, festive garnish for cheesecakes and other desserts during the Christmas season. To add this elegant garnish to your treat, just use this simple method:

Combine 1/3 cup cranberries, several orange peel strips and 1/2 cup sugar. Stir the ingredients gently to combine them. Cover and refrigerate them for 1 hour.

Just before serving, arrange the mixture of sugared cranberries and orange peel strips on top or alongside of your dessert.

BAKED to a golden brown and dotted with sweet-tart goodness, tender White Chocolate Cranberry Bread and Cream Cheese Cranberry Muffins (shown above) make lovely, jewel-toned treats for breakfast or brunch.

Cream Cheese Cranberry Muffins

(Pictured above and on page 81)

Prep: 15 min. **Bake:** 20 min. per batch

Moist and chock-full of pecans, these berry-dotted muffins are a specialty of mine during the holiday season. They also freeze very well...but you probably won't have any leftovers! —*Leonard Keszler, Bismarck, North Dakota*

1 cup butter, softened
1 package (8 ounces) cream cheese, softened
1-1/2 cups sugar
1-1/2 teaspoons vanilla extract
4 eggs
2 cups all-purpose flour
1-1/2 teaspoons baking powder
1/2 teaspoon salt
2 cups fresh *or* frozen cranberries
1/2 cup chopped pecans

In a large mixing bowl, cream the butter, cream cheese, sugar and vanilla. Add eggs, one at a time, beating well after each addition. Combine the flour, baking powder and salt; stir into creamed mixture just until moistened. Fold in cranberries and pecans.

Fill greased or paper-lined muffin cups three-fourths full. Bake at 350° for 20-25 minutes or until a toothpick comes out clean. Cool the muffins for 5 minutes before removing from pans to wire racks. Serve warm. **Yield:** 2 dozen.

Black Raspberry Bubble Ring

(Pictured above)

Prep: 35 min. + rising **Bake:** 25 min.

I first made this as a 4-H project. It helped me win Grand Champion at the fair. —Kila Frank, Reedsville, Ohio

- **1 package (1/4 ounce) active dry yeast**
- **1/4 cup warm water (110° to 115°)**
- **1 cup warm milk (110° to 115°)**
- **1/4 cup plus 2 tablespoons sugar, *divided***
- **1/2 cup butter, melted, *divided***
- **1 egg**
- **1 teaspoon salt**
- **4 cups all-purpose flour**
- **1 jar (10 ounces) seedless black raspberry preserves**

SYRUP:

- **1/3 cup corn syrup**
- **2 tablespoons butter, melted**
- **1/2 teaspoon vanilla extract**

In a large mixing bowl, dissolve yeast in warm water. Add the milk, 1/4 cup sugar, 1/4 cup butter, egg, salt and 3-1/2 cups flour. Beat until smooth. Stir in enough remaining flour to form a soft dough.

Turn onto a floured surface; knead until smooth and elastic, about 6-8 minutes. Place in a greased bowl, turning once to grease top. Cover and let rise in a warm place until doubled, about 1-1/4 hours.

Punch dough down. Turn onto a lightly floured surface; divide into 32 pieces. Flatten each into a 3-in. disk. Place about 1 teaspoon of preserves on the center of each piece; bring edges together and seal.

Place 16 dough balls in a greased 10-in. fluted tube pan. Brush with half of the remaining butter; sprinkle with 1 tablespoon sugar. Top with remaining balls, butter and sugar. Cover and let rise until doubled, about 35 minutes.

Bake at 350° for 25-30 minutes or until golden brown. Combine syrup ingredients; pour over warm bread. Cool for 5 minutes before inverting onto a serving plate. **Yield:** 1 loaf.

Date-Nut Bran Muffins

Prep: 30 min. **Bake:** 15 min. per batch

This spectacular recipe has been in my family for decades. The muffins are moist and chock-full of raisins, dates and nuts. —Dario Burcar, Rochester, Michigan

- **2 cups All-Bran**
- **1 cup crushed Shredded Wheat**
- **1 cup boiling water**
- **2-1/2 cups all-purpose flour**
- **2 cups sugar**
- **2-1/2 teaspoons baking soda**
- **1/2 teaspoon salt**
- **2 eggs**
- **2 cups buttermilk**
- **1/2 cup vegetable oil**
- **1 cup chopped dates**
- **1 cup raisins**
- **1 cup chopped walnuts**

In a bowl, combine cereals and boiling water; let stand for 5 minutes. In a large bowl, combine the flour, sugar, baking soda and salt. In another bowl, combine eggs, buttermilk and oil. Stir the cereal and buttermilk mixtures into dry ingredients just until blended. Fold in the dates, raisins and walnuts.

Fill paper-lined muffin cups two-thirds full. Bake at 400° for 15-20 minutes or until a toothpick comes out clean. Cool for 5 minutes before removing from pans to wire racks. Serve warm. **Yield:** 2-1/2 dozen.

Hambalaya Corn Bread

Prep: 25 min. **Bake:** 20 min.

Loaded with sausage and ham, this savory corn bread took the blue ribbon in the corn bread cook-off at the State Fair of Texas. A big slice is practically a meal in itself! —Donna Thomas, Dallas, Texas

- **1/2 cup chopped green pepper**
- **1/2 cup chopped green onions**

- **3 tablespoons butter**
- **1-1/4 cups all-purpose flour**
- **1 cup yellow cornmeal**
- **1-1/2 teaspoons baking soda**
- **1/2 teaspoon salt**
- **2 eggs, beaten**
- **2 cups buttermilk**
- **1 cup diced fully cooked ham**
- **1/3 pound bulk pork sausage, cooked and drained**

In a large skillet, saute green pepper and onions in butter until tender. In a bowl, combine the flour, cornmeal, baking soda and salt. In another bowl, combine eggs and buttermilk; stir into dry ingredients just until moistened. Fold in the ham, sausage and green pepper mixture.

Transfer to a greased 10-in. ovenproof skillet or 9-in. square baking pan. Bake at 425° for 20-25 minutes or until a toothpick inserted near the center comes out clean. Serve warm. Refrigerate leftovers. **Yield:** 8 servings.

Caraway Yeast Bread

Prep: 20 min. + rising **Bake:** 25 min. + cooling

Caraway blends beautifully with celery seed, nutmeg and sage in this tender bread. *—DeEtta Rasmussen, Fort Madison, Iowa*

- **1 package (1/4 ounce) active dry yeast**
- **1 cup warm milk (110° to 115°)**
- **2 tablespoons honey**
- **2 tablespoons butter, softened**
- **1 egg, beaten**
- **1 tablespoon caraway seeds**
- **1-1/2 teaspoons salt**
- **1 teaspoon celery seed**
- **1 teaspoon rubbed sage**
- **1/2 teaspoon ground nutmeg**
- **3 to 3-1/2 cups all-purpose flour**

In a large mixing bowl, dissolve yeast in warm milk. Add honey; let stand for 5 minutes. Add the butter, egg, caraway seeds, salt, celery seed, sage, nutmeg and 1 cup flour; beat until blended. Stir in enough remaining flour to form a soft dough.

Turn onto a floured surface; knead until smooth and elastic, about 6-8 minutes. Place in a greased bowl, turning once to grease top. Cover and let rise in a warm place until doubled, about 1 hour.

Punch dough down. Place in a greased 9-in. x 5-in. x 3-in. loaf pan. Cover and let rise until doubled, about 40 minutes. Bake at 375° for 22-27 minutes or until golden brown. Cool for 10 minutes before removing from pan to a wire rack. **Yield:** 1 loaf.

Simple Stollen

(Pictured below)

Prep: 25 min. **Bake:** 40 min. + cooling

When it comes to seasonal sweets, this easy Christmas stollen is a recipe I can count on. It's made with baking powder instead of yeast, so it requires no rising. *—Shirley Glaab, Hattiesburg, Mississippi*

- **2-1/4 cups all-purpose flour**
- **1/2 cup sugar**
- **1-1/2 teaspoons baking powder**
- **1/4 teaspoon salt**
- **7 tablespoons cold butter, *divided***
- **1 cup ricotta cheese**
- **1/2 cup chopped mixed candied fruit**
- **1/2 cup raisins**
- **1/3 cup slivered almonds, toasted**
- **1 teaspoon vanilla extract**
- **1/2 teaspoon almond extract**
- **1/2 teaspoon grated lemon peel**
- **1 egg**
- **1 egg yolk**

Confectioners' sugar

In a large bowl, combine the flour, sugar, baking powder and salt. Cut in 6 tablespoons butter until mixture resembles fine crumbs. Stir in the ricotta, candied fruit, raisins, almonds, extracts, lemon peel, egg and yolk just until moistened.

Turn the dough onto a floured surface; knead five times. Roll the dough into a 10-in. x 8-in. oval. Fold a long side over to within 1 in. of opposite side; press the edge lightly to seal. Place on a greased baking sheet; curve the ends slightly.

Bake at 350° for 40-45 minutes or until golden brown. Melt remaining butter; brush over loaf. Remove to a wire rack to cool completely. Dust with confectioners' sugar. **Yield:** 1 loaf.

Editor's Note: This recipe does not contain yeast.

Rhubarb Sticky Buns

(Pictured above)

Prep: 25 min. + rising **Bake:** 20 min.

These gooey cinnamon rolls don't last long at my house. The tangy rhubarb sauce is so pretty and makes these buns special. —*Kathy Kittell, Lenexa, Kansas*

- **1 package (16 ounces) hot roll mix**
- **4 tablespoons sugar, *divided***
- **1 cup warm water (120° to 130°)**
- **1 egg, beaten**
- **2 tablespoons plus 1/2 cup butter, softened, *divided***
- **2 cups sliced fresh *or* frozen rhubarb**
- **1/2 cup packed brown sugar**
- **1/2 cup light corn syrup**
- **2 teaspoons ground cinnamon**

In a large mixing bowl, combine the hot roll mix with the contents of the yeast packet and 2 tablespoons sugar. Stir in the warm water, egg and 2 tablespoons butter to form a soft dough. Turn onto a floured surface. Knead until smooth, about 5 minutes. Cover and let rest for 5 minutes.

Meanwhile, in a large saucepan, combine the rhubarb, brown sugar, corn syrup and remaining butter. Bring to a boil; cook and stir for 3 minutes. Pour into an ungreased 13-in. x 9-in. x 2-in. baking dish.

On a lightly floured surface, roll dough into a 15-in. x 10-in. rectangle. Combine cinnamon and remaining sugar; sprinkle over dough. Roll up jelly-roll style, starting with a long side; pinch seam to seal.

Cut the roll into 12 slices. Place slices cut side down over the rhubarb sauce. Cover and let rise in a warm place until doubled, about 30 minutes.

Bake at 375° for 20-25 minutes or until golden brown. Immediately invert onto a serving platter. Serve warm. **Yield:** 1 dozen.

Vermont Honey-Wheat Bread

Prep: 30 min. + rising **Bake:** 30 min. + cooling

Made with whole wheat flour, oats and wheat germ, this hearty bread gets its pleasant sweetness from honey and maple syrup. —*Roderick Crandall, Hartland, Vermont*

✓ **Uses less fat, sugar or salt. Includes Nutrition Facts and Diabetic Exchanges.**

- **2 packages (1/4 ounce *each*) active dry yeast**
- **3/4 cup warm water (110° to 115°)**
- **1 cup old-fashioned oats**
- **1 cup warm buttermilk (110° to 115°)**
- **1/3 cup butter, softened**
- **1-1/2 cups whole wheat flour**
- **1/2 cup maple syrup**
- **1/2 cup honey**
- **1/3 cup toasted wheat germ**
- **2 eggs, beaten**
- **1 teaspoon salt**
- **3 to 4 cups all-purpose flour**

TOPPING:

- **1 egg white**
- **1 tablespoon water**
- **1/4 cup old-fashioned oats**

In a small bowl, dissolve yeast in warm water. In a large mixing bowl, combine the oats, buttermilk and butter; add yeast mixture. Add the whole wheat flour, syrup, honey, wheat germ, eggs and salt; beat on low speed for 30 seconds. Beat on high for 3 minutes. Stir in enough all-purpose flour to form a firm dough.

Turn onto a floured surface; knead until smooth and elastic, about 6-8 minutes. Place in a bowl coated with nonstick cooking spray, turning once to coat top. Cover and let rise in a warm place until doubled, about 1 hour.

Punch the dough down. Shape into two loaves; place each in a 9-in. x 5-in. x 3-in. loaf pan coated with nonstick cooking spray. Cover and let rise in a warm place until doubled, about 45 minutes.

Beat the egg white and water; brush over loaves. Sprinkle with oats. Bake at 375° for 30-35 minutes or until golden brown. Cool for 10 minutes before removing from pans to wire racks to cool completely. **Yield:** 2 loaves (16 slices each).

Nutrition Facts: 1 slice equals 133 calories, 3 g fat (1 g saturated fat), 19 mg cholesterol, 108 mg sodium, 24 g carbohydrate, 2 g fiber, 4 g protein. **Diabetic Exchange:** 1-1/2 starch.

Fruit 'n' Nut Stollen

Prep: 40 min. + rising **Bake:** 15 min.

With four different kinds of nuts and lots of colorful fruit, these loaves are wonderful for the Christmas season. I top them off with a lemony sugar glaze.
—Rebekah Radewahn, Wauwatosa, Wisconsin

✓ **Uses less fat, sugar or salt. Includes Nutrition Facts and Diabetic Exchanges.**

- **4 to 4-1/2 cups all-purpose flour**
- **1/4 cup sugar**
- **3 teaspoons active dry yeast**
- **1 teaspoon ground cardamom**
- **1/2 teaspoon salt**
- **1-1/4 cups milk**
- **1/2 cup plus 3 tablespoons butter, softened, *divided***
- **1 egg**
- **1/4 cup *each* raisins and dried cranberries**
- **1/4 cup *each* chopped dried pineapple and apricots**
- **1/4 cup *each* chopped pecans, almonds, Brazil nuts and walnuts**
- **1/2 teaspoon lemon extract**

LEMON GLAZE:

- **1 cup confectioners' sugar**
- **4-1/2 teaspoons lemon juice**

In a large mixing bowl, combine 2 cups flour, sugar, yeast, cardamom and salt. In a small saucepan, heat milk and 1/2 cup butter to 120°-130°. Add to dry ingredients; beat just until moistened. Add egg; beat until smooth. Stir in enough remaining flour to form a soft dough (dough will be sticky).

Turn onto a floured surface; knead until smooth and elastic, about 6-8 minutes. Place in a bowl coated with nonstick cooking spray, turning once to coat top. Cover and let rise in a warm place until doubled, about 1 hour. In a small bowl, combine the dried fruits, nuts and extract; set aside.

Punch dough down. Turn onto a lightly floured surface; knead fruit mixture into dough. Divide into thirds. Roll each portion into a 10-in. x 8-in. oval. Melt remaining butter; brush over dough. Fold a long side over to within 1 in. of opposite side; press edges lightly to seal. Place on baking sheets coated with nonstick cooking spray. Cover and let rise until doubled, about 45 minutes.

Bake at 375° for 14-16 minutes or until golden brown. Remove the loaves to wire racks. Combine the glaze ingredients; drizzle glaze over the loaves. **Yield:** 3 loaves (12 slices each).

Nutrition Facts: 1 slice equals 142 calories, 6 g fat (3 g saturated fat), 16 mg cholesterol, 76 mg sodium, 19 g carbohydrate, 1 g fiber, 3 g protein. **Diabetic Exchanges:** 1 starch, 1 fat.

Golden Lemon Bread

(Pictured below)

Prep: 20 min. **Bake:** 55 min. + cooling

This wonderful bread from my grandmother won "Best of Show" at the New Mexico State Fair. My grandchildren love it. *—Marjorie Rose, Albuquerque, New Mexico*

- **1/2 cup shortening**
- **1 cup sugar**
- **2 eggs**
- **1-1/2 cups all-purpose flour**
- **1 teaspoon baking powder**
- **1/2 teaspoon salt**
- **3/4 cup milk**

GLAZE:

- **1/2 cup confectioners' sugar**
- **2 teaspoons grated lemon peel**
- **2 to 3 tablespoons lemon juice**

In a large mixing bowl, cream shortening and sugar. Add eggs, one at a time, beating well after each addition. Combine the flour, baking powder and salt; add to creamed mixture alternately with milk.

Pour the batter into a greased 8-in. x 4-in. x 2-in. loaf pan. Bake at 350° for 55-60 minutes or until a toothpick inserted near the center comes out clean. Place pan on a wire rack.

Combine the glaze ingredients; immediately pour over warm bread. Cool completely before removing from pan. **Yield:** 1 loaf.

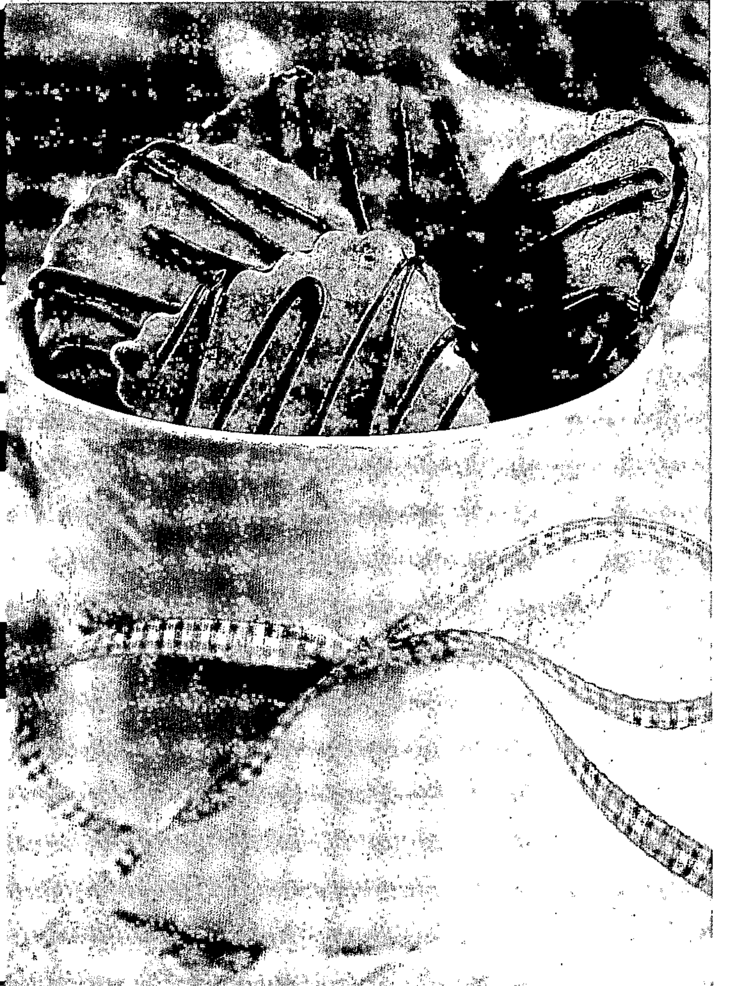

Cookies & Bars

When you make these nutty brownies, melt-in-your-mouth cutouts, rich caramel goodies and more, you'll have sweet treats well in hand!

DELIGHTS BY THE DOZEN. Clockwise from upper left: Iced Pumpkin Cookies (p. 94), Apricot Pastry Bars (p. 95), Hazelnut-Espresso Sandwich Cookies (p. 95), Chocolate Pecan Bars (p. 98) and Sledding Teddies (p. 96).

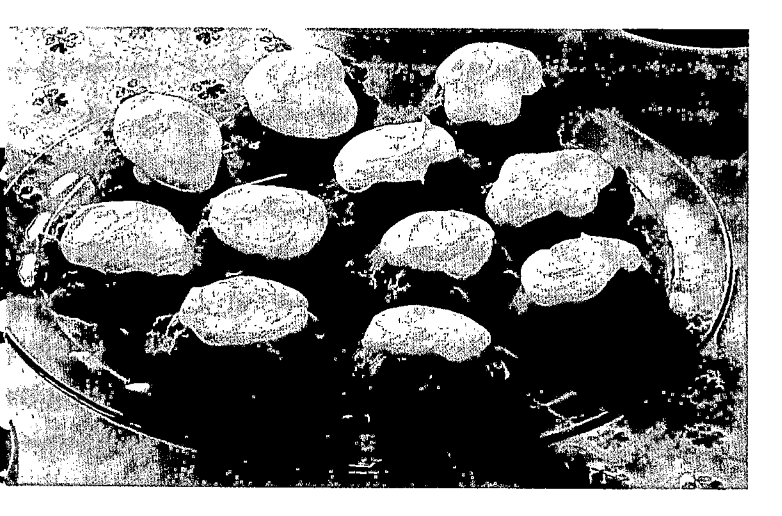

Iced Pumpkin Cookies

(Pictured above and on page 92)

Prep: 45 min. **Bake:** 15 min. per batch + cooling

My son, Joshua, likes to test—or should I say consume—these chunky treats! —*Johna Nilson, Vista, California*

1 cup butter, softened
1/2 cup sugar
1/2 cup packed brown sugar
1 egg
1 cup canned pumpkin
1 cup all-purpose flour
1 cup whole wheat flour
1-1/2 teaspoons ground cinnamon
1 teaspoon baking powder
1 teaspoon ground ginger
1/2 teaspoon salt
1/2 teaspoon baking soda
1/2 teaspoon ground nutmeg
1/4 teaspoon ground cloves
1 cup granola without raisins
1 cup chopped walnuts
1 cup vanilla *or* white chips
1 cup dried cranberries

ICING:

1/4 cup butter, softened
2 cups confectioners' sugar
3 tablespoons milk

In a large mixing bowl, cream butter and sugars. Beat in egg and pumpkin. Combine the flours, cinnamon, baking powder, ginger, salt, baking soda, nutmeg and cloves; gradually add to creamed mixture. Stir in the granola, walnuts, chips and cranberries.

Drop by tablespoonfuls 2 in. apart onto greased baking sheets. Bake at 350° for 15-18 minutes or until lightly browned. Remove to wire racks to cool.

In a small mixing bowl, combine icing ingredients until smooth. Frost cookies. Store in the refrigerator. **Yield:** 3 dozen.

Back-to-School Cookies

Prep: 30 min. **Bake:** 10 min. per batch

These nutty goodies are great for a classroom party or to share with friends. I like mine best with a glass of cold milk. —*Frances Pierce, Waddington, New York*

1 cup butter-flavored shortening
1 cup creamy peanut butter
2 cups packed brown sugar
4 egg whites
1 teaspoon vanilla extract
2 cups all-purpose flour
1 teaspoon baking soda
1/2 teaspoon baking powder
2 cups crisp rice cereal
1-1/2 cups chopped nuts
1 cup flaked coconut
1 cup quick-cooking oats

In a large mixing bowl, cream the shortening, peanut butter and brown sugar until light and fluffy. Beat in egg whites and vanilla. Combine the flour, baking soda and baking powder; gradually add to creamed mixture and mix well. Stir in the cereal, nuts, coconut and oats.

Drop by rounded tablespoonfuls 2 in. apart onto ungreased baking sheets. Flatten with a fork, forming a crisscross pattern. Bake at 375° for 7-8 minutes. Remove to wire racks. **Yield:** 12 dozen.

Rhubarb Cranberry Cookies

Prep: 30 min. **Bake:** 10 min. per batch + cooling

The sweetness of white chocolate in these cookies really complements the tart flavor of the rhubarb and cranberries. —*Elaine Scott, Lafayette, Indiana*

1 cup butter, softened
1 cup packed brown sugar
1/2 cup sugar
2 eggs
1 teaspoon vanilla extract
1-1/2 cups all-purpose flour
1 teaspoon baking soda
1/2 teaspoon salt
1/2 teaspoon ground cinnamon
2-1/2 cups old-fashioned oats
1-1/2 cups diced frozen rhubarb
1 cup vanilla *or* white chips
1 cup dried cranberries
4 squares (1 ounce *each*) white baking chocolate

In a large mixing bowl, cream the butter and sugars. Beat in eggs and vanilla. Combine the flour, baking soda, salt and cinnamon; gradually add to creamed mixture. Stir in the oats, rhubarb, chips and cranberries.

Drop by tablespoonfuls 2 in. apart onto parchment paper-lined baking sheets. Bake at 350° for 10-12 minutes or until set. Remove to wire racks to cool.

In a microwave-safe bowl, melt white chocolate; stir until smooth. Drizzle over cookies; let stand until set. Store in an airtight container. **Yield:** about 5-1/2 dozen.

Hazelnut-Espresso Sandwich Cookies

(Pictured below and on page 93)

Prep: 45 min. + chilling
Bake: 10 min. per batch + cooling

The inspiration for these came from a cookie my sister tried in Italy. —Cindy Beberman, Orland Park, Illinois

1 cup butter, softened
1-1/4 cups sugar
1 egg
1 egg yolk
4 teaspoons instant espresso granules
2 teaspoons vanilla extract
2-1/2 cups all-purpose flour
1/2 teaspoon salt
1/2 teaspoon baking powder
1 cup finely ground hazelnuts
FILLING:
1 cup heavy whipping cream
1-1/4 cups milk chocolate chips
1-3/4 cups semisweet chocolate chips, *divided*

In a large mixing bowl, cream butter and sugar. Beat in the egg, yolk, espresso granules and vanilla. Combine the flour, salt and baking powder; gradually add to creamed mixture and mix well. Stir in hazelnuts.

Divide dough into thirds; flatten each portion into a circle. Wrap each in plastic wrap; refrigerate for 1 hour or until easy to handle.

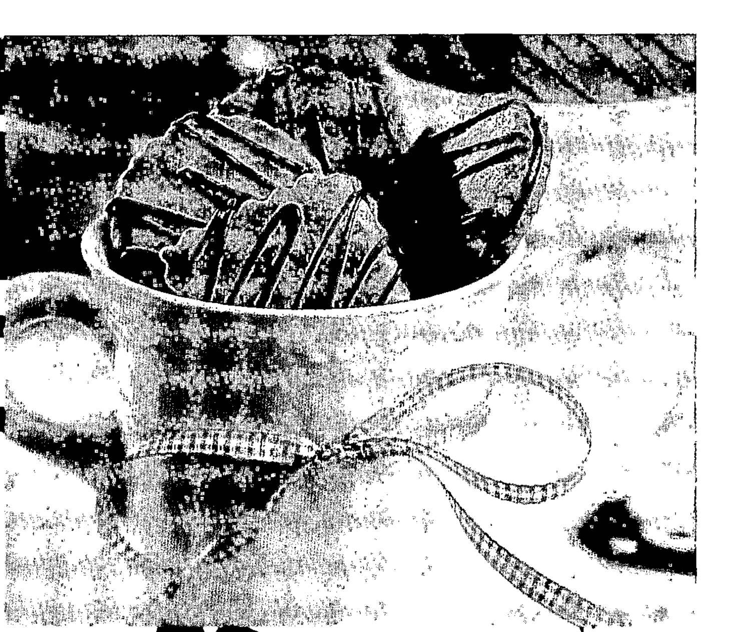

On a lightly floured surface, roll out one portion of dough to 1/8-in. thickness. Cut with a floured 1-1/2-in. cookie cutter; place 1/2 in. apart on ungreased baking sheets. Repeat with remaining dough; chill and reroll scraps. Bake at 375° for 6-8 minutes or until edges begin to brown. Remove to wire racks to cool.

For filling, in a small saucepan, bring cream to a boil. Remove from the heat; stir in milk chocolate chips and 3/4 cup semisweet chocolate chips until melted. Transfer to a bowl; refrigerate for 1-1/2 hours or until filling reaches spreading consistency, stirring occasionally.

Spread filling over the bottom of half of the cookies; top with remaining cookies. Melt remaining chips; drizzle over cookies. Let stand until set. Store in an airtight container in the refrigerator. **Yield:** 3 dozen.

Apricot Pastry Bars

(Pictured on page 93)

Prep: 45 min. **Bake:** 35 min. + cooling

Ideal for either a casual or a fancy event, this recipe is a crowd-pleaser. —Nancy Foust, Stoneboro, Pennsylvania

4 cups all-purpose flour
1 cup plus 2 tablespoons sugar, *divided*
3 teaspoons baking powder
1/2 teaspoon salt
1/4 teaspoon baking soda
1 cup shortening
3 eggs, *separated*
1/4 cup milk
1-1/2 teaspoons vanilla extract
4 cans (12 ounces *each*) apricot filling
1 cup chopped walnuts

In a large bowl, combine the flour, 1 cup sugar, baking powder, salt and baking soda. Cut in shortening until mixture resembles coarse crumbs.

In a small bowl, whisk the egg yolks, 2 egg whites, milk and vanilla; gradually add to crumb mixture, tossing with a fork until dough forms a ball. Divide in half, making one portion slightly larger.

Roll out larger portion of dough between two large sheets of waxed paper into a 17-in. x 12-in. rectangle. Transfer to an ungreased 15-in. x 10-in. x 1-in. baking pan. Press pastry onto the bottom and up the sides of pan; trim pastry even with top edges. Spread apricot filling over dough; sprinkle with walnuts.

Roll out remaining pastry to fit top of pan; place over filling. Trim, seal and flute edges. Cut slits in top. Whisk remaining egg white; brush over pastry. Sprinkle with remaining sugar.

Bake at 350° for 35-40 minutes or until golden brown. Cool on a wire rack. **Yield:** about 4 dozen.

Sledding Teddies

(Pictured below and on page 92)

Prep: 3 hours + chilling
Bake: 15 min. per batch + standing

I've been making these special holiday cookies at Christmastime for almost 30 years. They take a little extra time, but they're worth it because the teddy bears are so cute and yummy. —*Linda Nealley, Newburgh, Maine*

- **1 cup butter, cubed**
- **2/3 cup packed brown sugar**
- **2/3 cup molasses**
- **1 egg**
- **1-1/2 teaspoons vanilla extract**
- **4 cups all-purpose flour**
- **2 teaspoons ground cinnamon**
- **1 teaspoon ground ginger**
- **3/4 teaspoon baking soda**
- **3/4 teaspoon ground cloves**
- **2 tablespoons miniature semisweet chocolate chips**

FROSTING:

- **1/2 cup shortening**
- **2-1/2 cups confectioners' sugar**
- **2 tablespoons milk**
- **1/2 teaspoon vanilla extract**

SLEDS:

- **16 candy canes (about 5-1/4 inches)**
- **8 whole graham crackers**
- **8 red-hot candies**
- **8 cake decorator hearts**

In a small saucepan, cook the butter, brown sugar and molasses over medium heat until sugar is dissolved. Pour into a large mixing bowl; let stand for 10 minutes. Beat in egg and vanilla. Combine the flour, cinnamon, ginger, baking soda and cloves; gradually add to butter mixture and mix well. Cover and refrigerate for 4 hours or overnight.

For big bears, shape dough into eight 1-1/4-in. balls, eight 1-in. balls, sixteen 1/4-in. balls, sixteen 1-3/4-in. x 1/2-in. logs and sixteen 1-1/2-in. x 1/2-in. logs. Set remaining dough aside.

For bodies, place the 1-1/4-in. balls on three ungreased baking sheets; flatten to 1/2-in. thickness. Position the 1-in. balls for heads; flatten to 1/2-in. thickness. Attach two 1-3/4-in. logs for arms and 1/4-in. balls for ears. Do not attach 1-1/2-in. logs for legs; place separately on baking sheets. Add chocolate chips on each paw and on heads for eyes.

For small bears, shape the remaining dough into eight 1-in. balls, eight 3/4-in. balls, sixteen 1/4-in. balls, sixteen 1-1/4-in. x 3/8-in. logs and sixteen 1-in. x 3/8-in. logs. Position the small bears and logs on two ungreased baking sheets the same as for the big bears; add the chocolate chips.

Bake small bears and logs at 325° for 11-13 minutes, and big bears and logs for 14-16 minutes or until set. Cool for 10 minutes before carefully removing from pans to wire racks to cool completely.

In a small mixing bowl, combine frosting ingredients. For sleds, use frosting to attach two candy canes to the bottom of each graham cracker; let stand until set.

Trim bear bodies so the bottom edge is flat. Using frosting, attach a big bear to the back end of each sled. Attach small bears in front of big bears. For legs, attach big logs in front of bears and small logs on top. For noses, attach red-hots to big bears and hearts to small bears with a dab of frosting. Let stand until set. **Yield:** 8 servings.

Peanut Butter Chippers

Prep: 10 min. **Bake:** 15 min. per batch

The aroma of these chocolate chip-dotted cookies in the oven brings my family running to the kitchen. The recipe's so easy, I often stir up a batch while fixing dinner.
—*Pat Doerflinger, Centerview, Missouri*

✓ **Uses less fat, sugar or salt. Includes Nutrition Facts and Diabetic Exchanges.**

- **6 tablespoons butter, softened**
- **1/4 cup peanut butter**
- **1/2 cup sugar**
- **1/2 cup packed brown sugar**
- **1 egg**
- **1 teaspoon vanilla extract**

1-1/4 cups all-purpose flour
1/2 teaspoon baking soda
1/4 teaspoon salt
1 cup milk chocolate chips

In a small mixing bowl, cream the butter, peanut butter and sugars. Add egg; mix well. Beat in vanilla. Combine the flour, baking soda and salt; gradually add to creamed mixture. Stir in chocolate chips.

Drop by tablespoonfuls 2 in. apart onto ungreased baking sheets. Bake at 350° for 11-14 minutes or until golden brown. Remove to wire racks. **Yield:** 3-1/2 dozen.

Nutrition Facts: 1 cookie equals 78 calories, 4 g fat (2 g saturated fat), 10 mg cholesterol, 59 mg sodium, 10 g carbohydrate, trace fiber, 1 g protein. **Diabetic Exchanges:** 1 fat, 1/2 starch.

Hazelnut Mocha Brownies

(Pictured above)

Prep: 35 min. **Bake:** 25 min. + chilling

Flavored instant coffee perks up these brownies. I created the recipe while cooking with my young daughter.
—Anna Ginsberg, Austin, Texas

1-1/2 cups semisweet chocolate chips, *divided*
3/4 cup butter, cubed
1 tablespoon hazelnut-flavored instant coffee granules
2 eggs
2/3 cup packed brown sugar
1 teaspoon vanilla extract
3/4 cup all-purpose flour
1/2 teaspoon baking powder
1/4 teaspoon salt
1/4 cup chopped hazelnuts, toasted

FROSTING:

1/2 cup semisweet chocolate chips
1 tablespoon butter
1/4 cup heavy whipping cream
1-1/2 teaspoons instant hazelnut-flavored coffee granules
1/2 cup confectioners' sugar
1 teaspoon vanilla extract

In a microwave-safe bowl, melt 1 cup chocolate chips and the butter with coffee granules; stir until smooth. Cool slightly.

In a large mixing bowl, beat eggs and brown sugar; stir in chocolate mixture and vanilla. Combine the flour, baking powder and salt; gradually add to chocolate mixture. Stir in hazelnuts and remaining chips.

Pour into a greased 9-in. square baking pan. Bake at 325° for 22-26 minutes or until a toothpick inserted near the center comes out clean. Cool on a wire rack. For frosting, in a microwave-safe bowl, melt chocolate chips and butter with cream and coffee granules; stir until smooth. Cool to room temperature.

Transfer to a small mixing bowl; beat on high speed for 2 minutes. Beat in confectioners' sugar and vanilla until fluffy. Frost brownies; refrigerate for at least 30 minutes before cutting. **Yield:** 16 servings.

Coconut Drop Cookies

Prep: 25 min. **Bake:** 15 min. per batch

My mother always tried to add nutritious ingredients to recipes. With nuts and oats, these crispy-chewy treats are the perfect example. *—Cathy Wilson, Midvale, Utah*

1 cup shortening
1 cup sugar
1 cup packed brown sugar
2 eggs
1 teaspoon vanilla extract
2 cups all-purpose flour
2 cups old-fashioned oats
1 teaspoon baking powder
1 teaspoon baking soda
1/2 teaspoon salt
2 cups flaked coconut
1 cup chopped walnuts

In a large mixing bowl, cream the shortening and sugars. Add eggs, one at a time, beating well after each addition. Beat in vanilla. Combine the flour, oats, baking powder, baking soda and salt; gradually add to creamed mixture and mix well. Stir in coconut and walnuts.

Drop by rounded tablespoonfuls 3 in. apart onto greased baking sheets. Flatten slightly. Bake at 350° for 11-14 minutes or until golden brown. Cool for 2 minutes before removing to wire racks. Store in an airtight container. **Yield:** 5-1/2 dozen.

Chocolate Pecan Bars

(Pictured above and on page 92)

Prep: 25 min. **Bake:** 25 min. + cooling

These chewy, drizzled bars are great for Thanksgiving or Christmas...and always a hit. The recipe makes a large batch, too. —Carole Fraser, North York, Ontario

2/3 cup butter, softened
1/3 cup sugar
2 cups all-purpose flour
FILLING:
6 squares (1 ounce *each*) semisweet chocolate
1-1/4 cups light corn syrup
1-1/4 cups sugar
4 eggs, lightly beaten
1-1/4 teaspoons vanilla extract
2-1/4 cups chopped pecans
DRIZZLE:
4 squares (1 ounce *each*) semisweet chocolate
1-1/4 teaspoons shortening

In a small mixing bowl, cream butter and sugar until light and fluffy. Beat in flour. Press into a greased 15-in. x 10-in. x 1-in. baking pan. Bake at 350° for 12-15 minutes or until golden brown.

Meanwhile, in a large saucepan, melt chocolate with corn syrup over low heat; stir until smooth. Remove from the heat. Stir in the sugar, eggs and vanilla. Add pecans.

Spread evenly over hot crust. Bake for 25-30 minutes or until firm around the edges. Cool on a wire rack. Melt chocolate and shortening; stir until smooth. Drizzle over bars. **Yield:** 4 dozen.

Hazelnut Chocolate Cookies

Prep: 20 min. **Bake:** 15 min. per batch + cooling

Hazelnuts are the stars of these crisp cookies. Chocolate and nuts are big in our house, so I knew these would be popular. —Elisa Lochridge, Aloha, Oregon

1/2 cup butter, softened
6 tablespoons sugar
1 teaspoon vanilla extract
3/4 cup cake flour
1/4 cup baking cocoa
3/4 cup ground hazelnuts
24 whole hazelnuts, toasted and peeled
Confectioners' sugar

In a small mixing bowl, cream butter and sugar. Beat in vanilla. Combine flour and cocoa; gradually add to creamed mixture. Beat in ground hazelnuts.

Roll into 1-in. balls. Press one whole hazelnut into each. Place 2 in. apart on ungreased baking sheets.

Bake at 325° for 15-19 minutes or until firm to the touch. Let stand for 2 minutes before removing from pans to wire racks to cool completely. Sprinkle with confectioners' sugar. **Yield:** 2 dozen.

Double-Chocolate Espresso Cookies

Prep: 30 min. **Bake:** 10 min. per batch

Anyone who is a chocolate and espresso fan is sure to love these goodies. —Cindi Paulson, Anchorage, Alaska

3 squares (1 ounce *each*) unsweetened chocolate, chopped
2 cups (12 ounces) semisweet chocolate chips, *divided*
1/2 cup butter, cubed
1 tablespoon instant coffee granules
1 cup plus 2 tablespoons sugar
3 eggs
3/4 cup all-purpose flour
1/2 teaspoon baking powder
1/4 teaspoon salt

In a small heavy saucepan, melt unsweetened chocolate, 1 cup chocolate chips and butter with coffee granules; stir until smooth. Remove from the heat; set aside to cool.

In a small mixing bowl, beat sugar and eggs for 3 minutes or until thick and lemon-colored. Beat in the chocolate mixture. Combine the flour, baking powder and salt; add to chocolate mixture. Stir in remaining chips.

Drop by rounded teaspoonfuls 2 in. apart onto greased baking sheets. Bake at 350° for 10-12 minutes or until puffed and tops are cracked. Cool for 5 minutes before removing to wire racks. **Yield:** 3 dozen.

A Flock of Turkeys

(Pictured below)

Prep: 1 hour

Creating a gaggle of these gobblers is a fun project to do with kids. The cute turkey cookies make festive school treats or yummy favors for a Thanksgiving gathering.
—Zoraya Jennings, Fredericksburg, Virginia

48 sugar cookies (3 inches)
1 package (12 ounces) chocolate and marshmallow cookies
1 can (16 ounces) chocolate frosting
1 cup vanilla frosting
Yellow and red paste food coloring

Using a serrated knife, cut 1/2 in. from one side of 24 sugar cookies. Using a sharp knife, cut marshmallow cookies in half vertically. Spread chocolate frosting over the bottom of each marshmallow cookie half. Align cut edges of a marshmallow cookie and sugar cookie; press together to form each turkey body and feathers.

Spread chocolate frosting over the cut edges of each turkey; position and attach near the back edge of a whole sugar cookie.

Cut a small hole in the corner of a pastry or plastic bag. Insert a #12 round pastry tip; fill the bag with 3/4 cup chocolate frosting. Pipe the neck and head on each turkey.

Tint 1/2 cup vanilla frosting yellow and 1/2 cup red. With a #3 round pastry tip and yellow frosting, add eyes, beak and legs. With #3 round tip and red frosting, pipe snood and tail feathers. Store in the refrigerator. **Yield:** 2 dozen.

Editor's Note: This recipe was tested with Nabisco Pinwheels.

Rustic Nut Bars

Prep: 20 min. **Bake:** 35 min. + cooling

Everyone will crunch with joy when they bite into these gooey bars, which feature four different kinds of nuts. People rave about the shortbread-like crust and sweet-salty topping. These squares are so chock-full, just one will satisfy. *—Barbara Driscoll, West Allis, Wisconsin*

1 tablespoon plus 3/4 cup cold butter, *divided*
2-1/3 cups all-purpose flour
1/2 cup sugar
1/2 teaspoon baking powder
1/2 teaspoon salt
1 egg, lightly beaten

TOPPING:
2/3 cup honey
1/2 cup packed brown sugar
1/4 teaspoon salt
6 tablespoons butter, cubed
2 tablespoons heavy whipping cream
1 cup chopped hazelnuts, toasted
1 cup roasted salted almonds
1 cup salted cashews, toasted
1 cup pistachios, toasted

Line a 13-in. x 9-in. x 2-in. baking pan with foil; grease the foil with 1 tablespoon butter. Set aside.

In a large bowl, combine the flour, sugar, baking powder and salt; cut in remaining butter until mixture resembles coarse crumbs. Stir in egg until blended (mixture will be dry). Press firmly into prepared pan. Bake at 375° for 18-20 minutes or until edges are golden brown. Cool on a wire rack.

In a large heavy saucepan over medium heat, bring the honey, brown sugar and salt to a boil; stir until the brown sugar is dissolved. Boil without stirring for 2 minutes.

Add the butter and heavy whipping cream. Bring to a boil; cook and stir for 1 minute. Remove from the heat; stir in the hazelnuts, almonds, cashews and pistachios. Spread over the crust.

Bake at 375° for 15-20 minutes or until topping is bubbly. Cool completely on a wire rack. Using foil, lift bars out of pan. Discard foil; cut into squares. **Yield:** about 3 dozen.

Nutty Nutrition

Nuts are one of the world's oldest foods and a good source of protein. Some research even suggests that eating nuts can boost your brainpower. Nuts tend to be high in fat, but it's mostly the heart-healthy, unsaturated kind.

Cakes & Pies

Does your family love layer cakes topped with creamy frosting? How about golden, lattice-pastry pies filled with fruit? Whatever cake or pie variety they prefer, you're sure to find it here.

DELIGHTFUL DESSERTS. Clockwise from upper left: Red Velvet Heart Torte (p. 106), Raspberry Patch Cream Pie (p. 103), Caramel-Pecan Apple Pie (p. 103), Elegant Chocolate Torte (p. 107) and Barnyard Cupcakes (p. 116).

Fudgy Pudgy Cake

(Pictured above)

Prep: 30 min. **Bake:** 20 min. + cooling

This wonderful layer cake is definitely worth the effort. Enjoy a big piece with a glass of cold milk or a cup of hot coffee. —*Doris Jennings, Allen Park, Michigan*

1/2 cup baking cocoa
1 cup hot water
3/4 cup butter, softened
2-3/4 cups packed brown sugar
3 eggs
2-1/2 cups all-purpose flour
2 teaspoons baking soda
1 teaspoon baking powder
3/4 teaspoon salt
1 cup buttermilk
1-1/2 teaspoons vanilla extract

FILLING:

1 cup sugar
2 tablespoons cornstarch
1 cup milk
2 teaspoons vanilla extract
1 cup butter, softened

GLAZE:

1/4 cup butter, softened
2 tablespoons baking cocoa
1/4 cup milk
2 cups confectioners' sugar
Chocolate curls, optional

In a small bowl, combine cocoa and hot water until smooth; cool. In a large mixing bowl, cream butter and brown sugar. Add eggs, one at a time, beating well after each. Combine the flour, baking soda, baking powder and salt; add to creamed mixture alternately with buttermilk and cocoa mixture. Stir in vanilla.

Pour into three greased and waxed paper-lined 9-in. round baking pans. Bake at 350° for 20-25 minutes or until a toothpick comes out clean. Cool for 10 minutes; remove from pans to wire racks to cool completely.

In a small saucepan, combine sugar and cornstarch. Gradually stir in milk until smooth. Bring to a boil; cook and stir for 2 minutes or until thickened. Remove from the heat; stir in vanilla. Cool completely. In a small mixing bowl, cream butter. Gradually beat in cooled mixture.

Place one cake layer on a serving plate; spread with half of the filling. Repeat layers. Top with the remaining cake layer.

For glaze, in a small saucepan, melt butter; stir in cocoa until smooth. Add milk. Bring to a boil. Remove from the heat. With an electric mixer, beat in confectioners' sugar until blended. Cool for 15-20 minutes; spread over top of cake. Garnish with chocolate curls if desired. Store in the refrigerator. **Yield:** 12-14 servings.

Old-Fashioned Rhubarb Pudding Cake

Prep: 30 min. **Bake:** 30 min. + cooling

Homey and comforting, this is one of my mother's best desserts. It's a yummy way to use garden-grown rhubarb. —*Barbara Collins, Andover, Massachusetts*

4 cups diced fresh *or* frozen rhubarb
1-1/2 cups sugar, *divided*
1/4 cup shortening
1 egg
1/2 teaspoon vanilla extract
1 cup all-purpose flour
2 teaspoons baking powder
1/4 teaspoon salt
1/2 cup milk

In a large saucepan, combine rhubarb and 1 cup sugar. Cook over medium heat for 12-15 minutes or until rhubarb is tender.

Meanwhile, in a small mixing bowl, cream shortening and remaining sugar; beat in the egg and vanilla extract. Combine the flour, baking powder and salt; add to the creamed mixture alternately with the milk. Beat just until combined.

Pour the batter into a greased 9-in. square baking dish. Pour the rhubarb and sugar mixture over the batter. Bake at 350° for 30-35 minutes or until a toothpick inserted near the center comes out clean. Cool on a wire rack. **Yield:** 9 servings.

Raspberry Patch Cream Pie

(Pictured below and on page 100)

Prep: 35 min. + chilling

Our family loves raspberries, and this pie keeps the flavor and firmness of the berries intact. The added treat of gelatin and cream cheese guarantees requests for seconds.
—Allison Anderson, Raymond, Washington

1 cup graham cracker crumbs
1/2 cup sugar
5 tablespoons butter, melted

FILLING:
1 package (8 ounces) cream cheese, softened
1/4 cup confectioners' sugar
2 teaspoons milk
1 teaspoon vanilla extract

TOPPING:
3/4 cup sugar
3 tablespoons cornstarch
1-1/3 cups cold water
1/4 cup raspberry gelatin powder
3 cups fresh raspberries

In a small bowl, combine cracker crumbs, sugar and butter. Press onto the bottom and up the sides of an ungreased 9-in. pie plate. Bake at 350° for 9-11 minutes or until set. Cool on a wire rack.

For the filling, in a small mixing bowl, combine the cream cheese, confectioners' sugar, milk and vanilla. Carefully spread over crust.

For the topping, in a small saucepan, combine the sugar, cornstarch and cold water until smooth. Bring to a boil; cook and stir for 2 minutes or until thickened. Remove from the heat; stir in raspberry gelatin powder until dissolved. Cool to room temperature. Refrigerate until slightly thickened.

Arrange berries over filling. Spoon gelatin mixture over berries. Refrigerate until set. **Yield:** 6-8 servings.

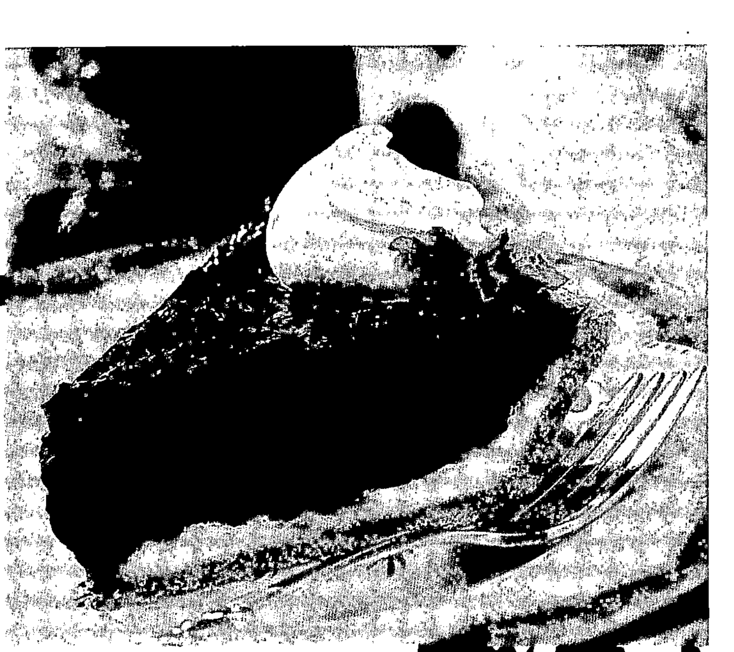

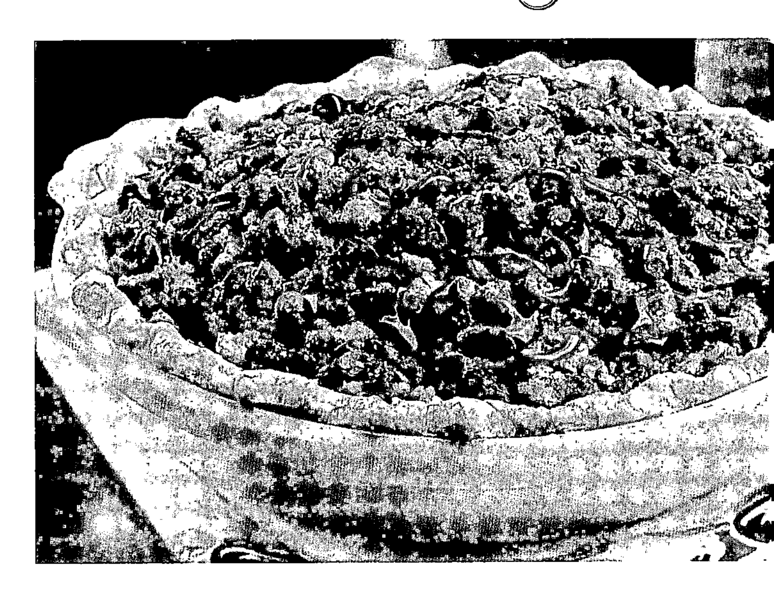

Caramel-Pecan Apple Pie

(Pictured above and on page 100)

Prep: 45 min. **Bake:** 55 min. + cooling

This scrumptious pie drizzled with caramel sauce always takes me back home to Virginia and my granny's table.
—Gloria Castro, Santa Rosa, California

7 cups sliced peeled tart apples
1 teaspoon lemon juice
1 teaspoon vanilla extract
3/4 cup chopped pecans
1/3 cup packed brown sugar
3 tablespoons sugar
4-1/2 teaspoons ground cinnamon
1 tablespoon cornstarch
1/4 cup caramel ice cream topping, room temperature
1 unbaked pastry shell (9 inches)
3 tablespoons butter, melted

STREUSEL TOPPING:
3/4 cup all-purpose flour
2/3 cup chopped pecans
1/4 cup sugar
6 tablespoons cold butter
1/4 cup caramel ice cream topping, room temperature

In a large bowl, toss apples with lemon juice and vanilla. Combine the pecans, sugars, cinnamon and cornstarch; add to apple mixture and toss to coat. Pour caramel topping over bottom of pastry shell; top with apple mixture (shell will be full). Drizzle with butter.

In a small bowl, combine the flour, pecans and sugar. Cut in butter until mixture resembles coarse crumbs. Sprinkle over filling.

Bake at 350° for 55-65 minutes or until filling is bubbly and topping is browned. Immediately drizzle with caramel topping. Cool on a wire rack. **Yield:** 8 servings.

Blue-Ribbon Peanut Butter Torte

(Pictured below)

Prep: 55 min. + chilling **Bake:** 20 min. + cooling

This impressive three-layer torte claimed the Grand Prize in a Taste of Home Cooking School contest.
—Carol Wilson, Ancho, New Mexico

1/2 cup plus 2 tablespoons butter, softened
1/2 cup chunky peanut butter
2 cups packed brown sugar
4 eggs
1 teaspoon vanilla extract
2-1/2 cups all-purpose flour
1 teaspoon baking soda
1 teaspoon baking powder
1/2 teaspoon salt
1 cup buttermilk
CHOCOLATE FILLING:
2-1/4 cups heavy whipping cream
1/2 cup packed brown sugar
12 squares (1 ounce *each*) bittersweet *or* semisweet chocolate, coarsely chopped
1/2 cup chunky peanut butter
CREAM CHEESE FROSTING:
12 ounces cream cheese, softened
6 tablespoons butter, softened
1 teaspoon vanilla extract
2 cups confectioners' sugar, *divided*
3/4 cup heavy whipping cream
2 Butterfinger candy bars (2.1 ounces *each*), coarsely chopped
1/3 cup honey roasted peanuts, coarsely chopped

Grease three 9-in. round baking pans; set aside. In a large mixing bowl, cream the butter, peanut butter and brown sugar until light and fluffy. Add eggs, one at a time, beating well after each. Beat in vanilla. Combine flour, baking soda, baking powder and salt; add to creamed mixture alternately with buttermilk.

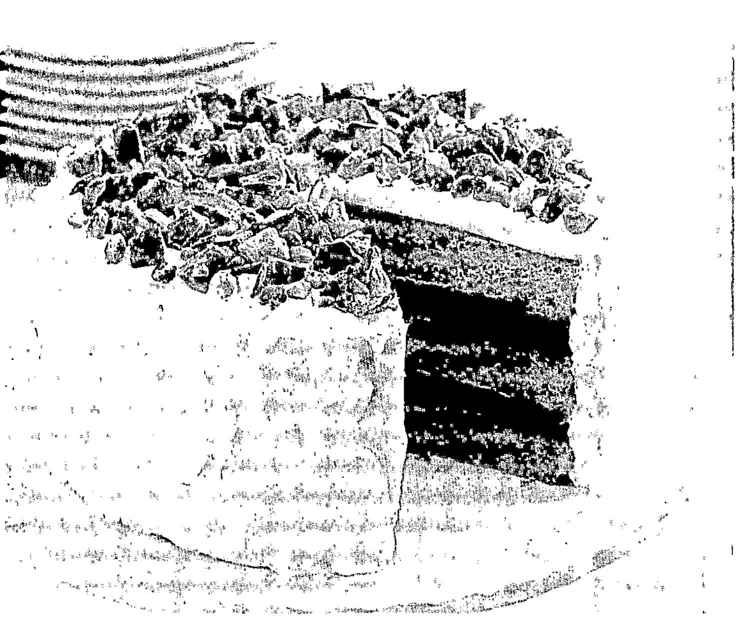

Pour into prepared pans (pans will have a shallow fill). Bake at 350° for 17-20 minutes or until a toothpick inserted near the center comes out clean. Cool for 10 minutes before removing from pans to wire racks.

For filling, in a large heavy saucepan, bring cream and brown sugar to a boil over medium heat. Reduce heat; cover and simmer for 1-2 minutes or until sugar is dissolved. Remove from the heat; stir in chocolate and peanut butter until blended. Transfer to a small bowl; chill until mixture achieves spreading consistency.

For frosting, in a large mixing bowl, beat cream cheese and butter until blended. Add vanilla; mix well. Gradually beat in 1-1/4 cups confectioners' sugar until light and fluffy. In a small mixing bowl, beat cream and remaining confectioners' sugar until stiff peaks form; fold into cream cheese mixture.

Spread filling between layers. Frost top and sides of cake. Garnish with chopped candy bars and peanuts. Store in the refrigerator. **Yield:** 14 servings.

Editor's Note: Reduced-fat or generic brands of peanut butter are not recommended for this recipe.

Chocolate Ganache Cake

Prep: 40 min. + chilling **Bake:** 20 min. + cooling

With a rich filling and glaze, this tantalizing dessert is a chocolate-lover's dream! —Kathy Kittell, Lenexa, Kansas

3/4 cup butter, softened
1-1/2 cups sugar
1 egg
1 teaspoon vanilla extract
1 cup buttermilk
3/4 cup sour cream
2 cups all-purpose flour
2/3 cup baking cocoa
1 teaspoon baking soda
1/4 teaspoon salt
FILLING:
4 squares (1 ounce *each*) semisweet chocolate
1 cup heavy whipping cream
1/2 teaspoon vanilla extract
GLAZE:
8 squares (1 ounce *each*) semisweet chocolate
1/4 cup butter, cubed
3/4 cup heavy whipping cream

In a large mixing bowl, cream butter and sugar until light and fluffy. Add egg and vanilla; beat for 2 minutes. Combine buttermilk and sour cream. Combine the flour, cocoa, baking soda and salt; add to creamed mixture alternately with buttermilk mixture, beating well after each addition.

Pour into two greased and waxed paper-lined 9-in.

round baking pans. Bake at 350° for 20-25 minutes or until a toothpick comes out clean. Cool for 10 minutes; remove from pans to wire racks to cool completely.

In a heavy saucepan, melt chocolate with cream over low heat. Remove from the heat; stir in vanilla. Transfer to a small mixing bowl; chill until slightly thickened, stirring occasionally. Beat on medium speed until light and fluffy. Chill until mixture achieves spreading consistency.

For glaze, in a heavy saucepan, melt chocolate and butter. Gradually add cream; heat until just warmed. Chill until slightly thickened.

Place one cake layer on a serving plate; spread with the filling. Top with the remaining cake layer. Slowly pour glaze over top of cake. Refrigerate until serving. **Yield:** 12-14 servings.

Cherry Gingerbread Cupcakes

Prep: 30 min. **Bake:** 20 min. + cooling

A sweet, lemony frosting complements these little spice cakes, each baked with a cherry in the center.
—Laura McAllister, Morganton, North Carolina

1/2 cup shortening
1 cup sugar
2 eggs
1 cup molasses
3 cups all-purpose flour
1 teaspoon baking soda
1 teaspoon ground ginger
1 teaspoon ground cinnamon
1 cup buttermilk
1/2 cup chopped walnuts
24 maraschino cherries, well drained
LEMON CREAM CHEESE FROSTING:
4 ounces cream cheese, softened
1/4 cup butter, softened
1 teaspoon vanilla extract
1 teaspoon grated lemon peel
1-3/4 to 2 cups confectioners' sugar

In a large mixing bowl, cream shortening and sugar until light and fluffy. Add eggs and molasses; mix well. Combine the flour, baking soda, ginger and cinnamon; add to creamed mixture alternately with buttermilk. Stir in walnuts.

Fill paper-lined muffin cups two-thirds full; place a cherry in the center of each. Bake at 375° for 20-24 minutes or until a toothpick comes out clean. Cool for 10 minutes before removing from pans to wire racks to cool completely.

For frosting, in a small mixing bowl, beat cream cheese and butter until smooth; add vanilla and lemon peel. Gradually beat in the confectioners' sugar. Frost cupcakes. **Yield:** 2 dozen.

Banana Nut Cake

(Pictured above)

Prep: 20 min. **Bake:** 30 min. + cooling

Topped with a sprinkling of chopped nuts, this yummy treat is loaded with banana and pecan flavor.
—Marlene Saunders, Lincoln, Nebraska

1/2 cup butter, softened
1 cup sugar
2 eggs
1/2 teaspoon vanilla extract
1-1/4 cups all-purpose flour
3/4 teaspoon baking soda
1/2 teaspoon salt
3/4 cup mashed ripe banana
1/2 cup chopped pecans, toasted
BUTTER PECAN FROSTING:
1/2 cup butter, softened
1/4 cup milk
1 teaspoon vanilla extract
Dash salt
2 to 2-1/2 cups confectioners' sugar
1 cup finely chopped pecans, toasted
Additional chopped pecans, optional

In a large mixing bowl, cream butter and sugar. Add eggs, one at a time, beating well after each addition. Stir in vanilla. Combine the flour, baking soda and salt; add to creamed mixture, beating just until combined. Fold in banana and pecans.

Pour into a greased 8-in. square baking dish. Bake at 350° for 30-35 minutes or until a toothpick inserted near the center comes out clean. Cool on a wire rack.

In a small mixing bowl, cream butter. Beat in the milk, vanilla, salt and enough confectioners' sugar to achieve spreading consistency. Stir in toasted pecans. Frost cake. Garnish with additional pecans if desired. **Yield:** 9 servings.

Red Velvet Heart Torte

(Pictured above and on page 100)

Prep: 25 min. **Bake:** 30 min. + cooling

I bake this scrumptious, fruit-topped layer cake every February 14 for my husband's birthday. The heart shape is really pretty for Valentine's Day, plus it's easier to make than it looks. —*Amy Freitag, Stanford, Illinois*

- **1 package (18-1/4 ounces) red velvet cake mix**
- **1 carton (6 ounces) raspberry yogurt**
- **1/3 cup confectioners' sugar**
- **1 carton (12 ounces) frozen whipped topping, thawed**
- **1 cup raspberry pie filling**

Splitting Cake Layers

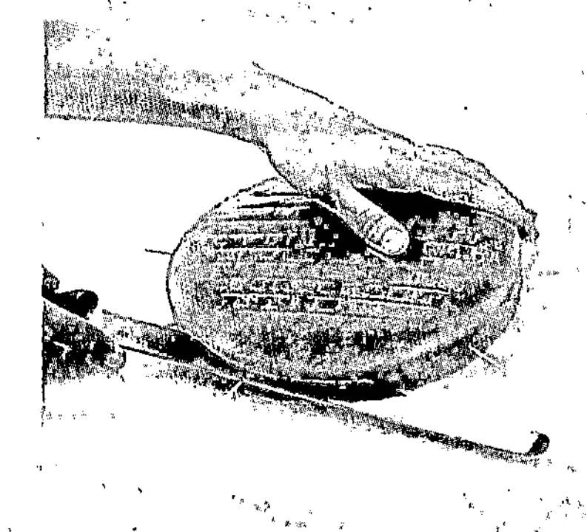

Use a ruler to locate the center of the cake layer, inserting toothpicks around the cake as a guide. Then cut the layer in half at the toothpicks using a long serrated knife.

Prepare cake batter according to package directions. Pour into two greased and floured 9-in. heart-shaped baking pans. Bake at 350° for 30-33 minutes or until a toothpick inserted near the centers comes out clean. Cool for 10 minutes before removing from pans to wire racks to cool completely.

In a large bowl, combine yogurt and confectioners' sugar; fold in whipped topping. Split each cake into two horizontal layers. Place one bottom layer on a serving plate; top with a fourth of the yogurt mixture. Repeat layers three times.

Spread the raspberry pie filling over the top of the cake to within 1 in. of the edges. Cover and refrigerate until serving. **Yield:** 14 servings.

Editor's Note: This recipe was tested with Duncan Hines red velvet cake mix.

Triple-Fruit Pie

Prep: 25 min. **Bake:** 50 min. + cooling

This golden-crusted pie is loaded with juicy slices of spiced peaches, apricots and nectarines. It's irresistible with a scoop of vanilla ice cream on a warm summer day. —*Janet Loomis, Terry, Montana*

- **2 cups all-purpose flour**
- **1 teaspoon salt**
- **3/4 cup shortening**
- **5 tablespoons cold water**
- **1-2/3 cups *each* sliced peeled peaches, nectarines and apricots**
- **1 tablespoon lemon juice**
- **1/2 cup packed brown sugar**
- **1/4 teaspoon ground ginger**
- **1/4 teaspoon ground cinnamon**
- **1 tablespoon butter**

In a small bowl, combine flour and salt; cut in shortening until crumbly. Gradually add water, tossing with a fork until dough forms a ball. Divide in half. On a lightly floured surface, roll out one portion to fit a 9-in. pie plate. Transfer pastry to plate; trim to 1/2 in. beyond edge.

In a large bowl, combine the peaches, nectarines, apricots and lemon juice. Combine the brown sugar, ginger and cinnamon; sprinkle over the fruit and toss gently to coat. Pour into crust; dot with butter.

Roll out remaining pastry to fit the top of pie; make decorative cutouts if desired. Set the cutouts aside. Place the top crust over the filling. Trim, seal and flute the edges. Moisten the cutouts with a small amount of water; place on top of pie.

Cover edges loosely with foil. Bake at 375° for 25 minutes. Uncover; bake 25-30 minutes longer or until crust is golden brown and filling is bubbly. Cool on a wire rack. **Yield:** 6-8 servings.

Elegant Chocolate Torte

(Pictured below and on page 100)

Prep: 50 min. **Bake:** 30 min. + cooling

When I want a really special dessert, I turn to this recipe. The four-layer torte has a yummy, pudding-like filling.
—Lois Gallup Edwards, Citrus Heights, California

1/3 cup all-purpose flour
3 tablespoons sugar
1 teaspoon salt
1-3/4 cups milk
1 cup chocolate syrup
1 egg, lightly beaten
1 tablespoon butter
1 teaspoon vanilla extract

BATTER:
1/2 cup butter, softened
1-1/4 cups sugar
4 eggs
1 teaspoon vanilla extract
1-1/4 cups all-purpose flour
1/3 cup baking cocoa
3/4 teaspoon baking soda
1/4 teaspoon salt
1-1/2 cups chocolate syrup
1/2 cup water

FROSTING:
2 cups heavy whipping cream
1/4 cup chocolate syrup
1/4 teaspoon vanilla extract

For filling, in a small saucepan, combine flour, sugar and salt. Stir in milk and syrup until smooth. Bring to a boil over medium heat, stirring constantly; cook and stir for 1-2 minutes or until thickened.

Remove from heat. Stir a small amount of hot mixture into egg; return all to the pan, stirring constantly. Bring to a gentle boil; cook and stir for 2 minutes. Remove from heat; stir in butter and vanilla. Cool to room temperature, stirring often.

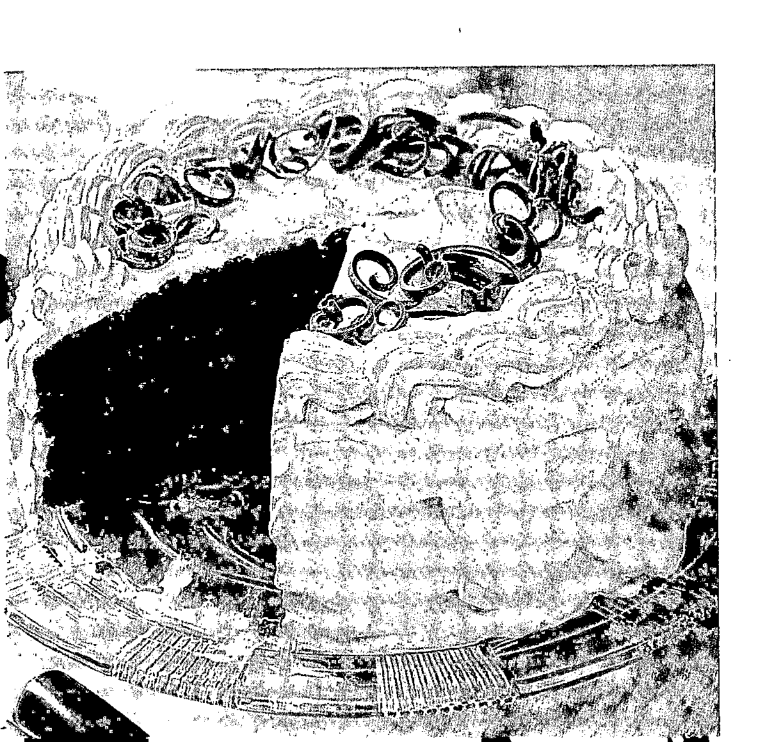

In a large mixing bowl, cream butter and sugar until light and fluffy. Add eggs, one at a time, beating well after each. Stir in vanilla. Combine dry ingredients; add to creamed mixture alternately with syrup and water. Beat just until combined.

Pour into two greased and floured 9-in. round baking pans. Bake at 350° for 30-35 minutes or until a toothpick comes out clean. Cool for 10 minutes; remove from pans to wire racks to cool.

Cut each cake in half horizontally. Place one bottom layer on a serving plate; spread with a third of the filling. Repeat layers twice. Top with remaining cake. In a mixing bowl, beat frosting ingredients until stiff peaks form; spread or pipe over the top and sides of cake. **Yield:** 16 servings.

Walnut Banana Cupcakes

Prep: 25 min. **Bake:** 20 min. + cooling

What makes these tender cupcakes extra special is the nutmeg. *—Rachel Krupp, Perkiomenville, Pennsylvania*

1/4 cup butter, softened
3/4 cup sugar
2 eggs
1/2 cup mashed ripe banana
1 teaspoon vanilla extract
1 cup all-purpose flour
1/2 teaspoon baking soda
1/2 teaspoon ground nutmeg
1/4 teaspoon salt
1/4 cup sour cream

CREAM CHEESE FROSTING:
4 ounces cream cheese, softened
1/2 teaspoon vanilla extract
1-3/4 cups confectioners' sugar
3 tablespoons chopped walnuts

In a small mixing bowl, cream butter and sugar. Add eggs, one at a time, beating well after each addition. Add banana and vanilla; mix well. Combine the flour, baking soda, nutmeg and salt; add to creamed mixture alternately with sour cream.

Fill paper-lined muffin cups half full. Bake at 350° for 18-22 minutes or until a toothpick comes out clean. Cool for 10 minutes before removing from pan to a wire rack to cool completely.

For frosting, in a small mixing bowl, combine cream cheese and vanilla. Gradually beat in confectioners' sugar. Frost cupcakes; sprinkle with walnuts. Store in the refrigerator. **Yield:** 1 dozen.

Dressed-Up Cupcakes

IRRESISTIBLE and yummy, cupcakes aren't just for kids anymore! From fluffy chocolate goodies to miniature upside-down cakes, the elegant treats here will delight young and old alike.

Special Mocha Cupcakes

(Pictured below)

Prep: 25 min. **Bake:** 20 min. + cooling

Topped with a fluffy frosting and chocolate sprinkles, these extra-rich cakes smell wonderful while baking...and taste even better! —*Mary Bilyeu, Ann Arbor, Michigan*

1-1/2 cups all-purpose flour
1 cup sugar
1/3 cup baking cocoa
1 teaspoon baking soda
1/2 teaspoon salt
2 eggs
1/2 cup cold brewed coffee
1/2 cup vegetable oil
3 teaspoons cider vinegar
3 teaspoons vanilla extract
MOCHA FROSTING:
3 tablespoons semisweet chocolate chips
3 tablespoons milk chocolate chips
1/3 cup butter, softened
2 cups confectioners' sugar
1 to 2 tablespoons brewed coffee
1/2 cup chocolate sprinkles

In a small mixing bowl, combine the flour, sugar, cocoa, baking soda and salt. In a small bowl, whisk the eggs, coffee, oil, vinegar and vanilla. Add to dry ingredients; mix well.

Fill paper-lined muffin cups three-fourths full. Bake at 350° for 20-25 minutes or until a toothpick comes out clean. Cool for 10 minutes before removing from pan to a wire rack to cool.

For frosting, in a small microwave-safe mixing bowl, melt chips; stir until smooth. Add butter; beat until blended. Gradually beat in confectioners' sugar and coffee. Pipe frosting onto cupcakes. Top with sprinkles; gently press down. **Yield:** 1 dozen.

Lemon Curd Cupcakes

Prep: 40 min. + chilling **Bake:** 20 min. + cooling

Homemade lemon curd gives these cakes great flavor. If you're a fan of lemon, you're sure to like these.
—*Kerry Barnett-Amundson, Ocean Park, Washington*

3 tablespoons plus 1-1/2 teaspoons sugar
3 tablespoons lemon juice
4-1/2 teaspoons butter
1 egg, lightly beaten
1 teaspoon grated lemon peel
BATTER:
3/4 cup butter, softened
1 cup sugar
2 eggs
1 teaspoon vanilla extract
1 teaspoon grated lemon peel
1-1/2 cups cake flour
1/2 teaspoon baking powder
1/4 teaspoon baking soda
1/4 teaspoon salt
2/3 cup buttermilk
FROSTING:
2 tablespoons butter, softened
1/2 teaspoon vanilla extract
Pinch salt
2 cups confectioners' sugar
2 to 4 tablespoons milk
Yellow food coloring, optional

For lemon curd, in a heavy saucepan, cook and stir sugar, lemon juice and butter until smooth. Stir a small amount into egg; return all to pan. Bring to a gentle boil, stirring constantly; cook 2 minutes longer. Stir in lemon peel. Cool 10 minutes. Cover and chill for 1-1/2 hours or until thickened.

In a mixing bowl, cream butter and sugar. Add eggs, one at a time, beating well after each. Add vanilla and lemon peel. Combine flour, baking powder, baking soda and salt; add to creamed mixture alternately with buttermilk.

Fill paper-lined muffin cups three-fourths full. Bake at 350° for 20-25 minutes or until a toothpick comes out clean. Cool 10 minutes; remove from pan to a wire rack. Cool completely.

Cut a small hole in the corner of a pastry or plastic bag; insert a small round pastry tip. Fill the bag with curd. Insert tip 1 in. into center of each cupcake; fill with curd just until tops of cupcakes begin to crack.

Combine frosting ingredients, tinting with food coloring if desired; frost cupcakes. Store in the refrigerator. **Yield:** 1 dozen.

Berry-Topped White Cupcakes

(Pictured above)

Prep: 30 min. **Bake:** 20 min. + cooling

Guests love these frosted cupcakes garnished with fresh berries. —Judy Kenninger, Brownsburg, Indiana

5 egg whites
1/2 cup plus 2 tablespoons butter, softened
1 cup sugar, *divided*
3/4 teaspoon vanilla extract
2-1/4 cups cake flour
2-1/4 teaspoons baking powder
1/2 teaspoon salt
3/4 cup milk
ICING:
4 ounces cream cheese, softened
1/3 cup butter, softened
2 cups confectioners' sugar
1/2 teaspoon lemon juice
Fresh blueberries, raspberries and sliced strawberries

Place egg whites in a large mixing bowl; let stand at room temperature for 30 minutes. In another mixing bowl, cream butter and 3/4 cup sugar until light and fluffy. Beat in vanilla. Combine flour, baking powder and salt; add to creamed mixture alternately with milk.

Beat egg whites on medium speed until soft peaks form. Gradually beat in remaining sugar, about 2 tablespoons at a time, on high until stiff glossy peaks form and sugar is dissolved. Fold a fourth of the egg whites into batter; fold in remaining whites.

With a spoon, gently fill foil- or paper-lined muffin cups two-thirds full. Bake at 350° for 18-22 minutes. Cool for 10 minutes before removing from pans to wire racks to cool completely.

For icing, in a small mixing bowl, beat cream cheese and butter until smooth. Gradually beat in confectioners' sugar and lemon juice. Spread over cupcakes. Top with berries. **Yield:** 22 cupcakes.

Pineapple Upside-Down Cupcakes

Prep: 30 min. **Bake:** 30 min. + cooling

These jumbo treats make an attractive dessert for special days or any day. —Barbara Hahn, Park Hills, Missouri

6 tablespoons butter, cubed
1 cup packed light brown sugar
2 tablespoons light corn syrup
1 small pineapple, peeled, cored and cut into 1/2-inch slices
12 maraschino cherries, well drained
3 eggs
2 cups sugar
1 cup vegetable oil
1 cup (8 ounces) sour cream
2 teaspoons vanilla extract
2-1/2 cups all-purpose flour
1/2 teaspoon baking powder
1/2 teaspoon baking soda
1/2 teaspoon salt
Whipped topping, optional

Line greased jumbo muffin cups with waxed paper; grease the paper and set aside. In a small saucepan, melt butter over low heat; stir in brown sugar and corn syrup. Cook and stir over medium heat until sugar is dissolved. Remove from the heat. Spoon 1 tablespoonful into each muffin cup; top each with a pineapple slice and a cherry.

In a large mixing bowl, beat eggs and sugar until thickened and lemon-colored. Beat in oil, sour cream and vanilla until smooth. Combine flour, baking powder, baking soda and salt. Add to egg mixture; mix well.

Fill muffin cups two-thirds full. Bake at 350° for 28-32 minutes or until a toothpick comes out clean. Cool for 5 minutes before inverting onto wire racks to cool completely. Garnish with whipped topping if desired. **Yield:** 1 dozen jumbo cupcakes.

Butterscotch Swirl Cake

(Pictured above)

Prep: 30 min. **Bake:** 65 min. + cooling

I was tickled when this swirled dessert took first place and "Best of Division" at the Los Angeles County Fair.
—Marina Castle-Henry, Burbank, California

1 cup butter, softened
2 cups sugar
6 eggs
3 teaspoons rum extract
1 teaspoon vanilla extract
3 cups all-purpose flour
1 teaspoon baking soda
1 teaspoon baking powder
1 cup (8 ounces) sour cream
1 package (3.4 ounces) instant butterscotch pudding mix
3/4 cup butterscotch ice cream topping
BUTTERSCOTCH GLAZE:
1/4 cup butter, cubed
1/4 cup packed brown sugar
2 tablespoons milk
1 cup confectioners' sugar
1 teaspoon vanilla extract
1/4 cup chopped pecans

In a large mixing bowl, cream butter and sugar until light and fluffy. Add 5 eggs, one at a time, beating well after each addition. Stir in extracts. Combine the flour, baking soda and baking powder; add to creamed mixture alternately with sour cream, beating well after each addition.

Transfer 2 cups of batter to another large mixing bowl; beat in the pudding mix, butterscotch topping and remaining egg until well blended. Pour half of the plain batter into a greased and floured 10-in. fluted tube pan. Top with half of the butterscotch batter; cut through with a knife to swirl. Repeat layers and swirl.

Bake at 350° for 65-70 minutes or until a toothpick inserted near the center comes out clean. Cool for 10 minutes before removing from pan to a wire rack to cool completely.

For glaze, in a small saucepan, combine the butter, brown sugar and milk. Bring to a boil. Remove from the heat; add confectioners' sugar and vanilla. Beat until smooth and creamy. Drizzle over cake; sprinkle with pecans. **Yield:** 12 servings.

Triple-Apple Pie

Prep: 30 min. **Bake:** 50 min. + cooling

When making this blue ribbon-winning pie, I like to use my homemade apple jelly. *—Louise Piper, Rolfe, Iowa*

5-1/2 cups thinly sliced peeled tart apples
1/4 cup apple cider *or* juice
1/3 cup apple jelly, melted
1 cup sugar
3 tablespoons all-purpose flour
1 tablespoon quick-cooking tapioca
1/8 teaspoon salt
Pastry for double-crust pie (9 inches)
2 tablespoons butter

In a large bowl, combine the apples, cider and jelly. Combine the sugar, flour, tapioca and salt; add to apple mixture and toss gently to coat. Let stand for 15 minutes.

Meanwhile, line a 9-in. pie plate with bottom pastry; trim pastry even with edge of plate. Add filling; dot with butter. Roll out remaining pastry to fit top of pie; place over filling. Trim, seal and flute edges. Cut slits in top. Cover edges loosely with foil.

Bake at 400° for 20 minutes. Remove foil; bake 30-35 minutes longer or until crust is golden brown and filling is bubbly. Cool on a wire rack. **Yield:** 6-8 servings.

Reusable Crust Cover

The top edges of a pie pastry often brown before the rest of the pie is thoroughly baked, which is why many recipes say to cover the edges with foil.

It's easy to make a foil cover that can be washed and reused. On a 12-inch disposable foil pizza pan, draw a 7-inch-diameter circle in the center. Cut out the 7-inch circle and discard it. Then just center the "O"-shaped foil over the pie to cover the edges of the crust.

Mocha Nut Torte

(Pictured below)

Prep: 40 min. + chilling **Bake:** 20 min. + cooling

My husband doesn't care for chocolate cake, but this spectacular three-layer torte is a favorite of his. I've been using this recipe for special occasions for many years.
—Megan Shepherdson, Winnipeg, Manitoba

7 eggs, *separated*
1 cup sugar, *divided*
1 teaspoon vanilla extract
1-1/4 cups ground walnuts
1-1/4 cups ground pecans
1/4 cup dry bread crumbs
1 teaspoon baking powder
3/4 teaspoon salt, *divided*

FILLING:
1 cup heavy whipping cream
1/2 cup confectioners' sugar
1 teaspoon vanilla extract

MOCHA FROSTING:
1/4 cup butter, cubed
4 squares (1 ounce *each*) unsweetened chocolate
1/2 cup brewed coffee
2 teaspoons vanilla extract
3 to 3-1/4 cups confectioners' sugar
Pecan halves, optional

Line three 9-in. round baking pans with waxed paper; set aside. Place egg whites in a large mixing bowl; let stand at room temperature for 30 minutes. Meanwhile, in another mixing bowl, beat egg yolks until slightly thickened. Gradually add 1/2 cup sugar, beating until thick and lemon-colored. Beat in vanilla. Combine nuts, crumbs, baking powder and 1/2 teaspoon salt; stir into yolk mixture until combined.

Add remaining salt to egg whites; beat on medium speed until soft peaks form. Gradually beat in remaining sugar, about 2 tablespoons at a time, on high until stiff glossy peaks form and sugar is dissolved. Gradually fold into batter just until blended. Divide among prepared pans. Bake at 375° for 20-25 minutes or until tops spring back when lightly touched. Invert pans; cool for 20 minutes. Remove from pans to wire racks; cool completely. Remove waxed paper.

In a small mixing bowl, beat cream until it begins to thicken. Add confectioners' sugar and vanilla; beat until stiff peaks form. Cover and refrigerate until assembling.

In a large saucepan, melt butter and chocolate over low heat. Remove from the heat. Stir in coffee, vanilla and enough confectioners' sugar to achieve frosting consistency. Spread filling between layers. Frost top and sides of cake. Garnish with pecans if desired. **Yield:** 12 servings.

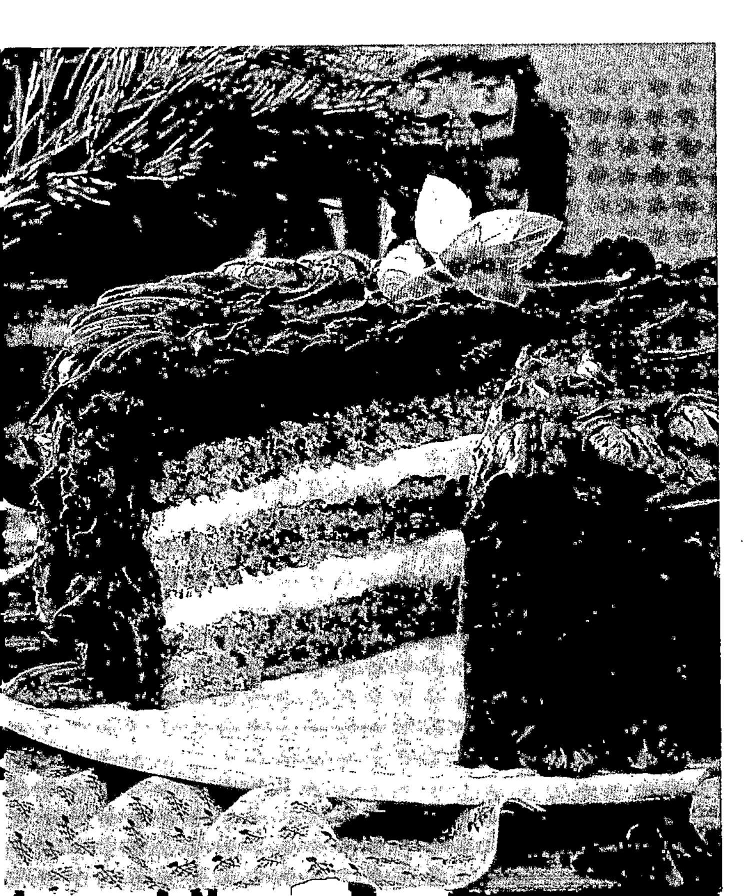

Yuletide Pound Cake

Prep: 15 min. **Bake:** 45 min. + cooling

We enjoy this Bundt cake at Christmastime with cups of hot chocolate. —Lorraine Caland, Thunder Bay, Ontario

1 cup butter, softened
1/2 cup shortening
3 cups sugar
5 eggs
1 teaspoon vanilla extract
1 teaspoon rum extract
3 cups all-purpose flour
1 teaspoon baking powder
1 cup (8 ounces) sour cream

GLAZE:
1 cup confectioners' sugar
2 to 3 teaspoons milk

In a large mixing bowl, cream the butter, shortening and sugar until light and fluffy, about 5 minutes. Add eggs, one at a time, beating well after each addition. Beat in extracts. Combine flour and baking powder; add to creamed mixture alternately with sour cream. Beat just until combined.

Pour into a greased and floured 10-in. fluted tube pan. Bake at 350° for 60-70 minutes or until a toothpick inserted near the center comes out clean. Cool for 10 minutes before removing from pan to a wire rack to cool completely. Combine glaze ingredients; drizzle over cake. **Yield:** 12-16 servings.

Strawberry-Mallow Cake Roll

(Pictured above)

Prep: 30 min. **Bake:** 10 min. + chilling

This stunning cake roll is so much fun to make and present to guests. —*Susan Olsen, Huntley, Montana*

- **4 eggs, *separated***
- **2/3 cup all-purpose flour**
- **1 teaspoon baking powder**
- **1/4 teaspoon salt**
- **3/4 cup sugar, *divided***
- **1/2 teaspoon vanilla extract**
- **2 cartons (8 ounces *each*) spreadable strawberry cream cheese**
- **1 jar (7 ounces) marshmallow creme**
- **3 cups sliced fresh strawberries, *divided***
- **Chocolate syrup, optional**

Let eggs stand at room temperature for 30 minutes. Line a greased 15-in. x 10-in. x 1-in. baking pan with waxed paper and grease the paper; set aside.

Sift together the flour, baking powder and salt; set aside. In a large mixing bowl, beat egg yolks until slightly thickened. Gradually add 1/4 cup sugar, beating until thick and lemon-colored. Beat in vanilla. Add dry ingredients; mix well.

In a small mixing bowl, beat egg whites on medium speed until soft peaks form. Gradually beat in remaining sugar, about 2 tablespoons at a time, on high until stiff glossy peaks form and sugar is dissolved. Fold a fourth of egg whites into batter; fold in remaining whites.

Gently spoon into prepared pan. Bake at 375° for 10-12 minutes or until cake springs back when lightly touched. Cool for 5 minutes. Turn cake onto a kitchen towel dusted with confectioners' sugar. Gently peel off waxed paper. Roll up cake in the towel, starting with a short side. Cool completely on a wire rack.

In a small mixing bowl, beat the cream cheese and marshmallow creme. Unroll the cake; spread cream cheese mixture to within 1/2 in. of edges. Top with 2-1/2 cups strawberries. Roll up again. Place seam side down on a platter. Refrigerate for at least 2 hours. Just before serving, garnish with remaining strawberries. Serve with chocolate syrup if desired. Refrigerate leftovers. **Yield:** 10 servings.

Coconut Pecan Cupcakes

Prep: 50 min. **Bake:** 20 min. + cooling

Pecan lovers have lots to cheer about with these luscious goodies. —*Tina Harrison, Prairieville, Louisiana*

- **5 eggs, *separated***
- **1/2 cup shortening**
- **1/2 cup butter, softened**
- **2 cups sugar**
- **3/4 teaspoon vanilla extract**
- **1/4 teaspoon almond extract**
- **1-1/2 cups all-purpose flour**
- **1/4 cup cornstarch**
- **1/2 teaspoon baking soda**
- **1/2 teaspoon salt**
- **1 cup buttermilk**
- **2 cups flaked coconut**
- **1 cup finely chopped pecans**

FROSTING:

- **1 package (8 ounces) cream cheese, softened**
- **1/4 cup butter, softened**
- **1/2 teaspoon vanilla extract**
- **1/4 teaspoon almond extract**
- **1-1/2 cups confectioners' sugar**
- **3/4 cup chopped pecans**

Let eggs stand at room temperature for 30 minutes. In a large mixing bowl, cream shortening, butter and sugar until light and fluffy. Add egg yolks, one at a time, beating well after each. Stir in extracts. Combine the flour, cornstarch, baking soda and salt; add to creamed mixture alternately with the buttermilk. Beat just until combined.

In a small mixing bowl, beat egg whites on high speed until stiff peaks form. Fold into batter. Stir in coconut and pecans.

Fill paper-lined muffin cups three-fourths full. Bake at 350° for 20-25 minutes or until a toothpick comes out clean. Cool 10 minutes; remove from pans to wire racks. Cool completely. Combine frosting ingredients; frost cupcakes. Store in refrigerator. **Yield:** 2 dozen.

Chocolate-Pecan Pudding Cakes

Prep: 15 min. **Bake:** 25 min. + cooling

Topped with syrup and whipped cream, these nutty chocolate mounds are both yummy and comforting.
—Cory Tower, Columbus, Nebraska

1 cup all-purpose flour
2/3 cup sugar
6 tablespoons baking cocoa, *divided*
2 teaspoons baking powder
1/4 teaspoon salt
1/2 cup milk
1/4 cup butter, melted
1 teaspoon vanilla extract
1/2 cup coarsely chopped pecans
2/3 cup packed brown sugar
3/4 cup hot water
Whipped cream, optional

In a large bowl, combine the flour, sugar, 3 tablespoons cocoa, baking powder and salt. Combine the milk, butter and vanilla; stir into dry ingredients just until combined. Stir in pecans. Spoon into six greased 6-oz. custard cups.

Combine brown sugar and remaining cocoa; sprinkle over batter. Pour 2 tablespoons hot water over each cup. Place cups on a baking sheet.

Bake at 350° for 25-30 minutes or until toothpick inserted in cake portion comes out clean. Cool on wire racks for 15 minutes. Run a knife around the edge of each cup; invert onto dessert plates. Serve warm with whipped cream if desired. **Yield:** 6 servings.

Editor's Note: This recipe does not use eggs.

Chip Lover's Cupcakes

(Pictured at right)

Prep: 30 min. **Bake:** 20 min. + cooling

My children's love of cookie dough inspired the recipe for these cupcakes. —Donna Scully, Middletown, Delaware

1 package (18-1/4 ounces) white cake mix
1/4 cup butter, softened
1/4 cup packed brown sugar
2 tablespoons sugar
1/3 cup all-purpose flour
1/4 cup confectioners' sugar
1/4 cup miniature semisweet chocolate chips
BUTTERCREAM FROSTING:
1/2 cup butter, softened
1/2 cup shortening
4-1/2 cups confectioners' sugar
4 tablespoons milk, *divided*
1-1/2 teaspoons vanilla extract
1/4 cup baking cocoa
18 miniature chocolate chip cookies

Prepare cake batter according to package directions; set aside. For filling, in a small mixing bowl, cream the butter and sugars. Beat in flour and confectioners' sugar until blended. Fold in chocolate chips.

Fill paper-lined muffin cups half full with cake batter. Drop filling by tablespoonfuls into the center of each; cover with remaining batter. Bake at 350° for 20-22 minutes or until a toothpick inserted in cake comes out clean. Cool for 10 minutes before removing from pans to wire racks to cool completely.

For frosting, in a large mixing bowl, cream butter, shortening and confectioners' sugar. Beat in 3 tablespoons milk and vanilla until creamy. Set aside 1 cup frosting; frost cupcakes with remaining frosting.

Stir baking cocoa and remaining milk into reserved frosting. Cut a small hole in a corner of a pastry or plastic bag; insert #21 star tip. Fill bag with chocolate frosting. Pipe a rosette on top of each cupcake; garnish with a cookie. **Yield:** 1-1/2 dozen.

Cupcake Dress-Up Ideas

- Top each cupcake with a confectionery rose, found in the supermarket's baking aisle.
- Sprinkle them with English toffee bits, crushed sandwich cookies, mini chocolate chips or mini M&M's baking bits.
- Add seasonal flair with crushed peppermint candies at Christmastime, candy corns for Halloween, jelly beans for Easter, etc.

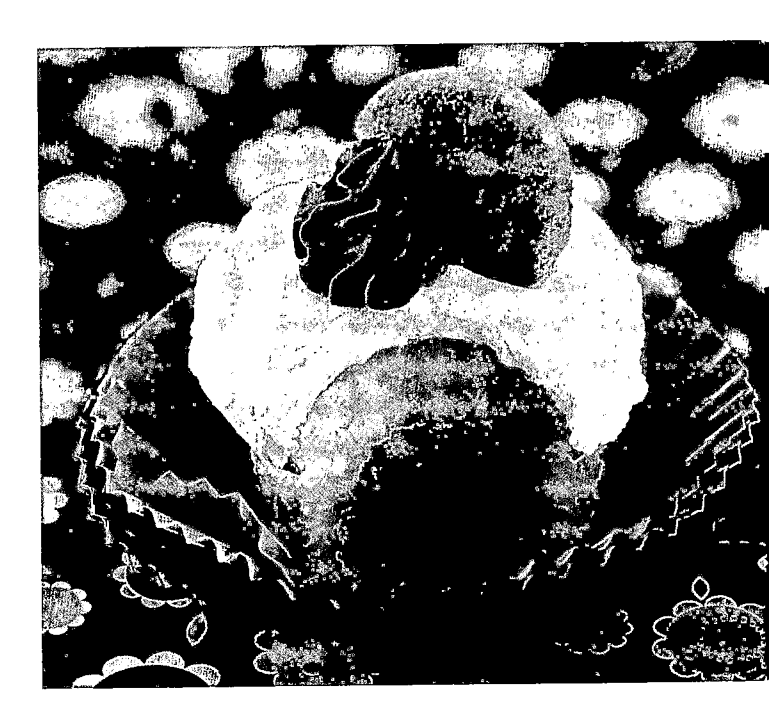

Chocolate Cream Cheese Cupcakes

(Pictured below)

Prep: 30 min. **Bake:** 25 min. + cooling

I got the recipe for these goodies from a dear friend of mine about 25 years ago. The moist, filled cupcakes are irresistible. —*Vivian Morris, Cleburne, Texas*

- **1 package (8 ounces) cream cheese, softened**
- **1-1/2 cups sugar, *divided***
- **1 egg**
- **1 teaspoon salt, *divided***
- **1 cup (6 ounces) semisweet chocolate chips**
- **1-1/2 cups all-purpose flour**
- **1/4 cup baking cocoa**
- **1 teaspoon baking soda**
- **1 cup water**
- **1/3 cup vegetable oil**
- **1 tablespoon white vinegar**

FROSTING:

- **3-3/4 cups confectioners' sugar**
- **3 tablespoons baking cocoa**
- **1/2 cup butter, melted**
- **6 tablespoons milk**
- **1 teaspoon vanilla extract**
- **1/3 cup chopped pecans**

For filling, in a small mixing bowl, beat cream cheese and 1/2 cup sugar until smooth. Beat in egg and 1/2 teaspoon salt until combined. Fold in chocolate chips; set aside.

In a bowl, combine flour, cocoa, baking soda, and remaining sugar and salt. In another bowl, whisk water, oil and vinegar; stir into dry ingredients just until moistened.

Fill paper-lined muffin cups half full with batter. Drop filling by heaping tablespoonfuls into the center of each. Bake at 350° for 24-26 minutes or until a toothpick inserted in cake comes out clean. Cool for 10 minutes before removing from pans to wire racks to cool completely.

For frosting, in a large mixing bowl, combine confectioners' sugar, cocoa, butter, milk and vanilla; beat until blended. Frost cupcakes; sprinkle with pecans. Store in the refrigerator. **Yield:** 20 cupcakes.

Marmalade Pudding Cakes

Prep: 20 min. **Bake:** 25 min.

For a light dessert, serve guests these cute custard treats accented with orange peel and glazed with marmalade sauce. —*Marian Platt, Sequim, Washington*

✓ **Uses less fat, sugar or salt. Includes Nutrition Facts and Diabetic Exchanges.**

- **2 tablespoons butter, softened**
- **3/4 cup sugar, *divided***
- **1/4 cup all-purpose flour**
- **4 eggs, *separated***
- **1 cup milk**
- **1/4 cup orange juice**
- **1/4 cup lemon juice, *divided***
- **1-1/2 teaspoons grated orange peel**
- **1/3 cup orange marmalade, warmed**

In a small mixing bowl, beat butter and 1/2 cup sugar until crumbly. Beat in flour and egg yolks until smooth. Gradually beat in the milk, orange juice, 2 tablespoons lemon juice and orange peel.

In another small mixing bowl, beat egg whites on high speed until soft peaks form. Add the remaining sugar; beat until stiff peaks form. Gently fold into orange mixture.

Pour into eight 6-oz. custard cups thoroughly coated with nonstick cooking spray. Place cups in two 13-in. x 9-in. x 2-in. baking pans; add 1 in. of boiling water to pans.

Bake at 325° for 25-30 minutes or until a knife inserted near the center comes out clean and tops are golden brown. Run a knife around the edges; carefully invert cakes onto dessert plates. Combine marmalade and remaining lemon juice; drizzle over warm cakes. **Yield:** 8 servings.

Nutrition Facts: 1 cake equals 206 calories, 6 g fat (3 g saturated fat), 118 mg cholesterol, 83 mg sodium, 34 g carbohydrate, trace fiber, 5 g protein. **Diabetic Exchanges:** 2 starch, 1 fat.

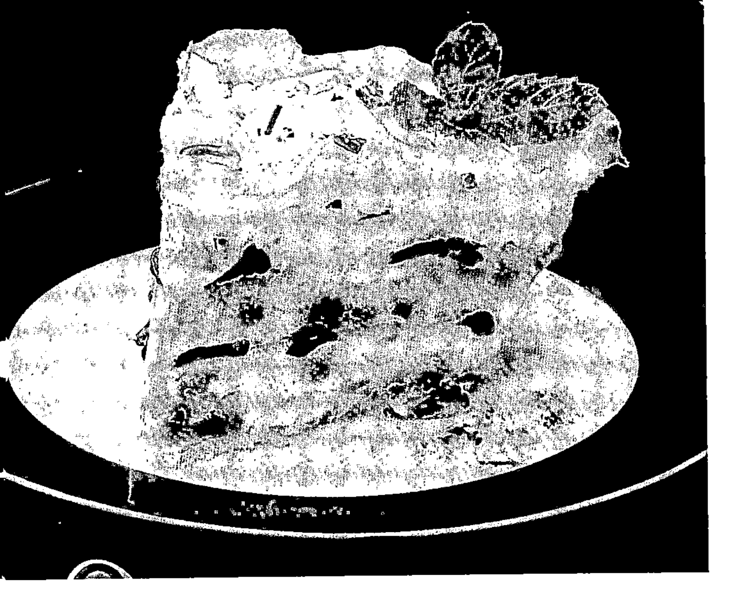

Peppermint Freezer Pie

(Pictured above)

Prep: 40 min. + freezing

Refreshing peppermint ice cream, fudgy chocolate plus fluffy meringue make this pie perfect for Christmastime. You can prepare and freeze the pie well in advance.
—Kelli Bucy, Massena, Iowa

- **2 squares (1 ounce *each*) unsweetened chocolate**
- **3 tablespoons butter**
- **1/2 cup sugar**
- **2/3 cup evaporated milk**
- **1 teaspoon vanilla extract**
- **1 quart peppermint ice cream, softened**
- **1 pastry shell (9 inches), baked**

MERINGUE:
- **3 egg whites**
- **6 tablespoons sugar**
- **1 tablespoon water**
- **1/4 teaspoon cream of tartar**
- **1/2 teaspoon vanilla extract**
- **3 tablespoons crushed peppermint-stick candy**

In a heavy saucepan, melt chocolate and butter over low heat. Stir in sugar and milk. Cook and stir for 8 minutes or until sugar is dissolved. Remove from the heat; stir in vanilla. Cool completely.

Spread half of the ice cream into pastry shell; freeze until firm. Spread half of the chocolate mixture over ice cream; freeze until set. Repeat layers. Freeze for several hours or overnight.

In a heavy saucepan, combine the egg whites, sugar, water and cream of tartar. With a portable mixer, beat on low speed for 1 minute. Continue beating over low heat until mixture reaches 160°, about 12 minutes.

Remove from the heat. Add vanilla; beat until stiff glossy peaks form and sugar is dissolved. Fold in peppermint candy. Spread over top of pie. Cover and freeze until serving. Pie may be frozen for up to 2 months. **Yield:** 6-8 servings.

Coconut-Filled Nut Torte

Prep: 45 min. **Bake:** 20 min. + cooling

This torte has five kinds of nuts, a coconut filling and lots of cream cheese frosting. It was fun creating this recipe because my family really enjoys desserts and nuts.
—Callie Barnum, Prattsville, New York

- **2/3 cup shortening**
- **1-2/3 cups sugar**
- **5 egg whites**
- **1 teaspoon almond extract**
- **2-1/4 cups all-purpose flour**
- **2-1/4 teaspoons baking powder**
- **1 teaspoon salt**
- **1-1/4 cups milk**
- **1/2 cup finely chopped almonds**
- **1/2 cup finely chopped hazelnuts**

FILLING:
- **1 can (5 ounces) evaporated milk**
- **3/4 cup sugar**
- **6 tablespoons butter, cubed**
- **3 egg yolks, lightly beaten**
- **1 teaspoon vanilla extract**
- **1-1/2 cups flaked coconut**
- **1/2 cup *each* chopped hazelnuts, pecans and walnuts**

FROSTING:
- **3/4 cup butter, softened**
- **2 packages (3 ounces *each*) cream cheese, softened**
- **1 teaspoon vanilla extract**
- **3 cups confectioners' sugar**
- **2 tablespoons milk**

Pecan halves, optional

In a large mixing bowl, cream the shortening and sugar until light and fluffy. Add egg whites and almond extract; mix well. Combine the flour, baking powder and salt; add to the creamed mixture alternately with the milk. Fold in the nuts.

Pour into three greased and floured 9-in. round baking pans. Bake at 350° for 20-25 minutes or until a toothpick inserted near the center comes out clean. Cool for 10 minutes; remove from pans to wire racks.

In a large saucepan, combine the evaporated milk, sugar, butter and egg yolks. Cook and stir over medium heat for 10-12 minutes or until mixture is thickened and reaches 160°. Remove from the heat; stir in the vanilla extract. Stir in the coconut and nuts. Cool to room temperature.

In a small mixing bowl, beat butter, cream cheese and vanilla until smooth. Gradually add confectioners' sugar. Beat in milk until light and fluffy. Spread filling between cake layers. Frost top and sides of cake. Garnish with pecans if desired. Store in the refrigerator. **Yield:** 12-14 servings.

Creative Party Cakes

WANT an extra-special dessert for your child's next birthday? Look no further than the fun monkey cupcakes, bright barnyard and playful game cake here. You're guaranteed to see smiles!

Candy Land Cake

(Pictured below)

Prep: 2 hours **Bake:** 30 min. + cooling

I was inspired by the ever-popular board game to create this whimsical cake. —*Pen Perez, Berkeley, California*

2 packages (18-1/4 ounces *each*) cake mix of your choice
Vanilla and chocolate frosting
Green mist food color spray, optional
Assorted decorations: Starburst candies, red Fruit Roll-Up, red coarse sugar, Dots, regular and miniature peanut butter cups, chocolate jimmies, large and small gumdrops, Dum Dum Pops, miniature candy canes, clear and blue rock candy, cake and waffle ice cream cones, multicolored sprinkles, green colored sugar, miniature marshmallows, round peppermints and conversation hearts

Line two 13-in. x 9-in. x 2-in. baking pans with waxed paper and grease the paper. Prepare cake batter; pour into prepared pans. Bake according to package directions. Cool for 15 minutes before removing from pans to wire racks to cool completely; remove waxed paper.

Level tops of cakes; place side by side on a covered board. Frost top and sides of cake with vanilla frosting; mist with food color spray if desired.

With a Candy Land game board as your guide, form a path using Starburst candies. With vanilla frosting, pipe "Happy Birthday" on Starburst candies. With chocolate frosting, make an arrow; pipe "Start" on the arrow with vanilla frosting.

For the Mountain/Gumdrop pass, use a red Fruit Roll-Up, red coarse sugar and Dots.

For forests, add peanut butter cups topped with piped chocolate frosting, chocolate jimmies, gumdrops, Dum Dum Pops, candy canes and rock candy.

For castle, pipe vanilla frosting into ice cream cones. Garnish with Dots and sprinkles.

Between pathways, add green sugar, sprinkles, marshmallows, peppermints and hearts. Pipe additional frosting to fill in spaces; top with sprinkles.

Pipe vanilla frosting around the base of the cake; place peppermints around the top edge of the cake. **Yield:** 30-40 servings.

Editor's Note: This cake is best eaten the day it's prepared. Do not refrigerate.

Barnyard Cupcakes

(Pictured above right and on page 100)

Prep: 1 hour 20 min. **Bake:** 55 min. + cooling

The sun, pig and chick cupcakes are adorable with this red barn. —*Colleen Palmer, Epping, New Hampshire*

1 package (18-1/2 ounces) butter recipe golden cake mix
1 package (18-1/4 ounces) devil's food cake mix
4 cans (16 ounces *each*) vanilla frosting
Red, blue, pink and yellow paste food coloring
Miniature semisweet chocolate chips
Black and red decorating gel
Miniature M&M's baking bits
Assorted candies (pink mint candy lozenges, Good & Plenty candies and Brach's white dessert mints)
Candy corn

Prepare golden cake batter according to package directions. Pour into a greased and waxed paper-lined

13-in. x 9-in. x 2-in. baking pan. Bake at 350° for 33-38 minutes or until a toothpick inserted near the center comes out clean. Cool for 15 minutes before removing from pan to a wire rack.

Prepare devil's food cake batter according to package directions. Fill 24 paper-lined muffin cups two-thirds full. Bake at 350° for 21-26 minutes or until a toothpick comes out clean. Cool for 10 minutes before removing from pans to wire racks.

Tint 1 cup frosting red. Frost two-thirds of the cake, forming a barn shape; spread frosting on sides of barn. For sky, tint 2/3 cup frosting blue; frost remaining top and sides of cake.

Outline barn roof with chocolate chips. Insert a #6 round tip into a pastry bag; fill bag with 1/4 cup white frosting. Outline barn doors and window. With black decorating gel and an M&M's baking bit, add a spiderweb and spider.

For pigs, tint 1 cup frosting pink; frost 12 cupcakes. Attach baking bits for eyes, pink candy lozenges for snouts, chocolate chips for nostrils and Good & Plenty candies for ears.

For chicks, tint 2 cups frosting yellow; transfer to a pastry bag. Insert a #17 star tip; pipe frosting onto 11 cupcakes. Pipe heads and wings. Decorate with dessert mints, baking bits and candy corn.

For sun, frost remaining cupcake with yellow frosting. Position candy corn around edges. Form a face with chocolate chips and red decorating gel. **Yield:** 1 cake (12-15 servings) plus 2 dozen cupcakes.

Editor's Note: Use of a coupler ring will allow you to easily change pastry tips for different designs.

Monkey Cupcakes

(Pictured below)

Prep: 30 min. **Bake:** 20 min. + cooling

Kids' eyes will light up when they see these cute jungle goodies. They never fail to make my grandkids giggle, and they're always a hit at bake sales.—Sandra Seaman
Greensburg, Pennsylvania

1 package (18-1/4 ounces) chocolate cake mix
1 can (16 ounces) chocolate frosting
24 vanilla wafers
Black and red decorating gel
48 light blue, dark blue *or* white milk chocolate M&M's
12 peanut butter cream-filled sandwich cookies

Prepare cake batter; bake according to package directions for cupcakes. Cool completely on wire racks.

Set aside 1/4 cup frosting. Frost cupcakes with remaining frosting. With a serrated knife, cut off and discard a fourth from each vanilla wafer. Place a wafer on each cupcake, with the rounded edge of wafer near edge of cupcake, for face. Add dots of black gel for nostrils. With red gel, pipe a mouth on each.

Place the M&M's above wafers for eyes; add dots of black gel for pupils. Using reserved frosting and a #16 star tip, pipe hair. Carefully separate sandwich cookies; cut each in half. Position one on each side of cupcakes for ears. **Yield:** 2 dozen.

Honey Pecan Pie

(Pictured above)

Prep: 25 min. **Bake:** 45 min. + cooling

Looking for a special ending to a holiday meal? This attractive pie is bound to please with its traditional sugary filling and honey-glazed pecans. A store-bought pastry crust helps this recipe go together with ease.

—Cathy Hudak, Wadsworth, Ohio

4 eggs
1 cup chopped pecans
1 cup light corn syrup
1/4 cup sugar
1/4 cup packed brown sugar
2 tablespoons butter, melted
1 teaspoon vanilla extract
1/2 teaspoon salt
1 unbaked pastry shell (9 inches)

TOPPING:

3 tablespoons butter
1/3 cup packed brown sugar
3 tablespoons honey
1-1/2 cups pecan halves

In a large bowl, combine the eggs, pecans, corn syrup, sugars, butter, vanilla and salt. Pour into pastry shell. Bake at 350° for 30 minutes.

In a small saucepan, melt butter over medium heat. Stir in the brown sugar and honey until combined. Stir in pecan halves until coated. Spoon over pie. Bake 15-20 minutes longer or until bubbly and golden brown. Cool completely on a wire rack. Refrigerate leftovers. **Yield:** 8 servings.

Buttermilk Cake With Caramel Icing

Prep: 35 min. **Bake:** 45 min. + cooling

I like to bring this family-favorite dessert to church potlucks. It's moist, tender and melt-in-your-mouth good!

—Anna Jean Allen, West Liberty, Kentucky

1 cup butter, softened
2-1/3 cups sugar
3 eggs
1-1/2 teaspoons vanilla extract
3 cups all-purpose flour
1 teaspoon baking soda
1 cup buttermilk

ICING:

1/4 cup butter, cubed
1/2 cup packed brown sugar
1/3 cup heavy whipping cream
1 cup confectioners' sugar

In a large mixing bowl, cream butter and sugar. Add eggs, one at a time, beating well after each addition. Beat in vanilla. Combine flour and baking soda; add to creamed mixture alternately with buttermilk, beating well after each addition (batter will be thick).

Pour into a greased and floured 10-in. fluted tube pan. Bake at 350° for 45-50 minutes or until a toothpick inserted near the center comes out clean. Cool for 10 minutes before removing from pan to a wire rack to cool completely.

For icing, in a small saucepan, combine the butter, brown sugar and cream. Bring to a boil over medium heat, stirring constantly. Remove from the heat; cool for 5-10 minutes. Whisk in confectioners' sugar. Drizzle over cake. **Yield:** 12-16 servings.

Raspberry Peach Cupcakes

Prep: 25 min. **Bake:** 15 min. + cooling

These easy treats start with a convenient cake mix and have an appealing combination of fresh fruit. The lemon buttercream frosting adds a delightful burst of citrus.

—Arlene Kay Butler, Ogden, Utah

1 cup vanilla *or* white chips
6 tablespoons butter, cubed
1 package (18-1/4 ounces) white cake mix
1 cup milk
3 eggs
1 teaspoon vanilla extract
1 cup fresh raspberries
1/2 cup chopped peeled fresh peaches *or* frozen unsweetened peach slices, thawed and chopped

LEMON FROSTING:
- **1/2 cup butter, softened**
- **3 cups confectioners' sugar**
- **2 tablespoons lemon juice**

Fresh raspberries and peach pieces, optional

In a microwave-safe bowl, combine the chips and butter. Microwave at 70% power until melted; stir until smooth. In a large mixing bowl, combine the cake mix, milk, eggs, vanilla and melted chips; beat on low speed for 30 seconds. Beat on medium for 2 minutes. Fold in the raspberries and peaches.

Fill paper-lined muffin cups three-fourths full. Bake at 350° for 15-20 minutes or until a toothpick comes out clean. Cool for 10 minutes before removing from pans to wire racks to cool completely.

For frosting, in a small mixing bowl, beat the butter, confectioners' sugar and lemon juice until smooth. Frost cupcakes. Top with fruit if desired. **Yield:** 2 dozen.

Editor's Note: This recipe was tested with Betty Crocker cake mix.

Peanut Butter Truffle Cupcakes

Prep: 40 min. **Bake:** 15 min. + cooling

Cupcakes come in so many varieties, and these have a hidden treasure inside—a rich homemade truffle.
—Marlene Schollenberger, Bloomington, Illinois

- **6 squares (1 ounce *each*) white baking chocolate, coarsely chopped**
- **1/4 cup creamy peanut butter**
- **2 tablespoons baking cocoa**

BATTER:
- **1/2 cup butter, softened**
- **3/4 cup sugar**
- **2 eggs**
- **1 teaspoon vanilla extract**
- **3/4 cup all-purpose flour**
- **1/2 cup baking cocoa**
- **1/2 teaspoon baking soda**
- **1/4 teaspoon salt**
- **1/2 cup buttermilk**
- **1/2 cup strong brewed coffee**

FROSTING:
- **3 squares (1 ounce *each*) semisweet chocolate, chopped**
- **1/3 cup heavy whipping cream**
- **3 tablespoons creamy peanut butter**

For truffles, in a microwave-safe bowl, melt white chocolate at 70% power; stir until smooth. Stir in peanut butter. Cover and refrigerate for 15-20 minutes or until firm enough to form into balls. Shape into twelve 1-in. balls; roll in cocoa. Set aside.

In a large mixing bowl, cream butter and sugar. Add eggs, one at a time, beating well after each. Beat in vanilla. Combine the flour, cocoa, baking soda and salt; add to creamed mixture alternately with buttermilk and coffee. Fill paper-lined muffin cups two-thirds full. Top each with a truffle (do not press down).

Bake at 350° for 15-20 minutes or until a toothpick inserted in cake portion comes out clean. Cool for 10 minutes before removing from pan to a wire rack to cool completely.

In a heavy saucepan over low heat, melt chocolate with cream, stirring constantly. Remove from the heat; stir in peanut butter. Transfer to a bowl; chill until mixture reaches spreading consistency. Frost cupcakes. Store in the refrigerator. **Yield:** 1 dozen.

Frosty Key Lime Pie

(Pictured below)

Prep: 20 min. + freezing

This creamy, frozen refresher has a fluffy-smooth texture and luscious flavor. *—Lisa Feld, Grafton, Wisconsin*

- **1 can (14 ounces) sweetened condensed milk**
- **6 tablespoons key lime juice**
- **2 cups heavy whipping cream, whipped, *divided***
- **1 graham cracker crust (9 inches)**

In a large bowl, combine milk and lime juice. Refrigerate 1/4 cup whipped cream for garnish. Fold a fourth of the remaining whipped cream into lime mixture; fold in remaining whipped cream. Spoon into crust. Cover and freeze overnight.

Remove from the pie from the freezer 10-15 minutes before serving. Garnish with the reserved whipped cream. **Yield:** 6-8 servings.

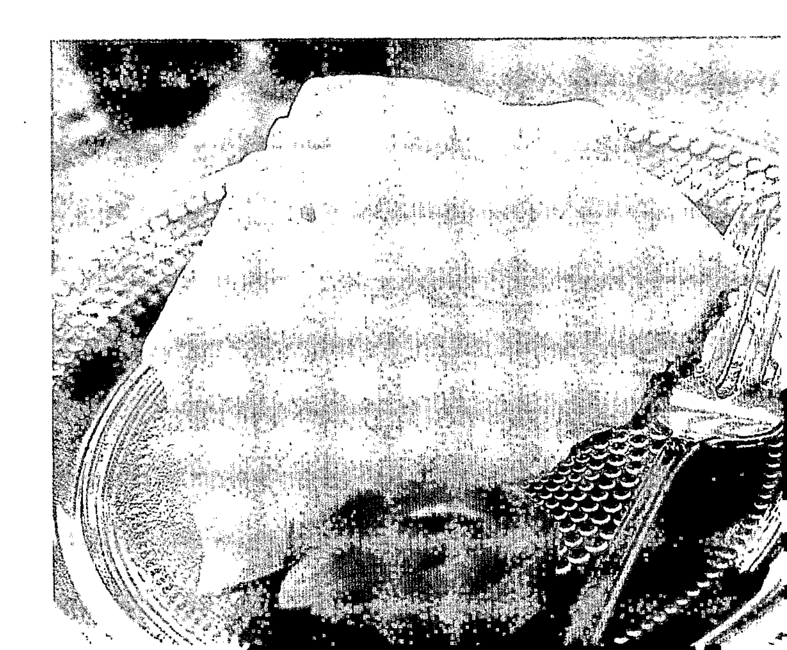

French Chocolate Cake

(Pictured above)

Prep: 35 min. **Bake:** 25 min. + cooling

A fudge filling and rum-flavored glaze are spread between these cake layers. —Claire Darby, New Castle, Delaware

- **1/2 cup butter, softened**
- **2 cups packed brown sugar**
- **2 egg yolks**
- **2 teaspoons vanilla extract**
- **6 squares (1 ounce *each*) semisweet chocolate, melted and cooled**
- **1 cup cold strong brewed coffee**
- **3 cups all-purpose flour**
- **1-1/2 teaspoons salt**
- **1-1/2 teaspoons baking soda**
- **1 cup buttermilk**

RUM GLAZE:

- **2/3 cup sugar**
- **1/2 cup milk**
- **1/4 cup butter, cubed**
- **1 teaspoon rum extract**

FILLING:

- **1 cup (6 ounces) semisweet chocolate chips**
- **1/4 cup milk**
- **1 carton (8 ounces) frozen whipped topping, thawed**

Chocolate curls

Line two greased 9-in. round baking pans with waxed paper and grease the paper; set aside. In a large mixing bowl, beat butter and brown sugar until crumbly. Beat in yolks and vanilla. Stir in chocolate and coffee. Combine flour, salt and baking soda; add to chocolate mixture alternately with buttermilk (batter will be thick).

Pour into prepared pans. Bake at 350° for 25-30 minutes or until a toothpick comes out clean. Cool for 10 minutes. Meanwhile, in a small saucepan, bring sugar, milk and butter to a gentle boil; cook and stir for 2 minutes or until sugar is dissolved. Remove from the heat; stir in extract.

Invert cakes onto wire racks; drizzle with glaze. Cool completely. In a microwave-safe bowl, melt chocolate chips with milk; stir until smooth. Refrigerate until spreadable, stirring occasionally.

Place one cake on a serving plate; spread with filling. Top with remaining cake. Spread whipped topping over top and sides of cake. Store in the refrigerator. Garnish with chocolate curls. **Yield:** 12 servings.

Apple Butter Cake Roll

Prep: 35 min. **Bake:** 15 min. + chilling

This spicy gingerbread cake is a new take on a classic pumpkin roll. —Debbie White, Williamson, West Virginia

- **3 eggs, *separated***
- **1 cup all-purpose flour, *divided***
- **2 tablespoons plus 1/2 cup sugar, *divided***
- **2 teaspoons ground cinnamon**
- **1 teaspoon baking powder**
- **1 teaspoon ground ginger**
- **1 teaspoon ground cloves**
- **1/4 teaspoon baking soda**
- **1/4 cup butter, melted**
- **1/4 cup molasses**
- **2 tablespoons water**
- **1 tablespoon confectioners' sugar**
- **2 cups apple butter**

Place egg whites in a small mixing bowl; let stand at room temperature for 30 minutes. Line a greased 15-

Making Chocolate Curls

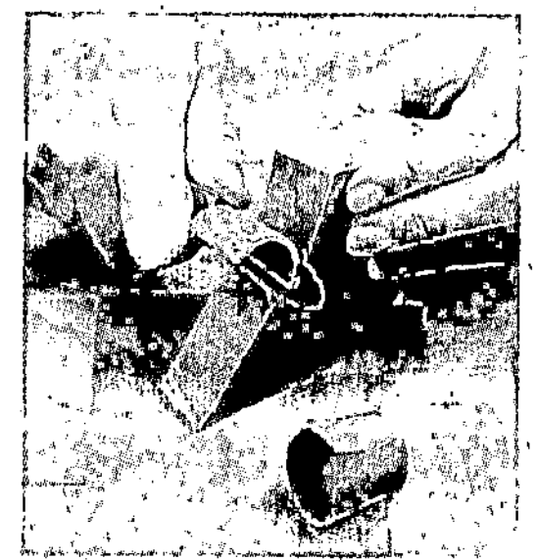

Using a vegetable peeler, peel off curls from a solid block of chocolate (see photo at left).

Allow curls to fall gently onto a work surface or plate in a single layer. If you get only shavings, try warming the chocolate block slightly.

ιking pan with waxed paper and
prinkle with 1 tablespoon flour
ıgar; set aside.
ɔwl, combine remaining flour and
ɪmon, baking powder, ginger,
a. In another bowl, whisk the egg
s and water. Add to dry ingredi-
ıded. Beat egg whites on medi-
...peaks form; fold into batter. Pour into prepared pan.

Bake at 375° for 12-14 minutes or until cake springs back when lightly touched. Cool for 5 minutes. Turn cake onto a kitchen towel dusted with confectioners' sugar. Gently peel off waxed paper. Roll up cake in the towel jelly-roll style, starting with a short side. Cool completely on a wire rack.

Unroll cake; spread apple butter to within 1/2 in. of edges. Roll up again. Cover and chill for 1 hour before serving. Refrigerate leftovers. **Yield:** 10 servings.

Editor's Note: This recipe was tested with commercially prepared apple butter.

Frosty Mocha Pie

Prep: 20 min. + freezing

No one guesses that this rich-tasting pie is on the lighter side. —Lisa Varner, Greenville, South Carolina

✓ **Uses less fat, sugar or salt. Includes Nutrition Facts and Diabetic Exchanges.**

4 ounces reduced-fat cream cheese
1/4 cup sugar
1/4 cup baking cocoa
1 tablespoon instant coffee granules
1/3 cup fat-free milk
1 teaspoon vanilla extract
1 carton (12 ounces) frozen reduced-fat whipped topping, thawed
1 extra-servings-size graham cracker crust (9 inches)
Reduced-calorie chocolate syrup, optional

In a large mixing bowl, beat the cream cheese, sugar and cocoa until smooth. Dissolve coffee granules in milk. Stir coffee mixture and vanilla into cream cheese mixture; fold in whipped topping.

Pour the filling into the crust. Cover and freeze for at least 4 hours. Remove from the freezer 10 minutes before serving. Drizzle the pie with chocolate syrup if desired. **Yield:** 10 servings.

Nutrition Facts: 1 piece (calculated without chocolate syrup) equals 259 calories, 13 g fat (7 g saturated fat), 8 mg cholesterol, 198 mg sodium, 31 g carbohydrate, 1 g fiber, 3 g protein. **Diabetic Exchanges:** 2 starch, 2 fat.

Peanut Butter Chocolate Cupcakes

(Pictured below)

Prep: 30 min. **Bake:** 25 min. + cooling

These peanut butter-filled chocolate cakes boast my favorite flavors. —Julie Small, Claremont, New Hampshire

1 package (3 ounces) cream cheese, softened
1/4 cup creamy peanut butter
2 tablespoons sugar
1 tablespoon milk

BATTER:
2 cups sugar
1-3/4 cups all-purpose flour
1/2 cup baking cocoa
1-1/2 teaspoons baking powder
1 teaspoon salt
1/4 teaspoon baking soda
2 eggs
1 cup water
1 cup milk
1/2 cup vegetable oil
2 teaspoons vanilla extract

FROSTING:
1/3 cup butter, softened
2 cups confectioners' sugar
6 tablespoons baking cocoa
3 to 4 tablespoons milk

In a small mixing bowl, beat cream cheese, peanut butter, sugar and milk until smooth; set aside.

In a bowl, combine sugar, flour, cocoa, baking powder, salt and baking soda. In another bowl, whisk the eggs, water, milk, oil and vanilla. Stir into dry ingredients just until moistened (batter will be thin).

Fill paper-lined jumbo muffin cups half full with batter. Drop scant tablespoonfuls of peanut butter mixture into center of each; cover with remaining batter. Bake at 350° for 25-30 minutes or until a toothpick inserted into cake comes out clean. Cool 10 minutes; remove from pans to wire racks. Cool completely.

Combine frosting ingredients; frost cupcakes. Store in the refrigerator. **Yield:** 1 dozen jumbo cupcakes.

Just Desserts

From creamy cheesecakes and chocolate mousse to comforting puddings and layered trifles, these treats are irresistible!

SWEET TEMPTATIONS: Clockwise from upper left: Festive White Chocolate Cheesecake (p. 135), Berry Pavlova (p. 126), Peach Cheesecake Ice Cream (p. 130), Java Cream Puffs (p. 131) and Pumpkin Pie Custard (p. 128).

Steamed Chocolate Pudding

(Pictured above)

Prep: 15 min. **Cook:** 2 hours + cooling

Warm and comforting, this old-fashioned dessert is timeless. You'll love its chocolaty goodness and moist, tender texture. —*Mary Kelley, Minneapolis, Minnesota*

2 tablespoons butter, softened
1 cup sugar
1 egg
2 squares (1 ounce *each*) unsweetened chocolate, melted and cooled
1-3/4 cups all-purpose flour
1 teaspoon salt
1/4 teaspoon cream of tartar
1/4 teaspoon baking soda
1 cup milk
VANILLA SAUCE:
1/2 cup sugar
1 tablespoon cornstarch
Dash salt
1 cup cold water
2 tablespoons butter
1 teaspoon vanilla extract
Dash ground nutmeg

In a large mixing bowl, beat butter and sugar until crumbly, about 2 minutes. Beat in egg. Stir in chocolate. Combine the flour, salt, cream of tartar and baking soda; add to creamed mixture alternately with milk. Beat just until combined.

Pour into a well-greased 7-cup pudding mold; cover. Place mold on a rack in a deep kettle; add 1 in. of hot water to pan. Bring to a gentle boil. Cover; steam for 2 to 2-1/4 hours or until top springs back when lightly touched, adding water as needed.

Remove mold to a wire rack; cool for 15 minutes. Meanwhile, in a small saucepan, combine the sugar, cornstarch and salt. Stir in water until smooth. Bring to a boil; cook and stir for 2 minutes or until thickened. Remove from heat; stir in butter, vanilla and nutmeg.

Unmold the pudding onto a serving plate; cut into wedges. Serve warm with sauce. **Yield:** 6-8 servings (1-1/2 cups sauce).

Gingered Strawberry Tart

Prep: 35 min. + chilling

This "very berry" tart is a favorite of mine. I serve it in spring or summer. —*Marie Rizzio, Interlochen, Michigan*

24 gingersnap cookies (about 1 cup)
2 tablespoons plus 1/3 cup sugar, *divided*
1/4 cup butter, melted
2 tablespoons cornstarch
1 teaspoon finely chopped crystallized ginger, optional
3 cups chopped fresh strawberries
1/4 cup water
TOPPING:
2 cups sliced fresh strawberries
5 tablespoons seedless strawberry jam

In a food processor, combine gingersnaps, 2 tablespoons sugar and butter. Cover and process until blended. Press onto bottom and up the sides of a 9-in. fluted tart pan with a removable bottom; set aside.

In a large saucepan, combine the cornstarch, crystallized ginger if desired and remaining sugar. Stir in the chopped strawberries and water. Bring to a boil; cook and stir for 2 minutes. Reduce the heat; simmer, uncovered, for 4-6 minutes or until thickened. Cool for 30 minutes. Pour into the crust. Cover and refrigerate 2 hours or until set.

Arrange the sliced strawberries over the filling. In a small microwave-safe bowl, heat the jam on high for 15-20 seconds or until pourable; brush over the berries. **Yield:** 8 servings.

Crystallized Ginger Clue

Crystallized, or candied, ginger is the root of the ginger plant that has been cooked in a sugar syrup. It's used mostly in fruit desserts, dips and sauces. Larger supermarkets carry candied ginger in the spice section.

Praline Chocolate Dessert

(Pictured below)

Prep: 25 min. + chilling **Bake:** 10 min. + cooling

A cookie crumb crust, lovely layers of praline and cream cheese, and a smooth chocolate glaze make this decadent dessert a real showstopper. It freezes well, too.
—Korrie Bastian, Clearfield, Utah

2 cups cream-filled chocolate sandwich cookie crumbs
1/2 cup butter, melted
1 cup chopped pecans

PRALINE:
1-1/2 cups butter, cubed
1 cup packed brown sugar
1 teaspoon vanilla extract

FILLING:
2 packages (8 ounces *each*) cream cheese, softened
1/2 cup confectioners' sugar
1/3 cup packed brown sugar

GANACHE:
1 cup (6 ounces) semisweet chocolate chips
1/2 cup heavy whipping cream
Pecan halves

In a small bowl, combine cookie crumbs and butter. Press onto the bottom of a greased 9-in. springform pan. Place on a baking sheet. Bake at 350° for 10 minutes. Cool on a wire rack. Sprinkle with pecans.

In a large saucepan over medium heat, bring butter and brown sugar to a boil, stirring constantly. Reduce heat; simmer, uncovered, for 10 minutes. Remove from the heat; stir in vanilla. Pour over pecans. Refrigerate for 1-2 hours or until set.

In a large mixing bowl, beat filling ingredients until smooth. Spread over praline layer. Refrigerate for 1-2 hours or until set.

For ganache, in a microwave-safe bowl, melt chocolate chips with cream; stir until smooth. Cool slightly; spread over filling. Refrigerate for 1-2 hours or until set. Carefully run a knife around edge of pan to loosen; remove sides of pan. Garnish with pecan halves. Refrigerate leftovers. **Yield:** 14-16 servings.

Maple-Walnut Ice Cream

Prep: 20 min. **Freeze:** 3 hours

An avid ice cream maker, I came up with this refreshing treat. The thick homemade fudge sauce is an excellent topping for the ice cream, which has a hint of maple flavor.
—Patricia Rauch, Kearny, New Jersey

1-1/2 cups half-and-half cream
3/4 cup sugar
2 cups heavy whipping cream
1 teaspoon walnut extract, optional
1/2 teaspoon vanilla extract
2 tablespoons maple syrup
1 cup chopped walnuts, toasted

HOT FUDGE SAUCE:
1 package (14 ounces) caramels
1/3 cup milk
1/2 cup semisweet chocolate chips
1/2 cup heavy whipping cream
1/4 cup chopped walnuts, toasted
1 teaspoon vanilla extract

In a small saucepan, heat the half-and-half cream to 175°; stir in the sugar. Cool quickly by placing the pan in a bowl of ice water; stir for 2 minutes. Stir in the whipping cream and walnut extract if desired. Stir in the vanilla extract.

Fill the cylinder of ice cream freezer two-thirds full; freeze according to manufacturer's directions. During the last 5 minutes, add half of the syrup and walnuts. Refrigerate remaining mixture until ready to freeze; repeat the process, adding remaining syrup and walnuts. Transfer to a freezer container; freeze for 2-4 hours before serving.

In a heavy saucepan over medium heat, melt caramels with milk. Stir in chocolate chips until melted. Stir in cream until blended. Remove from the heat. Stir in walnuts and vanilla. Serve warm over ice cream. Refrigerate leftover sauce. **Yield:** 1 quart ice cream (2 cups sauce).

Berry Pavlova

(Pictured above and on page 123)

Prep: 25 min. **Bake:** 1-1/4 hours + chilling

Guests are dazzled when they see the brightly colored berries nestled on a cloud of meringue in this heavenly dessert. —*Nancy Foust, Stoneboro, Pennsylvania*

- **6 egg whites**
- **1/2 teaspoon cream of tartar**
- **1 teaspoon cider vinegar**
- **1 teaspoon vanilla extract**
- **1-1/2 cups sugar**

FILLING:

- **2 packages (3 ounces *each*) cream cheese, softened**
- **1 cup sugar**
- **1 teaspoon vanilla extract**
- **2 cups heavy whipping cream, whipped**
- **2 cups miniature marshmallows**
- **1 can (21 ounces) blueberry pie filling**
- **1 package (16 ounces) frozen unsweetened strawberries, thawed**

Place egg whites in a large mixing bowl; let stand at room temperature for 30 minutes. Add cream of tartar; beat until foamy. Add vinegar and vanilla; beat until soft peaks form. Gradually beat in sugar, 1 tablespoon at a time, on high until stiff glossy peaks form and sugar is dissolved.

Spread evenly into a greased 13-in. x 9-in. x 2-in. baking dish. Bake at 225° for 1-1/4 hours; turn off oven and do not open door. Let dry in oven for 1 hour.

In a large mixing bowl, beat the cream cheese, sugar and vanilla until smooth; gently fold in the whipped cream and marshmallows. Spread over the meringue. Cover and chill for 12 hours.

Cut into squares; top with pie filling and strawberries. Refrigerate leftovers. **Yield:** 12 servings.

Rhubarb Raspberry Mousse

Prep: 30 min. + chilling

This pink mousse has a delightfully creamy texture, and bursts of rhubarb and raspberry add a tongue-tingling flavor. —*Linda Ahtiainen, Gibsons, British Columbia*

- **3 cups sliced fresh *or* frozen rhubarb**
- **1 cup plus 1 tablespoon cold water, *divided***
- **12 tablespoons sugar, *divided***
- **2 teaspoons lemon juice**
- **2 teaspoons unflavored gelatin**
- **1 cup fresh *or* frozen raspberries, thawed**
- **3/4 cup heavy whipping cream**

In a large saucepan, combine the rhubarb, 3/4 cup water, 6 tablespoons sugar and lemon juice. Bring to a boil. Reduce heat; cover and simmer for 10-12 minutes or until rhubarb is tender. Strain; return to the pan and set aside.

In a small bowl, sprinkle gelatin over 1/4 cup water; set aside. Place the raspberries and remaining water in a blender or food processor; cover and process until pureed. Strain; add to rhubarb mixture. Stir in softened gelatin. Cook over low heat for 3-5 minutes or until gelatin is dissolved, stirring occasionally.

In a small mixing bowl, beat cream and remaining sugar until stiff peaks form. Fold into rhubarb mixture. Spoon into dessert glasses or bowls. Cover and refrigerate for 4 hours or overnight. **Yield:** 4 servings.

Peaches 'n' Cream Tart

(Pictured on back cover)

Prep 30 min. **Bake:** 15 min. + cooling

Fresh peach slices and big, juicy berries crown this eye-catching tart. —*Brenda Harmon, Hastings, Minnesota*

- **2 cups crumbled soft macaroon cookies**
- **1 cup ground pecans**
- **3 tablespoons butter, melted**
- **1/2 cup heavy whipping cream**
- **1 package (8 ounces) cream cheese, softened**
- **1/3 cup sugar**
- **2 teaspoons orange juice**
- **1 teaspoon vanilla extract**
- **1/4 teaspoon almond extract**
- **4 medium fresh peaches, peeled and sliced *or* 3 cups frozen unsweetened sliced peaches, thawed**
- **2 tablespoons lemon juice**
- **1/2 cup fresh raspberries**
- **1/4 cup apricot preserves**
- **2 teaspoons honey**

In a food processor, combine the crumbled cookies, pecans and butter; cover and process until blended. Press onto the bottom and up the sides of an ungreased 11-in. fluted tart pan with removable bottom. Place pan on a baking sheet. Bake at 350° for 12-14 minutes or until golden brown. Cool completely on a wire rack.

In a small mixing bowl, beat cream until soft peaks form; set aside. In another small mixing bowl, beat cream cheese and sugar until smooth. Beat in juice and extracts. Fold in whipped cream. Spread over crust.

In a small bowl, combine peaches and lemon juice. Arrange peaches and raspberries over filling. In a small saucepan, combine preserves and honey. Cook and stir over low heat until melted; strain. Brush over fruit. Store in the refrigerator. **Yield:** 10 servings.

Mock Apple Pie Squares

Prep: 30 min. **Bake:** 25 min. + cooling

Your friends and family just might be fooled by these sweet "apple" slices, which are made with zucchini. In fact, there isn't a bit of apple in them! I like that this yummy recipe makes a lot using economical ingredients.
—Lynn Hamilton, Naperville, Illinois

4 cups all-purpose flour
2 cups sugar
1/2 teaspoon salt
1-1/2 cups cold butter
FILLING:
8 cups sliced peeled zucchini
2/3 cup lemon juice
1 cup sugar
1 teaspoon ground cinnamon
1/4 teaspoon ground nutmeg
1/2 cup chopped walnuts
1/2 cup golden raisins

In a large bowl, combine the flour, sugar and salt. Cut in the butter until the mixture resembles coarse crumbs. Press half of the crumb mixture into a greased 15-in. x 10-in. x 1-in. baking pan. Bake at 375° for 10-12 minutes or until lightly browned. Set the remaining crumb mixture aside.

Meanwhile, in a large saucepan, bring zucchini and lemon juice to a boil. Reduce heat; cover and simmer for 5-6 minutes or until tender. Drain. Stir in the sugar, cinnamon, nutmeg and 1/2 cup reserved crumb mixture. Cook and stir for 2-3 minutes. Stir in walnuts and raisins.

Spread filling evenly over crust. Sprinkle with remaining crumb mixture. Bake for 25-30 minutes or until golden brown. Cool on a wire rack. Cut into squares. **Yield:** about 2-1/2 dozen.

Fresh Plum Crumb Dessert

(Pictured below)

Prep: 25 min. **Bake:** 45 min. + cooling

This old-fashioned dessert has a terrific sweet-tart blend with its fresh-plum tang and crispy topping. Warm from the oven and served with ice cream, it's hard to resist.
—Janet Fahrenbruck-Lynch, Cincinnati, Ohio

7 large plums, pitted and quartered
1/2 cup packed brown sugar
3 tablespoons plus 1 cup all-purpose flour, *divided*
1 teaspoon ground cinnamon
1 cup sugar
1 teaspoon baking powder
1/4 teaspoon salt
1/4 teaspoon ground mace
1 egg, lightly beaten
1/2 cup butter, melted

In a large bowl, combine the plums, brown sugar, 3 tablespoons flour and cinnamon. Spoon into a greased 2-qt. baking dish.

In a small bowl, combine the sugar, baking powder, salt, mace and remaining flour. Add egg; stir with a fork until crumbly. Sprinkle over plum mixture. Drizzle with butter.

Bake at 375° for 40-45 minutes or until plums are tender and top is golden brown. Cool for 10 minutes before serving. Serve warm or at room temperature. **Yield:** 8 servings.

Mixed Nut Chocolate Tart

(Pictured below)

Prep: 20 min. **Bake:** 35 min. + cooling

For this impressive dessert, I start with a buttery shortbread crust. It holds a delectable filling made with chocolate and a mix of chopped pistachios, almonds and pecans.
—Debbie Cross, Sharon, Pennsylvania

1-1/2 cups all-purpose flour
1/4 cup sugar
1/2 cup plus 1 tablespoon cold butter
6 tablespoons heavy whipping cream
1-1/2 teaspoons vanilla extract
FILLING:
1 cup pistachios
1 cup pecan halves
3/4 cup unblanched almonds
3 eggs
3/4 cup light corn syrup
1/4 cup packed brown sugar
1/4 cup butter, melted
1-1/2 teaspoons almond extract
1 teaspoon vanilla extract
1 cup milk chocolate chips
Whipped cream

In a small bowl, combine flour and sugar; cut in butter until mixture resembles fine crumbs. Add cream and vanilla, tossing with a fork until dough forms a ball. Press onto the bottom and up the sides of an ungreased 11-in. tart pan with removable bottom; set aside.

Place nuts in a food processor; cover and process until chopped. In a large bowl, whisk the eggs, corn syrup, brown sugar, butter and extracts until smooth. Stir in chocolate chips and nut mixture; pour into crust.

Place pan on a baking sheet. Bake at 350° for 35-40 minutes or until center is set. Cool on a wire rack. Store in the refrigerator. Garnish with whipped cream. **Yield:** 12-14 servings.

Pumpkin Pie Custard

(Pictured on page 122)

Prep: 20 min. **Bake:** 40 min. + cooling

Next Thanksgiving, try this yummy light custard instead of the usual pumpkin pie. *—Nancy Zimmerman*
Cape May Court House, New Jersey

✓ **Uses less fat, sugar or salt. Includes Nutrition Facts and Diabetic Exchanges.**

2 cups canned pumpkin
1 can (12 ounces) fat-free evaporated milk
8 egg whites
1/2 cup fat-free milk
3/4 cup sugar
1 teaspoon ground cinnamon
1/2 teaspoon ground ginger
1/4 teaspoon salt
1/4 teaspoon ground cloves
1/4 teaspoon ground nutmeg
Fat-free whipped topping, optional

In a large mixing bowl, beat pumpkin, evaporated milk, egg whites and fat-free milk until smooth. Add sugar, cinnamon, ginger, salt, cloves and nutmeg; mix well.

Spoon into ten 6-oz. ramekins or custard cups coated with nonstick cooking spray. Place in a 15-in. x 10-in. x 1-in. baking pan.

Bake at 350° for 40-45 minutes or until a knife inserted near the center comes out clean. Cool on a wire rack. Refrigerate until serving. Garnish with whipped topping if desired. **Yield:** 10 servings.

Nutrition Facts: 1 serving (calculated without whipped topping) equals 120 calories, trace fat (trace saturated fat), 2 mg cholesterol, 151 mg sodium, 24 g carbohydrate, 2 g fiber, 7 g protein. **Diabetic Exchange:** 1-1/2 starch.

About Evaporated Milk

Evaporated milk is formed when milk is condensed to about half of its original volume in a heating process using a vacuum evaporator, which removes 60% of the water in milk.

Once opened, treat evaporated milk as you would fresh milk. Store it a sealed container in the refrigerator for up to 5 days.

Low-Fat Vanilla Ice Cream

Prep: 20 min. + chilling **Freeze:** 3 hours

This is my favorite recipe for "light" ice cream. It's smooth and creamy with wonderful vanilla flavor.
—Rebecca Baird, Salt Lake City, Utah

✓ **Uses less fat, sugar or salt. Includes Nutrition Facts.**

- **3/4 cup sugar**
- **3 tablespoons cornstarch**
- **1/8 teaspoon salt**
- **4 cups fat-free half-and-half**
- **2 egg yolks, beaten**
- **3 teaspoons vanilla extract**

In a large saucepan, combine the sugar, cornstarch and salt. Gradually add the half-and-half; stir until smooth. Bring to a boil over medium heat; cook and stir for 2 minutes or until thickened. Remove from the heat; cool slightly.

Whisk a small amount of hot mixture into egg yolks. Return all to the pan, whisking constantly. Cook and stir over medium heat for 2-3 minutes or until mixture reaches 160° and coats the back of a metal spoon.

Remove from heat. Stir in vanilla. Cool quickly by placing pan in a bowl of ice water; stir for 2 minutes. Transfer to a bowl. Press plastic wrap onto surface of custard. Refrigerate for several hours or overnight.

Fill cylinder of ice cream maker two-thirds full; freeze according to manufacturer's directions. Refrigerate remaining mixture until ready to freeze. Allow to ripen in ice cream freezer or firm up in the refrigerator freezer for 2-4 hours before serving. **Yield:** 1 quart.

Nutrition Facts: 1/2 cup (calculated without granola) equals 182 calories, 1 g fat (trace saturated fat), 53 mg cholesterol, 139 mg sodium, 34 g carbohydrate, trace fiber, 5 g protein.

Chocolate-Covered Cheesecake Squares

(Pictured above right)

Prep: 1 hour **Bake:** 35 min. + freezing

Dipped in chocolate, these rich, bite-sized delights are perfect for parties. But be warned...you won't be able to eat just one! *—Esther Neustaeter, La Crete, Alberta*

- **1 cup graham cracker crumbs**
- **1/4 cup finely chopped pecans**
- **1/4 cup butter, melted**

FILLING:

- **2 packages (8 ounces *each*) cream cheese, softened**

- **1/2 cup sugar**
- **1/4 cup sour cream**
- **2 eggs, lightly beaten**
- **1/2 teaspoon vanilla extract**

COATING:

- **24 squares (1 ounce *each*) semisweet chocolate**
- **3 tablespoons shortening**

Line a 9-in. square baking pan with foil and grease the foil. In a small bowl, combine the graham cracker crumbs, pecans and butter. Press into prepared pan; set aside.

In a large mixing bowl, beat the cream cheese, sugar and sour cream until smooth. Add the eggs; beat on low speed just until combined. Stir in the vanilla extract. Pour over the crust.

Bake at 325° for 35-40 minutes or until center is almost set. Cool on a wire rack. Refrigerate until chilled. Freeze overnight.

In a microwave-safe bowl, melt chocolate and shortening, stirring occasionally until smooth. Cool slightly. Using foil, lift cheesecake out of pan. Gently peel off foil; cut into 49 squares. Remove a few pieces at a time for dipping; keep remaining squares refrigerated until ready to dip.

Using a toothpick, completely dip squares, one at a time, in melted chocolate. Place on waxed paper-lined baking sheets; spoon about 1 teaspoon chocolate over each. (Reheat chocolate if needed to finish dipping.) Let stand for 20 minutes or until set. Store in an airtight container in the refrigerator or freezer. **Yield:** 49 servings.

Dreamy Whipping Cream

WHETHER you whisk it into fluffy peaks or stir it in for extra richness, whipping cream is a winning addition to recipes. Use that delightful ingredient to whip up any of the tempting treats here.

Peach Cheesecake Ice Cream

(Pictured below and on page 123)

Prep: 35 min. + freezing

I first served this at my grandparents' 50th-anniversary celebration. The ice cream was a wonderful finale for the event. —*Jenni Anderson, Bullhead City, Arizona*

- **1 cup milk**
- **1-1/4 cups plus 3 tablespoons sugar, *divided***
- **3 egg yolks, lightly beaten**
- **2 cups heavy whipping cream**
- **2 teaspoons vanilla extract**
- **2 packages (8 ounces *each*) cream cheese, cubed**
- **1/2 cup peach nectar**
- **4 teaspoons lemon juice**
- **4 medium fresh peaches, peeled and chopped *or* 1-1/2 cups frozen sliced peaches, chopped**

In a small saucepan, heat milk to 175°; stir in 1-1/4 cups sugar until dissolved. Whisk a small amount into the egg yolks. Return all to the pan, whisking constantly. Cook and stir over low heat until mixture reaches at least 160° and coats the back of a metal spoon.

Remove from the heat. Cool quickly by placing pan in a bowl of ice water; stir for 2 minutes. Transfer to a large bowl.

In a blender, combine the cream, vanilla and cream cheese; cover and process until smooth. Add to cooled milk mixture. Stir in peach nectar and lemon juice. Refrigerate for several hours or overnight.

Fill cylinder of ice cream freezer two-thirds full; freeze according to manufacturer's directions. Place peaches in a bowl; sprinkle with remaining sugar. Set aside, stirring several times.

Drain and discard juice from peaches. Add some of the peaches to each batch of ice cream during last 5 minutes of freezing.

Refrigerate remaining mixture and peaches until ready to freeze. Transfer ice cream to a freezer container; freeze for 2-4 hours before serving. May be frozen for up to 2 months. **Yield:** 2 quarts.

Cappuccino Truffles

Prep: 30 min. + chilling

Dark chocolate, coffee and cinnamon-sugar make a tantalizing trio in these rich, smooth truffles. I could eat them all myself—in one sitting! —*Ellen Swenson, Newport Center, Vermont*

- **1 tablespoon boiling water**
- **2 teaspoons instant coffee granules**
- **2-1/2 teaspoons ground cinnamon, *divided***
- **1/3 cup heavy whipping cream**
- **6 squares (1 ounce *each*) bittersweet chocolate, chopped**
- **2 tablespoons butter, softened**
- **3 tablespoons sugar**

In a small bowl, combine the water, coffee and 1 teaspoon cinnamon; set aside. In a small saucepan, bring cream just to a boil. Remove from the heat; whisk in chocolate and butter until smooth. Stir in coffee mixture. Press plastic wrap onto surface. Refrigerate for 1 hour or until easy to handle.

In a small bowl, combine sugar and remaining cinnamon. Shape chocolate into 1-in. balls; roll in cinnamon-sugar. Refrigerate for at least 2 hours or until firm. **Yield:** 1-1/2 dozen.

Java Cream Puffs

(Pictured above and on page 122)

Prep: 25 min. + chilling **Bake:** 30 min. + cooling

These fancy puffs have chopped pecans, a mocha cream filling and a drizzle of fudge topping. They're almost too beautiful to eat. —Iola Egle, Bella Vista, Arkansas

- **1/2 cup water**
- **1/4 cup butter, cubed**
- **1/8 teaspoon salt**
- **1/2 cup all-purpose flour**
- **2 eggs**
- **1/4 cup finely chopped pecans**

MOCHA CREAM FILLING:

- **1/2 cup strong brewed coffee**
- **24 large marshmallows**
- **1-1/2 cups heavy whipping cream**
- **1/4 cup hot fudge ice cream topping, warmed**

In a large saucepan, bring water, butter and salt to a boil. Add flour all at once and stir until a smooth ball forms. Remove from the heat; let stand for 5 minutes. Add eggs, one at a time, beating well after each addition. Continue beating until mixture is smooth and shiny. Stir in pecans.

Drop by rounded tablespoonfuls 3 in. apart onto a greased baking sheet. Bake at 400° for 30-35 minutes or until golden brown. Remove to a wire rack. Immediately split puffs open; remove tops and set aside. Discard soft dough from inside. Cool puffs.

For filling, in a large saucepan, combine coffee and marshmallows. Cook over low heat until marshmallows are melted. Remove from the heat. Transfer to a bowl; cover and chill just until thickened.

In a large mixing bowl, beat cream until soft peaks form. Whisk chilled coffee mixture until light in color; fold in whipped cream. Just before serving, fill each puff with about 1/3 cup filling. Replace tops and drizzle with fudge topping. **Yield:** 8 servings.

Elegant White Chocolate Mousse

Prep: 20 min. + chilling

"Simply elegant" is a fitting description for this decadent treat. —Laurinda Johnston, Belchertown, Massachusetts

- **12 squares (1 ounce *each*) white baking chocolate, coarsely chopped**
- **2 cups heavy whipping cream, *divided***
- **1 tablespoon confectioners' sugar**
- **1 teaspoon vanilla extract**

In a heavy saucepan over medium-low heat, combine chocolate and 2/3 cup cream. Cook and stir until chocolate is melted and mixture is smooth. Cool to room temperature.

In a small mixing bowl, beat remaining cream with confectioners' sugar and vanilla until soft peaks form. Fold about 1/4 cup into chocolate mixture. Fold in remaining whipped cream mixture. Spoon into dessert dishes. Cover and refrigerate for at least 2 hours. **Yield:** 8 servings.

The Biltmore's Bread Pudding

Prep: 15 min. **Bake:** 40 min.

This comforting yet special dessert comes from the Inn on Biltmore Estate in Asheville, North Carolina.

- **8 cups cubed day-old bread**
- **9 eggs**
- **2-1/4 cups milk**
- **1-3/4 cups heavy whipping cream**
- **1 cup sugar**
- **3/4 cup butter, melted**
- **3 teaspoons vanilla extract**
- **1-1/2 teaspoons ground cinnamon**

CARAMEL SAUCE:

- **1 cup sugar**
- **1/4 cup water**
- **1 tablespoon lemon juice**
- **2 tablespoons butter**
- **1 cup heavy whipping cream**

Place bread in a greased 13-in. x 9-in. x 2-in. baking dish. In a large bowl, whisk eggs, milk, cream, sugar, butter, vanilla and cinnamon. Pour evenly over bread.

Bake, uncovered, at 350° for 40-45 minutes or until a knife inserted near the center comes out clean. Let stand for 5 minutes before cutting.

Meanwhile, in a small saucepan, bring the sugar, water and lemon juice to a boil. Reduce heat to medium; cook until sugar is dissolved and mixture turns a golden amber color. Stir in butter until melted. Add cream. Remove from the heat. Serve with bread pudding. **Yield:** 12 servings.

Peanut Butter Pudding Dessert

(Pictured above)

Prep: 25 min. **Bake:** 25 min. + chilling

Here's a fun, layered dessert that will appeal to all ages. If you want it even nuttier, use chunky peanut butter.
—Barbara Schindler, Napoleon, Ohio

- **1 cup all-purpose flour**
- **1/2 cup cold butter, cubed**
- **1-1/2 cups chopped cashews, *divided***
- **1 package (8 ounces) cream cheese, softened**
- **1/3 cup creamy peanut butter**
- **1 cup confectioners' sugar**
- **1 carton (12 ounces) frozen whipped topping, thawed, *divided***
- **2-2/3 cups cold milk**
- **1 package (3.9 ounces) instant chocolate pudding mix**
- **1 package (3.4 ounces) instant vanilla pudding mix**
- **1 milk chocolate candy bar (1.55 ounces), coarsely chopped**

Place flour and butter in a food processor; cover and process until mixture resembles coarse crumbs. Add 1 cup cashews; pulse a few times until combined.

Press into a greased 13-in. x 9-in. x 2-in. baking dish. Bake at 350° for 25-28 minutes or until golden brown. Cool completely on a wire rack.

In a small mixing bowl, beat the cream cheese, peanut butter and confectioners' sugar until smooth. Fold in 1 cup whipped topping. Spoon over crust.

In another bowl, whisk milk and both pudding mixes for 2 minutes. Let stand for 2 minutes or until soft-set. Spread over cream cheese layer. Top with remaining whipped topping. Sprinkle with chopped candy bar and remaining cashews. Cover and refrigerate for at least 1 hour before serving. **Yield:** 12-16 servings.

Cranberry-Topped Lemon Tarts

Prep: 45 min. + chilling **Bake:** 20 min. + cooling

This recipe is a tempting combination of colors and tangy-sweet flavors. You'll receive lots of compliments when you present each person at the table with an elegant, individual tart. *—Ruth Lee, Troy, Ontario*

- **2 cups all-purpose flour**
- **3 tablespoons sugar**
- **3/4 teaspoon salt**
- **1 cup cold butter**

TOPPING:

- **3 cups fresh *or* frozen cranberries**
- **1-1/4 cups sugar**
- **1/4 cup water**

FILLING:

- **5 eggs**
- **1-1/2 cups sugar**
- **3/4 cup lemon juice**
- **1/3 cup butter, softened**
- **4 teaspoons grated lemon peel**

GARNISH:

- **1 medium lemon, cut into 1/4-inch slices**
- **1/2 cup sugar**
- **1/4 cup water**

In a bowl, combine the flour, sugar and salt; cut in butter until mixture resembles coarse crumbs. Stir until dough forms a ball. Divide into eight portions; press each onto the bottoms and up the sides of eight 4-in. tart pans.

Cover and refrigerate tart shells for 20 minutes. Bake at 350° for 20-25 minutes or until golden brown. Cool on wire racks.

In a large saucepan, combine the cranberries, sugar and water. Cook over medium heat until berries have popped, about 20 minutes. Meanwhile, in a small saucepan, whisk eggs. Stir in the sugar, lemon juice, butter and lemon peel. Cook and stir over medium heat for 20-25 minutes or until filling is thickened and reaches 160°.

Transfer filling to a small bowl; cover and refrigerate for 1 hour. Transfer berry topping to another bowl; refrigerate until serving.

Spoon filling into tart shells. Chill, uncovered, until set. For the garnish, in a small saucepan, bring lemon slices, sugar and water to a boil. Reduce heat; simmer, uncovered, for 20-25 minutes or until lemon is tender. Cut slices in half; chill.

Just before serving, spoon cranberry topping over tarts. Garnish with lemon slices. **Yield:** 8 servings.

Blueberry Cheesecake

(Pictured on back cover)

Prep: 1 hour **Bake:** 70 min. + chilling

Savor the best of summer's bounty with this rich blueberry cheesecake. I like to garnish it with edible pansies.
—Dick Deacon, Lawrenceville, Georgia

40 vanilla wafers, crushed
1 cup finely chopped pecans
1/3 cup butter, melted
FILLING:
2 cups 4% cottage cheese
2 packages (8 ounces each) cream cheese, softened
1/2 cup butter, softened
1-1/2 cups sugar
2 cups (16 ounces) sour cream
6 tablespoons cornstarch
6 tablespoons all-purpose flour
4 eggs, lightly beaten
4-1/2 teaspoons lemon juice
1 teaspoon vanilla extract
BLUEBERRY GLAZE:
3-1/2 cups fresh blueberries, *divided*
1 cup sugar
2 tablespoons cornstarch
Edible pansies or violas and fresh mint leaves, optional

In a bowl, combine the wafer crumbs, pecans and butter. Press onto the bottom and 2 in. up the sides of a greased 10-in. springform pan. Place on a baking sheet. Bake at 375° for 8 minutes. Cool on a wire rack. Reduce heat to 325°.

Process cottage cheese in a blender until smooth; pour into a large mixing bowl. Add cream cheese, butter and sugar; beat until smooth. Add sour cream, cornstarch and flour; beat well. Add eggs; beat on low speed just until combined. Stir in lemon juice and vanilla just until combined. Pour over crust.

Return pan to baking sheet. Bake for 70-80 minutes or until center is almost set. Cool on a wire rack for 10 minutes. Carefully run a knife around edge of pan to loosen; cool 1 hour longer. Refrigerate overnight.

For glaze, puree 2-1/2 cups berries in a food processor; press through a fine mesh sieve, reserving 1 cup juice. Discard pulp and seeds. In a small saucepan, combine sugar, cornstarch and reserved juice until smooth. Bring to a boil; cook and stir for 2 minutes or until thickened. Refrigerate until completely cooled. Remove pan sides. Spread glaze over cheesecake. Top with remaining berries; garnish with pansies and mint if desired. Refrigerate leftovers. **Yield:** 14-16 servings.

Editor's Note: Make sure to properly identify flowers before picking. Double-check that they're edible and have not been treated with chemicals.

Cheddar-Biscuit Peach Cobbler

(Pictured below)

Prep: 20 min. **Bake:** 35 min.

For this mouth-watering dessert, tender cheese biscuits top fragrant peaches bubbling in sweet juices. Yum!
—Marie Oliphant, Homer, Michigan

4 pounds fresh peaches, peeled and sliced *or* 8 cups frozen unsweetened sliced peaches
2 tablespoons lemon juice
1/2 teaspoon almond extract
1-1/2 cups sugar
2 tablespoons cornstarch
1/2 teaspoon salt
3 tablespoons cold butter
CHEDDAR BISCUITS:
2 cups biscuit/baking mix
1 cup (4 ounces) shredded cheddar cheese
2/3 cup milk
1 tablespoon butter, melted

In a large bowl, combine the peaches, lemon juice and extract. Transfer to a greased 13-in. x 9-in. x 2-in. baking dish. Combine the sugar, cornstarch and salt; sprinkle over peaches. Dot with butter. Bake, uncovered, at 400° for 15 minutes.

Meanwhile, in a bowl, combine biscuit mix and cheese. Combine milk and butter; stir into biscuit mixture just until blended. Drop by tablespoonfuls onto hot peach mixture. Bake 20-25 minutes longer or until biscuits are golden brown. Serve warm. **Yield:** 10-12 servings.

Pecan Caramel Clusters

(Pictured above)

Prep: 25 min. + chilling

A box of this mouth-watering homemade candy is the perfect way to show your affection on Valentine's Day or anytime. The crunchy clusters are yummy and so easy to make. —*Janice Price, Lexington, Kentucky*

1 package (14 ounces) caramels
2 tablespoons water
2 tablespoons butter
2 cups coarsely chopped pecans
4 ounces white candy coating, coarsely chopped
4 ounces semisweet chocolate candy coating, coarsely chopped

In a microwave-safe bowl, combine the caramels, water and butter. Microwave, uncovered, on high for 3 to 3-1/2 minutes, stirring every 30 seconds. Stir in the chopped pecans.

Drop by tablespoonfuls onto greased baking sheets. Freeze for 15-20 minutes or until set.

In a microwave-safe bowl, combine candy coatings. Microwave, uncovered, on high for 1-2 minutes, stirring every 15 seconds; stir until smooth. Dip caramel clusters in coating; place on waxed paper-lined baking sheets. Chill until firm. **Yield:** about 2 pounds.

Editor's Note: This recipe was tested in a 1,100-watt microwave.

Layered Pumpkin Dessert

Prep: 40 min. + cooling **Bake:** 15 min. + chilling

Pretty layers of cheesecake and pumpkin star in this prize-winning torte. Not too heavy, it's especially nice with a big holiday meal. There's never a morsel left!
—*Ruth Ann Stelfox, Raymond, Alberta*

1-1/2 cups graham cracker crumbs
1/3 cup sugar
1 teaspoon ground cinnamon
1/3 cup butter, melted

CREAM CHEESE FILLING:
12 ounces cream cheese, softened
1 cup sugar
3 eggs

PUMPKIN FILLING:
1 can (15 ounces) solid-pack pumpkin
3 eggs, *separated*
3/4 cup sugar, *divided*
1/2 cup milk
2 teaspoons ground cinnamon
1/2 teaspoon salt
1 envelope unflavored gelatin
1/4 cup cold water

TOPPING:
1 cup heavy whipping cream
3 tablespoons sugar
1/4 teaspoon vanilla extract

In a large bowl, combine the graham cracker crumbs, sugar and cinnamon; stir in butter. Press into an ungreased 13-in. x 9-in. x 2-in. baking dish. In a large mixing bowl, beat the cream cheese until smooth. Beat in the sugar and eggs until fluffy. Pour over the crust. Bake at 350° for 15-20 minutes or until set. Cool on a wire rack.

In a large saucepan, combine the pumpkin, egg yolks, 1/2 cup sugar, milk, cinnamon and salt. Cook and stir over low heat for 10-12 minutes or until mixture is thickened and reaches 160°. Remove from the heat.

In a small saucepan, sprinkle gelatin over cold water; let stand for 1 minute. Heat over low heat, stirring until gelatin is completely dissolved. Stir into pumpkin mixture; cool.

In a large heavy saucepan, combine egg whites and remaining sugar. With a portable mixer, beat on low speed for 1 minute. Continue beating over low heat until mixture reaches 160°, about 12 minutes.

Remove from the heat; beat until stiff glossy peaks form and sugar is dissolved. Fold into pumpkin mixture. Pour over cream cheese layer. Cover and refrigerate for at least 4 hours or until set.

Just before serving, in a large mixing bowl, beat cream until it begins to thicken. Add sugar and vanilla; beat until stiff peaks form. Spread over pumpkin layer. **Yield:** 15 servings.

Festive White Chocolate Cheesecake

(Pictured above and on page 122)

Prep: 30 min. **Bake:** 35 min. + chilling

I created this luscious dessert from a pie recipe, using a shortbread crust, white chocolate filling and berry sauce.
—Mary Alice Graves, Kempton, Indiana

2 cups crushed shortbread cookies
1/4 cup butter, melted

FILLING:

2 packages (8 ounces *each*) cream cheese, softened
1 cup vanilla *or* white chips, melted and cooled
2/3 cup sour cream
3/4 cup sugar
1 tablespoon grated orange peel
1 teaspoon vanilla extract
3 eggs, lightly beaten

SAUCE:

1 cup whole-berry cranberry sauce
1/2 cup seedless raspberry jam
1/2 teaspoon grated orange peel

TOPPING:

2 cups heavy whipping cream
1/4 cup confectioners' sugar

In a small bowl, combine cookie crumbs and butter. Press onto the bottom of a greased 9-in. springform pan; set aside.

In a large mixing bowl, beat cream cheese, chips, sour cream, sugar, orange peel and vanilla until smooth. Add eggs; beat on low speed just until combined. Pour over crust. Place pan on a baking sheet.

Bake at 350° for 35-40 minutes or until center is almost set. Cool on a wire rack for 10 minutes. Carefully run a knife around edge of pan to loosen; cool 1 hour longer. Refrigerate overnight.

In a small saucepan, combine sauce ingredients. Cook and stir over medium heat until blended. Cool.

Just before serving, remove sides of springform pan. Spoon sauce over cheesecake to within 1 in. of edges. In a small mixing bowl, beat cream and confectioners' sugar until stiff peaks form. Pipe over sauce. Refrigerate leftovers. **Yield:** 12 servings.

Cranberry Walnut Tart

Prep: 30 min. + chilling **Bake:** 20 min. + cooling

Both attractive and delicious, this is a holiday favorite at our house. *—Patricia Harmon, Baden, Pennsylvania*

2-1/2 cups all-purpose flour
1 cup cold butter, cubed
1/4 cup sugar
2 egg yolks
3 tablespoons cold water
1 tablespoon lemon juice
1/2 teaspoon grated lemon peel

FILLING:

1 cup sugar
1/4 cup butter, cubed
1/4 cup water
2/3 cup heavy whipping cream
3 tablespoons honey
1/2 teaspoon salt
2 cups chopped walnuts
1/2 cup dried cranberries
1 egg white, beaten
1 teaspoon coarse sugar

Place the flour, butter and sugar in a food processor; cover and process until the mixture resembles coarse crumbs. Add the egg yolks, water, lemon juice and lemon peel; cover and process until the dough forms a ball. Divide dough in half; wrap in plastic wrap. Refrigerate for 1 hour or until firm.

In a small saucepan, bring sugar, butter and water to a boil; cook and stir for 1 minute. Cook, without stirring, until mixture turns a golden amber color, about 7 minutes. Remove from heat; gradually stir in cream.

Return to heat; stir in honey and salt until smooth. Stir in walnuts and cranberries. Bring to a boil. Reduce heat; simmer, uncovered, for 5 minutes. Remove from the heat; cool to room temperature.

On a lightly floured surface, roll out one portion of dough into an 11-in. circle. Transfer to an ungreased 9-in. fluted tart pan with a removable bottom; trim pastry even with edge. Add filling. Roll out remaining dough to fit top of tart; place over filling. Trim and seal edges. Cut slits in pastry.

Brush with egg white; sprinkle with coarse sugar. Bake at 400° for 20-25 minutes or until filling is bubbly. Cool on a wire rack. **Yield:** 10-12 servings.

Potluck Pleasers

Plan on bringing home an empty dish when you bring any of these crowd-size specialties to your next event. This chapter offers a wide array of main courses, sides, salads, desserts and more.

COOKING TO CROWD AROUND. Clockwise from upper left: Barbecue Beef Taco Plate (p. 144); Cashew-Pear Tossed Salad (p. 152), Orange Gelatin Pretzel Salad (p. 139), Picnic Chicken (p. 147) and Corn Tortilla Chicken Lasagna (p. 138).

Corn Tortilla Chicken Lasagna

(Pictured below and on page 136)

Prep: 40 min. **Bake:** 35 min. + standing

This Southwest-style lasagna will satisfy a hungry crowd. It can be "stretched" with extra beans, and it's super easy to put together. —Susan Seymar, Valatie, New York

36 corn tortillas (6 inches)
6 cups cubed cooked chicken breast
2 cans (one 28 ounces, one 16 ounces) kidney beans, rinsed and drained
3 jars (16 ounces *each*) salsa
3 cups (24 ounces) sour cream
3 large green peppers, chopped
3 cans (3.8 ounces *each*) sliced ripe olives, drained
3 cups (12 ounces) shredded Monterey Jack cheese
3 cups (12 ounces) shredded cheddar cheese

In each of two greased 13-in. x 9-in. x 2-in. baking dishes, arrange six tortillas. Top each with 1 cup chicken, 2/3 cup kidney beans, 1 cup salsa, 1/2 cup sour cream, 1/2 cup green pepper, about 1/3 cup olives, 1/2 cup Monterey Jack cheese and 1/2 cup cheddar cheese. Repeat layers twice.

Cover and bake at 350° for 25 minutes. Uncover and bake 10-15 minutes longer or until the cheese is melted. Let stand for 10 minutes before serving. **Yield:** 2 casseroles (12 servings each).

Shrimp Macaroni Salad

Prep: 30 min. + chilling

Whenever I'm going to a potluck or other gathering, I'm asked to bring this tasty pasta salad. The recipe was a family secret for over 50 years. My husband really likes it as a main dish, served with garlic bread. —Barbara Robbins Cave Junction, Oregon

1 package (16 ounces) elbow macaroni
1 to 1-1/2 pounds cooked small shrimp
1 package (16 ounces) frozen peas, thawed
7 to 8 celery ribs, finely chopped
1 small onion, finely chopped

DRESSING:

1-3/4 cups mayonnaise
3/4 cup French salad dressing
2 tablespoons sugar
2 tablespoons white wine vinegar
1-1/2 to 2-1/2 teaspoons paprika
1 to 2 teaspoons salt
1 to 2 teaspoons garlic powder
1 to 2 teaspoons pepper

Cook macaroni according to package directions; drain and rinse in cold water. In a large bowl, combine macaroni, shrimp, peas, celery and onion.

In another bowl, whisk the dressing ingredients. Pour over salad and toss to coat. Cover and refrigerate for at least 8 hours before serving. **Yield:** 16 servings.

Strawberry Fluff

Prep: 20 min. + chilling

This pretty salad was a hit at our parents' 50th wedding anniversary. The fresh berries are complemented by a fluffy filling. —Cindy Borg, Newfolden, Minnesota

✓ **Uses less fat, sugar or salt. Includes Nutrition Facts.**

1 package (3 ounces) strawberry gelatin
2 cartons (13-1/2 ounces *each*) strawberry glaze
2 quarts buttermilk
4 packages (5.1 ounces *each*) instant vanilla pudding mix
4 cartons (16 ounces *each*) frozen whipped topping, thawed
4 quarts fresh strawberries, sliced
Additional fresh strawberries, optional

In a small bowl, combine gelatin and glaze; set aside. In two large bowls, whisk buttermilk and pudding mixes for 2 minutes or until slightly thickened. Stir in glaze mixture; fold in whipped topping and strawberries. Cover and refrigerate overnight. Garnish with additional strawberries if desired. **Yield:** 80 servings.

Nutrition Facts: 1/2 cup equals 126 calories, 4 g fat (4 g saturated fat), 1 mg cholesterol, 140 mg sodium, 19 g carbohydrate, 1 g fiber, 1 g protein.

Orange Gelatin Pretzel Salad

(Pictured below and on page 136)

Prep: 30 min. + chilling

I adapted this recipe from one that uses strawberries and strawberry gelatin. A pretzel crust gives the layered salad a nice crunch. —Peggy Boyd, Northport, Alabama

2 cups crushed pretzels
3 teaspoons plus 3/4 cup sugar, *divided*
3/4 cup butter, melted
2 packages (3 ounces *each*) orange gelatin
2 cups boiling water
2 cans (8 ounces *each*) crushed pineapple, drained
1 can (11 ounces) mandarin oranges, drained
1 package (8 ounces) cream cheese, softened
2 cups whipped topping
Additional whipped topping, optional

In a small bowl, combine pretzels and 3 teaspoons sugar; stir in butter. Press into an ungreased 13-in. x 9-in. x 2-in. baking dish. Bake at 350° for 10 minutes. Cool on a wire rack.

In a large bowl, dissolve gelatin in boiling water. Add pineapple and oranges. Chill until partially set, about 30 minutes.

In a small mixing bowl, beat cream cheese and remaining sugar until smooth. Fold in whipped topping. Spread over crust. Gently spoon gelatin mixture over cream cheese layer. Cover and refrigerate for 2-4 hours or until firm.

Cut into squares. Garnish with additional whipped topping if desired. **Yield:** 15 servings.

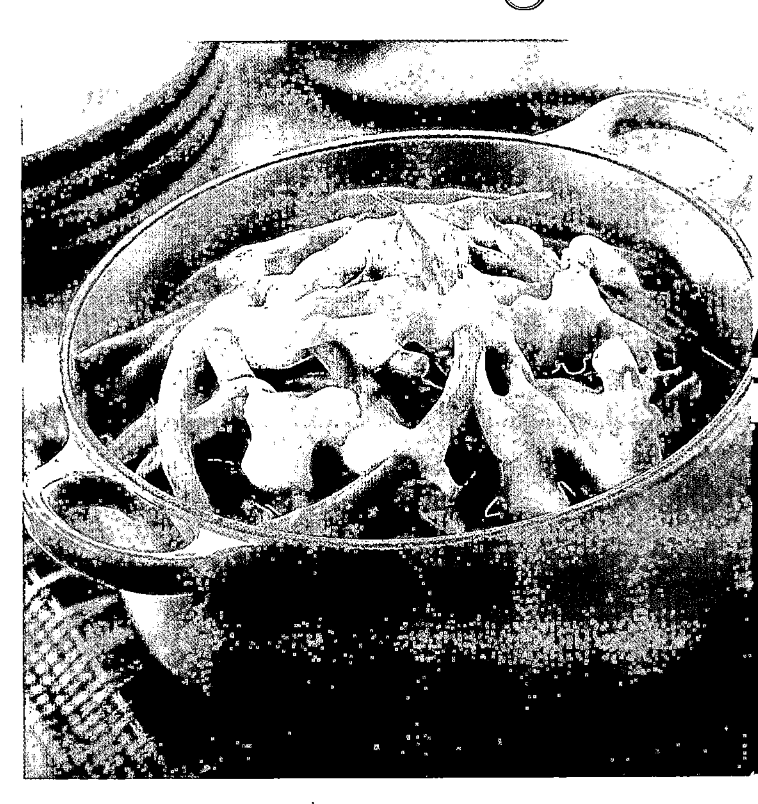

Green Beans in Lemon Chiffon Sauce

(Pictured above)

Prep: 15 min. **Cook:** 25 min.

A light citrus sauce drapes the fresh beans in this delicious side dish. It goes well with chicken, turkey or a roast. —Marian Platt, Sequim, Washington

3 pounds fresh green beans, trimmed
3 cups water
1 tablespoon cornstarch
1-1/2 cups chicken broth
6 egg yolks, beaten
1/4 cup grated Parmesan cheese
1/4 cup lemon juice
1/2 cup butter, cubed
2 teaspoons minced fresh parsley
2 teaspoons chopped green onion

Place beans and water in a Dutch oven. Bring to a boil; cover and cook for 8-10 minutes or until crisp-tender.

Meanwhile, in a small heavy saucepan, whisk the cornstarch, broth, egg yolks, Parmesan cheese and lemon juice until blended.

Cook and stir over low heat until mixture begins to thicken, bubbles around edges and reaches 160°, about 15 minutes.

Add butter, one piece at a time, whisking after each addition until butter is melted. Remove from the heat; stir in the parsley and onion. Drain beans; top with sauce. **Yield:** 12 servings.

Peppered Squash Relish

(Pictured above)

Prep: 15 min. + chilling **Cook:** 20 min. + cooling

I cook for a local middle school and have been making this fresh-tasting relish for more than 25 years. Whenever I take it to a dinner or picnic, it goes fast and brings lots of requests for the recipe. Sometimes I put it into gift baskets with other homemade goodies. —*Rose Cole Salem, West Virginia*

- **3 pounds yellow summer squash, finely chopped**
- **3 pounds zucchini, finely chopped**
- **6 large onions, finely chopped**
- **3 medium green peppers, finely chopped**
- **3 medium sweet red peppers, finely chopped**
- **1/4 cup salt**
- **2 cups sugar**
- **2 cups packed brown sugar**
- **2 cups white vinegar**
- **4 teaspoons celery seed**
- **1 teaspoon ground turmeric**
- **1 teaspoon ground mustard**
- **1/2 teaspoon alum**

In a large bowl, combine the first six ingredients. Cover and refrigerate overnight.

Drain vegetable mixture. Rinse in cold water and drain again. Place vegetables in a Dutch oven. Add the sugars, vinegar, celery seed, turmeric, mustard and alum. Bring to a boil. Reduce heat; simmer, uncovered, for 15-20 minutes or until liquid is clear.

Remove from the heat; cool. Spoon into containers. Cover and refrigerate for up to 3 weeks. **Yield:** 4 quarts.

Special Sesame Chicken Salad

Prep: 30 min. + chilling

With its mix of crunchy peanuts, tangy dried cranberries and mandarin oranges, this colorful pasta salad is a deliciously different choice on a buffet table. Water chestnuts and a teriyaki dressing give this main dish Oriental flair. —*Carolee Ewell, Santaquin, Utah*

- **1 package (16 ounces) bow tie pasta**
- **1 cup vegetable oil**
- **2/3 cup white wine vinegar**
- **2/3 cup teriyaki sauce**
- **1/3 cup sugar**
- **1/2 teaspoon pepper**
- **3 cans (11 ounces *each*) mandarin oranges, drained**
- **2 cans (8 ounces *each*) sliced water chestnuts, drained**
- **2 cups cubed cooked chicken**
- **1-1/3 cups honey roasted peanuts**
- **1 package (9 ounces) fresh spinach, torn**
- **1 package (6 ounces) dried cranberries**
- **6 green onions, chopped**
- **1/2 cup minced fresh parsley**
- **1/4 cup sesame seeds, toasted**

Cook pasta according to package directions; drain and place in a very large bowl. In a small bowl, combine the oil, vinegar, teriyaki sauce, sugar and pepper. Pour over pasta and toss to coat. Cover and refrigerate for 2 hours.

Just before serving, add the remaining ingredients; gently toss to coat. **Yield:** 22 servings (1 cup each).

Citrus Cranberry Tea

Prep/Total Time: 30 min.

Stir up some holiday spirit at Christmastime with this hot, tangy brew. It's full of fruit and cinnamon flavor. —*Pat Habiger, Spearville, Kansas*

- **4 quarts water**
- **1-1/2 cups sugar**
- **6 cinnamon sticks (3 inches)**
- **8 cups cranberry juice**

4 cups orange juice
1/3 cup lemon juice

In a Dutch oven or large kettle, combine water, sugar and cinnamon. Bring to a boil; reduce heat. Cover and simmer for 25 minutes.

Discard cinnamon sticks. Stir juices into syrup mixture. Serve warm. **Yield:** 32 servings (8 quarts).

Editor's Note: This recipe does not contain tea.

Frozen Peanut Parfait Pies

(Pictured below)

Prep: 20 min. + freezing

People will think you went to a lot of trouble to make these luscious pies, but with just six ingredients, they're quite simple to make. The crowd-pleasing dessert will be the hit of your potluck. —Anne Powers, Munford, Alabama

1 package (8 ounces) cream cheese, softened
1 can (14 ounces) sweetened condensed milk
1 carton (16 ounces) frozen whipped topping, thawed
2 pastry shells (9 inches), baked
1 jar (11-3/4 ounces) hot fudge ice cream topping, warmed
2 cups dry roasted peanuts

In a large mixing bowl, beat cream cheese and condensed milk until smooth; fold in whipped topping. Spread a fourth of the mixture into each pie shell. Drizzle each with a fourth of the fudge topping; sprinkle each with 1/2 cup peanuts. Repeat layers.

Cover and freeze for 4 hours or overnight. Remove from the freezer 5 minutes before cutting. **Yield:** 2 pies (6-8 servings each).

Texas Chocolate Cupcakes

Prep: 30 min. **Bake:** 15 min. + cooling

My husband remembers his mother making "little black cupcakes with caramel icing." I never thought caramel icing would taste good on chocolate cupcakes...boy, was I wrong! —Cathy Bodkins, Dayton, Virginia

2 cups all-purpose flour
2 cups sugar
1 teaspoon salt
1/2 teaspoon baking soda
1/4 cup baking cocoa
1 cup water
1 cup vegetable oil
1/2 cup butter, cubed
2 eggs
1/3 cup buttermilk
1 teaspoon vanilla extract

CARAMEL ICING:

1 cup packed brown sugar
1/2 cup butter, cubed
1/4 cup milk
2 to 2-1/4 cups confectioners' sugar

In a large mixing bowl, combine the flour, sugar, salt and baking soda. In a large saucepan over medium heat, bring cocoa, water, oil and butter to a boil. Gradually add to dry ingredients; mix well. Combine eggs, buttermilk and vanilla; gradually add to batter and mix well (batter will be very thin).

Fill paper-lined muffin cups three-fourths full. Bake at 350° for 15-20 minutes or until a toothpick comes out clean. Cool for 10 minutes before removing from pans to wire racks to cool completely.

For icing, in a heavy saucepan, combine the brown sugar, butter and milk. Cook and stir over low heat until sugar is dissolved. Increase heat to medium. Do not stir. Cook for 3-6 minutes or until bubbles form in center of mixture and syrup turns amber. Remove from the heat; transfer to a small mixing bowl. Cool to room temperature. Gradually beat in confectioners' sugar. Spread over cupcakes. **Yield:** 2 dozen.

Potluck Safety Tips

- Remember to keep hot foods hot and cold foods cold. Don't let food sit at room temperature for more than 2 hours.
- Keep food from spoiling by setting out smaller serving dishes. When replenishing your buffet table, wash or replace the empty bowls or platters first to prevent contamination.

Cheese Enchiladas

(Pictured below)

Prep: 25 min. **Bake:** 25 min.

You won't leave with leftovers when you make these easy enchiladas for a potluck. With a homemade tomato sauce and cheesy filling, they always go fast. You can substitute any type of cheese you wish. —Ashley Schackow Defiance, Ohio

- **2 cans (15 ounces *each*) tomato sauce**
- **1-1/3 cups water**
- **2 tablespoons chili powder**
- **2 garlic cloves, minced**
- **1 teaspoon dried oregano**
- **1/2 teaspoon ground cumin**
- **16 flour tortillas (8 inches), warmed**
- **4 cups (16 ounces) shredded Monterey Jack cheese**
- **2-1/2 cups (10 ounces) shredded cheddar cheese, *divided***
- **2 medium onions, finely chopped**
- **1 cup (8 ounces) sour cream**
- **1/4 cup minced fresh parsley**
- **1/2 teaspoon salt**
- **1/2 teaspoon pepper**
- **Shredded lettuce, sliced ripe olives and additional sour cream, optional**

In a large saucepan, combine the first six ingredients. Bring to a boil. Reduce heat; simmer, uncovered, for 4-5 minutes or until thickened, stirring occasionally. Spoon 2 tablespoons sauce over each tortilla.

In a large bowl, combine the Monterey Jack cheese, 2 cups cheddar cheese, onions, sour cream, parsley, salt and pepper. Place about 1/3 cup down the center of each tortilla. Roll up and place seam side down in two greased 13-in. x 9-in. x 2-in. baking dishes. Pour remaining sauce over top.

Bake, uncovered, at 350° for 20 minutes. Sprinkle with remaining cheddar cheese. Bake 4-5 minutes longer or until cheese is melted. Garnish with lettuce, olives and sour cream if desired. **Yield:** 16 enchiladas.

Biscuit-Topped Lemon Chicken

Prep: 40 min. **Bake:** 35 min.

This comforting baked dish combines two of my favorite things—hot, crusty biscuits and a flavorful lemon-pepper sauce. I've taken it to many get-togethers. Sometimes I add boiled, cubed potatoes and carrots to the recipe. —Pattie Ishee, Stringer, Mississippi

- **2 large onions, finely chopped**
- **4 celery ribs, finely chopped**
- **2 garlic cloves, minced**
- **1 cup butter, cubed**
- **8 green onions, thinly sliced**
- **2/3 cup all-purpose flour**
- **1/2 gallon milk**
- **12 cups cubed cooked chicken**
- **2 cans (10-3/4 ounces *each*) condensed cream of chicken soup, undiluted**
- **1/2 cup lemon juice**
- **2 tablespoons grated lemon peel**
- **2 teaspoons pepper**
- **1 teaspoon salt**

CHEDDAR BISCUITS:

- **5 cups self-rising flour**
- **2 cups milk**
- **2 cups (8 ounces) shredded cheddar cheese**
- **1/4 cup butter, melted**

In a Dutch oven, saute the onions, celery and garlic in butter. Add green onions. Stir in flour until blended; gradually add milk. Bring to a boil; cook and stir for 2 minutes or until thickened.

Add the chicken, soup, lemon juice and peel, pepper and salt; heat through. Pour into two greased 13-in. x 9-in. x 2-in. baking dishes; set aside.

In a large bowl, combine biscuit ingredients just until moistened. Turn onto a lightly floured surface; knead 8-10 times. Pat or roll out to 3/4-in. thickness. With a floured 2-1/2-in. biscuit cutter, cut out 30 biscuits.

Place the biscuits over the chicken mixture. Bake, uncovered, at 350° for 35-40 minutes or until golden brown. **Yield:** 15 servings (30 biscuits).

Editor's Note: As a substitute for each cup of self-rising flour, place 1-1/2 teaspoons baking powder and 1/2 teaspoon salt in a measuring cup. Add all-purpose flour to measure 1 cup.

Honey-Glazed Spiral Ham

Prep: 10 min. **Bake:** 1-1/4 hours

This big, savory ham is a wonderful way to celebrate a holiday. The spiraled slices taste so good drenched in a sweet and tangy glaze. If you like extra glaze, simply double the recipe and drizzle away...or serve some on the side. —*Connie Flechler, Louisville, Kentucky*

1 fully cooked spiral-sliced ham (7 to 8 pounds)
1/2 cup pear nectar
1/2 cup orange juice
1/2 cup packed brown sugar
1/2 cup honey

Line a roasting pan with heavy-duty foil. Place ham on a rack in pan. Combine pear nectar and orange juice; brush 1/3 cup over ham. Bake, uncovered, at 325° for 30 minutes, brushing twice with remaining juice mixture.

Combine brown sugar and honey; spread over ham. Bake 45-55 minutes longer or until ham is heated through, basting occasionally with the pan drippings. **Yield:** 20-24 servings.

Apple 'n' Pear Kabobs

Prep/Total Time: 30 min.

Instead of providing a fruit platter, offer these fun kabobs drizzled with a warm butter pecan sauce. The juicy skewers will definitely perk up your party spread. —*Robin Boynton, Harbor Beach, Michigan*

5 medium apples, cut into 1-inch chunks
4 medium pears, cut into 1-inch chunks
1 tablespoon lemon juice

BUTTER PECAN SAUCE:

1/3 cup packed brown sugar
2 tablespoons sugar
4 teaspoons cornstarch
3/4 cup heavy whipping cream
1 tablespoon butter
1/2 cup chopped pecans

Toss apples and pears with lemon juice. Thread fruit alternately onto 12 metal or soaked wooden skewers; place on an ungreased baking sheet. Bake at 350° for 15-20 minutes or until tender.

Meanwhile, in a small saucepan, combine sugars and cornstarch. Gradually stir in cream until smooth. Bring to a boil, stirring constantly; cook and stir for 2-3 minutes or until slightly thickened. Remove from the heat; stir in butter until smooth. Add the pecans. Serve warm with kabobs. **Yield:** 12 servings.

Hearty Beef Soup

(Pictured below)

Prep: 1 hour **Cook:** 40 min.

While this delicious soup is great for large groups, I also like to freeze it in meal-sized portions for busy days. —*Marcia Severson, Hallock, Minnesota*

✓ **Uses less fat, sugar or salt. Includes Nutrition Facts and Diabetic Exchanges.**

4 pounds boneless beef top sirloin steak, cut into 1/2-inch cubes
4 cups chopped onions
1/4 cup butter
4 quarts hot water
4 cups sliced carrots
4 cups cubed peeled potatoes
2 cups chopped cabbage
1 cup chopped celery
1 large green pepper, chopped
8 teaspoons beef bouillon granules
1 tablespoon seasoned salt
1 teaspoon dried basil
1 teaspoon pepper
4 bay leaves
6 cups tomato juice

In two Dutch ovens or one large soup kettle, brown beef and onions in butter in batches; drain. Add the water, vegetables and seasonings; bring to a boil. Reduce heat; cover and simmer for 20 minutes.

Add tomato juice; cover and simmer 10 minutes longer or until the beef and vegetables are tender. Discard bay leaves. **Yield:** 32 servings (8 quarts).

Nutrition Facts: 1 cup (prepared with reduced-fat butter and reduced-sodium bouillon) equals 123 calories, 4 g fat (2 g saturated fat), 34 mg cholesterol, 428 mg sodium, 10 g carbohydrate, 2 g fiber, 12 g protein. **Diabetic Exchanges:** 2 vegetable, 1 lean meat, 1/2 fat.

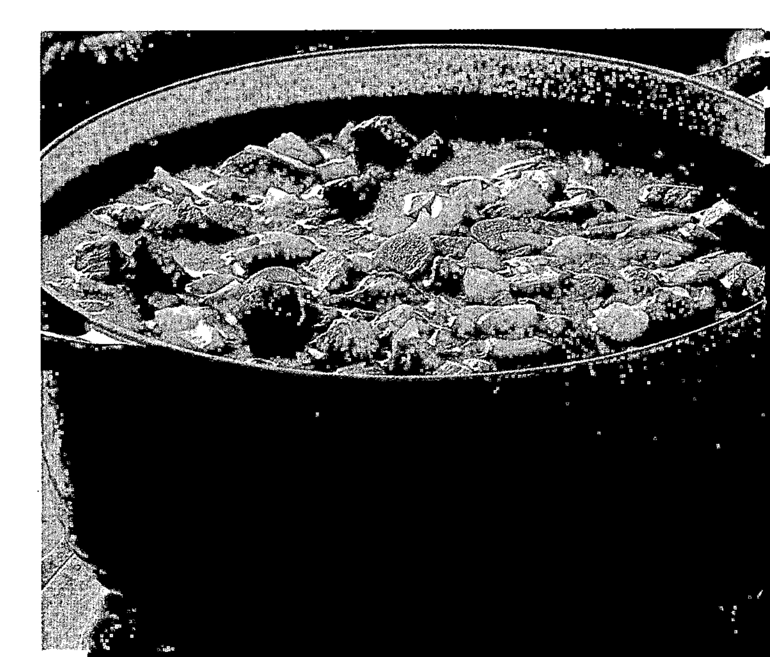

Barbecue Beef Taco Plate

(Pictured above and on page 136)

Prep: 20 min. **Cook:** 20 min.

I prepared this hearty appetizer for 200 guests who came to our beef barbecue cookout. There wasn't a morsel left over! —Iola Egle, Bella Vista, Arkansas

4 pounds ground beef
2 envelopes taco seasoning
1 cup water
4 packages (8 ounces *each*) cream cheese, softened
1 cup milk
2 envelopes ranch salad dressing mix
4 cans (4 ounces *each*) chopped green chilies, drained
1 cup chopped green onions
3 to 4 cups shredded romaine
2 cups (8 ounces) shredded cheddar cheese
4 medium tomatoes, seeded and chopped
2 to 3 cups honey barbecue sauce
2 to 3 packages (13-1/2 ounces *each*) tortilla chips

In a Dutch oven, cook the beef over medium heat until no longer pink; drain. Stir in the taco seasoning and water. Bring to a boil. Reduce heat; simmer, uncovered, for 15 minutes.

In a large mixing bowl, beat the cream cheese, milk and dressing mixes until blended. Spread over two 14-in. plates. Layer with beef mixture, chilies, onions, romaine, cheese and tomatoes. Drizzle with barbecue sauce. Arrange some tortilla chips around the edge; serve with remaining chips. **Yield:** 40-50 servings.

Layered Tortellini-Spinach Salad

Prep/Total Time: 25 min.

Layers of red cabbage, spinach, cherry tomatoes and tortellini make this attractive salad a real centerpiece. —Genise Krause, Sturgeon Bay, Wisconsin

1 package (19 ounces) frozen cheese tortellini
2 packages (6 ounces *each*) fresh baby spinach
6 cups shredded red cabbage
1 pint cherry tomatoes, quartered
3 tablespoons thinly sliced green onions
1 package (1 pound) sliced bacon, cooked and crumbled
1 bottle (16 ounces) ranch salad dressing

Cook tortellini according to package directions. Meanwhile, in a large salad bowl, layer spinach, cabbage, tomatoes and onions.

Drain the tortellini and rinse in cold water; place over the onions. Top with the bacon. Drizzle with ranch salad dressing; do not toss. Cover and refrigerate until serving. **Yield:** 18 servings.

Italian Sausage Egg Bake

Prep: 20 min. + chilling **Bake:** 50 min.

This hearty, make-ahead dish warms up any breakfast or brunch menu with lots of sausage and cheese. —Darlene Markham, Rochester, New York

8 slices white bread, cubed
1 pound Italian sausage links, casings removed and sliced
2 cups (8 ounces) shredded sharp cheddar cheese
2 cups (8 ounces) shredded part-skim mozzarella cheese
9 eggs
3 cups milk
1 teaspoon dried basil
1 teaspoon dried oregano
1 teaspoon fennel seed, crushed

Place bread cubes in a greased 13-in. x 9-in. x 2-in. baking dish; set aside. In a large skillet, cook sausage over medium heat until no longer pink; drain. Spoon sausage over bread; sprinkle with cheeses.

In a large bowl, whisk the eggs, milk and seasonings; pour over casserole. Cover and refrigerate overnight.

Remove from the refrigerator 30 minutes before baking. Bake, uncovered, at 350° for 50-55 minutes or until golden brown. Let stand for 5 minutes before cutting. **Yield:** 12 servings.

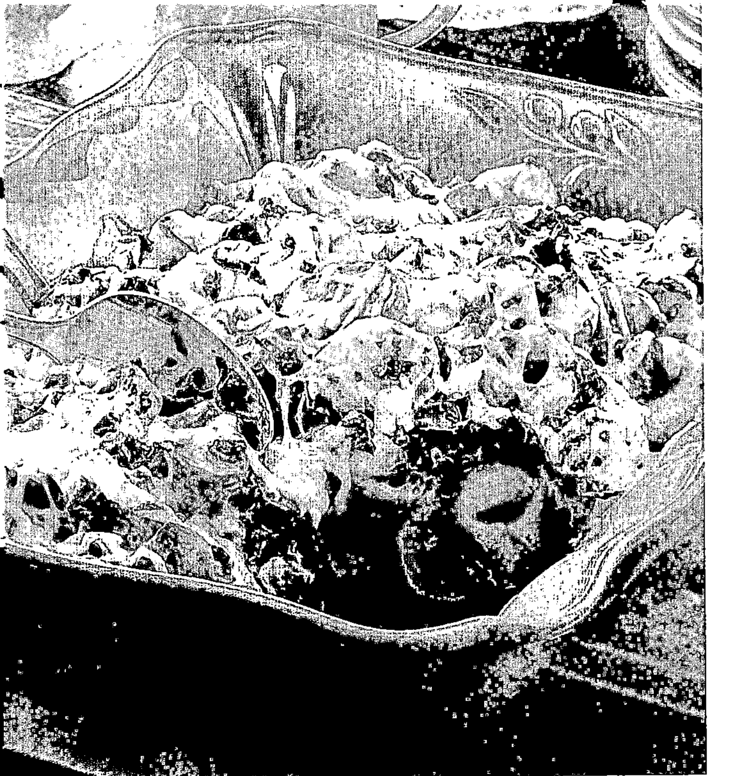

Tortellini Spinach Casserole

(Pictured above)

Prep: 20 min. **Bake:** 20 min.

This popular casserole has a fresh taste that delights even those who say they don't care for spinach.
—Barbara Kellen, Antioch, Illinois

- **2 packages (10 ounces *each*) frozen cheese tortellini**
- **1 pound sliced fresh mushrooms**
- **1 teaspoon garlic powder**
- **1/4 teaspoon onion powder**
- **1/4 teaspoon pepper**
- **1/2 cup butter, *divided***
- **1 can (12 ounces) evaporated milk**
- **1 block (8 ounces) brick cheese, cubed**
- **3 packages (10 ounces *each*) frozen chopped spinach, thawed and squeezed dry**
- **2 cups (8 ounces) shredded part-skim mozzarella cheese**

Cook tortellini according to package directions. Meanwhile, in a large skillet, saute the mushrooms, garlic powder, onion powder and pepper in 1/4 cup butter until mushrooms are tender. Remove and keep warm.

In the same skillet, combine milk and remaining butter. Bring to a gentle boil; stir in brick cheese. Cook and stir until smooth. Drain tortellini; place in a large bowl. Stir in the mushroom mixture and spinach. Add cheese sauce and toss to coat.

Transfer to a greased 3-qt. baking dish; sprinkle with mozzarella cheese. Cover and bake at 350° for 15 minutes. Uncover; bake 5-10 minutes longer or until heated through and cheese is melted. **Yield:** 12 servings.

Cheddar Loaves

(Pictured below)

Prep: 25 min. + rising **Bake:** 35 min. + cooling

Shredded cheddar cheese gives these homemade loaves great flavor. Try slices for sandwiches, toasted for breakfast or served on the side with a Caesar salad.
—Agnes Ward, Stratford, Ontario

- **3 teaspoons active dry yeast**
- **1/2 cup warm water (110° to 115°)**
- **2 cups warm milk (110° to 115°)**
- **2 tablespoons butter, melted**
- **2 eggs**
- **3 teaspoons sugar**
- **2 teaspoons salt**
- **6 to 6-1/2 cups all-purpose flour**
- **2 cups (8 ounces) shredded sharp cheddar cheese**

In a large mixing bowl, dissolve yeast in warm water. Add the milk, butter, eggs, sugar, salt and 6 cups flour. Beat on medium speed for 3 minutes. Stir in enough remaining flour to form a soft dough.

Turn onto a lightly floured surface; knead until smooth and elastic, about 6-8 minutes. Place in a greased bowl, turning once to grease top. Cover and let rise in a warm place until doubled, about 1 hour.

Punch dough down. Turn onto a lightly floured surface; knead cheese into the dough. Divide in half; shape each portion into a 6-in. round loaf. Place on greased baking sheets. Cover and let rise until doubled, about 45 minutes.

Bake at 350° for 35-40 minutes or until golden brown. Remove from pans to wire racks to cool. Refrigerate leftovers. **Yield:** 2 loaves.

Sausage Ham Loaves

(Pictured below)

Prep: 30 min. **Bake:** 1 hour 20 min.

These savory ham and pork loaves are drizzled with a sweet mustard glaze. When our pastor got married, he asked me to make them for his rehearsal dinner. A big fan of the special glaze, he requested triple the amount!
—Esther Martin, Goshen, Indiana

12 eggs, lightly beaten
6 cups milk
6 cups soft bread crumbs
1-1/2 teaspoons pepper
8 pounds bulk pork sausage
4 pounds ground fully cooked ham

GLAZE:
4 cups packed brown sugar
6 tablespoons ground mustard
1-1/2 cups white vinegar

In a large bowl, combine the eggs, milk, bread crumbs and pepper. Crumble pork and ham over mixture and mix well.

Shape into four loaves; place each in a greased 13-in. x 9-in. x 2-in. baking dish. Bake, uncovered, at 350° for 45 minutes.

In a large saucepan, combine the glaze ingredients. Bring to a boil; cook and stir until the brown sugar is dissolved. Pour 1/2 cup over each ham loaf; bake for 30 minutes. Pour the remaining glaze over the loaves; bake 5-10 minutes longer or until a meat thermometer reads 160°. Let stand for 5 minutes before slicing. **Yield:** 4 loaves (12 servings each).

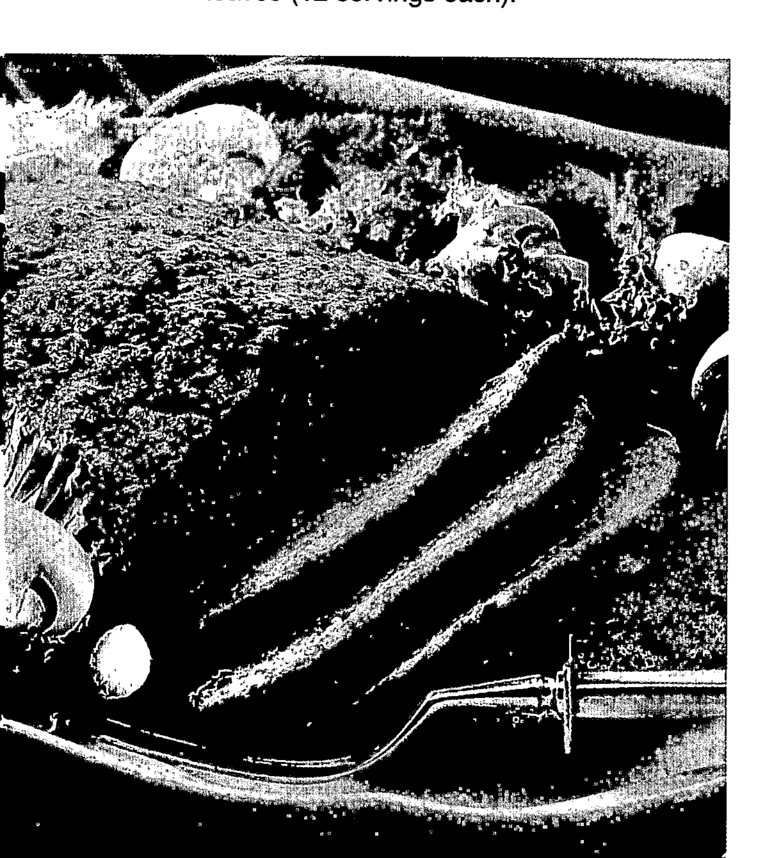

Breakfast Patties

Prep/Total Time: 30 min.

While looking for lower-fat, high-protein breakfast dishes, I took an old sausage recipe and made it new again.
—Jo Ann Honey, Longmont, Colorado

✓ **Uses less fat, sugar or salt. Includes Nutrition Facts and Diabetic Exchanges.**

2 pounds lean ground turkey
1-1/2 teaspoons salt
1 teaspoon dried sage leaves
1 teaspoon pepper
1/2 teaspoon ground ginger
1/2 teaspoon cayenne pepper

Crumble turkey into a large bowl. Add the salt, sage, pepper, ginger and cayenne. Shape into sixteen 2-1/2-in. patties.

In a large skillet, cook patties over medium heat for 4-6 minutes on each side or until meat is no longer pink. **Yield:** 16 patties.

Nutrition Facts: 2 patties equals 85 calories, 5 g fat (1 g saturated fat), 45 mg cholesterol, 275 mg sodium, trace carbohydrate, trace fiber, 10 g protein. **Diabetic Exchanges:** 2 very lean meat, 1/2 fat.

Mac 'n' Cheese for a Bunch

Prep: 30 min. **Bake:** 35 min.

They'll come running when you set this rich, comforting bake on the buffet table. The easy-to-fix macaroni tastes wonderful covered in a creamy, homemade cheese sauce.
—Dixie Terry, Goreville, Illinois

3 packages (two 16 ounces, one 7 ounces) elbow macaroni
1-1/4 cups butter, *divided*
3/4 cup all-purpose flour
2 teaspoons salt
3 quarts milk
3 pounds sharp cheddar cheese, shredded
1-1/2 cups dry bread crumbs

Cook macaroni according to package directions until almost tender. Meanwhile, in a large soup kettle, melt 1 cup butter. Stir in flour and salt until smooth. Gradually stir in milk. Bring to a boil; cook and stir for 2 minutes or until thickened. Reduce heat. Add cheese, stirring until melted. Drain macaroni; stir into sauce.

Transfer to three greased 13-in. x 9-in. x 2-in. baking dishes. Melt remaining butter; toss with bread crumbs. Sprinkle over casseroles. Bake, uncovered, at 350° for 35-40 minutes or until golden brown. **Yield:** 36 servings (1 cup each).

Picnic Chicken

(Pictured below and on page 136)

Prep: 20 min. **Bake:** 1 hour + chilling

I made this well-seasoned chicken for dinner one night and served it hot from the oven. While raiding the fridge the next day, I discovered how delicious it was cold and created the yogurt dip to go with it. —Ami Okasinski Memphis, Tennessee

3 eggs
3 tablespoons water
1-1/2 cups dry bread crumbs
2 teaspoons paprika
1 teaspoon salt
1/2 teaspoon *each* dried marjoram, thyme and rosemary, crushed
1/2 teaspoon pepper
1 cup butter, melted
12 chicken drumsticks
12 chicken thighs

CREAMY LEEK DIP:

1 cup heavy whipping cream
1-1/2 cups plain yogurt
1 envelope leek soup mix
1 cup (4 ounces) shredded Colby cheese

In a shallow bowl, whisk eggs and water. In another shallow bowl, combine bread crumbs and seasonings. Divide the butter between two 13-in. x 9-in. x 2-in. baking dishes.

Dip chicken pieces in egg mixture, then coat with crumb mixture. Place in prepared pans. Bake, uncovered, at 375° for 1 hour or until juices run clear, turning once. Cool for 30 minutes; refrigerate until chilled.

For the dip, in a small mixing bowl, beat cream until stiff peaks form. In another bowl, combine the yogurt, soup mix and cheese; fold in whipped cream. Cover and refrigerate until serving. Serve with cold chicken. **Yield:** 24 servings (4 cups dip).

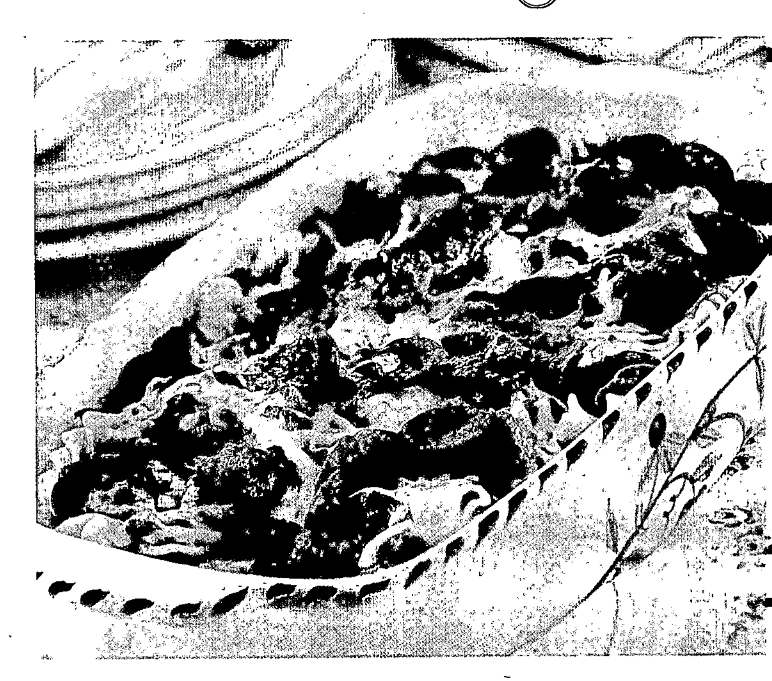

Spiral Pepperoni Pizza Bake

(Pictured above)

Prep: 30 min. **Bake:** 40 min.

My grandmother used to fix this popular dish for my Girl Scout troop when I was growing up. Now, I make the casserole for my stepdaughters' troop. —Kimberly Howland, Fremont, Michigan

1 package (16 ounces) spiral pasta
2 pounds ground beef
1 large onion, chopped
1 teaspoon salt
1/2 teaspoon pepper
2 cans (15 ounces *each*) pizza sauce
1/2 teaspoon garlic salt
1/2 teaspoon Italian seasoning
2 eggs
2 cups milk
1/2 cup shredded Parmesan cheese
4 cups (16 ounces) shredded part-skim mozzarella cheese
1 package (3-1/2 ounces) sliced pepperoni

Cook pasta according to package directions. Meanwhile, in a Dutch oven, cook the beef, onion, salt and pepper over medium heat until meat is no longer pink; drain. Stir in the pizza sauce, garlic salt and Italian seasoning; remove from the heat and set aside.

In a small bowl, combine the eggs, milk and Parmesan cheese. Drain pasta; toss with egg mixture. Transfer to a greased 3-qt. baking dish. Top with beef mixture, mozzarella cheese and pepperoni.

Cover and bake at 350° for 20 minutes. Uncover; bake 20-25 minutes longer or until golden brown. **Yield:** 12 servings.

Cherry Gelatin Supreme

(Pictured below)

Prep: 20 min. + chilling

When I was growing up, this yummy dessert was always on the menu for special occasions. Years ago, my aunt gave me the recipe, and now I think of her every time I fix it.
—Janice Rathgeb, Brighton, Illinois

- **2 cups water, *divided***
- **1 package (3 ounces) cherry gelatin**
- **1 can (21 ounces) cherry pie filling**
- **1 package (3 ounces) lemon gelatin**
- **1 package (3 ounces) cream cheese, softened**
- **1/3 cup mayonnaise**
- **1 can (8 ounces) crushed pineapple, undrained**
- **1 cup miniature marshmallows**
- **1/2 cup heavy whipping cream, whipped**
- **2 tablespoons chopped pecans**

In a large saucepan, bring 1 cup water to a boil. Stir in cherry gelatin until dissolved. Add pie filling; mix well. Pour into an 11-in. x 7-in. x 2-in. dish. Cover and refrigerate for 2 hours or until set.

In a saucepan, bring remaining water to a boil. Stir in lemon gelatin until dissolved. In a small mixing bowl, beat the cream cheese and mayonnaise until smooth. Beat in lemon gelatin and pineapple. Cover and refrigerate for 45 minutes.

Fold in the marshmallows and whipped cream. Spoon over cherry layer; sprinkle with pecans. Cover and refrigerate for 2 hours or until set. **Yield:** 12 servings.

Frozen Fruit Cups

Prep: 30 min. + freezing

Add some sparkle to your next get-together or church supper with these sunny citrus treats. They're bursting with refreshing fruit and look so cute served in shiny foil cups.
—Sue Ross, Casa Grande, Arizona

✓ **Uses less fat, sugar or salt. Includes Nutrition Facts and Diabetic Exchanges.**

- **5 packages (3 ounces *each*) lemon gelatin**
- **10 cups boiling water**
- **5 cans (20 ounces *each*) unsweetened pineapple tidbits, undrained**
- **5 cans (11 ounces *each*) mandarin oranges, drained**
- **5 cans (6 ounces *each*) frozen orange juice concentrate, partially thawed**
- **5 large firm bananas, sliced**

In a very large bowl, dissolve the lemon gelatin in the boiling water; cool for 10 minutes. Stir in the remaining ingredients. Spoon the mixture into foil cups. Freeze until firm. Remove from the freezer 30 minutes before serving. **Yield:** 9-1/2 dozen.

Nutrition Facts: 1 fruit cup equals 48 calories, trace fat (trace saturated fat), 0 cholesterol, 11 mg sodium, 12 g carbohydrate, 1 g fiber, 1 g protein. **Diabetic Exchange:** 1 fruit.

Duo Tater Bake

Prep: 40 min. **Bake:** 20 min. + chilling

I made this creamy and comforting potato dish for Thanksgiving, and it was an instant winner with my family. In fact, they asked me to include it at every holiday dinner.
—Joan McCulloch, Abbotsford, British Columbia

- **4 pounds russet *or* Yukon Gold potatoes, peeled and cubed**
- **3 pounds sweet potatoes, peeled and cubed**
- **2 cartons (8 ounces *each*) spreadable chive and onion cream cheese**
- **1 cup (8 ounces) sour cream**
- **1/4 cup shredded Colby-Monterey Jack cheese**
- **1/3 cup milk**
- **1/4 cup shredded Parmesan cheese**
- **1/2 teaspoon salt**
- **1/2 teaspoon pepper**

TOPPING:

- **1 cup (4 ounces) shredded Colby-Monterey Jack cheese**
- **1/2 cup chopped green onions**
- **1/4 cup shredded Parmesan cheese**

Place russet potatoes in a Dutch oven and cover with water. Bring to a boil. Reduce heat; cover and cook for 15-20 minutes or until tender.

Meanwhile, place sweet potatoes in a large saucepan; cover with water. Bring to a boil. Reduce heat; cover and cook for 15-20 minutes or until tender. Drain; mash with half of the cream cheese and sour cream and all of the Colby cheese.

Drain russet potatoes; mash with the remaining cream cheese and sour cream. Add the milk, Parmesan cheese, salt and pepper; mix well.

Spread 2-2/3 cups russet potato mixture into each of two greased 11-in. x 7-in. x 2-in. baking dishes. Layer with 4 cups sweet potato mixture. Repeat layers. Spread with remaining russet potato mixture.

Bake, uncovered, at 350° for 15 minutes or until heated through. Combine topping ingredients; sprinkle over casseroles. Bake 2-3 minutes longer or until cheese is melted. **Yield:** 2 casseroles (10 servings each).

Rhubarb Biscuit Coffee Cakes

Prep: 45 min. **Bake:** 30 min.

This recipe makes 10 scrumptious coffee cakes—perfect for a church breakfast or other large gathering. Starting with refrigerated biscuits makes preparation a snap!
—Carla Hodenfield, Ray, North Dakota

10 tubes (12 ounces *each*) refrigerated buttermilk biscuits
20 cups sliced fresh *or* frozen rhubarb (about 6 pounds)
2-1/2 cups sugar
5 teaspoons cornstarch
10 eggs, beaten
5 cartons (16 ounces *each*) sour cream
1 pint heavy whipping cream
2-1/2 teaspoons vanilla extract
TOPPING:
3 tablespoons sugar
1-3/4 teaspoons ground cinnamon

Divide biscuits among 10 ungreased 9-in. pie plates; top each with 2 cups rhubarb. In a large mixing bowl, combine the sugar, cornstarch, eggs, sour cream, cream and vanilla. Beat on high for 2 minutes; pour over rhubarb.

Combine sugar and cinnamon; sprinkle over filling. Bake, uncovered, at 350° for 30-35 minutes or until golden brown and set. Remove to wire racks. Serve warm. Refrigerate leftovers. **Yield:** 10 coffee cakes (8 servings each).

Editor's Note: If using frozen rhubarb, measure rhubarb while still frozen, then thaw completely. Drain in a colander, but do not press liquid out.

Glazed Peanut Butter Bars

(Pictured above)

Prep: 15 min. **Bake:** 20 min. + cooling

Memories of lunchtime at school and my Aunt Shelly's kitchen come to mind whenever I bite into one of these sweet, chewy bars. My husband is their biggest fan.
—Janis Luedtke, Thornton, Colorado

3/4 cup butter, softened
3/4 cup creamy peanut butter
3/4 cup sugar
3/4 cup packed brown sugar
2 teaspoons water
2 eggs
1-1/2 teaspoons vanilla extract
1-1/2 cups all-purpose flour
1-1/2 cups quick-cooking oats
3/4 teaspoon baking soda
1/2 teaspoon salt
GLAZE:
1-1/4 cups milk chocolate chips
1/2 cup butterscotch chips
1/2 cup creamy peanut butter

In a large mixing bowl, cream the butter, peanut butter, sugars and water. Beat in eggs and vanilla. Combine the flour, oats, baking soda and salt; gradually add to creamed mixture.

Spread into a greased 15-in. x 10-in. x 1-in. baking pan. Bake at 325° for 18-22 minutes or until lightly browned.

For glaze, in a microwave-safe bowl, melt chips and peanut butter; pour over warm bars and spread evenly. Cool completely on a wire rack before cutting. **Yield:** 4 dozen.

Ham Bundles

(Pictured above)

Prep: 55 min. + rising **Bake:** 20 min.

Whenever I serve ham, I can't wait for the leftovers so I can make these tasty ham buns. I like to pair them with homemade bean or split pea soup. My husband warms them up for breakfast, too. —*Chris Sendelbach, Henry, Illinois*

1 package (1/4 ounce) active dry yeast
1/4 cup warm water (110° to 115°)
3/4 cup warm milk (110° to 115°)
1/2 cup shortening
3 eggs, lightly beaten
1/2 cup sugar
1-1/2 teaspoons salt
4-1/2 to 4-3/4 cups all-purpose flour

FILLING:

1 large onion, finely chopped
5 tablespoons butter, *divided*
4 cups cubed fully cooked ham, coarsely ground
4 bacon strips, cooked and crumbled, optional
1/4 to 1/3 cup sliced pimiento-stuffed olives, optional
1/2 to 3/4 cup shredded cheddar cheese, optional

In a large mixing bowl, dissolve the yeast in the warm water. Add the milk, shortening, eggs, sugar, salt and 2 cups flour; beat until smooth. Add enough remaining flour to form a soft dough.

Turn onto a lightly floured surface; knead until smooth and elastic, about 8 minutes. Place in a greased bowl, turning once to grease top. Cover and let rise in a warm place until doubled, about 1 hour.

Meanwhile, in a large skillet, saute onion in 2 tablespoons butter until tender. Add ham and mix well; set aside.

Punch the dough down. Turn onto a lightly floured surface; divide into thirds. Roll each portion into a 16-in. x 8-in. rectangle. Cut each rectangle into eight squares. Place a tablespoonful of ham mixture in the center of each square. Add bacon, olives and/or cheese if desired. Fold up corners to center of dough; seal edges.

Place 2 in. apart on greased baking sheets. Cover and let rise in a warm place until doubled, about 45 minutes. Melt remaining butter; brush over dough. Bake at 350° for 16-20 minutes or until golden brown and filling is heated through. Refrigerate leftovers. **Yield:** 2 dozen.

Spicy Nacho Bake

Prep: 1 hour **Bake:** 20 min.

I fixed this hearty, layered Southwestern casserole for a dinner meeting once, and now I'm asked to bring it every time we have a potluck. People love the ground beef and bean filling along with the crunchy, cheesy topping.
—*Anita Wilson, Mansfield, Ohio*

2 pounds ground beef
2 large onions, chopped
2 large green peppers, chopped
2 cans (28 ounces *each*) diced tomatoes, undrained
2 cans (15-1/2 ounces *each*) hot chili beans
2 cans (15 ounces *each*) black beans, rinsed and drained
2 cans (11 ounces *each*) whole kernel corn, drained
2 cans (8 ounces *each*) tomato sauce
2 envelopes taco seasoning
2 packages (13 ounces *each*) spicy nacho tortilla chips
4 cups (16 ounces) shredded cheddar cheese

In a Dutch oven or large kettle, cook the beef, onions and green peppers over medium heat until meat is no longer pink; drain. Stir in the tomatoes, beans, corn, tomato sauce and taco seasoning. Bring to a boil. Reduce heat; simmer, uncovered, for 30 minutes (mixture will be thin).

In each of two greased 13-in. x 9-in. x 2-in. baking dishes, layer 5 cups of chips and 4-2/3 cups of meat mixture. Repeat layers. Top each with 4 cups of chips and 2 cups of cheese. Bake, uncovered, at 350° for 20-25 minutes or until golden brown. **Yield:** 2 casseroles (15 servings each).

Bread Pudding for 40

(Pictured below)

Prep: 20 min. **Bake:** 50 min.

Comforting and easy, this is one of my favorite recipes for bread pudding. Its texture is neither soggy nor dry, and everyone praises the lemony sauce drizzled on top.
—Loreta Dressel, Nathrop, Colorado

16 eggs, lightly beaten
12 cups milk
1 cup butter, melted
3 cups sugar
4 teaspoons salt
4 teaspoons vanilla extract
2 teaspoons ground cinnamon
4 loaves (1 pound *each*) day-old white bread, cubed
4 cups raisins
LEMON SAUCE:
3 cups sugar
6 tablespoons cornstarch
Dash salt
2 cups cold water
1-1/2 cups lemon juice
6 tablespoons butter, cubed
3 tablespoons grated lemon peel

In a very large bowl, combine eggs, milk and butter. Whisk in sugar, salt, vanilla and cinnamon until combined. Gently stir in bread cubes and raisins. Transfer to four well-greased 13-in. x 9-in. x 2-in. baking dishes. Bake, uncovered, at 350° for 50-55 minutes or until a knife inserted 1 in. from edges comes out clean.

Meanwhile, in a large saucepan, combine the sugar, cornstarch and salt. Gradually stir in the cold water and lemon juice until smooth. Bring to a boil; cook and stir for 10 minutes or until thickened. Remove from the heat. Stir in the butter and lemon peel until blended. Serve the sauce warm with the bread pudding. **Yield:** 40 servings (7-1/2 cups sauce).

Scalloped Potatoes and Ham

(Pictured above)

Prep: 30 min. **Bake:** 1 hour 20 min.

A friend of mine served this scrumptious dish at her wedding. I enjoyed these potatoes so much, I asked her for the recipe. *—Ruth Ann Stelfox, Raymond, Alberta*

2 cans (10-3/4 ounces *each*) condensed cream of mushroom soup, undiluted
2 cans (10-3/4 ounces *each*) condensed cream of celery soup, undiluted
1 can (10-3/4 ounces) condensed cheddar cheese soup, undiluted
1 can (12 ounces) evaporated milk
10 pounds medium potatoes, peeled and thinly sliced
5 pounds fully cooked ham, cubed
4 cups (16 ounces) shredded cheddar cheese

In two large bowls, combine soups and milk. Add potatoes and ham; toss to coat. Divide among four greased 13-in. x 9-in. x 2-in. baking dishes.

Cover and bake at 325° for 1-1/4 hours or until potatoes are tender. Uncover; sprinkle with cheese. Bake 5-10 minutes longer or until cheese is melted. **Yield:** 4 casseroles (10 servings each).

Hungarian Strawberry Pastry Bars

(Pictured below)

Prep: 45 min. + chilling **Bake:** 25 min. + cooling

Golden pastry crisscrosses a fruity, nutty filling in this family favorite. The recipe makes a big batch—ideal for a party or special occasion. —Ron Roth, Wyoming, Michigan

- **5 cups all-purpose flour**
- **1 cup plus 3 tablespoons sugar, *divided***
- **4 teaspoons baking powder**
- **2 teaspoons baking soda**
- **1/8 teaspoon salt**
- **1-1/4 cups shortening**
- **4 egg yolks**
- **1/2 cup sour cream**
- **1/4 cup water**
- **1 teaspoon vanilla extract**
- **2-1/2 cups chopped walnuts, *divided***
- **1 jar (18 ounces) seedless strawberry jam**

In a large bowl, combine the flour, 1 cup sugar, baking powder, baking soda and salt. Cut in shortening until mixture resembles coarse crumbs. In a bowl, whisk the egg yolks, sour cream, water and vanilla; gradually add to crumb mixture, tossing with a fork until dough forms a ball. Divide into thirds. Chill for 30 minutes.

Between two large sheets of waxed paper, roll out one portion of dough into a 15-in. x 10-in. rectangle. Transfer to an ungreased 15-in. x 10-in. x 1-in. baking pan. Sprinkle with 1-1/4 cups walnuts and 2 tablespoons sugar. Roll out another portion of dough into a 15-in. x 10-in. rectangle; place over walnuts. Spread with jam; sprinkle with remaining walnuts and sugar.

Roll out remaining pastry; cut into strips. Arrange in a crisscross pattern over filling. Trim and seal edges. Bake at 350° for 25-30 minutes or until golden brown. Cool on a wire rack. Cut into bars. **Yield:** 2 dozen.

Frosted Chocolate Chip Brownies

Prep: 25 min. **Bake:** 35 min. + cooling

These moist, cake-like brownies are so chocolaty-good that they go fast at any get-together. A homemade sour cream frosting gives the yummy squares their pleasing tang. —Connie Poto, Canton, Ohio

- **1/2 cup butter, softened**
- **1 cup sugar**
- **4 eggs**
- **1 can (16 ounces) chocolate syrup**
- **2 teaspoons vanilla extract**
- **1 cup all-purpose flour**
- **1 cup (6 ounces) semisweet chocolate chips**
- **1 cup chopped nuts**

CHOCOLATE SOUR CREAM FROSTING:

- **1 cup (6 ounces) semisweet chocolate chips**
- **1/4 cup butter**
- **1/2 cup sour cream**
- **2-1/4 cups confectioners' sugar**

In a large mixing bowl, cream butter and sugar. Add eggs, one at a time, beating well after each addition. Beat in chocolate syrup and vanilla. Stir in the flour, chocolate chips and nuts.

Pour into a greased 13-in. x 9-in. x 2-in. baking pan. Bake at 350° for 35-40 minutes or until a toothpick inserted near the center comes out clean. Cool on a wire rack.

For frosting, in a microwave-safe bowl, melt chocolate chips and butter; stir until smooth. Cool for 5 minutes. Whisk in sour cream. Gradually stir in confectioners' sugar until smooth. Frost brownies. Cut into bars. Store in the refrigerator. **Yield:** 1 dozen.

Cashew-Pear Tossed Salad

(Pictured on page 136)

Prep/Total Time: 15 min.

A friend who does a lot of catering prepared this salad for our staff Christmas party several years ago, and we all asked for the recipe. The unexpected sweet-salty combination and lovely dressing make it a hit with everyone who tastes it. —Arlene Muller, Kingwood, Texas

1 bunch romaine, torn
1 cup (4 ounces) shredded Swiss cheese
1 cup salted cashews
1 medium pear, thinly sliced
1/2 cup dried cranberries

POPPY SEED VINAIGRETTE:

2/3 cup olive oil
1/2 cup sugar
1/3 cup lemon juice
2 to 3 teaspoons poppy seeds
2 teaspoons finely chopped red onion
1 teaspoon prepared mustard
1/2 teaspoon salt

In a large salad bowl, combine the romaine, shredded Swiss cheese, cashews, pear slices and dried cranberries. In a jar with a tight-fitting lid, combine the poppy seed vinaigrette ingredients; shake well. Drizzle vinaigrette over the salad and toss to coat. Serve immediately. **Yield:** 15 servings.

Cranberry-Nut Couscous Salad

Prep: 25 min. + chilling

If you're looking for something different to present at a carry-in dinner, try this pretty salad featuring couscous, dried cranberries and toasted almonds. It goes together quickly...and is on the lighter side, too! —*Jean Ecos Hartland, Wisconsin*

Uses less fat, sugar or salt. Includes Nutrition Facts and Diabetic Exchanges.

1 package (10 ounces) plain couscous
1 cup dried cranberries
3/4 cup chopped green onions
3/4 cup chopped sweet yellow *or* red pepper
3/4 cup slivered almonds, toasted
1/3 cup lemon juice
1/4 cup olive oil
1/2 teaspoon paprika
1/4 teaspoon salt
1/8 teaspoon pepper

Prepare couscous according to package directions. Transfer to a large bowl; fluff with a fork. Cover and refrigerate for 30 minutes or until chilled.

Stir in the dried cranberries, onions, yellow pepper and almonds. In a small bowl, whisk the lemon juice, oil, paprika, salt and pepper; pour over the salad and toss to coat. Cover salad and refrigerate until serving. **Yield:** 14 servings.

Nutrition Facts: 1/2 cup equals 170 calories, 7 g fat (1 g saturated fat), 0 cholesterol, 45 mg sodium, 25 g carbohydrate, 2 g fiber, 4 g protein. **Diabetic Exchanges:** 1-1/2 fat, 1 starch, 1/2 fruit.

Pumpkin Cheesecake Dessert

(Pictured above)

Prep: 25 min. **Bake:** 40 min. + chilling

With its gingersnap cookie crust and topping of maple syrup, this spiced pumpkin dessert never fails to get rave reviews, especially during autumn months. It also cuts nicely into squares. —*Cathy Hall, Phoenix, Arizona*

32 gingersnap cookies, crushed (about 1-1/2 cups)
1/4 cup butter, melted
5 packages (8 ounces *each*) cream cheese, softened
1 cup sugar
1 can (15 ounces) solid-pack pumpkin
1 teaspoon ground cinnamon
1 teaspoon vanilla extract
5 eggs, lightly beaten
Dash ground nutmeg
Maple syrup

In a small bowl, combine the gingersnap crumbs and butter. Press into a greased 13-in. x 9-in. x 2-in. baking dish; set aside.

In a large mixing bowl, beat the cream cheese and sugar until smooth. Beat in the pumpkin, cinnamon and vanilla. Add eggs; beat on low speed just until combined. Pour over crust; sprinkle with nutmeg.

Bake at 350° for 40-45 minutes or until center is almost set. Cool on a wire rack for 10 minutes. Carefully run a knife around edge of baking dish to loosen; cool 1 hour longer. Refrigerate overnight.

Cut into squares; serve with syrup. Refrigerate leftovers. **Yield:** 24 servings.

Chocolate Chip Cinnamon Rolls

(Pictured below)

Prep: 45 min. + rising **Bake:** 25 min. per batch

I tried this chocolate-cinnamon combination when I noticed that several kids in my Sunday school class didn't like raisins. The class loves this twist, and so does my family.
—Patty Wynn, Pardeeville, Wisconsin

4 packages (1/4 ounce *each*) active dry yeast
2-1/2 cups warm water (110° to 115°)
3 cups warm milk (110° to 115°)
1/2 cup butter, softened
2 eggs
3/4 cup honey
4 teaspoons salt
14 cups all-purpose flour

FILLING:

6 tablespoons butter, softened
2-1/4 cups packed brown sugar
1 package (12 ounces) miniature semisweet chocolate chips
3 teaspoons ground cinnamon

GLAZE:

6 tablespoons butter, softened
3 cups confectioners' sugar
1 teaspoon vanilla extract
6 to 8 tablespoons milk

In a bowl, dissolve yeast in warm water; let stand 5 minutes. Add milk, butter, eggs, honey, salt and 3 cups flour; beat on low for 3 minutes. Stir in enough remaining flour to form a soft dough. Turn onto a floured surface; knead until smooth and elastic, 6-8 minutes. Place in a greased bowl, turning once. Cover and let rise in a warm place until doubled, about 1 hour.

Punch dough down. Turn onto a floured surface; divide into four pieces. Roll each into a 14-in. x 8-in. rectangle; spread with butter. Combine brown sugar, chips and cinnamon; sprinkle over dough to within 1/2 in. of edges and press into dough. Roll up each jelly-roll style, starting with a long side; pinch seam to seal. Cut each into 12 slices. Place cut side down in four greased 13-in. x 9-in. x 2-in. baking dishes. Cover and let rise until doubled, about 30 minutes.

Bake at 350° for 25-30 minutes or until golden brown. Cool for 5 minutes; remove from pans to wire racks. In a small mixing bowl, combine glaze ingredients. Drizzle over warm rolls. **Yield:** 4 dozen.

Editor's Note: This recipe can be halved to fit into a mixing bowl.

Floret Salad

(Pictured above)

Prep: 30 min. + chilling

Colorful and crunchy, this crowd-pleasing salad can be made a day ahead. Everyone likes the zip it gets from a dash of hot sauce. *—Denise Elder, Hanover, Ontario*

6 cups fresh broccoli florets
6 cups fresh cauliflowerets
3 medium red onions, halved and sliced
2 cups mayonnaise
1 cup (8 ounces) sour cream
1/4 cup packed brown sugar
1/4 cup cider vinegar
1 tablespoon Worcestershire sauce
3 teaspoons dill weed
2 teaspoons salt
Dash Louisiana-style hot sauce

In a large bowl, combine the broccoli, cauliflower and onions. In a small bowl, combine the remaining ingredients. Pour over vegetables and toss to coat. Cover and refrigerate salad for 4 hours before serving. **Yield:** 25 servings (2/3 cup each).

Chocolate Chip Oatmeal Cookies

(Pictured below)

Prep: 20 min. **Bake:** 10 min. per batch

People will come back for more of these treats, so the big batch is perfect. —Diane Neth, Menno, South Dakota

✓ **Uses less fat, sugar or salt. Includes Nutrition Facts and Diabetic Exchanges.**

- **1 cup butter, softened**
- **3/4 cup sugar**
- **3/4 cup packed brown sugar**
- **2 eggs**
- **1 teaspoon vanilla extract**
- **3 cups quick-cooking oats**
- **1-1/2 cups all-purpose flour**
- **1 package (3.4 ounces) instant vanilla pudding mix**
- **1 teaspoon baking soda**
- **1 teaspoon salt**
- **2 cups (12 ounces) semisweet chocolate chips**
- **1 cup chopped nuts**

In a large mixing bowl, cream butter and sugars. Beat in eggs and vanilla. Combine the oats, flour, pudding mix, baking soda and salt; gradually add to creamed mixture. Stir in chocolate chips and nuts.

Drop by rounded teaspoonfuls 2 in. apart onto ungreased baking sheets. Bake at 375° for 10-12 minutes or until lightly browned. Remove to wire racks. **Yield:** about 7 dozen.

Nutrition Facts: 1 cookie equals 87 calories, 5 g fat (2 g saturated fat), 11 mg cholesterol, 84 mg sodium, 11 g carbohydrate, 1 g fiber, 1 g protein. **Diabetic Exchanges:** 1 starch, 1 fat.

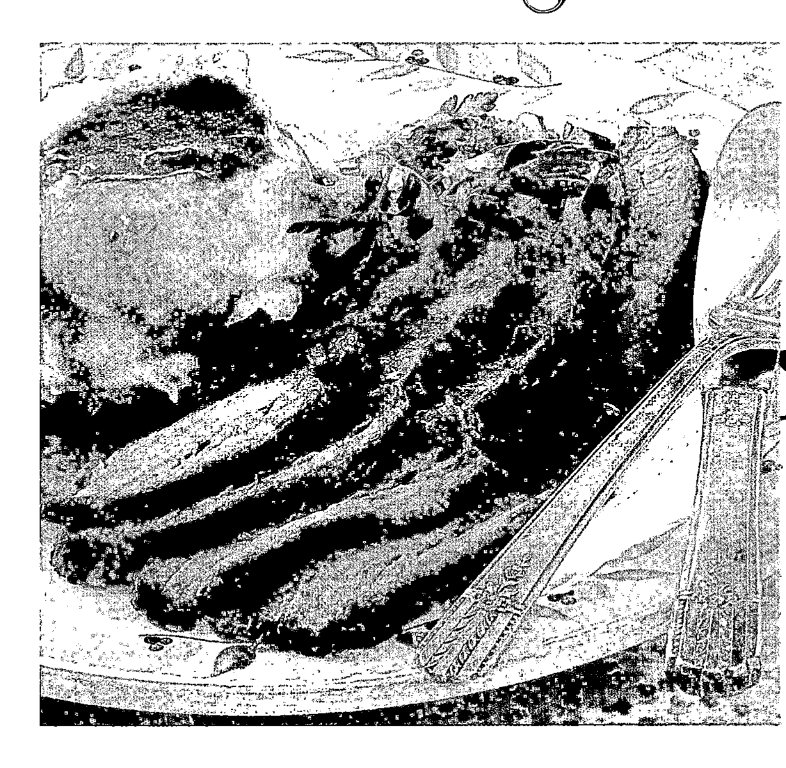

Brisket in a Bag

(Pictured above)

Prep: 15 min. **Bake:** 2-1/2 hours

This tender brisket is served with a savory cranberry gravy made right in the oven roasting bag. You'll want to serve the meat with mashed potatoes so you can drizzle the gravy over them. —Peggy Stigers, Fort Worth, Texas

- **3 tablespoons all-purpose flour, *divided***
- **1 oven roasting bag (17-1/2 inches x 16 inches)**
- **1 fresh beef brisket (5 pounds), trimmed**
- **1 can (16 ounces) whole-berry cranberry sauce**
- **1 can (10-3/4 ounces) condensed cream of mushroom soup, undiluted**
- **1 can (8 ounces) tomato sauce**
- **1 envelope onion soup mix**

Place 1 tablespoon flour in the roasting bag; shake to coat. Place bag in an ungreased 13-in. x 9-in. x 2-in. baking pan; place brisket in bag.

Combine the cranberry sauce, soup, tomato sauce, soup mix and remaining flour; pour over beef. Seal the bag. Cut slits in top of the bag according to package directions.

Bake at 325° for 2-1/2 to 3 hours or until meat is tender. Let stand for 5 minutes. Carefully remove brisket from bag. Thinly slice meat across the grain; serve with gravy. **Yield:** 12 servings.

Editor's Note: This is a fresh beef brisket, not corned beef. The meat comes from the first cut of the brisket.

Hot Wings

(Pictured below)

Prep: 15 min. **Bake:** 1 hour

These appetizers are great for a party—and have just the right amount of "heat." My family enjoys the wings so much that I'll even serve them as our main course for dinner. —*Coralie Begin, Fairfield, Maine*

- **7 to 8 pounds fresh *or* frozen chicken wingettes, thawed**
- **4 cups ketchup**
- **2-1/2 cups packed brown sugar**
- **1-1/3 cups water**
- **1 cup Louisiana-style hot sauce**
- **1/3 cup Worcestershire sauce**
- **2-1/2 teaspoons chili powder**
- **2 teaspoons garlic powder**
- **1/2 teaspoon onion powder**

Place the chicken wings in two greased 15-in. x 10-in. x 1-in. baking pans. In a large bowl, combine the remaining ingredients. Pour over wings.

Bake, uncovered, at 350° for 1 hour or until chicken juices run clear. Spoon sauce from pans over wings if desired. **Yield:** about 6 dozen.

Classic Turkey Tetrazzini

Prep: 30 min. **Bake:** 30 min.

Popular and easy to make, this casserole works well with leftover turkey or fresh turkey cutlets. —*Shannon Weddle Berryville, Virginia*

- **1 package (16 ounces) spaghetti**
- **2 medium onions, chopped**
- **9 tablespoons butter, *divided***
- **1 pound sliced fresh mushrooms**

- **1 large sweet red pepper, chopped**
- **1/2 cup all-purpose flour**
- **1 teaspoon salt**
- **6 cups milk**
- **1 tablespoon chicken bouillon granules**
- **6 cups cubed cooked turkey breast**
- **1 cup grated Parmesan cheese**
- **1-1/2 cups dry bread crumbs**
- **4 teaspoons minced fresh parsley**

Cook spaghetti according to package directions. Meanwhile, in a Dutch oven, saute onions in 6 tablespoons butter until tender. Add mushrooms and red pepper; saute 4-5 minutes longer.

Stir in flour and salt until blended. Gradually whisk in milk and bouillon. Bring to a boil; cook and stir for 2 minutes or until thickened. Stir in turkey and Parmesan cheese; heat through. Remove from the heat.

Drain spaghetti; add to turkey mixture and mix well. Transfer to one greased 13-in. x 9-in. x 2-in. baking dish and one greased 11-in. x 7-in. x 2-in. baking dish.

Melt the remaining butter; toss with bread crumbs. Sprinkle over casseroles. Bake, uncovered, at 350° for 30-35 minutes or until heated through. Sprinkle with parsley. **Yield:** 16 servings.

Chicken Salad with a Twist

Prep: 30 min. + chilling

I got the recipe for this colorful salad from my cousin, who always has excellent dishes at her parties...and this one is no exception! —*Valerie Holt, Cartersville, Georgia*

- **8 ounces uncooked spiral pasta**
- **2-1/2 cups cubed cooked chicken**
- **1 medium onion, chopped**
- **2 celery ribs, chopped**
- **1 medium cucumber, seeded and chopped**
- **1/2 cup sliced ripe olives**
- **1/3 cup zesty Italian salad dressing**
- **1/3 cup mayonnaise**
- **2 teaspoons spicy brown *or* horseradish mustard**
- **1 teaspoon lemon juice**
- **1/2 teaspoon salt**
- **1/4 teaspoon pepper**
- **3 plum tomatoes, chopped**

Cook pasta according to package directions; drain and rinse in cold water. In a large bowl, combine the pasta, chicken, onion, celery, cucumber and olives.

In a small bowl, whisk the Italian dressing, mayonnaise, mustard, lemon juice, salt and pepper. Pour over salad and toss to coat. Cover and refrigerate for 2 hours or until chilled. Just before serving, fold in the tomatoes. **Yield:** 12 servings.

Chocolate-Berry Cream Pies

(Pictured at right)

Prep: 15 min. + freezing

These crowd-pleasing pies feature a chilled chocolate filling and tangy raspberry garnish. Sometimes I lighten up the recipe by using frozen yogurt and reduced-fat whipped topping. —*Cleo Miller, Mankato, Minnesota*

1/2 gallon chocolate ice cream, softened
1 can (11-1/2 ounces) frozen cranberry-raspberry juice concentrate, thawed
1 carton (16 ounces) frozen whipped topping, thawed, *divided*
3 chocolate crumb crusts (9 inches)
1 can (21 ounces) raspberry pie filling

In a large mixing bowl, combine ice cream and juice concentrate. Fold in 4 cups whipped topping. Spoon into crusts. Cover and freeze for 4 hours or until firm.

Remove pies from the freezer 15 minutes before serving. Garnish with pie filling and remaining whipped topping. **Yield:** 3 pies (8 servings each).

Warm Dressing for Spinach

Prep/Total Time: 25 min.

You'll satisfy everyone with this flavorful dressing. I transfer it to a slow cooker to keep it warm on the buffet table. —*Victoria Hahn, Northampton, Pennsylvania*

1-1/2 cups sugar
1/2 cup all-purpose flour
3/4 teaspoon salt
4 cups cold water
1 cup cider vinegar
4 eggs, beaten
Torn fresh spinach, sliced fresh mushrooms and salad croutons
1 package (1 pound) sliced bacon, cooked and crumbled

In a large saucepan, combine the sugar, flour and salt. Gradually stir in water and vinegar. Cook and stir over medium-high heat until thickened and bubbly. Reduce the heat; cook and stir 2 minutes longer. Remove from the heat.

Stir a small amount of hot dressing into eggs; return all to the pan, stirring constantly. Bring to a gentle boil; cook and stir 2 minutes longer. Remove from the heat. Serve immediately over spinach with mushrooms and croutons; sprinkle with bacon. Or transfer dressing to a slow cooker and keep warm on low. Refrigerate leftovers. **Yield:** 6-1/2 cups.

Soft Breadsticks

Prep: 25 min. + rising **Bake:** 15 min.

Once you taste these chewy breadsticks, you'll never want store-bought again. I like to give potluck-goers a choice by sprinkling some of the breads with different seeds and seasonings. —*Linda Craig, Edmonton, Alberta*

3 teaspoons active dry yeast
2 cups warm water (110° to 115°)
1-1/2 teaspoons sugar
1/4 cup olive oil
1-1/2 teaspoons salt
3/4 cup whole wheat flour
4-1/2 to 5 cups all-purpose flour
1 egg
1 tablespoon water
Poppy seeds, sesame seeds, Italian seasoning, coarse salt *and/or* shredded Parmesan cheese, optional

In a large mixing bowl, dissolve yeast in warm water. Add sugar; let stand for 5 minutes. Stir in the oil, salt, whole wheat flour and enough all-purpose flour to form a soft dough.

Turn onto a floured surface; knead until smooth and elastic, about 6-8 minutes. Place in a greased bowl, turning once to grease top. Cover and let rise in a warm place until doubled, about 1 hour.

Punch dough down. Let rest for 10 minutes. Divide into 14 portions. Roll each portion into a 9-in. x 1-in. rope. Place 1 in. apart on greased baking sheets. Cover and let rise for 20 minutes or until doubled.

Whisk egg and water; brush over dough. Sprinkle with toppings if desired. Bake at 375° for 15-20 minutes. Serve warm. **Yield:** 14 breadsticks.

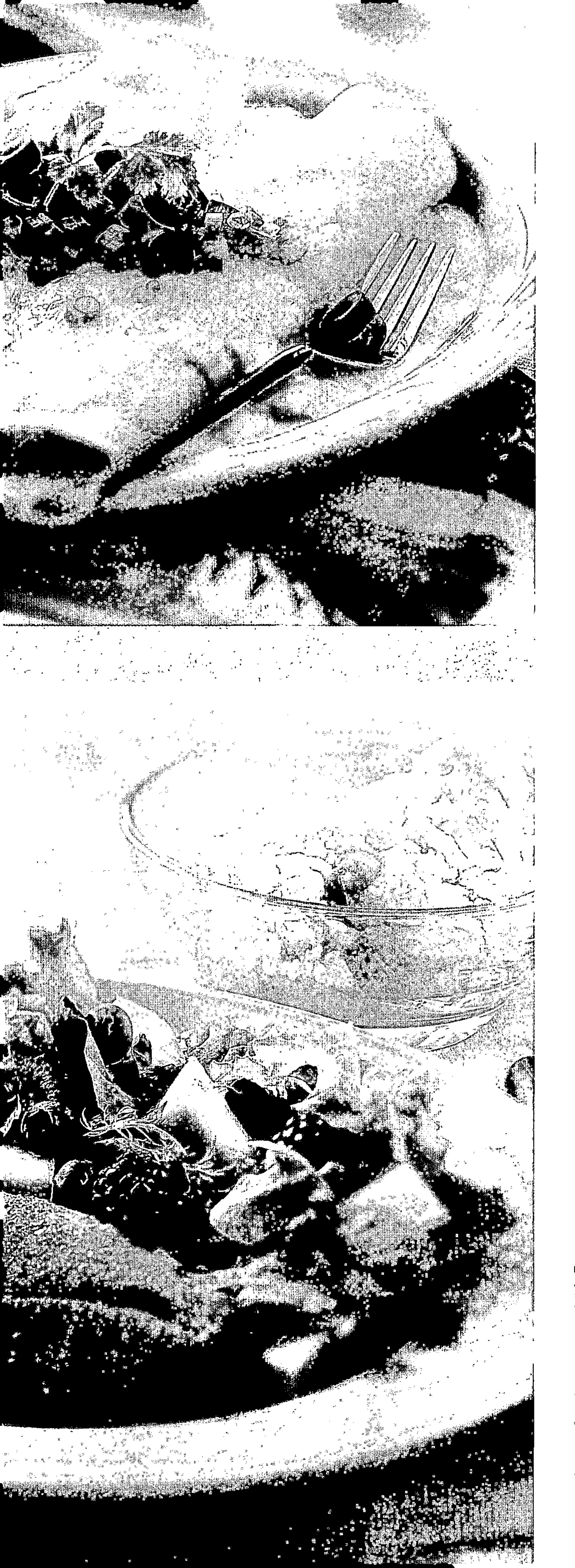

Cooking for One or Two

Why make enough for a crowd when you're cooking for just a few? These small-size dishes are big on taste and won't leave lots of leftovers.

DOWN-SIZED SPECIALTIES. Clockwise from upper left: Primo Pasta Salad (p. 161), Barbecued Pork Sandwiches (p. 162), Chili Chicken Enchiladas (p. 166), Pear Pandowdy (p. 167), Apple Juice Vinaigrette (p. 166), Herb-Crusted Red Snapper (p. 166), Crab-Stuffed Portobellos (p. 162), Asparagus Fettuccine (p. 162), Cinnamon-Raisin Bread Pudding (p. 163) and Pepperoni Veggie Pizzas (p. 162).

YOU CAN COOK for only one or two people and still enjoy fancy, restaurant-style fare...just consider tempting Seafood Potpies and Broccoli Chicken Fettuccine (shown above).

Seafood Potpies

(Pictured above)

Prep: 15 min. **Bake:** 30 min.

If you like crab and shrimp, you'll really like this family-favorite recipe. The individual potpies taste gourmet, even though they're super easy to make. —*Carol Hickey*
Lake St. Louis, Missouri

- **1 sheet refrigerated pie pastry**
- **1 can (6 ounces) crabmeat, drained, flaked and cartilage removed**
- **1 can (4-1/4 ounces) tiny shrimp, rinsed and drained**
- **1/2 cup chopped celery**
- **1/2 cup mayonnaise**
- **1/4 cup chopped green pepper**
- **2 tablespoons diced pimientos**
- **1 tablespoon lemon juice**
- **1-1/2 teaspoons chopped onion**
- **1/4 teaspoon seafood seasoning**
- **1/4 cup shredded cheddar cheese**

On a lightly floured surface, roll out pastry to 1/8-in. thickness. Cut out two 7-in. circles (discard scraps or save for another use). Press pastry circles onto the bottom and up the sides of two ungreased 10-oz. custard cups.

Place on a baking sheet. Bake at 425° for 7-10 minutes or until golden brown. Reduce heat to 375°.

In a small bowl, combine the crab, shrimp, celery, mayonnaise, green pepper, pimientos, lemon juice, onion and seafood seasoning. Spoon into hot shells. Sprinkle with cheese. Bake for 20-25 minutes or until bubbly and cheese is melted. **Yield:** 2 servings.

Broccoli Chicken Fettuccine

(Pictured at left)

Prep/Total Time: 25 min.

I served this creamy, comforting pasta dish with garlic bread, and my finicky-eating grandson absolutely loved it! —Elaine Mizzles, Ben Wheeler, Texas

- **4 ounces uncooked fettuccine**
- **1/2 pound boneless skinless chicken breasts, cut into 1-inch pieces**
- **1 small onion, halved and sliced**
- **4 garlic cloves, minced**
- **2 tablespoons butter**
- **1 can (10-3/4 ounces) condensed cream of chicken soup, undiluted**
- **1 cup chicken broth**
- **1-1/2 cups frozen broccoli florets, thawed**
- **1 can (4 ounces) mushroom stems and pieces, drained**
- **1 teaspoon onion powder**
- **1/2 teaspoon pepper**
- **1/4 cup shredded Parmesan cheese**

Cook fettuccine according to package directions. Meanwhile, in a large skillet, saute the chicken, onion and garlic in butter until no longer pink. Stir in the soup, broth, broccoli, mushrooms, onion powder and pepper. Bring to a boil.

Drain fettuccine; add to chicken mixture. Reduce heat; cover and simmer for 5 minutes or until heated through. Sprinkle with Parmesan cheese. **Yield:** 2 servings.

Primo Pasta Salad

(Pictured on page 158)

Prep/Total Time: 25 min.

When I needed to make a pasta salad, I decided to try something a bit different. This one blends Italian and ranch dressings. —Teresa McGhee, Luttrell, Tennessee

- **3/4 cup uncooked spiral pasta**
- **1/4 cup *each* finely chopped onion, celery, green pepper and sweet red pepper**
- **1 hard-cooked egg, diced**
- **1 tablespoon chopped ripe olives**
- **1/8 teaspoon garlic powder**
- **1/8 teaspoon dried basil**
- **Dash salt and pepper**
- **2 tablespoons ranch salad dressing**
- **1 tablespoon Italian salad dressing**

Cook pasta according to package directions; drain and rinse in cold water. In a bowl, combine the pasta, onion, celery, peppers, egg, olives, garlic powder, basil, salt and pepper. Add dressings and toss to coat. Cover and refrigerate until serving. **Yield:** 2 servings.

Apple-Cherry Pork Chops

(Pictured below)

Prep/Total Time: 30 min.

You'll never want pork chops any other way once you taste this recipe. I season the juicy chops with an herb rub and top them with a lovely apple-cherry sauce. —Doris Heath, Franklin, North Carolina

- **2 boneless pork loin chops (1/2 inch thick and 5 ounces *each*)**
- **1/4 teaspoon dried thyme**
- **1/8 teaspoon salt**
- **1 tablespoon olive oil**
- **2/3 cup apple juice**
- **1 small apple, sliced**
- **2 tablespoons dried cherries *or* cranberries**
- **2 tablespoons chopped onion**
- **1 teaspoon cornstarch**
- **1 tablespoon cold water**

Rub pork chops with thyme and salt. In a skillet, cook pork in oil for 3-4 minutes on each side or until juices run clear. Remove and keep warm.

In the same skillet, combine the apple juice, apple slices, dried cherries and onion. Bring to a boil. In a small bowl, combine the cornstarch and cold water until smooth; stir into the skillet. Bring to a boil; cook and stir for 1-2 minutes or until thickened. Spoon the sauce over the pork chops. **Yield:** 2 servings.

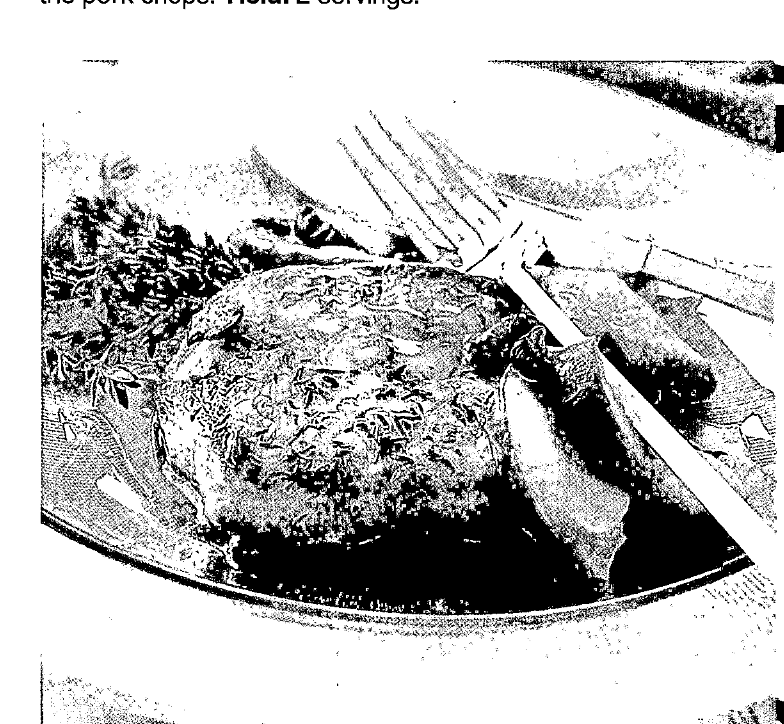

Barbecued Pork Sandwiches

(Pictured on page 158)

Prep/Total Time: 30 min.

A sweet yet tangy barbecue sauce flavors the juicy, shredded pork in these classic sandwiches. —*Mary Goff*
Rowlesburg, West Virginia

3 tablespoons chopped onion
1 tablespoon butter
1 can (8 ounces) tomato sauce
2 tablespoons brown sugar
1-1/2 teaspoons Worcestershire sauce
1 teaspoon lemon juice
1 teaspoon prepared mustard
1 cup shredded *or* diced cooked pork
2 hamburger buns, split

In a small saucepan, saute onion in butter until tender. Stir in tomato sauce, brown sugar, Worcestershire sauce, lemon juice and mustard. Bring to a boil. Reduce heat; simmer, uncovered, for 20 minutes.

Add pork; cook and stir until heated through. Serve on buns. **Yield:** 2 servings.

Pepperoni Veggie Pizzas

(Pictured below and on page 158)

Prep/Total Time: 20 min.

These mini pizzas start with a tortilla crust and are loaded with veggies. For a meat lover's version, I add browned ground round and Canadian bacon. —*Vicki Milam*
Huntsville, Alabama

2 whole wheat tortillas (8 inches)
2 tablespoons prepared pesto
12 slices turkey pepperoni
12 thin slices zucchini
4 fresh mushrooms, thinly sliced
2 thin slices red onion, separated into rings
2 tablespoons *each* chopped green, red and yellow pepper
2 thin slices provolone cheese, cut into strips
1/4 cup shredded part-skim mozzarella cheese

Coat both sides of tortillas with nonstick cooking spray; place on a greased baking sheet. Spread with pesto; top with remaining ingredients. Bake at 350° for 5-10 minutes or until cheese is melted. **Yield:** 2 servings.

Crab-Stuffed Portobellos

(Pictured at right and on page 158)

Prep/Total Time: 25 min.

Fans of portobello mushrooms will make these delectable stuffed caps again and again. I fill them with a tasty blend of crabmeat, cheese and roasted sweet red pepper.
—*Pat Ford, Southampton, Pennsylvania*

2 portobello mushrooms (5 ounces *each*)
2 tablespoons olive oil
1 garlic clove, minced
1 can (6 ounces) crabmeat, drained, flaked and cartilage removed
5 teaspoons mayonnaise
2 roasted sweet red pepper halves, drained
2 slices provolone cheese

Remove and discard stems from mushrooms. Place caps on a greased baking sheet. Combine oil and garlic; brush over mushrooms. Broil 4-6 in. from the heat for 4-5 minutes or until tender.

In a small bowl, combine crab and mayonnaise. Place a red pepper half on each mushroom; top with crab mixture. Broil for 2-3 minutes or until heated through. Top with cheese; broil 1-2 minutes longer or until cheese is melted. **Yield:** 2 servings.

Asparagus Fettuccine

(Pictured above right and on page 158)

Prep/Total Time: 25 min.

When asparagus is in season, celebrate with this satisfying meatless dish. The tender spears pair beautifully with the creamy white sauce, which is tossed with fettuccine right in the skillet. —*Genise Krause*
Sturgeon Bay, Wisconsin

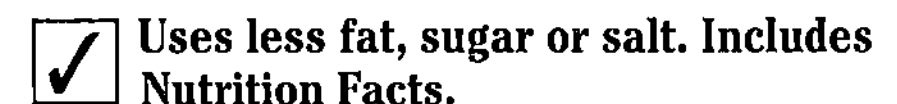

Uses less fat, sugar or salt. Includes Nutrition Facts.

TRY SOMETHING NEW for dinner tonight with this memorable menu featuring scrumptious Crab-Stuffed Portobellos, Asparagus Fettuccine and Cinnamon-Raisin Bread Pudding (shown above).

- **4 ounces uncooked fettuccine**
- **1/2 pound fresh asparagus, trimmed and cut into 1-inch pieces**
- **1/4 cup chopped onion**
- **1 garlic clove, minced**
- **1 tablespoon butter**
- **2 ounces cream cheese, cubed**
- **1/4 cup milk**
- **1/4 cup shredded Parmesan cheese**
- **1-1/2 teaspoons lemon juice**
- **1 teaspoon grated lemon peel**
- **1/4 teaspoon salt**
- **1/8 teaspoon pepper**

Cook fettuccine according to package directions. Meanwhile, in a large skillet, saute the asparagus, onion and garlic in butter until tender.

Add the remaining ingredients. Cook and stir over medium heat for 5 minutes or until cheese is melted and sauce is blended. Drain fettuccine; toss with asparagus mixture. **Yield:** 2 servings.

Nutrition Facts: 1-1/4 cups (prepared with 2% milk) equals 396 calories, 17 g fat (10 g saturated fat), 45 mg cholesterol, 388 mg sodium, 47 g carbohydrate, 4 g fiber, 17 g protein.

Cinnamon-Raisin Bread Pudding

(Pictured above and on page 158)

Prep: 5 min. **Bake:** 35 min.

This rich bread pudding recipe goes together in minutes and has plenty of old-fashioned cinnamon flavor. It's sure to become a favorite. —Edna Hoffman, Hebron, Indiana

- **1 cup cubed cinnamon-raisin bread**
- **1 egg**
- **2/3 cup milk**
- **3 tablespoons brown sugar**
- **1 tablespoon butter, melted**
- **1/2 teaspoon ground cinnamon**
- **1/4 teaspoon ground nutmeg**
- **Dash salt**
- **1/3 cup raisins**

Place bread cubes in a greased 2-cup baking dish. In a bowl, whisk the egg, milk, brown sugar, butter, cinnamon, nutmeg and salt until blended. Stir in raisins. Pour over bread.

Bake at 350° for 35-40 minutes or until a knife inserted 1/2 in. from the edge comes out clean. Serve warm. **Yield:** 2 servings.

ON ST. PATRICK'S DAY or any time at all, you'll feel lucky as a leprechaun when you sit down to Corned Beef Supper, Irish Soda Muffins and Minty Ice Cream Shamrocks (shown above).

Corned Beef Supper

(Pictured above)

Prep: 25 min. **Cook:** 3 hours 50 min.

What better way to celebrate St. Patrick's Day than with this hearty one-pot meal featuring tender corned beef, potatoes and carrots? I often fix this during March, but it's good any time of the year. —*Dawn Fagerstrom Warren, Minnesota*

- **1 small onion, sliced**
- **4 small carrots, cut into chunks**
- **2 medium potatoes, cut into chunks**
- **1 corned beef brisket with spice packet (1 pound)**
- **1/3 cup unsweetened apple juice**
- **2 whole cloves**
- **1 tablespoon brown sugar**
- **1/2 teaspoon grated orange peel**
- **1/2 teaspoon prepared mustard**
- **2 cabbage wedges**

Place onion in a 3-qt. slow cooker. Top with carrots, potatoes and brisket. Combine the apple juice, cloves, brown sugar, orange peel, mustard and contents of spice packet; pour over brisket. Cover and cook on high for 3-1/2 to 4 hours.

Add cabbage; cover and cook 20-30 minutes longer or until meat and vegetables are tender. Strain and discard cloves; serve pan juices with corned beef and vegetables. **Yield:** 2 servings.

Irish Soda Muffins

(Pictured above)

Prep/Total Time: 30 min.

Who says muffins have to be boring? These little gems are especially fun for Irish heritage celebrations. I like to use blueberries when they are in season instead of the raisins. —*Camille Wisniewski, Jackson, New Jersey*

1 cup plus 2 tablespoons all-purpose flour
1/4 cup plus 1 teaspoon sugar, *divided*
1 teaspoon baking powder
1/4 teaspoon baking soda
1/4 teaspoon salt
1/2 cup sour cream
2 tablespoons vegetable oil
2 tablespoons beaten egg
1/3 cup raisins

In a small bowl, combine the flour, 1/4 cup sugar, baking powder, baking soda and salt. In another bowl, whisk the sour cream, oil and egg; stir into dry ingredients just until moistened. Fold in raisins.

Fill six greased or paper-lined muffin cups half full. Sprinkle with remaining sugar. Bake at 400° for 15-18 minutes or until a toothpick comes out clean. Cool for 5 minutes before removing to a wire rack. Serve warm. **Yield:** 6 muffins.

Minty Ice Cream Shamrocks

(Pictured at left)

Prep: 30 min. + chilling **Bake:** 10 min. + freezing

With chocolate and mint flavor, these festive treats are sure to get smiles. —*Beverly Coyde, Gasport, New York*

6 tablespoons butter, softened
3/4 cup sugar
1 egg
1-1/2 teaspoons milk
3/4 teaspoon vanilla extract
1-1/3 cups all-purpose flour
1/4 cup baking cocoa
1-1/4 teaspoons baking powder
1/4 teaspoon salt
1-1/2 to 2 cups mint chocolate chip ice cream, softened

In a small mixing bowl, cream butter and sugar. Add egg, milk and vanilla; mix well. Combine the flour, cocoa, baking powder and salt; gradually add to creamed mixture just until blended. Divide dough into two portions; flatten. Wrap in plastic wrap and refrigerate for 1 hour or until firm.

On a lightly floured surface, roll out dough to 1/8- to 1/4-in. thickness. Cut with a floured 3-in. shamrock cookie cutter. Place 2 in. apart on parchment paper-lined baking sheets. Prick with a fork if desired. Bake at 350° for 7-10 minutes or until set. Cool for 2 minutes before removing from pans to wire racks to cool completely.

Spread 1/4 to 1/3 cup ice cream over the bottoms of six cookies; top with remaining cookies. Wrap individually in plastic wrap; freeze. May be frozen for up to 2 months. **Yield:** 6 servings.

Simple Salsa Chicken

(Pictured below)

Prep: 10 min. **Bake:** 25 min.

My husband and I prefer our food a little spicier than our children do, so one evening I baked plain chicken for them and created this dish for us. My husband liked it so much that it's now a regular menu item at our house.
—*Jan Cooper, Troy, Alabama*

✓ **Uses less fat, sugar or salt. Includes Nutrition Facts and Diabetic Exchanges.**

2 boneless skinless chicken breast halves (5 ounces *each*)
1/8 teaspoon salt
1/3 cup salsa
2 tablespoons taco sauce
1/3 cup shredded Mexican cheese blend

Place chicken in a shallow 2-qt. baking dish coated with nonstick cooking spray. Sprinkle with salt. Combine salsa and taco sauce; drizzle over chicken. Sprinkle with cheese.

Cover and bake at 350° for 25-30 minutes or until chicken juices run clear. **Yield:** 2 servings.

Nutrition Facts: 1 chicken breast half (prepared with reduced-fat cheese) equals 226 calories, 7 g fat (3 g saturated fat), 92 mg cholesterol, 628 mg sodium, 3 g carbohydrate, trace fiber, 34 g protein. **Diabetic Exchanges:** 5 very lean meat, 1 fat.

Chili Chicken Enchiladas

(Pictured below and on page 159)

Prep: 20 min. **Bake:** 25 min.

Rich and saucy, this south-of-the-border entree makes a hearty meal. When I take dinner to friends who aren't feeling well, these enchiladas always seem to hit the spot.
—Alicia Johnson, Hillsboro, Oregon

- **1/4 cup chopped onion**
- **2 garlic cloves, minced**
- **1 tablespoon butter**
- **2 tablespoons all-purpose flour**
- **1 cup chicken broth**
- **1 can (4 ounces) chopped green chilies**
- **1/4 teaspoon ground coriander**
- **1/8 teaspoon pepper**
- **1 cup (4 ounces) shredded Monterey Jack cheese, *divided***
- **1/2 cup sour cream**
- **2 cups chopped cooked chicken**
- **4 flour tortillas (8 inches)**
- **Sliced ripe olives, chopped tomatoes and green onions, optional**

In a small saucepan, saute onion and garlic in butter until tender. Combine flour and broth until smooth; gradually add to pan. Stir in the chilies, coriander and pepper. Bring to a boil; cook and stir for 2 minutes or until thickened.

Remove from the heat; stir in 1/2 cup cheese and sour cream until cheese is melted. Combine chicken and 3/4 cup sauce. Place about 1/2 cup chicken mixture down the center of each tortilla. Roll up and place seam side down in a greased 11-in. x 7-in. x 2-in. baking dish. Pour remaining sauce over enchiladas.

Bake, uncovered, at 350° for 20 minutes. Sprinkle with remaining cheese. Bake 5-10 minutes longer or until cheese is melted. Garnish with olives, tomatoes and green onions if desired. **Yield:** 2 servings.

Herb-Crusted Red Snapper

(Pictured at right and on page 159)

Prep/Total Time: 25 min.

An appetizing blend of herbs complements the mild flavor of these flaky fillets. *—Nella Parker, Hersey, Michigan*

☑ **Uses less fat, sugar or salt. Includes Nutrition Facts and Diabetic Exchanges.**

- **1 tablespoon dry bread crumbs**
- **1 teaspoon dried basil**
- **1 teaspoon paprika**
- **1/2 teaspoon salt**
- **1/2 teaspoon fennel seeds**
- **1/2 teaspoon dried thyme**
- **1/2 teaspoon dried oregano**
- **1/4 teaspoon pepper**
- **1/4 teaspoon crushed red pepper flakes**
- **2 red snapper fillets (5 ounces *each*), skin removed**
- **2 teaspoons canola oil**

In a food processor, combine the first nine ingredients; cover and process until fennel is finely ground. Transfer to a shallow bowl; dip fillets in herb mixture, coating both sides.

In a heavy skillet over medium-high heat, cook fillets in oil for 3-4 minutes on each side or until fish flakes easily with a fork. **Yield:** 2 servings.

Nutrition Facts: 1 fillet equals 200 calories, 7 g fat (1 g saturated fat), 50 mg cholesterol, 681 mg sodium, 4 g carbohydrate, 1 g fiber, 29 g protein. **Diabetic Exchanges:** 4 very lean meat, 1 fat.

Apple Juice Vinaigrette

(Pictured above right and on page 159)

Prep/Total Time: 15 min.

You can make this tart vinaigrette, just enough for two, in mere minutes. *—Shirley Glaab, Hattiesburg, Mississippi*

- **2 teaspoons unsweetened apple juice**
- **1 teaspoon water**
- **1 teaspoon rice wine vinegar**
- **1 teaspoon lemon juice**
- **1 teaspoon olive oil**
- **1/4 teaspoon sugar**
- **1/4 teaspoon Dijon mustard**
- **1/8 teaspoon salt**
- **Spring mix salad greens**
- **Assorted fresh fruit**

In a small bowl, whisk the first eight ingredients. Drizzle over greens and fruit. **Yield:** 2 tablespoons.

A SEAFOOD SPREAD featuring flaky Herb-Crusted Red Snapper, salad greens with Apple Juice Vinaigrette and warm Pear Pandowdy (shown above) is sure to catch compliments.

Pear Pandowdy

(Pictured above and on page 159)

Prep: 20 min. **Bake:** 20 min.

One night, my husband was craving something sweet after dinner, so I searched through my file of recipes and pulled out this one. He couldn't have been happier! A great last-minute dessert, the pandowdy almost melts in your mouth...and is especially good with ice cream.
—*Jennifer Class, Kirkland, Washington*

2 medium firm pears, peeled and sliced
2 tablespoons brown sugar
4-1/2 teaspoons butter
1-1/2 teaspoons lemon juice
1/8 teaspoon ground cinnamon
1/8 teaspoon ground nutmeg
TOPPING:
1/2 cup all-purpose flour
2 tablespoons plus 1/2 teaspoon sugar, *divided*
1/2 teaspoon baking powder
1/8 teaspoon salt
1/4 cup cold butter
2 tablespoons water
Vanilla ice cream, optional

In a small saucepan, combine the first six ingredients. Cook and stir over medium heat for 5 minutes or until the pear slices are tender. Pour into a greased 3-cup baking dish.

In a small bowl, combine the flour, 2 tablespoons sugar, baking powder and salt; cut in butter until crumbly. Stir in water. Sprinkle over pear mixture. Sprinkle with remaining sugar.

Bake, uncovered, at 375° for 20-25 minutes or until a toothpick inserted into topping comes out clean and topping is lightly browned. Serve warm with ice cream if desired. **Yield:** 2 servings.

happy holidays
to: Mom
from: Santa
Taste of Home

Holiday & Seasonal Favorites

Merry cookie recipes for Christmas...a patriotic dessert for the Fourth of July...spooky snacks for Halloween...these and many other festive foods for special occasions are all right here.

TIMELY TREATS. Clockwise from upper left: Pistachio Cranberry Biscotti (p. 199), Apple Snack Mix (p. 199), Turkey with Sausage-Corn Bread Stuffing (p. 187), Thanksgiving Gravy (p. 189), Golden Harvest Potato Bake (p. 188), Green Beans Supreme (p. 188), Fruity Cranberry Chutney (p. 188), Sour Cream Cutouts (p. 195), Lemon Butter Spritz (p. 195), Wicked Witch Cupcakes (p. 184), Garlic Pepper Corn (p. 179), Fiesta Rib Eye Steaks (p. 179) and Fruit 'n' Cake Kabobs (p. 179).

Happy, Healthy New Year

CELEBRATING New Year's Eve with festive food is easy with the party menu here. From an elegant main course to a delectable dessert, these delights will satisfy everyone right up until midnight.

What's more, each delicious dish is on the lighter side. So if you've made a resolution to eat better in the coming year, you'll be well on your way!

Chicken Taco Cups

(Pictured at left)

Prep: 20 min. **Bake:** 20 min.

For holiday parties and summer picnics, I stuff cute wonton cups with a Southwest-style chicken filling. You can freeze them, too—just reheat after thawing.
—Lee Ann Lowe, Gray, Maine

✓ **Uses less fat, sugar or salt. Includes Nutrition Facts and Diabetic Exchanges.**

1 pound boneless skinless chicken breasts, cut into 1-inch pieces
1 envelope reduced-sodium taco seasoning
1 small onion, chopped
1 jar (16 ounces) salsa, *divided*
2 cups (8 ounces) shredded reduced-fat cheddar cheese, *divided*
36 wonton wrappers
Sour cream, chopped green onions and chopped ripe olives, optional

Sprinkle chicken with taco seasoning. In a large skillet coated with nonstick cooking spray, cook and stir the chicken over medium heat for 5 minutes or until juices run clear. Transfer chicken to a food processor; cover and process until chopped. In a bowl, combine the chicken, onion, half of the salsa and 1 cup cheese.

Press wonton wrappers into miniature muffin cups coated with nonstick cooking spray. Bake at 375° for 5 minutes or until lightly browned.

Spoon rounded tablespoonfuls of chicken mixture into cups; top with remaining salsa and cheese. Bake 15 minutes longer or until heated through. Serve warm. Garnish with sour cream, green onions and olives if desired. **Yield:** 3 dozen.

Nutrition Facts: 2 appetizers (calculated without garnishes) equals 124 calories, 3 g fat (2 g saturated fat), 24 mg cholesterol, 408 mg sodium,12 g carbohydrate, trace fiber, 10 g protein. **Diabetic Exchanges:** 1 starch, 1 very lean meat, 1/2 fat.

Roasted Eggplant Spread

(Pictured at far left)

Prep: 20 min. **Bake:** 45 min. + cooling

Black pepper and garlic perk up this out-of-the-ordinary spread. —Barbara McCalley, Allison Park, Pennsylvania

✓ **Uses less fat, sugar or salt. Includes Nutrition Facts and Diabetic Exchanges.**

2 large sweet red peppers, cut into 1-inch pieces
1 medium eggplant, cut into 1-inch pieces
1 medium red onion, cut into 1-inch pieces
3 garlic cloves, minced
3 tablespoons olive oil
1/2 teaspoon salt
1/2 teaspoon pepper
1 tablespoon tomato paste
Toasted bread slices *or* assorted crackers

In a large bowl, combine the red peppers, eggplant, onion and garlic. Drizzle with the oil; sprinkle with the salt and pepper. Toss to coat.

Transfer to a 15-in. x 10-in. x 1-in. baking pan coated with nonstick cooking spray. Bake, uncovered, at 400° for 45-50 minutes or until lightly browned and tender, stirring once. Cool slightly.

Place the vegetables and tomato paste in a food processor; cover and process until chopped and blended. Transfer to a serving bowl; cool to room temperature. Serve with bread or crackers. **Yield:** 2 cups.

Nutrition Facts: 1/4 cup (calculated without bread or crackers) equals 83 calories, 5 g fat (1 g saturated fat), 0 cholesterol, 153 mg sodium, 9 g carbohydrate, 3 g fiber, 1 g protein. **Diabetic Exchanges:** 1 vegetable, 1 fat.

Healthier Habits

Resolving to eat healthier in the coming year? Here are some easy-to-handle tips:

- Switch from whole milk to 2% (reduced fat), 1% (low-fat) or skim (fat-free).
- Instead of sipping colas or sweetened juices, choose water or sparkling water.
- Bring healthy snacks to work, such as low-fat yogurt, granola, fruit or raisins.

AN EXCELLENT ENDING to the old year on New Year's Eve starts with Tomato and Onion Salmon, Broccoli with Mock Hollandaise Sauce and Two-Tone Potato Wedges (shown above).

Tomato and Onion Salmon

(Pictured above)

Prep/Total Time: 30 min.

Tomatoes, onion and lemon juice make this flaky salmon something special. —Lillian Denchick, Peru, New York

✓ Uses less fat, sugar or salt. Includes Nutrition Facts and Diabetic Exchanges.

4 salmon fillets (5 ounces *each*)
2 teaspoons olive oil
1/4 teaspoon dill weed
1/4 teaspoon pepper
2 medium tomatoes, thinly sliced
1 medium onion, thinly sliced
4 garlic cloves, minced
1/2 cup reduced-sodium chicken broth
1 tablespoon lemon juice
2 tablespoons minced fresh parsley

Place the salmon fillets in a 13-in. x 9-in. x 2-in. baking dish coated with nonstick cooking spray. Brush with oil; sprinkle with dill and pepper. Top with the tomato slices; set aside.

In a small skillet coated with nonstick cooking spray, saute onion and garlic. Add the broth, lemon juice and parsley. Bring to a boil; cook for 2-3 minutes or until most of the liquid has evaporated.

Spoon over salmon. Cover and bake at 350° for 13-18 minutes or until fish flakes easily with a fork. **Yield:** 4 servings.

Nutrition Facts: 1 serving equals 318 calories, 18 g fat (3 g saturated fat), 84 mg cholesterol, 171 mg sodium, 9 g carbohydrate, 2 g fiber, 30 g protein. **Diabetic Exchanges:** 4 lean meat, 1-1/2 fat, 1/2 starch.

Two-Tone Potato Wedges

(Pictured above)

Prep: 10 min. **Bake:** 40 min.

I think these tasty wedges are even better than french fries. —Maria Nicolau Schumacher, Larchmont, New York.

✓ Uses less fat, sugar or salt. Includes Nutrition Facts and Diabetic Exchanges.

2 medium potatoes
1 medium sweet potato
1 tablespoon olive oil
1/4 teaspoon salt
1/4 teaspoon pepper
1 tablespoon grated Parmesan cheese
2 garlic cloves, minced

Cut each potato into eight wedges; place in a large resealable plastic bag. Add oil, salt and pepper; seal bag and shake to coat. Arrange in a single layer in a 15-in. x 10-in. x 1-in. baking pan coated with nonstick cooking spray. Bake, uncovered, at 425° for 20 minutes. Turn the potatoes; sprinkle with cheese and garlic. Bake 20-25 minutes longer or until golden brown, turning once. **Yield:** 4 servings.

Nutrition Facts: 6 wedges equals 169 calories, 4 g fat (1 g saturated fat), 1 mg cholesterol, 182 mg sodium, 31 g carbohydrate, 3 g fiber, 4 g protein. **Diabetic Exchange:** 2 starch.

Broccoli with Mock Hollandaise Sauce

(Pictured at left)

Prep/Total Time: 20 min.

This broccoli has a lemony sauce that complements the veggie flavor. —Roxana Quarles, Ralph, Alabama

✓ **Uses less fat, sugar or salt. Includes Nutrition Facts and Diabetic Exchanges.**

2 packages (9 ounces *each*) frozen broccoli spears
4 ounces reduced-fat cream cheese, cubed
1 egg
2 tablespoons lemon juice
1/4 teaspoon salt
1/4 teaspoon pepper

Place the broccoli in a steamer basket; place in a saucepan over 1 in. of water. Bring to a boil; cover and steam for 5-7 minutes or until crisp-tender.

Meanwhile, in a small saucepan, combine the cream cheese, egg, lemon juice, salt and pepper. Cook and stir over low heat for 3-5 minutes or until sauce is thickened and reaches 160°. Serve with broccoli. **Yield:** 4 servings.

Nutrition Facts: 1 cup broccoli with 3 tablespoons sauce equals 127 calories, 8 g fat (4 g saturated fat), 73 mg cholesterol, 313 mg sodium, 8 g carbohydrate, 3 g fiber, 9 g protein. **Diabetic Exchanges:** 1-1/2 vegetable, 1-1/2 fat.

Hot Crab Pinwheels

(Pictured on page 170)

Prep: 15 min. + chilling **Bake:** 10 min.

My husband, who doesn't like most seafood, couldn't stop eating these! —Kitti Boesel, Woodbridge, Virginia

✓ **Uses less fat, sugar or salt. Includes Nutrition Facts and Diabetic Exchanges.**

1 package (8 ounces) reduced-fat cream cheese
1 can (6 ounces) crabmeat, drained, flaked and cartilage removed
3/4 cup finely chopped sweet red pepper
1/2 cup shredded reduced-fat cheddar cheese
2 green onions, finely chopped
3 tablespoons minced fresh parsley
1/4 to 1/2 teaspoon cayenne pepper
6 flour tortillas (6 inches)

In a small mixing bowl, beat the cream cheese until smooth. Stir in the crab, red pepper, cheese, onions, parsley and cayenne. Spread 1/3 cupful over one side of each tortilla; roll up tightly. Wrap in plastic wrap and refrigerate for at least 2 hours.

Cut and discard ends of roll-ups. Cut each into six slices. Place on baking sheets coated with nonstick cooking spray. Bake at 350° for 10 minutes or until bubbly. Serve warm. **Yield:** 3 dozen.

Nutrition Facts: 3 appetizers equals 123 calories, 7 g fat (3 g saturated fat), 29 mg cholesterol, 270 mg sodium, 8 g carbohydrate, trace fiber, 8 g protein. **Diabetic Exchanges:** 1 lean meat, 1/2 starch, 1/2 fat.

Black Cherry Cake

(Pictured below)

Prep: 10 min. **Bake:** 30 min. + cooling

A friend baked this pretty pink cake for a birthday party, and I loved it. —Judy Lentz, Emmetsburg, Iowa

✓ **Uses less fat, sugar or salt. Includes Nutrition Facts.**

1 package (18-1/4 ounces) white cake mix
1-1/4 cups water
1/3 cup canola oil
4 egg whites
2 cartons (6 ounces *each*) fat-free reduced-sugar black cherry yogurt, *divided*
1 carton (8 ounces) frozen fat-free whipped topping, thawed

In a large mixing bowl, combine the cake mix, water, oil and egg whites just until moistened; beat on low speed for 2 minutes. Fold in one carton of yogurt.

Pour into a 13-in. x 9-in. x 2-in. baking dish coated with nonstick cooking spray. Bake at 350° for 30-35 minutes or until a toothpick inserted near the center comes out clean. Cool on a wire rack.

Place remaining yogurt in a bowl; fold in whipped topping. Spread over cake. Store in the refrigerator. **Yield:** 15 servings.

Nutrition Facts: 1 piece equals 230 calories, 8 g fat (2 g saturated fat), trace cholesterol, 255 mg sodium, 34 g carbohydrate, 1 g fiber, 3 g protein.

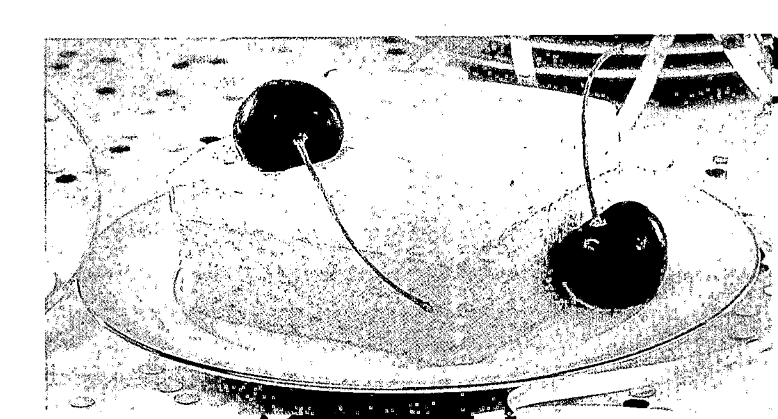

Beautiful Bridal Shower

ORDER sunshine and blue sky...then use these delightful recipes to host a memorable celebration for the bride-to-be! You'll have a full buffet table with a crunchy chicken salad, elegant crepes, refreshing punch, a pretty lemon loaf and luscious desserts.

Strawberry Rhubarb Crepes

(Pictured at left)

Prep: 35 min. + chilling **Cook:** 20 min.

With a ruby-red filling and whipped cream on top, this lovely dessert is ideal for a special brunch or luncheon.
—Susan McDonald, Germantown, Tennessee

2 eggs
1 cup milk
1 tablespoon butter, melted
1/2 cup all-purpose flour
1/8 teaspoon salt
Dash ground nutmeg
Dash ground cinnamon
FILLING:
3/4 cup sugar
1/3 cup cornstarch
1/4 teaspoon ground cinnamon
Dash ground nutmeg
1 cup orange juice
2 tablespoons grated orange peel
1-1/2 teaspoons lemon juice
3 cups chopped fresh rhubarb
3 cups halved fresh strawberries
Whipped cream

In a small mixing bowl, beat eggs, milk and butter. Combine the flour, salt, nutmeg and cinnamon; add to egg mixture. Cover and refrigerate for 30 minutes.

Meanwhile, in a large saucepan, combine the sugar, cornstarch, cinnamon and nutmeg. Add the orange juice, orange peel and lemon juice; stir in rhubarb and strawberries. Bring to a boil. Reduce heat; simmer, uncovered, for 15-20 minutes or until rhubarb is tender, stirring occasionally. Set aside.

Heat a lightly greased 8-in. nonstick skillet; pour 2 tablespoons batter into the center of skillet. Lift and tilt pan to evenly coat bottom. Cook until top appears dry; turn and cook 15-20 seconds longer. Remove to a wire rack. Repeat with remaining batter, greasing skillet as needed. Stack cooled crepes with waxed paper or paper towels in between.

To serve, spoon 1/4 cup filling over each crepe; roll up. Top with whipped cream. **Yield:** 1 dozen.

Special-Occasion White Cake

(Pictured at far left)

Prep: 1 hour 50 min. **Bake:** 20 min. + cooling

My grandma used this recipe for her famous wedding cakes. *—Sue Gronholz, Beaver Dam, Wisconsin*

3/4 cup shortening
1-1/2 cups sugar
1/2 teaspoon vanilla extract
1/2 teaspoon almond extract
2-1/2 cups cake flour
2 teaspoons baking powder
1/4 teaspoon salt
1 cup cold water
4 egg whites
FROSTING:
2 cups shortening
6 cups confectioners' sugar
1 teaspoon almond extract
1 teaspoon vanilla extract
1/4 teaspoon salt
2 to 3 tablespoons milk
Lilac paste food coloring
Edible pansies

In a large mixing bowl, cream shortening and sugar until light and fluffy, about 5 minutes. Beat in extracts. Combine the flour, baking powder and salt; add to creamed mixture alternately with water. Beat just until combined. In another mixing bowl, beat egg whites on high speed until stiff peaks form; fold into batter.

Pour into two greased and floured 9-in. round baking pans. Bake at 350° for 20-25 minutes or until a toothpick comes out clean. Cool for 10 minutes; remove from pans to wire racks to cool completely.

In a large mixing bowl, beat shortening, confectioners' sugar, extracts and salt until light and fluffy. Add enough milk to achieve frosting consistency.

Place one cake on a plate; spread with 1 cup frosting. Top with remaining cake. Tint 2 cups frosting with lilac food coloring; spread over top and sides of cake.

Cut a small hole in the corner of a pastry or plastic bag; insert #10 tip and fill with remaining white frosting. Pipe a border of "pearls" around bottom and top of cake. With #3 tip, pipe smaller "pearls" on sides of cake. Garnish with pansies. **Yield:** 14-16 servings.

Editor's Note: Use of a coupler ring will allow you to easily change pastry tips for different designs. Make sure to properly identify flowers before picking. Double-check that they're edible and have not been treated with chemicals.

SPECIAL SWEETS. Lavender Cookies and Spritz Butter Blossoms (shown above) prove irresistible.

Lavender Cookies

(Pictured above)

Prep: 30 min. **Bake:** 10 min. per batch

I am a wedding and event planner. One of my brides served these unusual cookies, which contain dried lavender flowers, at her reception. I just had to have the recipe. You can guess what her wedding color was!

—Glenna Tooman, Boise, Idaho

✓ **Uses less fat, sugar or salt. Includes Nutrition Facts and Diabetic Exchanges.**

- **1/2 cup shortening**
- **1/2 cup butter, softened**
- **1-1/4 cups sugar**
- **2 eggs**
- **1 teaspoon vanilla extract**
- **1/2 teaspoon almond extract**
- **2-1/4 cups all-purpose flour**
- **4 teaspoons dried lavender flowers**
- **1 teaspoon baking powder**
- **1/2 teaspoon salt**

In a large mixing bowl, cream the shortening, butter and sugar. Add eggs, one at a time, beating well after each addition. Beat in extracts. Combine the flour, lavender, baking powder and salt; gradually add to creamed mixture and mix well.

Drop by rounded teaspoonfuls 2 in. apart onto baking sheets lightly coated with nonstick cooking spray. Bake at 375° for 8-10 minutes or until golden brown. Cool for 2 minutes before removing to wire racks. Store in an airtight container. **Yield:** about 7 dozen.

Nutrition Facts: 1 cookie equals 46 calories, 2 g fat (1 g saturated fat), 8 mg cholesterol, 31 mg sodium, 6 g carbohydrate, trace fiber, 1 g protein. **Diabetic Exchanges:** 1/2 starch, 1/2 fat.

Editor's Note: Dried lavender flowers are available from Penzeys Spices. Call 1-800/741-7787 or visit *www.tasteofhome.com* for a Web site link.

Spritz Butter Blossoms

(Pictured at left)

Prep: 30 min. **Bake:** 10 min. per batch

These pretty cookies can be customized to the bride's chosen colors with food coloring or colored sugars.

—Christine Omar, Harwich Port, Massachusetts

✓ **Uses less fat, sugar or salt. Includes Nutrition Facts and Diabetic Exchanges.**

- **3/4 cup sugar**
- **1/2 cup blanched almonds, toasted**
- **1 cup butter, softened**
- **1 egg**
- **1 teaspoon almond extract**
- **1 teaspoon vanilla extract**
- **2 cups all-purpose flour**
- **1/4 teaspoon salt**
- **Sprinkles *or* colored sugar**

Place sugar and almonds in a food processor; cover and process until almonds are finely ground. In a large mixing bowl, cream butter and sugar mixture. Beat in the egg and extracts. Combine flour and salt; gradually add to creamed mixture and mix well.

Using a cookie press fitted with the disk of your choice, press dough 2 in. apart onto ungreased baking sheets. Decorate with sprinkles or colored sugar. Bake at 375° for 7-9 minutes. Remove to wire racks. Store in an airtight container. **Yield:** about 6-1/2 dozen.

Nutrition Facts: 1 cookie equals 44 calories, 3 g fat (1 g saturated fat), 9 mg cholesterol, 31 mg sodium, 4 g carbohydrate, trace fiber, 1 g protein. **Diabetic Exchange:** 1 fat.

Lemony Zucchini Bread

(Pictured on page 174)

Prep: 25 min. **Bake:** 50 min. + cooling

Flecks of zucchini give a third dimension to the popular lemon and poppy seed combination in this moist quick bread.

—Carol Funk, Richard, Saskatchewan

4 cups all-purpose flour
1-1/2 cups sugar
1 package (3.4 ounces) instant lemon pudding mix
1-1/2 teaspoons baking soda
1 teaspoon baking powder
1 teaspoon salt
4 eggs
1-1/4 cups milk
1 cup vegetable oil
3 tablespoons lemon juice
1 teaspoon lemon extract
2 cups shredded zucchini
1/4 cup poppy seeds
2 teaspoons grated lemon peel

In a large bowl, combine the flour, sugar, pudding mix, baking soda, baking powder and salt. In another bowl, whisk the eggs, milk, oil, lemon juice and extract. Stir into dry ingredients just until moistened. Fold in the zucchini, poppy seeds and lemon peel.

Pour into two greased 9-in. x 5-in. x 3-in. loaf pans. Bake at 350° for 50-55 minutes or until a toothpick inserted near the center comes out clean. Cool for 10 minutes before removing from pans to wire racks to cool completely. **Yield:** 2 loaves.

Sweetheart Punch

(Pictured on page 174)

Prep: 20 min. + chilling

A berry-filled ice ring adds elegance to this punch. If you like, use a heart-shaped mold to make the ice ring.
—Gretchen Montgomery, Marietta, Ohio

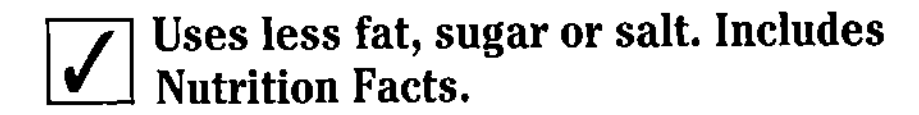

2 cups sliced fresh strawberries, *divided*
6 cups water, *divided*
1 can (12 ounces) frozen pink lemonade concentrate, thawed
1 package (10 ounces) frozen sweetened sliced strawberries, thawed
1 can (6 ounces) frozen orange juice concentrate, thawed
2 liters lemon-lime soda, chilled

Arrange 1 cup fresh berries in a 4-1/2-cup ring mold; add 2 cups water. Freeze until solid. Top with remaining fresh berries. Slowly pour 1 cup water into mold to almost cover berries. Freeze until solid.

In a punch bowl, combine the lemonade concentrate, thawed berries, orange juice concentrate and remaining water. Refrigerate until chilled.

Just before serving, stir in lemon-lime soda. Unmold ice ring by wrapping the bottom of the mold in a damp hot dishcloth; invert onto a baking sheet. Place fruit side up in punch bowl. **Yield:** 14 servings (3-1/2 quarts).

Nutrition Facts: 1 cup equals 149 calories, trace fat (trace saturated fat), 0 cholesterol, 18 mg sodium, 38 g carbohydrate, 1 g fiber, 1 g protein.

Luncheon Chicken Salad

(Pictured below)

Prep/Total Time: 30 min.

For a bridal shower or summer buffet, this crunchy salad is a perfect choice. *—Pat Stevens, Granbury, Texas*

6 cups cubed cooked chicken
2 celery ribs, finely chopped
1 large green pepper, chopped
2/3 to 1 cup sweet pickle relish
1/4 cup shredded carrot
1 jar (4 ounces) sliced pimientos, drained
1 cup mayonnaise
2 tablespoons lemon juice
2 teaspoons sugar
1/2 teaspoon salt
1/2 teaspoon pepper
4 cups *each* torn Bibb, leaf and iceberg lettuce
2 cups (8 ounces) shredded Colby cheese
2 cups green grapes, halved
2 cans (11 ounces *each*) mandarin oranges, drained
1 cup slivered almonds, toasted

In a large bowl, combine chicken, celery, green pepper, relish, carrot and pimientos. In another bowl, combine mayonnaise, lemon juice, sugar, salt and pepper; spoon over chicken mixture and toss to coat.

Combine lettuces; divide among 12 salad plates. Top each with 1 cup chicken salad, cheese, grapes, oranges and almonds. **Yield:** 12 servings.

Sizzlin' Summer Cookout

SAVOR warm summer days by cooking up fresh, satisfying fare in the great outdoors. With grilled steaks, corn on the cob and sweet kabobs, this spread is ideal to enjoy with friends on the patio or deck.

You'll even find a star-spangled dessert featuring fresh seasonal berries, sugared pastry cutouts and a creamy dip. It's the perfect finale to a feast on Memorial Day, July Fourth or any summer day.

Fiesta Rib Eye Steaks

(Pictured at left and on page 168)

Prep/Total Time: 30 min.

This is a great recipe for grilling out or for camping trips. You can adapt it for indoor cooking by making the steaks in a skillet and heating the tortillas in a warm oven.
—Jodee Harding, Granville, Ohio

- **8 flour tortillas (6 inches)**
- **8 boneless beef rib eye steaks (3/4 inch thick)**
- **1/4 cup lime juice**
- **1 cup (4 ounces) shredded Colby-Monterey Jack cheese**
- **2 cups salsa**

Place the tortillas on a sheet of heavy-duty foil (about 18 in. x 12 in.). Fold the foil around tortillas and seal tightly; set aside.

Drizzle both sides of steaks with lime juice. Grill, covered, over medium-hot heat for 7-9 minutes on each side or until meat reaches desired doneness (for medium-rare, a meat thermometer should read 145°; medium, 160°; well-done, 170°).

Place the tortillas on the outer edge of the grill; heat for 5-6 minutes, turning once. Sprinkle shredded cheese over the steaks; serve with salsa and warmed tortillas. **Yield:** 8 servings.

Garlic Pepper Corn

(Pictured at left and on page 168)

Prep/Total Time: 25 min.

I've loved corn served with this simple seasoning since I was a child. It makes corn on the cob extra special.
—Anna Minegar, Zolfo Springs, Florida

✓ **Uses less fat, sugar or salt. Includes Nutrition Facts and Diabetic Exchanges.**

- **1 tablespoon dried parsley flakes**
- **1 tablespoon garlic pepper blend**
- **1/2 teaspoon paprika**
- **1/4 teaspoon salt**
- **8 medium ears sweet corn, husked**
- **1/4 cup butter, melted**

In a small bowl, combine the parsley, garlic pepper, paprika and salt; set aside. Place corn in a Dutch oven or kettle; cover with water. Bring to a boil; cover and cook for 3 minutes or until tender. Drain.

Brush corn with butter; sprinkle with seasoning mixture. Serve immediately. **Yield:** 8 servings.

Nutrition Facts: 1 ear of corn equals 128 calories, 7 g fat (4 g saturated fat), 15 mg cholesterol, 251 mg sodium, 17 g carbohydrate, 2 g fiber, 3 g protein. **Diabetic Exchanges:** 1 starch, 1 fat.

Fruit 'n' Cake Kabobs

(Pictured at far left and on page 168)

Prep/Total Time: 25 min.

A neighbor brought us this grilled dessert to sample. I was pleasantly surprised with the unusual kabobs.
—Mary Ann Dell, Phoenixville, Pennsylvania

✓ **Uses less fat, sugar or salt. Includes Nutrition Facts.**

- **1/2 cup apricot preserves**
- **1 tablespoon water**
- **1 tablespoon butter**
- **1/8 teaspoon ground cinnamon**
- **1/8 teaspoon ground nutmeg**
- **3 medium nectarines, pitted and quartered**
- **3 medium plums, pitted and quartered**
- **3 medium peaches, pitted and quartered**
- **1 loaf (10-3/4 ounces) frozen pound cake, thawed and cut into 2-inch cubes**

In a small saucepan over medium heat, combine preserves, water, butter, cinnamon and nutmeg until blended. On eight metal or soaked wooden skewers, alternately thread nectarines, plums, peaches and cake cubes. Grill, uncovered, over medium heat for 1-2 minutes on each side or until cake is golden brown and fruit is tender, brushing occasionally with apricot mixture. **Yield:** 8 servings.

Nutrition Facts: 1 kabob equals 259 calories, 8 g fat (4 g saturated fat), 58 mg cholesterol, 161 mg sodium, 46 g carbohydrate, 3 g fiber, 4 g protein.

Curried Chutney Spread

(Pictured above)

Prep/Total Time: 15 min.

This unusual blend of ingredients really creates a winner of an appetizer. It appeals to people of all ages.
—Kate McSoley, Indianapolis, Indiana

- **1 package (8 ounces) cream cheese, softened**
- **2/3 cup sour cream**
- **1 cup mango chutney**
- **1-1/2 teaspoons curry powder**
- **1 cup salted cashew halves**
- **Assorted crackers**

In a small mixing bowl, beat the cream cheese and sour cream until smooth; beat in chutney and curry until blended. Just before serving, stir in cashews. Serve with crackers. Refrigerate leftovers. **Yield:** 3 cups.

Keeping Cool at Picnics

When serving a cold dip or spread on a hot day, it's important to keep it cool.

Fill a large glass or plastic serving bowl with ice cubes, crushed ice or ice packs. Fill a smaller bowl with dip and set it on top of the ice. Replace the ice as it melts.

If you're taking the dip to an outing, put the dip in a small bowl (plastic is best for traveling because it won't break), cover it with plastic wrap and put it in a cooler. Assemble the ice-filled serving bowl when you get to the picnic.

Perky Parsleyed Tomatoes

(Pictured below)

Prep: 25 min. + chilling

My mother gave me this delightful recipe many years ago with a notation, "Oh, so good!" You can make these tomatoes ahead of time, and they're always so attractive.
—Marie Melchert, Las Vegas, Nevada

✓ **Uses less fat, sugar or salt. Includes Nutrition Facts and Diabetic Exchanges.**

- **8 medium tomatoes**
- **1/4 cup olive oil**
- **1/4 cup minced fresh parsley**
- **2 tablespoons tarragon *or* cider vinegar**
- **2 teaspoons Dijon mustard**
- **1 garlic clove, minced**
- **1 teaspoon salt**
- **1 teaspoon sugar**
- **1/4 teaspoon pepper**

Cut a thin slice off the bottom of each tomato so it sits flat. Cut each tomato into 1/2-in. horizontal slices; reassemble tomatoes, stacking slices on top of each other. Place the stacks in a 13-in. x 9-in. x 2-in. dish.

In a jar with a tight-fitting lid, combine the remaining ingredients; shake well. Pour over tomatoes. Cover and refrigerate for 4 hours or overnight. Remove from the refrigerator 20 minutes before serving. **Yield:** 8 servings.

Nutrition Facts: 1 tomato equals 96 calories, 7 g fat (1 g saturated fat), 0 cholesterol, 341 mg sodium, 8 g carbohydrate, 2 g fiber, 1 g protein. **Diabetic Exchanges:** 1-1/2 fat, 1 vegetable.

Grilled Three-Potato Salad

(Pictured above)

Prep: 25 min. **Grill:** 10 min.

Everyone in our extended family loves to cook, so I put all of our favorite recipes in a cookbook to be handed down from generation to generation. This recipe from the book is a delicious twist on traditional potato salad.
—Suzette Jury, Keene, California

- **3/4 pound Yukon Gold potatoes (about 3 medium)**
- **3/4 pound red potatoes (about 3 medium)**
- **1 medium sweet potato, peeled**
- **1/2 cup thinly sliced green onions**
- **1/4 cup vegetable oil**
- **2 to 3 tablespoons white wine vinegar**
- **1 tablespoon Dijon mustard**
- **1 teaspoon salt**
- **1/2 teaspoon celery seed**
- **1/4 teaspoon pepper**

Place all of the potatoes in a Dutch oven; cover with water. Bring to a boil. Reduce heat; cover and simmer for 15-20 minutes or until tender. Drain and rinse in cold water. Cut into 1-in. chunks.

Place the potatoes in a grill wok or basket. Grill, uncovered, over medium heat for 8-12 minutes or browned, stirring frequently. Transfer to a large salad bowl; add onions.

In a small bowl, whisk the oil, vinegar, mustard, salt, celery seed and pepper. Drizzle over potato mixture and toss to coat. Serve warm or at room temperature. **Yield:** 6 servings.

Editor's Note: If you do not have a grill wok or basket, use a disposable foil pan. Poke holes in the bottom of the pan with a meat fork to allow the liquid to drain.

Stars and Stripes Forever Dessert

(Pictured below)

Prep: 30 min. **Bake:** 10 min. + cooling

Sugared puff pastry stars and a medley of fresh berries, served with a creamy dip, make up this festive dessert.
—Gail Sykora, Menomonee Falls, Wisconsin

- **1 sheet frozen puff pastry, thawed**
- **1 to 2 tablespoons water**
- **1 tablespoon coarse sugar**
- **2 cups sliced fresh strawberries**
- **1-1/2 cups fresh raspberries**
- **1 cup fresh blueberries**
- **1/4 cup plus 1 tablespoon sugar, *divided***
- **1/2 cup heavy whipping cream**
- **1 cup (8 ounces) sour cream**

On a lightly floured surface, roll out pastry to 1/8-in. thickness. Cut with floured star-shaped cookie cutters. Place 1 in. apart on parchment paper-lined baking sheets. Bake at 400° for 8-10 minutes or until golden brown. Remove to wire racks. Brush lightly with water and sprinkle with coarse sugar. Cool.

In a large bowl, combine berries and 1/4 cup sugar; set aside. In a small mixing bowl, beat whipping cream until it begins to thicken. Add remaining sugar; beat until soft peaks form. Place sour cream in a small serving bowl; fold in whipped cream. Place bowl on a serving platter. Spoon berries onto platter; top with pastry stars. **Yield:** 8 servings.

Halloween Tricks and Treats

TO SCARE UP a spine-tingling good time next October 31, try this frightfully fun party fare. From jack-o'-lantern turnovers to cute witch cakes and bone-shaped cookies, they'll fit the eerie evening.

Goblin Gorp

(Pictured at left)

Prep/Total Time: 20 min.

Ghosts, goblins and ghouls alike can't keep their hands out of this snack mix. —*Renae Moncur, Burley, Idaho*

1 package (3.3 ounces) butter-flavored microwave popcorn, popped
1 cup salted pumpkin seeds
1 cup corn chips
2 tablespoons butter, melted
1 tablespoon taco sauce
1 teaspoon Mexican seasoning

In a large bowl, combine the popcorn, pumpkin seeds and corn chips. Combine the butter, taco sauce and Mexican seasoning; drizzle over popcorn mixture and toss to coat.

Transfer to an ungreased 13-in. x 9-in. x 2-in. baking pan. Bake at 275° for 10 minutes, stirring once. Serve warm, or cool before storing in an airtight container. **Yield:** 3 quarts.

Roasted Pumpkin Seeds

Prep: 20 min. **Bake:** 1 hour 35 min.

To enjoy the seeds from a pumpkin you hollow out, just spice them and bake them using this easy recipe. —*Dawn Fagerstrom, Warren, Minnesota*

2 cups fresh pumpkin seeds
3 tablespoons butter, melted
1 teaspoon salt
1 teaspoon Worcestershire sauce

Line a 15-in. x 10-in. x 1-in. baking pan with foil and grease the foil. In a small bowl, combine all ingredients; spread into prepared pan. Bake at 225° for 1-1/2 hours, stirring occasionally.

Increase heat to 325°. Bake 5 minutes longer or until seeds are dry and lightly browned. Serve warm, or cool before storing in an airtight container. **Yield:** 2 cups.

Peanut Butter Popcorn Balls

(Pictured at far left)

Prep: 20 min. + standing

Trick-or-treaters are always happy to receive these yummy balls. —*Betty Claycomb, Alverton, Pennsylvania*

5 cups popped popcorn
1 cup dry roasted peanuts
1/2 cup sugar
1/2 cup light corn syrup
1/2 cup chunky peanut butter
1/2 teaspoon vanilla extract

Place popcorn and peanuts in a large bowl; set aside. In a large heavy saucepan over medium heat, bring sugar and corn syrup to a rolling boil, stirring occasionally. Remove from the heat; stir in peanut butter and vanilla. Quickly pour over popcorn mixture and mix well.

When cool enough to handle, quickly shape into ten 2-1/2-in. balls. Let stand at room temperature until firm; wrap in plastic wrap. **Yield:** 10 popcorn balls.

Trick-or-Treat Turnovers

(Pictured at far left)

Prep: 30 min. **Bake:** 10 min.

I "carved" these beefy jack-o'-lantern pastries to feed my hungry bunch. —*Marge Free, Brandon, Mississippi*

1/2 pound ground beef
1 tablespoon finely chopped onion
4 ounces part-skim mozzarella cheese, cubed
1/4 cup prepared mustard
2 tubes (16.3 ounces *each*) large refrigerated flaky biscuits
1 egg, lightly beaten

In a large skillet, cook beef and onion over medium heat until meat is no longer pink; drain. Add cheese and mustard; cook and stir until cheese is melted. Cool slightly.

Flatten each biscuit into a 4-in. circle; place four biscuits in each of two greased 15-in. x 10-in. x 1-in. baking pans. Spoon 2 heaping tablespoons of meat mixture onto each. Using a sharp knife or cookie cutters, cut out faces from remaining biscuits; place over meat mixture and pinch edges to seal tightly. Reroll scraps if desired and cut out stems for pumpkins.

Brush with egg. Bake at 350° for 10-15 minutes or until golden brown. **Yield:** 8 servings.

Wicked Witch Cupcakes

(Pictured below and on page 168)

Prep: 1-1/4 hours **Bake:** 20 min. + cooling

More fun than frightening, these colorful witches cast a yummy spell. —Joan Antneon, Arlington, South Dakota

1/2 cup butter, softened
1 cup sugar
2 eggs
1-1/2 cups all-purpose flour
1/2 cup baking cocoa
1 teaspoon baking soda
1/2 teaspoon baking powder
1/2 teaspoon salt
1-1/4 cups milk
1 can (16 ounces) vanilla frosting
Green food coloring
WITCH HATS:
1 can (16 ounces) vanilla frosting
2 teaspoons milk
Assorted food coloring of your choice
12 to 16 ice cream sugar cones
Fruit Roll-Ups, licorice and assorted candies of your choice

In a large mixing bowl, cream butter and sugar until light and fluffy. Add eggs, one at a time, beating well after each addition. Combine the flour, cocoa, baking soda, baking powder and salt; add to creamed mixture alternately with milk. Beat just until combined.

Fill paper- or foil-lined muffin cups two-thirds full. Bake at 350° for 18-22 minutes or until a toothpick comes out clean. Cool for 10 minutes before removing from pan to a wire rack to cool completely. Tint frosting green; frost cupcakes. For hats, combine frosting and milk; tint with food coloring. Frost ice cream cones. Using small cookie cutters, cut out shapes from Fruit Roll-Ups; arrange on hats. Add licorice for hair and candies for faces. Place a hat on each witch. **Yield:** about 1 dozen.

Editor's Note: Chocolate frosting and jimmies may also be used to decorate witch hats.

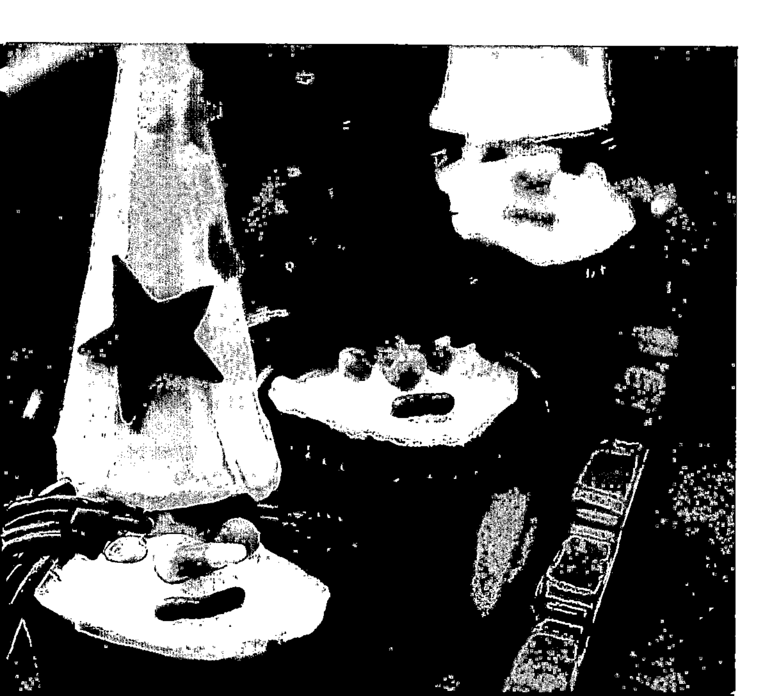

Pumpkin Cheese Ball

(Pictured on page 182)

Prep: 20 min. + chilling

Everyone gets a kick out of this creamy, pumpkin-shaped spread. —Suzanne Cleveland, Lyons, Georgia

1 package (8 ounces) cream cheese, softened
1 carton (8 ounces) spreadable chive and onion cream cheese
2 cups (8 ounces) shredded sharp cheddar cheese
2 teaspoons paprika
1/2 teaspoon cayenne pepper
1 celery rib *or* broccoli stalk
Sliced apples and assorted crackers

In a small mixing bowl, beat cream cheeses until smooth. Stir in the cheddar cheese, paprika and cayenne. Shape into a ball; wrap in plastic wrap. Refrigerate for 4 hours or until firm.

With a knife, add vertical lines to ball to resemble a pumpkin; insert celery rib or broccoli stalk for the stem. Serve with apples and crackers. **Yield:** 2-1/2 cups.

Mulled Cider

(Pictured on page 182)

Prep: 5 min. **Cook:** 45 min.

I dress up this apple cider using lemonade, orange juice, honey and spices. —Glenna Tooman, Boise, Idaho

4 cups water
2 teaspoons ground allspice
1 cinnamon stick (3 inches)
Dash ground cloves
1 gallon apple cider *or* juice
1 can (12 ounces) frozen lemonade concentrate, thawed
3/4 cup orange juice
1/3 cup honey
1 individual tea bag

In a large kettle, combine the water, allspice, cinnamon stick and cloves. Bring to a boil. Reduce heat; simmer, uncovered, for 30 minutes.

Add the remaining ingredients. Return just to a boil. Discard cinnamon stick and tea bag. Stir and serve warm. **Yield:** 18 servings (4-1/2 quarts).

Sausage-Stuffed Pumpkins

Prep: 50 min. **Bake:** 1-1/4 hours

This is a festive fall meal. To serve it, cut the pumpkin into wedges, giving each person both pumpkin and stuffing.
—Rebecca Baird, Salt Lake City, Utah

- **2 cups water**
- **1 cup uncooked brown rice**
- **2 teaspoons chicken bouillon granules**
- **1/2 teaspoon curry powder**
- **1 pound bulk Italian sausage**
- **4 cups sliced fresh mushrooms**
- **1 small onion, chopped**
- **2 shallots, minced**
- **1 garlic clove, minced**
- **5 tablespoons dried currants**
- **1/4 cup chicken broth**
- **1 teaspoon poultry seasoning**
- **1/2 teaspoon rubbed sage**
- **1/4 teaspoon dried marjoram**
- **2 medium pie pumpkins (2-1/2 pounds *each*)**
- **1/2 teaspoon salt**
- **1/2 teaspoon garlic powder**

In a large saucepan, bring the water, rice, bouillon and curry powder to a boil. Reduce heat; cover and simmer for 45-50 minutes or until tender.

Meanwhile, in a large skillet, cook sausage over medium heat until no longer pink; drain and set aside. In the same skillet, saute the mushrooms, onion, shallots and garlic for 3-5 minutes or until tender. Reduce heat; add the currants, broth, poultry seasoning, sage and marjoram. Return sausage to the pan. Cook and stir for 5-7 minutes or until liquid is absorbed. Remove from the heat; stir in rice.

Wash pumpkins; cut a 3-in. circle around each stem. Remove tops and set aside. Remove and discard loose fibers; save seeds for another use. Prick inside each pumpkin with a fork; sprinkle with salt and garlic powder. Stuff with sausage mixture; replace tops.

Place in a 13-in. x 9-in. x 2-in. baking dish; add 1/2 in. of water. Bake, uncovered, at 350° for 30 minutes. Cover loosely with foil; bake 45-50 minutes longer or until tender. Cut each pumpkin into four wedges to serve. **Yield:** 8 servings.

Boneyard Cookies

(Pictured above right)

Prep: 1 hour + chilling

Bake: 10 min. per batch + cooling

With a pinch of "spook," these bone-shaped sandwich cookies are eye-catching at any Halloween party.
—Celena Cantrell-Richardson, Eau Claire, Michigan

✓ **Uses less fat, sugar or salt. Includes Nutrition Facts and Diabetic Exchanges.**

- **1 cup confectioners' sugar**
- **1/2 cup cornstarch**
- **1/2 cup cold butter**
- **2 eggs**
- **1 teaspoon almond extract**
- **2 cups all-purpose flour**
- **1/8 teaspoon salt**
- **2 to 3 tablespoons seedless raspberry jam**
- **16 to 18 squares (1 ounce *each*) white baking chocolate**

In a small mixing bowl, combine confectioners' sugar and cornstarch. Cut in butter until mixture resembles coarse crumbs. Add eggs, one at a time, beating well after each addition. Beat in extract. Combine flour and salt; gradually add to sugar mixture.

Shape the dough into a ball; flatten into a disk. Wrap the dough in plastic wrap; refrigerate for 30 minutes or until easy to handle.

On a lightly floured surface, roll dough to 1/8-in. thickness. Cut with a floured 3-1/2-in. bone-shaped cookie cutter. Place 1 in. apart on parchment paper-lined baking sheets. Bake at 350° for 8-10 minutes or until edges begin to brown. Remove to wire racks to cool.

On the bottom of half of the cookies, spread 1/8 to 1/4 teaspoon jam down the center; top with remaining cookies. In a microwave-safe bowl, melt white chocolate at 70% power. Dip each cookie in chocolate, allowing excess to drip off. Place on waxed paper; let stand until set. **Yield:** 34 sandwich cookies.

Nutrition Facts: 1 cookie equals 155 calories, 8 g fat (5 g saturated fat), 23 mg cholesterol, 54 mg sodium, 19 g carbohydrate, trace fiber, 2 g protein. **Diabetic Exchanges:** 1-1/2 fat, 1 starch.

Three Thanksgiving Menus

WHAT KIND of Thanksgiving dinner appeals to you most? A traditional, just-like-Mom-made feast? A menu with dishes you can prepare ahead? How about a meal that's extra easy, so you're out of the kitchen fast?

You'll find all three types of menus here. Simply choose the one that suits you best! See the list of recipes in each menu in the sidebar below right.

Turkey with Sausage-Corn Bread Stuffing

(Pictured at left and on page 168)

Prep: 1 hour + cooling

Bake: 3-3/4 hours + standing

Turkey is so good with this meaty stuffing. I've made it for many get-togethers, and even those who usually skip stuffing enjoy this. —June Kathrein, Delta, Colorado

2-1/2 cups yellow cornmeal
1-1/2 cups all-purpose flour
1/4 cup sugar
6 teaspoons baking powder
1-1/2 teaspoons salt
2 eggs
2 cups milk
6 tablespoons vegetable oil

STUFFING:

1 pound bulk sage pork sausage
1 bunch celery, chopped
2 medium onions, chopped
1 cup chopped fresh mushrooms
1/2 cup chopped sweet red pepper
1/2 cup butter, cubed
1 can (49-1/2 ounces) chicken broth
2 to 3 tablespoons rubbed sage
1 tablespoon poultry seasoning
1/2 teaspoon pepper
1 turkey (14 to 16 pounds)
Additional butter, melted

In a large bowl, combine the first five ingredients. Combine eggs, milk and oil; stir into dry ingredients just until moistened. Pour into a greased 13-in. x 9-in. x 2-in. baking pan. Bake at 425° for 18-20 minutes or until a toothpick comes out clean. Cool on a wire rack.

In a Dutch oven, cook sausage over medium heat until no longer pink; drain and set aside. In the same pan, saute celery, onions, mushrooms and red pepper in butter until tender. Stir in the broth, seasonings and sausage. Bring to a boil. Reduce heat; simmer, uncovered, for 10 minutes. Remove from the heat. Cut corn bread into 1/2-in. cubes; fold into sausage mixture.

Just before baking, loosely stuff the turkey. Place remaining stuffing in a greased 13-in. x 9-in. x 2-in. baking dish; cover and refrigerate. Remove from the refrigerator 30 minutes before baking.

Skewer turkey openings; tie drumsticks together with kitchen string. Place breast side up on a rack in a roasting pan. Brush with melted butter. Bake, uncovered, at 325° for 3-3/4 to 4-1/2 hours or until a meat thermometer reads 180° for the turkey and 165° for the stuffing, basting occasionally with pan drippings (cover loosely with foil if turkey browns too quickly).

Bake the additional stuffing, covered, for 45-50 minutes. Uncover; bake 10 minutes longer or until lightly browned. Cover the turkey with foil and let stand for 20 minutes before removing the stuffing and carving. **Yield:** 12 servings (16 cups stuffing).

A Trio of Meals to Choose From

TRADITIONAL DINNER

Turkey with Sausage-Corn Bread Stuffing (p. 187)
Thanksgiving Gravy (p. 189)
Green Beans Supreme (p. 188)
Golden Harvest Potato Bake (p. 188)
Fruity Cranberry Chutney (p. 188)

MAKE-AHEAD DINNER

(Roasted and sliced turkey)
Apple Pumpkin Soup (p. 189)
Applesauce-Berry Gelatin Mold (p. 191)
Baked Vegetable Medley (p. 190)
Freezer Crescent Rolls (p. 190)
White Chocolate Pumpkin Cheesecake (p. 191)

EASY DINNER

Turkey Breast with Vegetables (p. 192)
Green Apple Spinach Salad (p. 192)
Maple Butternut Squash (p. 193)
Sweet Potato Bread (p. 193)
Nantucket Cranberry Tart (p. 193)

Golden Harvest Potato Bake

(Pictured below and on pages 168 and 186)

Prep: 40 min. **Bake:** 25 min.

This brightly colored casserole features two kinds of potatoes. It goes well with nearly any holiday entree.
—Pat Tomkins, Sooke, British Columbia

5 pounds Yukon Gold potatoes, peeled and cubed
2 medium sweet potatoes, peeled and cubed
5 large carrots, cut into 2-inch pieces
3 garlic cloves, minced
1 teaspoon dried tarragon
1 cup (4 ounces) shredded cheddar cheese, *divided*
1/2 cup milk
2 eggs
2 tablespoons cider vinegar
1 tablespoon butter
1 teaspoon salt
1 teaspoon dried parsley flakes
1/2 to 1 teaspoon pepper

Place the potatoes, carrots, garlic and tarragon in a Dutch oven; cover with water. Bring to a boil. Reduce heat; cover and cook for 15-20 minutes or until tender. Drain.

In a large mixing bowl, mash the vegetables. Stir in 1/2 cup cheese, milk, eggs, vinegar, butter, salt, parsley and pepper until blended.

Transfer to a greased 13-in. x 9-in. x 2-in. baking dish; sprinkle with remaining cheese. Bake, uncovered, at 325° for 25-30 minutes or until a thermometer reads 160°. **Yield:** 16 servings.

Fruity Cranberry Chutney

(Pictured on pages 168 and 186)

Prep: 15 min. **Cook:** 55 min. + chilling

An enticing blend of fruits and spices comes alive in this versatile chutney recipe. It complements turkey and ham nicely, but we also enjoy it spooned over cream cheese and served with crackers. *—Pat Stevens, Granbury, Texas*

2-1/4 cups packed brown sugar
1-1/2 cups cranberry juice
1/2 cup cider vinegar
1/2 teaspoon ground ginger
1/4 teaspoon ground allspice
2 packages (12 ounces *each*) fresh *or* frozen cranberries
2 medium oranges, peeled and sectioned
1 medium tart apple, peeled and coarsely chopped
1/2 cup dried currants *or* golden raisins
1/2 cup dried apricots, coarsely chopped
2 tablespoons finely grated orange peel

In a large saucepan, combine the brown sugar, cranberry juice, vinegar, ginger and allspice. Bring to a boil. Reduce heat; simmer, uncovered, for 2 minutes or until sugar is dissolved.

Stir in the cranberries, oranges, apple, currants, apricots and orange peel. Return to a boil. Reduce heat; simmer, uncovered, for 45 minutes or until thickened, stirring occasionally. Cool to room temperature. Transfer to a serving bowl; cover and refrigerate until chilled. **Yield:** about 5 cups.

Green Beans Supreme

(Pictured on pages 168 and 186)

Prep/Total Time: 25 min.

Here's a pleasing substitute for plain green bean casserole. I prepare a well-seasoned cheese sauce to add zip to this familiar dish. *—Heather Campbell, Lawrence, Kansas*

4 packages (16 ounces *each*) frozen cut green beans
1/4 cup finely chopped onion
1/4 cup butter, cubed
2 tablespoons all-purpose flour
1 teaspoon salt
1 teaspoon paprika
1 teaspoon Worcestershire sauce
1/2 teaspoon ground mustard
2 cups evaporated milk
8 ounces process cheese (Velveeta), shredded

TOPPING:

1/4 cup dry bread crumbs
2 teaspoons butter, melted

Cook green beans according to package directions. Meanwhile, in a Dutch oven, saute the onion in butter until tender. Remove from the heat; whisk in the flour, salt, paprika, Worcestershire sauce and mustard until blended.

Gradually stir in milk. Bring to a boil; cook and stir for 2 minutes or until thickened and bubbly. Remove from the heat; stir in cheese.

Drain beans; gently fold into cheese sauce. Transfer to a large serving bowl. Toss bread crumbs and butter; sprinkle over beans. **Yield:** 12-16 servings.

Apple Pumpkin Soup

(Pictured above right)

Prep: 50 min. + chilling **Cook:** 10 min.

This creamy, golden soup lets you relish autumn's color and flavors. For an extra treat, serve it in hollowed-out small pumpkins. —Pat Habiger, Spearville, Kansas

 **Uses less fat, sugar or salt. Includes Nutrition Facts and Diabetic Exchanges.**

2 cups finely chopped peeled tart apples
1/2 cup finely chopped onion
2 tablespoons butter
1 tablespoon all-purpose flour
4 cups chicken broth
3 cups canned pumpkin
1/4 cup packed brown sugar
1/2 teaspoon *each* ground cinnamon, nutmeg and ginger
1 cup unsweetened apple juice
1/2 cup half-and-half cream
1/4 teaspoon salt
1/4 teaspoon pepper

In a large saucepan, saute the apples and onion in butter for 3-5 minutes or until tender. Stir in the flour until blended. Gradually whisk in the chicken broth. Stir in the pumpkin, brown sugar, cinnamon, nutmeg and ginger. Bring to a boil. Reduce heat; cover and simmer for 25 minutes. Cool slightly.

In a blender, cover and process soup in batches until smooth. Pour into a bowl; cover and refrigerate for 8 hours or overnight.

Just before serving, transfer soup to a saucepan. Cook over medium heat for 5-10 minutes. Stir in the apple juice, cream, salt and pepper; heat through. **Yield:** 12 servings (about 2 quarts).

Nutrition Facts: 3/4 cup equals 100 calories, 3 g fat (2 g saturated fat), 10 mg cholesterol, 389 mg sodium, 16 g carbohydrate, 3 g fiber, 2 g protein. **Diabetic Exchanges:** 1 starch, 1/2 fat.

Thanksgiving Gravy

(Pictured on pages 168 and 186)

Prep: 10 min. **Cook:** 1-1/2 hours

Each of my granddaughters asked for my Thanksgiving and Christmas recipes when they got married. Here's the flavorful, traditional gravy recipe I shared with them. —Iola Egle, Bella Vista, Arkansas

Turkey giblets and neck bone
2 cups chicken broth
2 medium carrots, finely chopped
1 celery rib, chopped
2 large shallots, finely chopped
1/3 cup cornstarch
3 cups cold water
1/4 cup turkey drippings
2 teaspoons chicken bouillon granules
Pepper to taste

Place the giblets, neck bone, broth, carrots, celery and shallots in a large saucepan; bring to a boil. Reduce heat; cover and simmer for 1-1/4 hours.

Strain and discard giblets, neck bone and vegetables; set cooking juices aside. In another large saucepan, combine cornstarch and water until smooth; stir in the drippings and bouillon until smooth. Gradually stir in the reserved cooking juices. Bring to a boil; cook and stir for 2 minutes or until thickened. Season with pepper. **Yield:** 4 cups.

ROUND OUT roast turkey slices with Applesauce-Berry Gelatin Mold, Baked Vegetable Medley, Freezer Crescent Rolls, White Chocolate Pumpkin Cheesecake (all shown above) and Apple Pumpkin Soup (shown on page 189).

Baked Vegetable Medley

(Pictured above)

Prep: 35 min. + chilling **Bake:** 40 min.

If you finish chopping all of the vegetables for this dish the night before, it's smooth sailing when company arrives the next day. Just prepare this casserole in advance, keep it in the fridge and bake it at serving time. —*Linda Vail, Ballwin, Missouri*

- **1 medium head cauliflower, broken into florets**
- **1 bunch broccoli, cut into florets**
- **6 medium carrots, sliced**
- **1 pound sliced fresh mushrooms**
- **1 bunch green onions, sliced**
- **1/4 cup butter, cubed**
- **1 can (10-3/4 ounces) condensed cream of chicken soup, undiluted**
- **1/2 cup milk**
- **1/2 cup process cheese sauce**

Place the cauliflower, broccoli and carrots in a steamer basket; place in a large saucepan over 1 in. of water. Bring to a boil; cover and steam for 7-9 minutes or until crisp-tender. Meanwhile, in a large skillet, saute mushrooms and onions in butter until tender.

Drain vegetables. In a large bowl, combine the soup, milk and cheese sauce. Add vegetables and mushroom mixture; toss to coat. Transfer to a greased 2-qt. baking dish. Cover and refrigerate overnight.

Remove from the refrigerator 30 minutes before baking. Bake, uncovered, at 350° for 40-45 minutes or until bubbly. **Yield:** 12 servings.

Freezer Crescent Rolls

(Pictured above)

Prep: 30 min. + freezing **Bake:** 15 min.

I really love the convenience of this homemade, freezer-friendly crescent dough. —*Kristine Buck, Payson, Utah*

☑ Uses less fat, sugar or salt. Includes Nutrition Facts and Diabetic Exchanges.

3 teaspoons active dry yeast
2 cups warm water (110° to 115°)
1/2 cup butter, softened
2/3 cup nonfat dry milk powder
1/2 cup sugar
1/2 cup mashed potato flakes
2 eggs
1-1/2 teaspoons salt
6 to 6-1/2 cups all-purpose flour

In a large mixing bowl, dissolve yeast in warm water. Add the butter, milk powder, sugar, potato flakes, eggs, salt and 3 cups flour. Beat until smooth. Stir in enough remaining flour to form a firm dough.

Turn onto a heavily floured surface; knead 8-10 times. Divide dough in half. Roll each portion into a 12-in. circle; cut each circle into 16 wedges. Roll up wedges from the wide end and place point side down 2 in. apart on waxed paper-lined baking sheets. Curve ends to form crescent shape.

Cover and freeze. When firm, transfer to a large resealable plastic freezer bag. Freeze for up to 4 weeks.

To use frozen rolls: Arrange frozen rolls 2 in. apart on baking sheets coated with nonstick cooking spray. Cover and thaw in the refrigerator overnight.

Let rise in a warm place for 1 hour or until doubled. Bake at 350° for 15-17 minutes or until golden brown. Serve warm. **Yield:** 32 rolls.

Nutrition Facts: 1 roll equals 141 calories, 3 g fat (2 g saturated fat), 21 mg cholesterol, 160 mg sodium, 23 g carbohydrate, 1 g fiber, 4 g protein. **Diabetic Exchanges:** 1-1/2 starch, 1/2 fat.

White Chocolate Pumpkin Cheesecake

(Pictured above left)

Prep: 30 min. **Bake:** 55 min. + chilling

This luscious cheesecake belongs on a pedestal! The dessert looks so pretty with its crunchy almond topping.
—Phyllis Schmalz-Eismann, Kansas City, Kansas

1-1/2 cups crushed gingersnap cookies (about 32 cookies)
1/4 cup butter, melted
3 packages (8 ounces *each*) cream cheese, softened
1 cup sugar
3 eggs, lightly beaten
1 teaspoon vanilla extract
5 squares (1 ounce *each*) white baking chocolate, melted and cooled
3/4 cup canned pumpkin
1 teaspoon ground cinnamon
1/4 teaspoon ground nutmeg
ALMOND TOPPING:
1/2 cup chopped almonds
2 tablespoons butter, melted
1 teaspoon sugar

In a small bowl, combine gingersnap crumbs and butter. Press onto the bottom of a greased 9-in. springform pan; set aside.

In a large mixing bowl, beat cream cheese and sugar until smooth. Add eggs and vanilla; beat on low speed just until combined. Stir in melted white chocolate. Combine pumpkin and spices; gently fold into cream cheese mixture. Pour over crust. Place pan on a baking sheet.

Bake at 350° for 55-60 minutes or until center is just set. Cool on a wire rack for 10 minutes. Meanwhile, combine the topping ingredients; spread in a shallow baking pan. Bake for 10 minutes or until golden brown, stirring twice. Cool.

Carefully run a knife around edge of pan to loosen; cool 1 hour longer. Refrigerate overnight. Transfer topping to an airtight container; store in refrigerator.

Just before serving, remove sides of pan; sprinkle topping over cheesecake. Refrigerate leftovers. **Yield:** 12 servings.

Applesauce-Berry Gelatin Mold

(Pictured above far left)

Prep: 10 min. + chilling

Want a head start on your Thanksgiving meal? Try this uncomplicated yet attractive apple-berry mold. Fresh cranberries and mint leaves are a beautiful garnish.
—Gloria Coates, Madison, Connecticut

☑ Uses less fat, sugar or salt. Includes Nutrition Facts and Diabetic Exchanges.

2 packages (3 ounces *each*) strawberry gelatin
2 cups boiling water
1 can (16 ounces) whole-berry cranberry sauce
1-3/4 cups chunky applesauce

In a large bowl, dissolve gelatin in boiling water. Stir in cranberry sauce and applesauce. Pour into a 6-cup ring mold coated with nonstick cooking spray. Cover and refrigerate overnight. Unmold onto a serving platter. **Yield:** 12 servings.

Nutrition Facts: 1/2 cup equals 134 calories, trace fat (trace saturated fat), 0 cholesterol, 41 mg sodium, 34 g carbohydrate, 1 g fiber, 1 g protein. **Diabetic Exchanges:** 1-1/2 starch, 1/2 fruit.

KEEP THINGS EASY with a low-fuss dinner of Turkey Breast with Vegetables, Green Apple Spinach Salad, Maple Butternut Squash, Sweet Potato Bread (all shown above) and Nantucket Cranberry Tart (shown at far right).

Turkey Breast with Vegetables

(Pictured above)

Prep: 20 min. **Bake:** 2-1/4 hours + standing

This recipe offers the convenient option of baking just the turkey breast. —*Doris Russell, Fallston, Maryland*

- **1/4 cup plus 1 tablespoon olive oil, *divided***
- **1 tablespoon minced fresh rosemary**
- **2 teaspoons fennel seed, crushed**
- **3 garlic cloves, minced**
- **1 package (16 ounces) fresh baby carrots**
- **3 large onions, cut into eighths**
- **8 small red potatoes, cut in half**
- **1 bone-in turkey breast (6 pounds)**
- **1/2 teaspoon salt**
- **1/4 teaspoon pepper**
- **1/8 teaspoon garlic powder**
- **1/2 cup chicken broth**

In a large resealable plastic bag, combine 1/4 cup oil, rosemary, fennel seed and garlic. Add vegetables; shake to coat. Place turkey in a shallow roasting pan. Rub skin with remaining oil; sprinkle with salt, pepper and garlic powder. Arrange vegetables around turkey.

Bake, uncovered, at 325° for 2-1/4 to 2-3/4 hours or until a meat thermometer reads 170°, basting occasionally with broth. Cover and let stand for 10 minutes before carving. **Yield:** 6 servings.

Green Apple Spinach Salad

(Pictured above)

Prep/Total Time: 20 min.

Here's a refreshing salad that's ready in 20 minutes, so it's nice for busy days. —*Rae Sorensen, Granby, Quebec*

- **1 package (9 ounces) fresh baby spinach, torn**
- **2 medium tart green apples, chopped**
- **1/4 cup golden raisins**
- **1/4 cup sugar**
- **1/4 cup cider vinegar**
- **1/4 cup vegetable oil**
- **1/4 teaspoon garlic salt**
- **1/4 teaspoon celery salt**
- **1/2 cup chopped cashews**

In a large salad bowl, combine the spinach, apples and raisins. In a small bowl, whisk the sugar, vinegar, oil, garlic salt and celery salt. Drizzle over salad and toss to coat. Sprinkle with cashews. Serve immediately. **Yield:** 6 servings.

Maple Butternut Squash

(Pictured at left)

Prep: 20 min. + cooling **Bake:** 30 min.

For a simple yet unique Thanksgiving side dish, try this spicy-sweet bake. —Rene Powell, Annapolis, Maryland

✓ **Uses less fat, sugar or salt. Includes Nutrition Facts.**

1 medium butternut squash
1 cup maple syrup
2 tablespoons raisins
1 tablespoon butter, melted
1/2 teaspoon ground cardamom
1/2 teaspoon ground allspice

Cut squash in half lengthwise; discard seeds. Place cut side down in a microwave-safe dish; add 1/2 in. of hot water. Microwave, uncovered, on high for 12-14 minutes or until crisp-tender.

Meanwhile, in a small bowl, combine the syrup, raisins, butter, cardamom and allspice; set aside. When squash is cool enough to handle, remove rind and cut into 1-in. pieces. Place in a 13-in. x 9-in. x 2-in. baking dish. Drizzle with syrup mixture.

Cover and bake at 350° for 20 minutes. Uncover; bake 10-15 minutes longer or until squash is tender. Serve with a slotted spoon. **Yield:** 6 servings.

Nutrition Facts: 3/4 cup equals 224 calories, 2 g fat (1 g saturated fat), 5 mg cholesterol, 30 mg sodium, 53 g carbohydrate, 5 g fiber, 1 g protein.

Editor's Note: This recipe was tested in a 1,100-watt microwave.

Sweet Potato Bread

(Pictured above left)

Prep: 15 min. **Bake:** 1 hour + cooling

My family won't eat traditional sweet potatoes, so I fix this instead. —Rebecca Cook Jones, Henderson, Nevada

1-3/4 cups all-purpose flour
1-1/2 cups sugar
1 teaspoon baking soda
1 teaspoon ground cinnamon
1 teaspoon ground nutmeg
3/4 teaspoon salt
1/4 teaspoon ground allspice
1/4 teaspoon ground cloves
2 eggs
1-1/2 cups mashed sweet potatoes (about 2 medium)
1/2 cup vegetable oil
6 tablespoons orange juice
1/2 cup chopped pecans

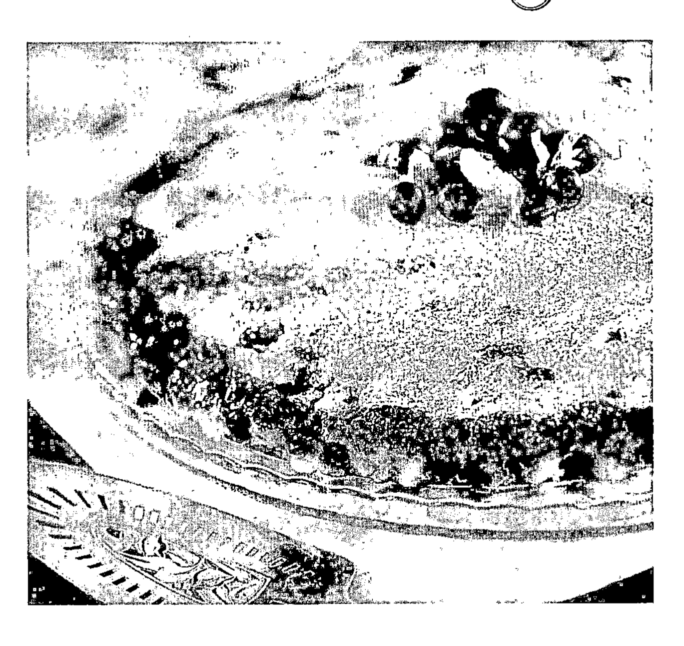

In a large bowl, combine the first eight ingredients. In a small bowl, whisk the eggs, sweet potatoes, oil and orange juice. Stir into dry ingredients just until moistened. Fold in pecans.

Transfer to a greased 9-in. x 5-in. x 3-in. loaf pan. Bake at 350° for 60-65 minutes or until a toothpick inserted near the center comes out clean. Cool for 10 minutes before removing from the pan to a wire rack. **Yield:** 1 loaf.

Nantucket Cranberry Tart

(Pictured above)

Prep: 15 min. **Bake:** 40 min. + cooling

This pretty dessert requires few ingredients, and it's a snap to assemble. —Jackie Zack, Riverside, Connecticut

1 package (12 ounces) fresh *or* frozen cranberries, thawed
1 cup sugar, *divided*
1/2 cup sliced almonds
2 eggs
3/4 cup butter, melted
1 teaspoon almond extract
1 cup all-purpose flour
1 tablespoon confectioners' sugar

In a small bowl, combine berries, 1/2 cup sugar and almonds. Transfer to a greased 11-in. fluted tart pan with a removable bottom. Place on a baking sheet.

In a small mixing bowl, beat the eggs, butter, extract and remaining sugar. Beat in flour just until moistened (batter will be thick). Spread evenly over berries.

Bake at 325° for 40-45 minutes or until a toothpick inserted near the center comes out clean. Cool on a wire rack. Dust with confectioners' sugar. Refrigerate leftovers. **Yield:** 12 servings.

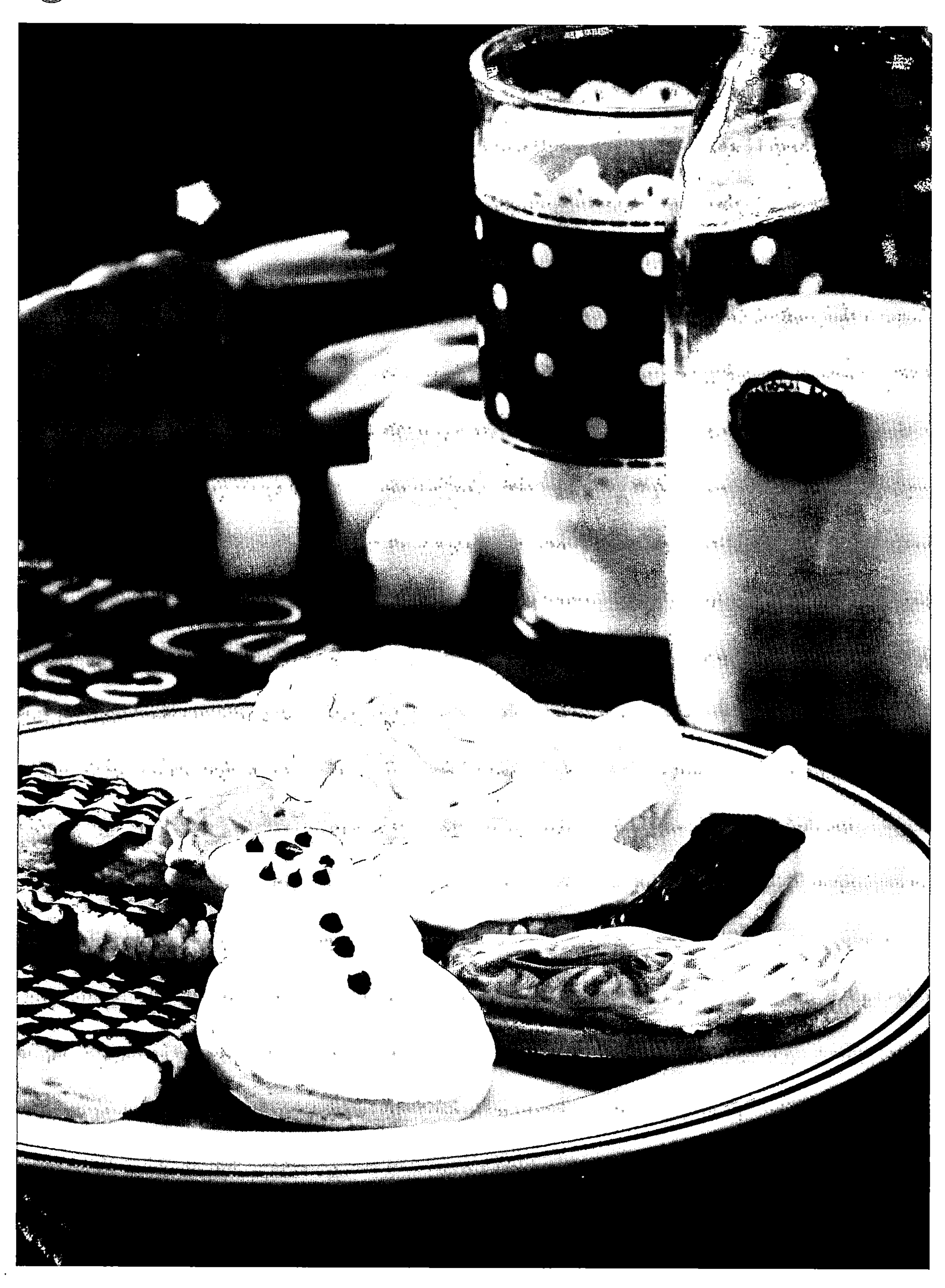

Festive Christmas Cookies

OLD-FASHIONED FAVORITES bring back wonderful memories of Christmas past. From ovens across America, these frosted cutouts, glazed spritz and more will make a merry assortment on your cookie tray.

Lemon Butter Spritz

(Pictured at left and on page 168)

Prep: 20 min. **Bake:** 5 min. per batch + cooling

As a veteran baker, I count this lemony spritz variation among my best-loved recipes. —*Iola Egle*
Bella Vista, Arkansas

1 cup butter, softened
1-1/4 cups confectioners' sugar
1 egg
1 tablespoon lemon juice
1 tablespoon grated lemon peel
2-1/2 cups all-purpose flour
1/4 teaspoon salt
GLAZE:
1 cup confectioners' sugar
2 tablespoons baking cocoa
1/2 teaspoon vanilla extract
2 to 3 tablespoons milk

In a large mixing bowl, cream butter and confectioners' sugar. Beat in the egg, lemon juice and peel. Combine flour and salt; gradually add to creamed mixture and mix well.

Using a cookie press fitted with a bar disk, form dough into long strips on ungreased baking sheets. Cut each strip into 2-1/4-in. pieces (do not separate pieces). Bake at 400° for 5-7 minutes or until set (do not brown). Cut into pieces again if necessary. Remove to wire racks to cool.

For glaze, in a small bowl, combine confectioners' sugar, cocoa, vanilla and enough milk to achieve desired consistency. Drizzle over cookies. Let stand until set. **Yield:** 4-1/2 dozen.

Sour Cream Cutouts

(Pictured at left and on page 168)

Prep: 30 min. + chilling
Bake: 15 min. per batch + cooling

Festively decorated, these vanilla-flavored cutouts are sure to catch Santa's eye! It just wouldn't be Christmas without them. —*Jane Grosvold, Holcombe, Wisconsin*

1 cup butter, softened
2 cups sugar
3 eggs
1 cup (8 ounces) sour cream
1 teaspoon vanilla extract
5-3/4 cups all-purpose flour
2 teaspoons baking powder
1/2 teaspoon baking soda
1/2 teaspoon salt
2 cans (16 ounces *each*) vanilla frosting
Gel food coloring of your choice

In a large mixing bowl, cream butter and sugar. Beat in eggs, sour cream and vanilla. Combine flour, baking powder, baking soda and salt; gradually add to creamed mixture and mix well. Cover; refrigerate overnight.

On a lightly floured surface, roll out dough to 1/8-in. thickness. Cut with floured cookie cutters. Place 1 in. apart on ungreased baking sheets. Bake at 375° for 12-15 minutes or until lightly browned. Remove to wire racks to cool.

Tint some of the frosting with food coloring; decorate cookies as desired. **Yield:** about 6-1/2 dozen.

Mincemeat Cookies

(Pictured on page 197)

Prep: 20 min. + chilling **Bake:** 10 min. per batch

After people taste this tender, chewy old-time cookie, they want to know what the "secret ingredient" is.
—*Lucie Fitzgerald, Spring Hill, Florida*

1/2 cup butter, softened
1 cup sugar, *divided*
1 egg
1 teaspoon vanilla extract
1-3/4 cups all-purpose flour
1-1/2 teaspoons baking powder
1/4 teaspoon salt
1 package (9 ounces) condensed mincemeat, cut into small pieces
1 egg white, lightly beaten

In a large mixing bowl, cream butter and 3/4 cup sugar. Beat in egg and vanilla. Combine flour, baking powder and salt; gradually add to creamed mixture and mix well. Stir in mincemeat. Cover; refrigerate for 2 hours.

Roll dough into 1-in. balls; dip into egg white and remaining sugar. Place sugar side up 2 in. apart on greased baking sheets. Bake at 375° for 10-12 minutes or until set. Remove to wire racks. **Yield:** 4 dozen.

Festive Stars

(Pictured below)

Prep: 50 min. + chilling
Bake: 10 min. per batch + cooling

These clever interlocking stars will add delightful dimension to your Christmas cookie assortment. Cutting a slit in each star makes them easy to assemble.
—Caren Zimmerman, Franklin, Wisconsin

1/2 cup butter, softened
1/4 cup shortening
1 cup sugar
2 eggs
1 teaspoon vanilla extract
2-1/2 cups all-purpose flour
1 teaspoon baking powder
1/4 teaspoon salt
3/4 cup red colored sugar

In a large mixing bowl, cream the butter, shortening and sugar. Beat in the eggs and vanilla extract. Combine the flour, baking powder and salt; gradually add to the creamed mixture and mix well. Chill for 1 hour or until easy to handle.

On a lightly floured surface, roll out dough to 1/8-in. thickness. Cut with a floured 2-1/2-in. five-point star-shaped cookie cutter. Cut a vertical slit between two points on each star to just above the center; spread dough apart to form a 1/4-in. opening.

Place 1 in. apart on ungreased baking sheets. Sprinkle with colored sugar. Bake at 400° for 6-7 minutes or until edges begin to brown. Remove to wire racks to cool. Assemble by placing two stars together at slits. **Yield:** about 3 dozen.

Chocolate Linzer Cookies

(Pictured at right)

Prep: 30 min. + chilling
Bake: 10 min. per batch + cooling

My mom and I used to bake these together. They remind me of home. *—Heather Peters, North Pole, Alaska*

3/4 cup butter, softened
1 cup sugar
2 eggs
1/2 teaspoon almond extract
2-1/3 cups all-purpose flour
1 teaspoon baking powder
1/2 teaspoon salt
1/2 teaspoon ground cinnamon
1 cup (6 ounces) semisweet chocolate chips, melted
Confectioners' sugar
6 tablespoons seedless raspberry jam

In a small mixing bowl, cream butter and sugar. Add eggs, one at a time, beating well after each addition. Beat in extract. Combine flour, baking powder, salt and cinnamon; gradually add to creamed mixture and mix well. Refrigerate for 1 hour or until easy to handle.

Divide dough in half. On a lightly floured surface, roll out one portion to 1/8-in. thickness; cut with a 2-1/2-in. round cookie cutter. Roll out the remaining dough; cut with a 2-1/2-in. doughnut cutter so the center is cut out of each cookie.

Place 1 in. apart on ungreased baking sheets. Bake at 350° for 8-10 minutes or until edges are lightly browned. Remove to wire racks to cool.

Spread melted chocolate over the bottoms of solid cookies. Place cookies with cutout centers over chocolate. Sprinkle with confectioners' sugar. Spoon 1/2 teaspoon jam in center of each cookie. **Yield:** 2 dozen.

Cherry Almond Cups

(Pictured above right)

Prep: 25 min. + cooling
Bake: 20 min. per batch + cooling

To create these attractive mini tarts, I adapted a bar cookie recipe. *—Christine Kneebone, Hilbert, Wisconsin*

1/2 cup butter, softened
1/2 cup sugar
1/2 teaspoon vanilla extract

YOUR HOLIDAY GUESTS will find it impossible to resist tempting Christmas treats such as old-fashioned Mincemeat Cookies, Chocolate Linzer Cookies and Cherry Almond Cups (shown above).

1-1/4 cups all-purpose flour
1/4 teaspoon salt
FILLING:
1 can (8 ounces) almond paste, cut into cubes
1/4 cup butter, softened
1/4 cup sugar
2 eggs
18 *each* red and green maraschino cherries, patted dry
TOPPING:
1 tablespoon butter
1 square (1 ounce) unsweetened chocolate
1/2 cup confectioners' sugar
1/2 teaspoon vanilla extract
1 to 2 tablespoons milk

In a small mixing bowl, cream butter and sugar. Beat in vanilla. Combine the flour and salt; gradually add to creamed mixture. Roll into 1-in. balls.

Press dough onto the bottoms and up the sides of well-greased miniature muffin cups. Bake at 350° for 8-10 minutes or until lightly browned. Cool in pans on wire racks.

In a small mixing bowl, beat almond paste, butter and sugar until smooth. Beat in eggs. Spoon 2 teaspoonfuls into each cooled cup; top each with a cherry.

Bake at 350° for 16-18 minutes or until lightly browned and filling is set. Cool for 10 minutes before removing from pans to wire racks to cool completely.

In a microwave-safe bowl, melt butter and chocolate; stir until smooth. Add the confectioners' sugar, vanilla and enough milk to achieve desired consistency. Drizzle over cookies. **Yield:** 3 dozen.

Decorative Ginger Cookies

Prep: 45 min. + chilling
Bake: 10 min. per batch + cooling

I add a ribbon loop to these cutouts so they can hang on the tree. —Cheryl Maczko, Reedsville, West Virginia

1 cup sugar
1 cup molasses
1/2 cup water
1/3 cup vegetable oil
1 egg
5-1/2 cups all-purpose flour
3 teaspoons baking soda
3 teaspoons cream of tartar
3 teaspoons ground ginger
1 teaspoon ground cinnamon
Frosting of your choice, optional

In a large mixing bowl, combine the sugar, molasses, water, oil and egg. Combine the flour, baking soda, cream of tartar, ginger and cinnamon; add to molasses mixture and beat until combined. Cover and refrigerate for 1 hour or until easy to handle.

On a well-floured surface, roll out dough to 1/8-in. thickness. Cut with floured cookie cutters. Using a floured spatula, place cookies 1 in. apart on greased baking sheets. If cookies will be hung, use a plastic straw to make a hole 1/2 in. from the top of each.

Bake at 375° for 10-12 minutes or until edges are lightly browned. Use straw to reopen holes. Remove to wire racks to cool completely.

Decorate with frosting if desired. Let dry completely. Thread ribbon through holes. **Yield:** 4 dozen.

happy holiday
to: Mom
from: Santa
Taste of Home

Gifts from the Kitchen

AT CHRISTMASTIME, do you struggle with the question of what to give the person who has everything? Why not get busy in the kitchen? For holiday gifts of good taste, try the colorful candy, fruit-flavored snack mix, rich fudge and more here.

You'll even find a special treat you can make especially for Fido! And the recipe calls for no unusual ingredients. (See the sidebar on page 201.)

Apple Snack Mix

(Pictured at left and on page 168)

Prep/Total Time: 15 min.

With its apple-cinnamon cereal and apple chips, this fun mix makes a cute gift for a teacher...or anyone. I put it in clear, cone-shaped plastic bags and tie them with ribbon.

—Rosemary Pacha, Brighton, Iowa

3 cups French Toast Crunch cereal
2 cups miniature pretzels
2 cups dry roasted peanuts
1-1/2 cups Frosted Cheerios
1-1/2 cups Apple Cinnamon Cheerios
1-1/2 cups yogurt-covered raisins
1-1/2 cups small apple-flavored red and green jelly beans
2 packages (3 ounces *each*) dried apple chips
2/3 cup sunflower kernels

In a large bowl, combine all ingredients. Store in an airtight container. **Yield:** 4 quarts.

Pistachio Cranberry Biscotti

(Pictured at left and on page 168)

Prep: 15 min. + chilling **Bake:** 35 min. + cooling

This fruit-and-nut biscotti mix always goes over big. Just layer the dry ingredients in a clear jar, add a bow and attach the instructions for preparation and baking.

—Dawn Fagerstrom, Warren, Minnesota

2 cups all-purpose flour
2 teaspoons baking powder
1/2 teaspoon ground cinnamon
2/3 cup sugar, *divided*
3/4 cup pistachios
3/4 cup dried cranberries *or* cherries

ADDITIONAL INGREDIENTS:
1/3 cup butter, softened
2 eggs

Combine flour and baking powder; pour into a wide-mouth 1-qt. glass container with a tight-fitting lid. Sprinkle cinnamon around edge of container. Layer with 1/3 cup sugar, pistachios and cranberries, packing each layer tightly (do not mix). Pour remaining sugar down the center. Cover and store in a cool, dry place for up to 6 months.

To prepare biscotti: In a large mixing bowl, beat butter and eggs. Gradually stir in biscotti mix (dough will be sticky). Chill for 30 minutes.

Divide dough in half. On an ungreased baking sheet, shape each half into a 10-in. x 2-in. rectangle. Bake at 350° for 25-30 minutes or until firm. Cool for 5 minutes.

Transfer to a cutting board; cut diagonally with a serrated knife into 3/4-in.-thick slices. Place cut side down on an ungreased baking sheet. Bake for 5 minutes. Turn and bake 5-6 minutes longer or until golden brown. Remove to wire racks to cool. Store in an airtight container. **Yield:** 16 cookies.

Peanut Butter Pretzel Bites

Prep: 1-1/2 hours + standing

My daughter and I came up with this recipe. Every bite gives you great peanut flavor and a salty pretzel crunch.

—Lois Farmer, Logan, West Virginia

1 package (14 ounces) caramels
1/4 cup butter, cubed
2 tablespoons water
5 cups miniature pretzels
1 jar (18 ounces) chunky peanut butter
26 ounces milk chocolate candy coating

In a microwave-safe bowl, melt caramels with butter and water; stir until smooth. Spread one side of each pretzel with 1 teaspoon peanut butter; top with 1/2 teaspoon caramel mixture. Place on waxed paper-lined baking sheets. Refrigerate until set.

In a microwave-safe bowl, melt candy coating. Using a small fork, dip each pretzel into coating until completely covered; shake off excess. Place on waxed paper. Let stand until set. Store in an airtight container in a cool, dry place. **Yield:** 8-1/2 dozen.

Editor's Note: This recipe was tested in a 1,100-watt microwave.

Layered Mint Fudge

(Pictured above)

Prep: 15 min. **Cook:** 20 min. + chilling

I've brought this festive fudge to many get-togethers...and am asked for copies of the recipe every time.
—Denise Hanson, Anoka, Minnesota

- **1-1/2 teaspoons butter, softened**
- **2 cups (12 ounces) semisweet chocolate chips**
- **1 can (14 ounces) sweetened condensed milk, *divided***
- **2 teaspoons vanilla extract**
- **1 cup vanilla *or* white chips**
- **3 teaspoons peppermint extract**
- **1 to 2 drops green food coloring**

Line a 9-in. square pan with foil; grease the foil with butter and set aside. In a heavy saucepan, melt chocolate chips and 1 cup milk over low heat; cook and stir for 5-6 minutes or until smooth. Remove from the heat. Add vanilla; stir for 3-4 minutes or until creamy. Spread half of the mixture into prepared pan. Refrigerate for 10 minutes or until firm. Set remaining chocolate mixture aside.

In a heavy saucepan, melt vanilla chips and remaining milk over low heat; cook and stir for 5-6 minutes or until smooth (mixture will be thick). Remove from the heat. Add peppermint extract and food coloring; stir for 3-4 minutes or until creamy. Spread evenly over chocolate layer. Refrigerate for 10 minutes or until firm.

Heat reserved chocolate mixture over low heat until mixture achieves spreading consistency; spread over mint layer. Cover; refrigerate overnight or until firm.

Using foil, lift fudge out of pan. Gently peel off foil; cut fudge into 1-in. squares. Store in the refrigerator. **Yield:** 1-3/4 pounds.

Pecan Caramels

(Pictured below)

Prep: 20 min. **Cook:** 35 min. + cooling

I altered the original recipe for these creamy caramels by substituting condensed milk for part of the whipping cream and cutting back on the sugar. Everybody raves about them, and they make an ideal holiday gift. You can't eat just one! *—Patsy Howell, Peru, Indiana*

- **1 tablespoon butter, softened**
- **1 cup sugar**
- **1 cup light corn syrup**
- **2 cups heavy whipping cream, *divided***
- **1 can (14 ounces) sweetened condensed milk**
- **2 cups chopped pecans**
- **1 teaspoon vanilla extract**

Line a 13-in. x 9-in. x 2-in. pan with foil; grease the foil with butter. Set aside.

In a large heavy saucepan, combine the sugar, corn syrup and 1 cup cream. Bring to a boil over medium heat. Cook and stir until smooth and blended, about 10 minutes. Stir in milk and remaining cream. Bring to a boil over medium-low heat, stirring constantly. Cook and stir until a candy thermometer reads 238° (soft-ball stage), about 25 minutes.

Remove from the heat; stir in pecans and vanilla. Pour into prepared pan (do not scrape saucepan). Cool. Using foil, lift candy out of pan; cut into 1-in. squares. Wrap individually in waxed paper. **Yield:** about 2-1/2 pounds.

Editor's Note: We recommend that you test your candy thermometer before each use by bringing water to a boil; the thermometer should read 212°. Adjust your recipe temperature up or down based on your test.

Christmas Hard Candy

(Pictured above)

Prep: 10 min. **Cook:** 30 min. + cooling

A dusting of confectioners' sugar gives a frosty look to this old-fashioned candy. The color is beautiful, and people are surprised by the wonderful watermelon flavor.

—Amy Short, Lesage, West Virginia

3-3/4 cups sugar
1-1/2 cups light corn syrup
1 cup water
2 to 3 drops red food coloring *or* color of your choice
1/4 teaspoon watermelon flavoring *or* flavoring of your choice
1/2 cup confectioners' sugar

Butter two 15-in. x 10-in. x 1-in. pans; set aside. In a large heavy saucepan, combine the sugar, corn syrup, water and food coloring. Cook and stir over medium heat until sugar is dissolved. Bring to a boil. Cook, without stirring, until a candy thermometer reads 300° (hard-crack stage).

Remove from the heat; stir in flavoring. Immediately pour into prepared pans; cool. Dust with confectioners' sugar; break into pieces. Store in airtight containers. **Yield:** 2 pounds.

Editor's Note: We recommend that you test your candy thermometer before each use by bringing water to a boil; the thermometer should read 212°. Adjust your recipe temperature up or down based on your test.

Remember Man's Best Friend

YOUR POOCH and other pet friends will be smacking their lips when they get a whiff of these yummy homemade dog treats.

A recipe from Lori Kimble of Mascoutah, Illinois, these healthy, meat-flavored biscuits make perfect gifts for pets. You may want to leave a few for Santa's reindeer, too!

Beefy Dog Treats

Prep: 20 min. **Bake:** 25 min. + cooling

1 package (1/4 ounce) active dry yeast
1/4 cup warm water (110° to 115°)
1 teaspoon beef bouillon granules
2 tablespoons boiling water
2-1/2 cups all-purpose flour
1 cup nonfat dry milk powder
1 cup whole wheat flour
1 cup cooked long grain rice
1 envelope unflavored gelatin
1 jar (4 ounces) vegetable beef dinner baby food
1 egg
2 tablespoons vegetable oil

In a small bowl, dissolve yeast in warm water. In another small bowl, dissolve bouillon in boiling water. In a large bowl, combine the all-purpose flour, milk powder, wheat flour, rice and gelatin. Stir in the baby food, egg, oil, yeast mixture and bouillon mixture until combined; knead until mixture forms a ball.

Turn onto a lightly floured surface; roll to 1/4-in. thickness. Cut with a floured 2-in. bone-shaped cookie cutter. Place 1 in. apart on ungreased baking sheets.

Bake at 300° for 25-30 minutes or until set. Remove to wire racks to cool. Let stand for 24 hours or until hardened. Store in an airtight container. **Yield:** about 3 dozen.

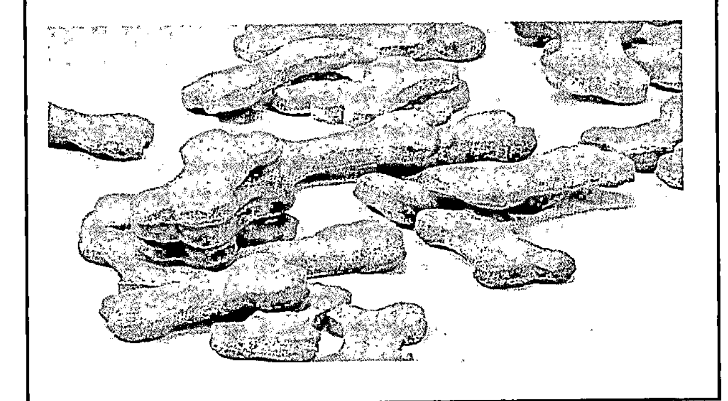

'My Mom's Best Meal'

Six family cooks fondly recall their mothers' cooking...and share the recipes for their favorite made-by-mom feast.

MEMORABLE MENUS. Clockwise from upper left: Treasured Turkey Dinner (p. 204), Meat-and-Potatoes Perfection (p. 220), Old-Fashioned Favorites (p. 212) and Family-Pleasing Feast (p. 208).

Her mother's succulent stuffed turkey—served up with all the finest fixings—made for a holiday-worthy feast that the whole family has treasured for years.

By June Blomquist, Eugene, Oregon

MY MOM, Ruby Kehoe (above), is the hardest worker I know. These days, she is as busy as ever... sewing, gardening and decorating her home in nearby Florence, Oregon. But of all the rooms in her house, the kitchen is where Mom spends most of her time, preparing appetizing meals, treats and more.

All of Mom's meals are five-star, but the one I will always cherish is this turkey dinner that's perfect for the holidays.

Whenever I think about her impressive Rolled-Up Turkey, I can almost smell its wonderful aroma. It starts with a deboned turkey that's flattened and rolled up with a scrumptious Southern-style corn bread stuffing.

For side dishes, Mom serves a colorful trio. The festive Cranberry Gelatin Mold is a tangy holiday favorite. The crunchy Sweet-Sour Red Cabbage gets its sweet-tart flavor from vinegar and maraschino cherry juice. And her Peas in Cheese Sauce is creamy and comforting.

It's difficult to leave room for dessert, but you can't turn down a slice of her yummy Caramel-Crunch Pumpkin Pie! A dollop of real whipped cream is the perfect finishing touch.

Mom's had plenty of practice in the kitchen. She started cooking at the age of 12 for a family of 15 as well as relatives, friends and farmhands during threshing season. As a wife and mother, she cooked nourishing meals every day for our family—my dad; me; my sister, Gretchen; and my brother, Colin.

Dad was a carpenter, and Mom often worked alongside him. She also sewed clothes for my sister and me...with matching outfits for our dolls. But she always found time to make cinnamon rolls, breads, cakes and pies. There were even baked goods on camping trips because the tent had a gas stove with an oven.

Mom still loves to cook and bake, especially homemade soups and desserts, and until a couple years ago, she canned fruit, vegetables and meat. But now the big, festive dinners are held at someone else's house or at my niece's restaurant.

The rest of the family must have caught the cooking "bug" from Mom. My brother and sister are both good cooks, and I worked as a cook at the University of Oregon for 15 years. Even my son, Daren, married a wonderful cook, Karen, who works as a food service manager.

When Mom got married, she told my dad that a 50-pound bag of flour had top priority when they moved their belongings west from Oklahoma. I'm sure glad cooking was so important to her because we all ended up reaping the rewards! I hope you'll reward your family by serving her meal.

Tips for Whipping Cream

When you want real whipped cream to dollop on Caramel-Crunch Pumpkin Pie (pictured at far left) or another dessert, try this recipe:

For 1 cup of whipped cream, place 1/2 cup heavy whipping cream in a chilled small mixing bowl. Using chilled beaters, beat the cream until it begins to thicken. Add 1-1/2 tablespoons confectioners' sugar and 1/4 teaspoon vanilla, then beat until soft peaks form. Store whipped cream in the refrigerator.

Here are some helpful hints to make the process easy and successful every time:

- Keep cream refrigerated until you're ready to whip it. Refrigerate the bowl and beaters for about 30 minutes before whipping.
- You have soft peaks in the cream if the points of the peaks curl over when you lift the beaters from the mixture. If the points stand straight up, you have stiff peaks, making the cream more suitable to use as frosting than for garnishing pies or other desserts.
- If you'd like to make whipped cream ahead of time, slightly underwhip the cream, then cover and refrigerate it for several hours. Beat the cream again briefly just before serving.

PICTURED AT LEFT: Rolled-Up Turkey, Cranberry Gelatin Mold, Peas in Cheese Sauce, Sweet-Sour Red Cabbage and Caramel-Crunch Pumpkin Pie (recipes are on the next page).

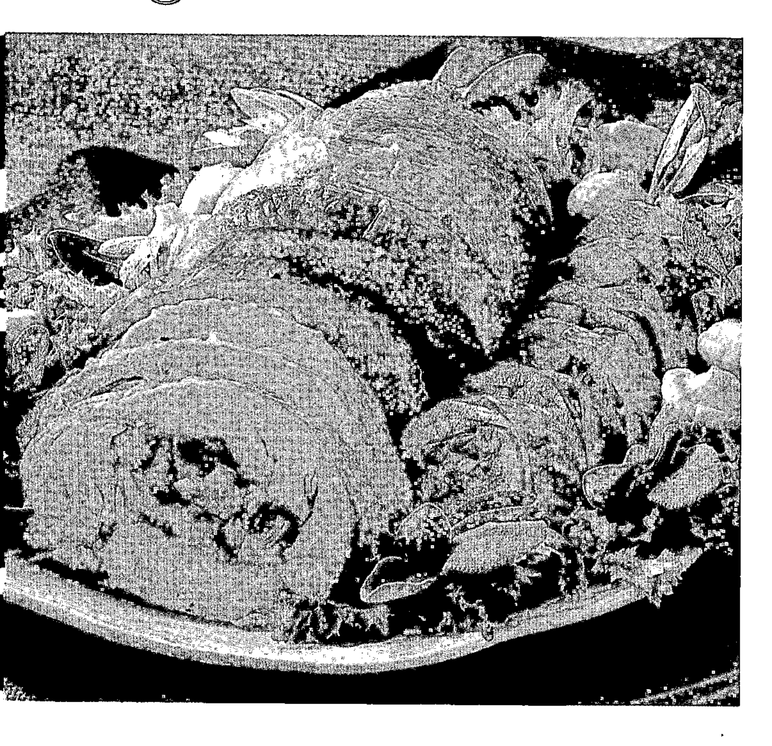

Rolled-Up Turkey

(Also pictured on front cover)

Prep: 30 min. **Bake:** 2 hours + standing

- **1 turkey (12 pounds), deboned and giblets removed**
- **1 cup chopped celery**
- **1 medium onion, chopped**
- **1/2 cup butter, cubed**
- **5 cups cubed white bread**
- **1-1/2 cups coarsely crumbled corn bread**
- **1 teaspoon salt, *divided***
- **1/2 teaspoon rubbed sage**
- **3/4 to 1 cup chicken broth**
- **3 tablespoons vegetable oil**
- **1/4 teaspoon pepper**

Unroll turkey on a large cutting board. With a sharp knife, remove the wings (save for another use). Flatten turkey to 3/4-in. thickness. Cut between the turkey breast and thighs to separate into three sections.

In a large skillet, saute celery and onion in butter until tender. In a large bowl, combine bread and corn bread; add celery mixture, 1/2 teaspoon salt and sage. Stir in enough broth to moisten.

Spoon 2 cups stuffing over turkey breast to within 1 in. of edges. Roll up jelly-roll style, starting with a long side. Tie with kitchen string at 1-1/2-in. intervals. Place on a rack in a large shallow roasting pan.

Spoon 1 cup stuffing over each thigh section to within 1 in. of edges. Roll up each jelly-roll style, starting with a short side. Tie with kitchen string; place on a rack in another shallow roasting pan.

Brush turkey with oil; sprinkle with pepper and remaining salt. Bake turkey breast at 325° for 2 to 2-1/2 hours or until a meat thermometer reads 170°; bake thighs for 1-1/2 to 1-3/4 hours or until a meat thermometer reads 180°. Let turkey breast stand for 15 minutes before slicing. **Yield:** 8 servings.

Editor's Note: Ask the butcher to debone the turkey for you.

Cranberry Gelatin Mold

Prep: 15 min. + chilling

✓ **Uses less fat, sugar or salt. Includes Nutrition Facts and Diabetic Exchanges.**

- **2 packages (3 ounces *each*) raspberry gelatin**
- **3 cups boiling water**
- **1 can (16 ounces) whole-berry cranberry sauce**
- **2 tablespoons lemon juice**
- **1 can (8 ounces) unsweetened crushed pineapple, drained**
- **1/2 cup finely chopped celery**

In a large bowl, dissolve gelatin in boiling water. Stir in cranberry sauce and lemon juice until blended. Chill until partially set.

Stir in pineapple and celery. Pour into a 6-cup ring mold coated with nonstick cooking spray. Refrigerate until firm. Unmold onto a platter. **Yield:** 8 servings.

Nutrition Facts: 3/4 cup (prepared with sugar-free gelatin) equals 174 calories, trace fat (trace saturated fat), 0 cholesterol, 479 mg sodium, 33 g carbohydrate, 1 g fiber, 12 g protein. **Diabetic Exchanges:** 2 fruit, 1/2 starch.

Peas in Cheese Sauce

Prep/Total Time: 20 min.

4-1/2 teaspoons butter
4-1/2 teaspoons all-purpose flour
1/4 teaspoon salt
1/8 teaspoon white pepper
1-1/2 cups milk
3/4 cup cubed process cheese (Velveeta)
2 packages (10 ounces *each*) frozen peas, thawed

In a large saucepan, melt butter over low heat. Stir in the flour, salt and pepper until smooth. Gradually add milk. Bring to a boil; cook and stir for 2 minutes or until thickened. Add the cheese; stir until melted. Stir in peas; cook 1-2 minutes longer or until heated through. **Yield:** 8 servings.

Sweet-Sour Red Cabbage

(Pictured on page 204)

Prep: 15 min. **Cook:** 35 min.

✓ **Uses less fat, sugar or salt. Includes Nutrition Facts and Diabetic Exchanges.**

1 medium head red cabbage, shredded
1 large onion, chopped
1/4 cup canola oil
2 medium apples, peeled and thinly sliced
1/4 cup cider vinegar
1/2 teaspoon salt
1/4 teaspoon pepper
2 tablespoons maraschino cherry juice, optional

In a large skillet, saute cabbage and onion in oil for 5-8 minutes or until crisp-tender. Add the apples, vinegar, salt, pepper and cherry juice if desired. Bring to a boil. Reduce heat; cover and simmer for 25 minutes or until cabbage is tender. **Yield:** 8 servings.

Nutrition Facts: 2/3 cup (calculated without cherry juice) equals 112 calories, 7 g fat (1 g saturated fat), 0 cholesterol, 160 mg sodium, 13 g carbohydrate, 3 g fiber, 2 g protein. **Diabetic Exchanges:** 1 vegetable, 1 fat, 1/2 fruit.

Caramel-Crunch Pumpkin Pie

Prep: 15 min. **Bake:** 50 min. + cooling

3/4 cup packed brown sugar, *divided*
1/2 cup finely chopped walnuts
2 tablespoons butter, melted
1 unbaked pastry shell (9 inches)
3 eggs
1 cup canned pumpkin
1 teaspoon rum extract
3/4 teaspoon ground cinnamon
1/2 teaspoon salt
1/2 teaspoon ground mace
1/4 teaspoon ground ginger
1-1/2 cups heavy whipping cream
Whipped cream and additional chopped walnuts, optional

In a small bowl, combine 1/4 cup brown sugar, walnuts and butter. Press onto the bottom of pastry shell. In a large bowl, whisk the eggs, pumpkin, extract, cinnamon, salt, mace, ginger and remaining brown sugar until blended; stir in cream.

Pour into pastry shell. Cover edges loosely with foil. Bake at 400° for 10 minutes. Reduce heat to 350°; bake 40-45 minutes longer or until a knife inserted near the center comes out clean. Remove foil. Cool on a wire rack.

Garnish with whipped cream and additional walnuts if desired. Refrigerate leftovers. **Yield:** 8 servings.

Every entree, dessert and other home-cooked dish that comes out of her mother's bustling, geared-for-a-crowd kitchen is big on two things—flavor and family.

By Heather Ahrens, Avon, Ohio

MY MOTHER, Tammy Ahrens (above), learned early on how to cook for a large family. Growing up on a small farm, she was the oldest of seven children. So you can imagine the responsibility that fell on her shoulders.

Her mom taught her how to make pies and how to can fruits and vegetables. Grandma let Mom make meals for company, too.

Those kitchen skills didn't go to waste. As a wife and mother, Mom has baked and cooked even more. She and my dad, Leslie, have 11 children. I have six brothers and four sisters—and there are two sets of twins. We all still live at home.

With such a large family, Mom relied on delicious casseroles, soups, stews and breads while we were growing up. Most of what she made was from scratch, and she'd always use the freshest herbs and vegetables from our garden. For each of our birthdays, she'd make our favorite meal.

My number-one choice is always Tender Flank Steak, Cheddar Twice-Baked Potatoes, crisp Romaine with Oranges and luscious Pistachio Ice Cream Dessert. For years, Mom has also prepared this special menu on Father's Day. If we're lucky, she makes it during the "off-season," too.

When she prepares her juicy flank steak, I often get my nose too close, trying to get a better whiff of the wonderful aroma! Cheddar Twice-Baked Potatoes are the perfect complement, so smooth and creamy...and easy, too. I've been making them since I was 12.

The refreshing Romaine with Oranges salad is also simple to prepare. I love the combination of sweet oranges and tangy olive oil and vinegar dressing. It's colorful, too.

Pistachio Ice Cream Dessert is a fabulous treat, with its crisp crust and crunchy toffee topping. It's a tradition with us to have just desserts on Sunday evenings, so this yummy dish might appear on our table two or three times a year...and we always look forward to it.

To feed our big family, a vegetable garden was an absolute necessity, and we helped Mom with the weeding. It was one of my least-liked tasks, but we knew it had to be done if we wanted a good supply of food available during the fall and winter.

Mom once said that even the toughest, most time-consuming job could be fun if we did it together. And that's the truth! Canning tomatoes and making sauerkraut weren't exactly enjoyable, but most of the time, we'd be laughing, singing or playing games like I Spy while we worked.

My mother believes that raising us kids is more important than anything. But that doesn't mean she didn't do other things. She sewed, played softball and took daily walks. She also has home-schooled all of us kids.

I recently graduated and hope to become a chef someday. I love practicing on my family and entering baked goods at our county fair. Last year, I was thrilled when I received a "best of show," a blue ribbon and 13 other ribbons. My oldest sister says I should open my own cafe!

The credit has to go to God and to my mom, who taught me to follow recipes, read directions carefully and start a meal soon enough so we don't end up eating at midnight!

Now that all but three of the younger kids are cooking, Mom doesn't cook as much anymore. But she still puts together Sunday lunch and gets up early on Saturday mornings to make pancakes. Her labors of love are always appreciated!

I'm sure you'll enjoy these family-favorite foods as much as we do.

Ice Cream Ease

Ice cream is often softened before being used in recipes, such as Pistachio Ice Cream Dessert (pictured at left). To soften it in the refrigerator, transfer the ice cream from the freezer to the refrigerator 20-30 minutes before using it. Or let it stand at room temperature for 10-15 minutes. Hard ice cream can also be microwaved at 30% power for about 30 seconds.

PICTURED AT LEFT: Tender Flank Steak, Cheddar Twice-Baked Potatoes, Romaine with Oranges and Pistachio Ice Cream Dessert (recipes are on the next page).

Tender Flank Steak

Prep: 10 min. + marinating **Cook:** 20 min.

1 cup reduced-sodium soy sauce
1/4 cup lemon juice
1/4 cup honey
6 garlic cloves, minced
1 beef flank steak (1-1/2 pounds)

In a large resealable plastic bag, combine the soy sauce, lemon juice, honey and garlic; add steak. Seal bag and turn to coat; refrigerate for 6-8 hours.

Drain and discard marinade. Broil 4-6 in. from the heat or grill over medium heat for 8-10 minutes on each side or until meat reaches desired doneness (for medium-rare, a meat thermometer should read 145°; medium, 160°; well-done, 170°). Thinly slice steak across the grain. **Yield:** 6 servings.

Broiling Beef

Broiling is a recommended cooking method for marinated flank steak and other tender cuts of beef, such as sirloin, T-bones, porterhouse steaks, rib and rib eye steaks, marinated top round or chuck shoulder steaks and ground beef patties.

Place the meat on a broiler pan (3 to 4 inches from the heat for cuts 3/4 inch thick; 4 to 6 inches from the heat for cuts 1 to 1-1/2 inches thick). Broil, turning once, until the meat is browned and cooked to the desired doneness.

Cheddar Twice-Baked Potatoes

Prep: 20 min. **Bake:** 1 hour 20 min.

6 large baking potatoes
8 tablespoons butter, *divided*
1/4 pound sliced bacon, diced
1 medium onion, finely chopped
1/2 cup milk
1 egg
1/2 teaspoon salt
1/8 teaspoon white pepper
1 cup (4 ounces) shredded cheddar cheese

Scrub and pierce potatoes; rub each with 1 teaspoon butter. Place on a baking sheet. Bake at 375° for 1 hour or until tender.

Meanwhile, in a small skillet, cook bacon over medium heat until crisp. Remove to paper towels; drain, reserving 1 tablespoon drippings. In the drippings, saute onion until tender; set aside.

When potatoes are cool enough to handle, cut a thin slice off the top of each and discard. Scoop out pulp, leaving a thin shell. In a bowl, mash pulp with remaining butter. Stir in the milk, egg, salt and pepper. Stir in the cheese, bacon and onion.

Spoon filling into the potato shells. Place on a baking sheet. Bake at 375° for 20-25 minutes or until heated through. **Yield:** 6 servings.

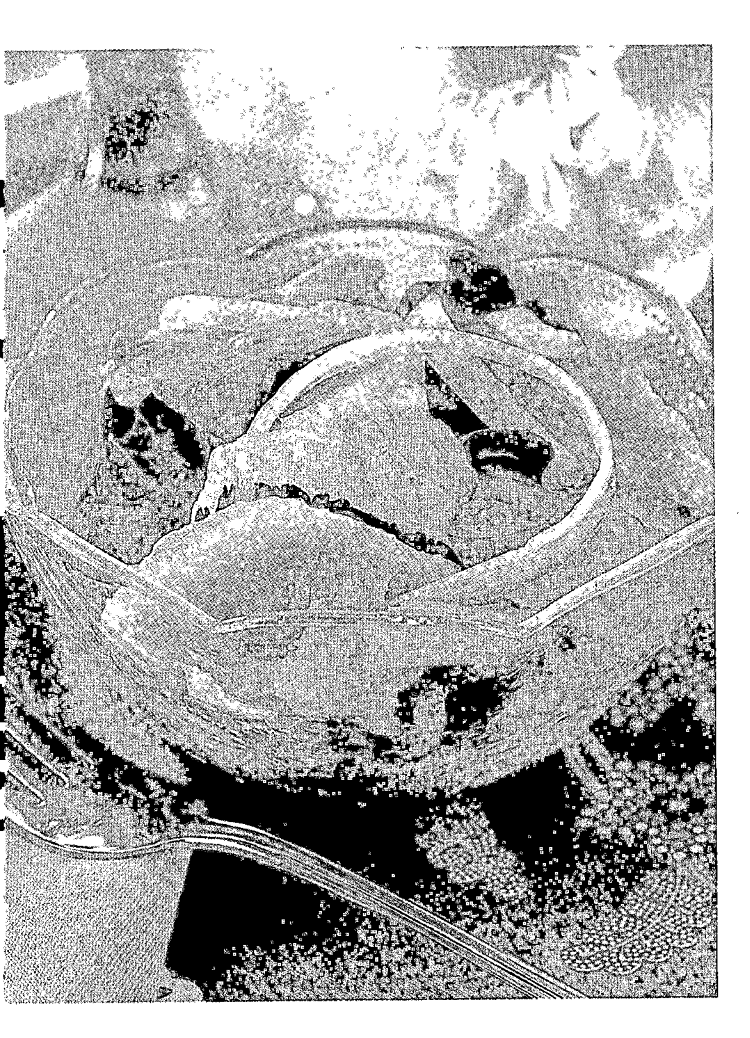

Romaine with Oranges

Prep: 20 min. + chilling

 Uses less fat, sugar or salt. Includes Nutrition Facts and Diabetic Exchanges.

6 cups torn romaine
2 medium navel oranges, peeled and sectioned
6 slices red onion, separated into rings
DRESSING:
1/4 cup olive oil
2 tablespoons orange juice
2 teaspoons cider vinegar
1/4 teaspoon salt
Dash ground mustard

In a large salad bowl, toss the romaine, oranges and onion. In a jar with a tight-fitting lid, combine the dressing ingredients; shake well. Drizzle over salad and toss to coat. Cover and refrigerate until chilled; toss before serving. **Yield:** 6 servings.

Nutrition Facts: 1 cup equals 115 calories, 9 g fat (1 g saturated fat), 0 cholesterol, 104 mg sodium, 8 g carbohydrate, 2 g fiber, 2 g protein. **Diabetic Exchanges:** 2 fat, 1/2 fruit.

Pistachio Ice Cream Dessert

Prep: 20 min. + freezing

1 cup crushed butter-flavored crackers
1/4 cup butter, melted
3/4 cup cold milk
1 package (3.4 ounces) instant pistachio pudding mix
1 quart vanilla ice cream, softened
1 carton (8 ounces) frozen whipped topping, thawed
2 packages (1.4 ounces *each*) Heath candy bars, crushed

In a bowl, combine the cracker crumbs and butter. Press into an ungreased 9-in. square baking pan. Bake at 325° for 7-10 minutes or until lightly browned. Cool on a wire rack.

Meanwhile, in a large bowl, whisk milk and pudding mix for 2 minutes. Let stand for 2 minutes or until soft-set. Stir in ice cream; pour over crust. Cover and freeze for 2 hours or until firm.

Spread with whipped topping; sprinkle with crushed candy bars. Cover and freeze for 1 hour or until firm. **Yield:** 9 servings.

These unforgettable dinner dishes lovingly made and served by her mom aren't your everyday chicken and dumplings...and that's exactly how family members like it.

By Pamela Eaton, Lambertville, Michigan

WHO NEEDS to go to a restaurant when you have a mom who can cook like mine does? My mom, Barbara Barocsi (above), specializes in what restaurants like to call "home-cooked" meals.

On Sundays, you'll often find me and my family dining at my parents' house in Oregon, Ohio, a suburb of Toledo. We're often Mom's "testers" for the new recipes she tries out from *Taste of Home*.

When my older sister, Cheryl, and I were growing up, Sunday dinner was typically meat, potatoes, salad, vegetables and rolls. Sunday was also the day Mom would write down her menus for the week ahead. She had a repertoire of meals that she repeated every 2 to 3 weeks.

Although Mom worked full-time as a customer service rep when we were kids, she still enjoyed cooking dinner from scratch. And after dinner, she'd start the next night's meal.

Her best meal—and my favorite—starts with Hungarian Chicken Paprikash, served with Spaetzle Dumplings. I always request it for my birthday, which is in August. Some years I have to wait because Mom doesn't want to cook over a hot stove in the summer heat. Last year, she served it for Christmas.

My mother learned to make the chicken and dumplings from the Hungarian women at her church. They serve it to about 300 people at an annual church dinner. To fix it at home, Mom had to cut down the recipe substantially!

It's been a tradition at our house to serve Sour Cream Cucumbers with the paprikash and spaetzle. This side dish is especially good during the summer when the cucumbers are fresh from the garden.

My favorite meal is not complete without a slice of Tart Cherry Lattice Pie, topped with vanilla ice cream. Whenever Mom is invited to a party or potluck, everyone requests her homemade, double-crust fruit pies. My dad, Donald, grows apple, cherry and apricot trees in their backyard, so fruit is always plentiful. Mom freezes some of the fruit so she can make pies during the winter, too.

She taught my dad, who's retired from an Ohio gas company, to make the pie crust and fillings. Sometimes, I don't know if my mom or my dad made a pie, because now they are both excellent pie makers.

Born and raised in Toledo, my mom learned her cooking skills from her mother. According to Mom, the secret to making food taste good is using onion, green pepper and celery for seasoning. Onion is our family's favorite vegetable, and we especially enjoy green onions from Dad's spring garden. We add onion to almost all of our salads, soups and main dishes.

Mom has passed on her cooking wisdom to my sister and me. She prepares a lot of our favorite dishes from memory, so when I started my own family, I asked her to write down the ingredients and instructions for many of those dishes.

Like my mom, I make a meal plan on Sunday so I can be ready for the upcoming week, since I work full-time. I'm a computer analyst at a local hospital. My husband, Robert, is a shipping manager at a competing hospital. We have two children, Matt and Kate.

We have Sunday dinner with my folks at their home once or twice a month during the winter. In the summer, however, the tables are turned, and my parents visit us at our cottage in Indiana. Mom still likes to do the cooking...and starts planning her meal a week before she and Dad arrive.

I'm happy to turn over kitchen duties to her because we all love her home-cooked meals.

Tart Cherry Tips

Peak season for tart cherries is June through July. Select cherries that are plump and firm with a shiny skin. Avoid fruit that is soft, bruised or shriveled, or that has browned near the stem.

Store cherries, unwashed, in a sealed plastic bag in the refrigerator for up to 2 days. When ready to use the cherries, gently wash them, then pit them using a cherry pitter, the tip of a vegetable peeler or the end of a clean bobby pin.

PICTURED AT LEFT: Hungarian Chicken Paprikash, Spaetzle Dumplings, Sour Cream Cucumbers and Tart Cherry Lattice Pie (recipes are on the next page).

Spaetzle Dumplings

Prep/Total Time: 15 min.

- **2 cups all-purpose flour**
- **4 eggs, lightly beaten**
- **1/3 cup milk**
- **2 teaspoons salt**
- **2 quarts water**
- **1 tablespoon butter**

In a large bowl, stir the flour, eggs, milk and salt until smooth (dough will be sticky). In a large saucepan, bring water to a boil. Pour dough into a colander or spaetzle maker coated with nonstick cooking spray; place over boiling water.

With a wooden spoon, press the dough until small pieces drop into boiling water. Cook for 2 minutes or until dumplings are tender and float. Remove with a slotted spoon; toss with butter. **Yield:** 6 servings.

All About Spaetzle

Spaetzle is a traditional German noodle or dumpling made of flour, eggs, water or milk, salt and sometimes a little nutmeg. The dough is either firm enough to roll out and cut into narrow strips or soft enough to force through a colander or spaetzle maker.

The noodles are then boiled in water or broth for a very short time before being tossed with butter or used in soups, sides or main dishes.

Hungarian Chicken Paprikash

Prep: 20 min. **Bake:** 1-1/2 hours

- **1 large onion, chopped**
- **1/4 cup butter, cubed**
- **4 to 5 pounds broiler/fryer chicken pieces**
- **2 tablespoons paprika**
- **1 teaspoon salt**
- **1/2 teaspoon pepper**
- **1-1/2 cups hot water**
- **2 tablespoons cornstarch**
- **2 tablespoons cold water**
- **1 cup (8 ounces) sour cream**

In a large skillet, saute onion in butter until tender. Sprinkle chicken with paprika, salt and pepper; place in an ungreased roasting pan. Spoon onion mixture over chicken. Add hot water. Cover and bake at 350° for 1-1/2 hours or until chicken juices run clear.

Remove the chicken and keep warm. In a small saucepan, combine cornstarch and cold water until smooth. Whisk in pan juices with onion. Bring to a boil over medium heat; cook and stir for 2 minutes or until thickened. Remove from the heat. Stir in sour cream. Serve with chicken. **Yield:** 6 servings.

Tart Cherry Lattice Pie

(Also pictured on front cover)

Prep: 20 min. **Bake:** 40 min. + cooling

1-1/3 cups sugar
1/3 cup all-purpose flour
4 cups fresh *or* frozen unsweetened pitted tart cherries, thawed and drained
1/4 teaspoon almond extract
Pastry for double-crust pie (9 inches)
2 tablespoons butter, cut into small pieces

In a large bowl, combine sugar and flour; stir in cherries and extract. Line a 9-in. pie plate with bottom pastry; trim to 1 in. beyond edge of plate. Pour filling into crust. Dot with butter.

Roll out remaining pastry; make a lattice crust. Seal and flute edges. Cover edges loosely with foil.

Bake at 425° for 20 minutes. Reduce heat to 375°. Remove foil; bake 20-25 minutes longer or until crust is golden brown and filling is bubbly. Cool on a wire rack. **Yield:** 6-8 servings.

Sour Cream Cucumbers

Prep: 15 min. + chilling

✓ **Uses less fat, sugar or salt. Includes Nutrition Facts and Diabetic Exchanges.**

1/2 cup sour cream
3 tablespoons white vinegar
1 tablespoon sugar
Pepper to taste
4 medium cucumbers, thinly sliced
1 small sweet onion, thinly sliced and separated into rings

In a serving bowl, combine the sour cream, vinegar, sugar and pepper. Stir in cucumbers and onion. Cover and refrigerate for at least 4 hours. Serve with a slotted spoon. **Yield:** 8 servings.

Nutrition Facts: 3/4 cup (prepared with reduced-fat sour cream) equals 52 calories, 1 g fat (1 g saturated fat), 5 mg cholesterol, 10 mg sodium, 8 g carbohydrate, 2 g fiber, 3 g protein. **Diabetic Exchange:** 1-1/2 vegetable.

Proving that necessity is indeed the mother of invention, her improvising mom worked wonders at mealtime for decades while serving as a missionary in Africa.

By Susan Hansen, Chicago, Illinois

IMAGINE COOKING without the basic ingredients—or even electricity—in 120° heat. My mom, Shirley Randall (above), faced such challenges daily when our family lived in Africa.

For 30 years, my mom and dad, Maurice, served as medical missionaries in Zimbabwe (formerly Rhodesia). One of my mother's main responsibilities was feeding and caring for the many visitors to the mission hospital and school...as well as tending to my brother, two sisters and me.

With no restaurants or stores nearby, visitors knew that they would be welcome in our home. The only visitor Mom turned away was a cobra that fell from the dining room light fixture just before dinner one day. Everyone else who came to our home, regardless of their cultural background, enjoyed my mom's cooking.

One meal I remember fondly is Cheese-Filled Meat Loaf with Butternut Squash Casserole, Parker House Dinner Rolls and Company Chocolate Cake.

Because flavorful Cheese-Filled Meat Loaf was one of my favorites, Mom would make it when I returned from boarding school in Kenya. Now I make it for special occasions.

Mom, who grew up in rural Georgia, was famous for her light, fluffy Parker House Dinner Rolls. Her comforting Butternut Squash Casserole filled in for sweet potatoes when we couldn't get them; it was a favorite of guests.

She made Company Chocolate Cake for birthday celebrations. One guest from New Zealand was so impressed with this cake that he included the recipe in his travel article!

Creating memorable meals was an accomplishment. Getting even the basic supplies wasn't always easy. The hospital lorry (a flatbed truck) picked up our groceries in town. When we heard the truck returning, we'd rush to it with our wheelbarrow...hoping the eggs and milk had survived the trip.

Electrical outages happened often during the rainy season, so Mom couldn't rely on mixers and other appliances. We also lacked air-conditioning, and some foods didn't hold up well in the heat.

Cooking was more than a necessity, though. It was our main form of entertainment. We'd invite friends for a fast-food night or a pizza party. We often held large social gatherings for people who worked at the hospital and school. The single nurses and teachers considered our house a second home and always helped in the kitchen.

Although the atmosphere was casual, Mom's table was set elegantly. We couldn't get paper plates and plastic utensils, so Mom used her china and crocheted tablecloths made by local women. And she always decorated with her homegrown roses.

Some of my best memories are of preparing foods with Mom and my sisters. She taught us to be resourceful and use what we have available to create satisfying meals.

Now that I have three children of my own (and another on the way), I appreciate all my mother did even more. My husband, Hunter, is director of behavioral health at a private clinic. I stay at home with our boys (Luke, J.P. and Rayne). I still cook most of our foods using the basic ingredients that were available in Zimbabwe. But on hectic nights, when we end up at the drive-thru or ordering pizza, I think of how my mom managed without such luxuries.

My parents live in Georgia but visit Zimbabwe every year. The grandkids (they have 13 scattered across the globe) call her "Gogo," the Ndebele word for grandmother. It suits her perfectly because she's always on the go, and she can still whip up a meal faster than anyone I know! I hope you'll enjoy this memorable one.

Selecting Squash

Butternut, acorn, hubbard, spaghetti and turban are the most common types of winter squash. Look for squash which feel heavy for their size and have hard, deep-colored, blemish-free rinds. Store unwashed winter squash in a dry, cool, well-ventilated place for up to 1 month.

PICTURED AT LEFT: Cheese-Filled Meat Loaf, Butternut Squash Casserole, Parker House Dinner Rolls and Company Chocolate Cake (recipes are on the next page).

Cheese-Filled Meat Loaf

Prep: 35 min. **Bake:** 65 min. + standing

- **1/2 cup milk**
- **2 eggs, lightly beaten**
- **1 tablespoon Worcestershire sauce**
- **1 cup crushed cornflakes**
- **1/2 cup finely chopped onion**
- **3 tablespoons finely chopped celery**
- **1 teaspoon salt**
- **1/2 teaspoon ground mustard**
- **1/2 teaspoon rubbed sage**
- **1/4 teaspoon pepper**
- **1-1/2 pounds ground beef**
- **1 cup (8 ounces) sour cream**
- **1 cup (4 ounces) finely shredded cheddar cheese**
- **1/2 cup sliced pimiento-stuffed olives**

In a large bowl, combine the first 10 ingredients. Crumble beef over mixture and mix well. On a large piece of heavy-duty foil, pat beef mixture into a 14-in. x 10-in. rectangle. Spread sour cream to within 1/2 in. of edges. Sprinkle with cheese and olives.

Roll up, jelly-roll style, starting with a short side and peeling away foil while rolling. Seal seam and ends. Place seam side down in a greased 13-in. x 9-in. x 2-in. baking dish.

Bake, uncovered, at 350° for 65-75 minutes or until the meat is no longer pink and a meat thermometer reads 160°. Let meat loaf stand for 10 minutes before slicing. **Yield:** 6 servings.

Butternut Squash Casserole

Prep: 30 min. **Bake:** 30 min.

✓ **Uses less fat, sugar or salt. Includes Nutrition Facts.**

- **2 medium butternut squash, peeled and cut into chunks**
- **1/2 cup sugar**
- **2 eggs**
- **1/4 cup milk**
- **2 tablespoons butter**
- **1 teaspoon vanilla extract**
- **1/4 teaspoon ground cinnamon**
- **1/4 teaspoon ground nutmeg**

Place squash in a large saucepan and cover with water; bring to a boil. Reduce heat; cover and simmer for 12-16 minutes or until tender. Drain.

In a small mixing bowl, beat squash until smooth. Add the remaining ingredients; beat well. Spoon into a 1-1/2-qt. baking dish coated with nonstick cooking spray. Cover and bake at 350° for 30-35 minutes or until a thermometer inserted near the center reads 160°. **Yield:** 6 servings.

Nutrition Facts: 3/4 cup equals 246 calories, 6 g fat (3 g saturated fat), 82 mg cholesterol, 75 mg sodium, 47 g carbohydrate, 9 g fiber, 5 g protein.

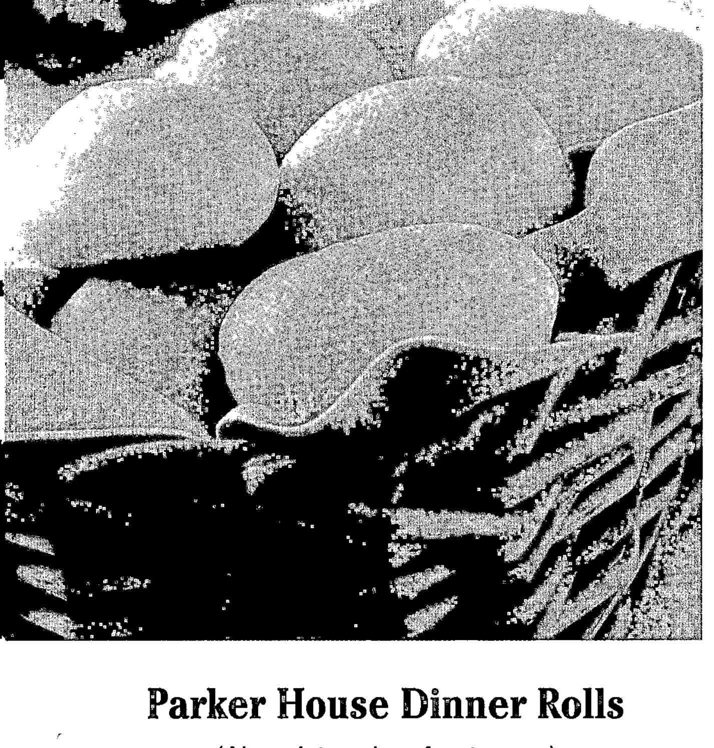

Parker House Dinner Rolls

(Also pictured on front cover)

Prep: 1 hour + rising **Bake:** 10 min.

✓ Uses less fat, sugar or salt. Includes Nutrition Facts and Diabetic Exchanges.

1/2 cup shortening
1/4 cup sugar
2 teaspoons salt
1-1/2 cups boiling water
2 tablespoons active dry yeast
1/2 cup warm water (110° to 115°)
3 eggs
6-3/4 to 7-1/4 cups all-purpose flour
1/4 cup butter, melted

In a large mixing bowl, combine the shortening, sugar and salt. Stir in boiling water. Cool to 110°-115°. Dissolve yeast in warm water. Add yeast mixture, eggs and 3 cups flour to shortening mixture; mix well. Stir in enough remaining flour to form a soft dough.

Turn onto a floured surface; knead until smooth and elastic, about 6-8 minutes. Place in a greased bowl, turning once to grease top. Cover and let rise in a warm place until doubled, about 45 minutes.

Punch dough down; turn onto a lightly floured surface. Roll dough to 1/2-in. thickness. Cut with a 2-1/2-in. biscuit cutter. Fold circles in half; press edges to seal. Place 2 in. apart on baking sheets coated with nonstick cooking spray. Cover and let rise until doubled, about 30 minutes.

Bake at 400° for 10-12 minutes or until golden brown. Remove to wire racks. Brush with the butter. Serve warm. **Yield:** 3 dozen.

Nutrition Facts: 1 roll equals 134 calories, 5 g fat (2 g saturated fat), 21 mg cholesterol, 150 mg sodium, 20 g carbohydrate, 1 g fiber, 3 g protein. **Diabetic Exchanges:** 1 starch, 1 fat.

Company Chocolate Cake

Prep: 35 min. **Bake:** 30 min. + cooling

2 cups cake flour
2 cups sugar
1 teaspoon baking soda
1/4 teaspoon salt
1/4 cup baking cocoa
1 cup water
1 cup butter, cubed
2 eggs
1/2 cup buttermilk
1 teaspoon vanilla extract
CHOCOLATE SAUCE:
1-3/4 cups sugar
2 tablespoons baking cocoa
2/3 cup milk
2 tablespoons butter
1 teaspoon vanilla extract
Vanilla ice cream
Maraschino cherries, optional

In a large bowl, combine the flour, sugar, baking soda and salt; set aside. In a small saucepan, combine cocoa and water until smooth; add butter. Bring to a boil over medium heat, stirring constantly; add to dry ingredients and stir well. Combine the eggs, buttermilk and vanilla; stir into chocolate mixture.

Pour into a greased 13-in. x 9-in. x 2-in. baking pan. Bake at 350° for 30-35 minutes or until a toothpick inserted near the center comes out clean. Cool on a wire rack.

For the chocolate sauce, in a small saucepan, combine the sugar, cocoa, milk and butter. Bring to a boil; boil for 1 minute. Remove from the heat; stir in vanilla. Serve sauce warm with cake and ice cream. Garnish with cherries if desired. **Yield:** 20 servings.

Comforting and classic, her mom's best-loved dinner menu of grilled roast, creamy mashed potatoes, crisp tossed salad and luscious cake was a summertime favorite.

By Krista Smith Kliebenstein, Broomfield, Colorado

WHEN I was 11 years old, I decided to become a vegetarian. We were a very meat-and-potatoes kind of family, but my mother respected my decision and adapted her recipes so I could follow my special diet...which I did until I graduated from college.

That's typical of my mom, Sherry Smith (above). She has always put family first. I have fond memories of bringing the best homemade goodies to school for birthdays and bake sales. Her willingness to cook and bake for my events (and those of my sister, Stacey, and brother, Craig) made me feel special.

A few years ago, I found Mom's list for cooking at Christmastime. It included a dozen cookies and candies, seven side dishes and an entree. Seeing that list made me appreciate how hard she worked to make the holiday season for our entire extended family so memorable.

The five of us—Mom; my dad, Jeff; and us kids—ate dinner together every night, and she was the sole chef. As I got older, she'd let me help in the kitchen. Of all the scrumptious dinners she has prepared, I think one of her very best is Grilled Sirloin Roast, Sour Cream Potatoes, Berry Tossed Salad and Coconut Pineapple Cake.

Her saucy grilled roast is always a hit, especially at summer gatherings. It's easy to put together and doesn't require a lot of attention.

With their mild garlic-herb flavor, Sour Cream Potatoes go so well with the roast. This rich, tasty side dish is real comfort food!

A creamy raspberry dressing highlights Berry Tossed Salad, a pretty combination of strawberries, kiwifruit, red onion, feta cheese and almonds. It fits in nicely at sit-down holiday dinners as well as casual backyard barbecues.

Coconut Pineapple Cake is a delight with chunks of pineapple, chopped walnuts and fluffy coconut frosting. Sometimes my mom will leave out the walnuts, particularly when she's taking the cake to a function, where someone might have an allergy.

Mom stayed at home while raising her three children but still did the bookkeeping for my father's business. He recently retired after more than 40 years as a financial adviser. They live nearby, in Fort Collins, Colorado.

My husband, Nick, is also a financial adviser. We have two daughters—Evelyn and Nora. I'm pursuing a master's degree in counseling psychology and also do freelance writing.

I'm not much of a baker, so every year my mom makes Christmas cookies with my daughters, who love that special time with their "nana."

Mom, who grew up in Wisconsin, has baking in her blood. Her father was a baker, and she often cooked with her mother.

My mother, sister and I share a special bond through cooking. We often call and E-mail each other with new and interesting recipes we've discovered. (Stacey and her family live in Illinois; my brother and his family are in Wisconsin.)

When Stacey and I had our babies, Mom came to our homes to help, making sure we had home-cooked dinners. It was a lifesaver! She's always there for us, whether she's taking care of my daughters so I can attend class or flying to Illinois if my sister needs her.

She and Dad are active in their church, and you might say Mom "ministers" with her cooking. Anytime there's a special occasion or event, she bakes delicious goodies to share.

I'm happy I could pass along this menu to you and your family. Enjoy!

Slow Grilling

Grilling over indirect heat is a cooking method used for foods that need to be cooked for a long time over medium or medium-low heat. It is recommended for roasts, ribs, chicken, turkey and other large cuts of meat.

Grilled Sirloin Roast (pictured at left) cooks for 3 hours over indirect heat, giving you plenty of time to prepare the rest of your meal.

PICTURED AT LEFT: Grilled Sirloin Roast, Sour Cream Potatoes, Berry Tossed Salad and Coconut Pineapple Cake (recipes are on the next page).

Sour Cream Potatoes

Prep: 30 min. **Bake:** 30 min.

- **10 medium red potatoes, peeled and quartered**
- **1 package (8 ounces) cream cheese, cubed**
- **1 cup (8 ounces) sour cream**
- **1/4 cup milk**
- **2 tablespoons butter, *divided***
- **1 tablespoon dried parsley flakes**
- **1-1/4 teaspoons garlic salt**
- **1/4 teaspoon paprika**

Place potatoes in a large saucepan and cover with water. Bring to a boil. Reduce heat; cover and cook for 15-20 minutes or until tender. Drain.

In a large mixing bowl, mash the potatoes. Add the cream cheese, sour cream, milk, 1 tablespoon butter, parsley and garlic salt; beat until smooth.

Spoon into a greased 2-qt. baking dish. Dot with remaining butter; sprinkle with paprika. Bake, uncovered, at 350° for 30-40 minutes or until heated through. **Yield:** 6-8 servings.

Grilled Sirloin Roast

Prep: 20 min. **Grill:** 3 hours

- **3 tablespoons all-purpose flour**
- **3/4 cup ketchup**
- **4-1/2 teaspoons Worcestershire sauce**
- **1 tablespoon brown sugar**
- **1 tablespoon cider vinegar**
- **1-1/4 teaspoons salt, *divided***
- **1/2 teaspoon prepared mustard**
- **1/4 teaspoon pepper**
- **1 boneless beef sirloin tip roast (3 pounds)**
- **1 pound fresh baby carrots**
- **2 medium tomatoes, quartered**
- **1 medium onion, quartered**

In a small bowl, combine the flour, ketchup, Worcestershire sauce, brown sugar, vinegar, 1 teaspoon salt and mustard until smooth; set aside. Sprinkle pepper and remaining salt over roast.

Grill roast over medium heat for 5-10 minutes or until browned on all sides. Transfer to a heavy-duty 13-in. x 9-in. disposable foil pan. Pour reserved sauce over roast. Top with carrots, tomatoes and onion. Cover pan with foil. Grill over indirect medium heat for 3 hours or until meat is tender. **Yield:** 6-8 servings.

Berry Tossed Salad

Prep/Total Time: 30 min.

- **1 package (10 ounces) ready-to-serve salad greens**
- **1 cup sliced fresh strawberries**
- **1 kiwifruit, peeled and sliced**
- **1/4 cup chopped red onion**
- **1/4 cup crumbled feta cheese**
- **2 tablespoons slivered almonds**

CREAMY RASPBERRY DRESSING:

- **1/2 cup mayonnaise**
- **2 tablespoons plus 2 teaspoons sugar**
- **1 tablespoon raspberry vinegar**
- **1 tablespoon milk**
- **2-1/2 teaspoons poppy seeds**
- **2-1/2 teaspoons seedless raspberry jam**

In a large salad bowl, combine the greens, strawberries, kiwi, onion, feta cheese and almonds. In a small bowl, whisk the dressing ingredients. Drizzle desired amount over salad and toss to coat. Serve immediately. Refrigerate any leftover dressing. **Yield:** 8 servings.

Kinds of Kiwi

Kiwifruit is available in both green and gold varieties. The green kiwi is egg-shaped with a fuzzy brown exterior and emerald-green flesh.

The golden kiwi is sweeter than the green. It has a pointed end, smooth brown skin and golden flesh. The flesh of both varieties of kiwifruit contains tiny black, edible seeds.

Coconut Pineapple Cake

Prep: 25 min. **Bake:** 35 min. + cooling

- **2 eggs**
- **2 cups sugar**
- **1 teaspoon vanilla extract**
- **2 cups all-purpose flour**
- **1/2 teaspoon baking soda**
- **1/2 teaspoon baking powder**
- **1/2 teaspoon salt**
- **1 can (20 ounces) crushed pineapple, undrained**
- **1/2 cup chopped walnuts**

FROSTING:

- **1 package (8 ounces) cream cheese, softened**
- **1/2 cup butter, softened**
- **2 cups confectioners' sugar**
- **1/2 cup flaked coconut**

In a large mixing bowl, beat the eggs, sugar and vanilla until fluffy. Combine the flour, baking soda, baking powder and salt; add to egg mixture alternately with pineapple. Stir in walnuts.

Pour into a greased 13-in. x 9-in. x 2-in. baking pan. Bake at 350° for 35-40 minutes or until a toothpick inserted near the center comes out clean. Cool on a wire rack.

In a small mixing bowl, beat the cream cheese, butter and confectioners' sugar until smooth. Frost cake. Sprinkle with coconut. Store in the refrigerator. **Yield:** 12 servings.

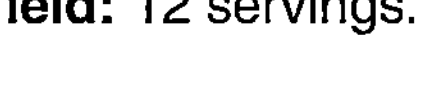

From the succulent main course to the delightful cake for dessert, her mom's Pennsylvania Dutch dinner is chock-full of fruit flavor...and memories.

By Shirley Joan Helfenbein, Lapeer, Michigan

WHEN I think about my mother's cooking, I remember how different it was back then to put food on the table. Her meals were fruits of love, prepared from scratch. Nearly everything was homegrown. Apples from the orchard, nuts from the woods, veggies from the garden, flour from the mill and meat from the butcher made up her storehouse of ingredients.

My mom, Elsie Hart (above), was born in Ohio in 1915. She and my dad, Verlin, raised six children, plus two daughters from my dad's first marriage. We all helped out, from baking to picking apples. I still recall Mom saying, "Many hands make lighter work."

Mom shared her memories with us as we helped her make the Pennsylvania Dutch meals from her childhood. I especially enjoyed her Apple-Raisin Pork Chops, Buttered Poppy Seed Noodles, Dutch Apple Salad and Spice Cake served with Hard Sauce.

The kitchen was a hubbub of activity when we helped Mom prepare this meal. While she browned the pork chops, we took turns peeling and slicing the apples. (Apples showed up in many of our entrees, salads and desserts.) We brought her the spices she needed, and Mom shook in what seemed right. The result was always delicious!

When I was growing up, we couldn't afford to have meat very often, so it was a special evening when we had these tender chops. The topping looked so impressive and gave this dish a sweet touch.

Mom had been making comforting Buttered Poppy Seed Noodles since she was a child, but she started with homemade noodles instead of packaged. She'd roll out the dough on the table, then let us cut the long noodles and hang them to dry.

Crisp apple chunks, celery and grapes give Dutch Apple Salad its refreshing flavor. Our family would sit around the table in the evening and crack nuts for Mom to use in recipes such as this salad.

Its simple milk and sugar dressing allows the fruit and veggies to really shine through. You could also use vanilla yogurt in place of the cooked dressing.

Grandma wrote down the recipe for Spice Cake on a piece of brown paper bag, and Mom often relied on this dessert when she wanted an extra-special treat. The raisin-dotted cake is loaded with popular spices, so the aroma during baking is out of this world!

A topping of Hard Sauce makes this dessert extra moist...and special. Today, many cooks spread cream cheese frosting on spice cake instead, but I think you're missing something if you bypass this warm, old-time topping. Any leftover sauce will keep in the refrigerator about a week.

Mom also loved to make date-nut pudding, and her pumpkin and mincemeat-apple pies were wonderful additions to our holiday and Sunday meals. My dad most enjoyed her molasses and sugar cookies. When he walked home from his job at the gas station, Dad would sniff the kitchen air to guess what goodies she had baked that day.

Once some of us were older, Mom went to work as a nurse. By that time, I was on my own. After working my way through school, I volunteered at a Navajo Indian mission in New Mexico. I met my husband, Roy, in New Mexico, where he worked in the oil fields and construction. We're now retired with three grown children and five grandchildren.

Because I taught school, I had to rely on simple dishes to feed our family. But for special days, I'd make meals like Mom used to cook. Some of those Pennsylvania Dutch recipes creep into our menus now, and I still love them. I hope you will, too!

Combining Cake Ingredients

Combining the dry ingredients of a cake before adding them to the creamed mixture helps to evenly distribute the leavener throughout the flour. This ensures that the leavener is evenly incorporated into the batter.

Once the combined dry ingredients are added to the creamed mixture, the ingredients should be beaten just until blended. Beating too much could give the cake a tough texture.

PICTURED AT LEFT: Apple-Raisin Pork Chops, Buttered Poppy Seed Noodles, Dutch Apple Salad, Spice Cake and Hard Sauce (recipes are on the next page).

Apple-Raisin Pork Chops

Prep: 25 min. **Cook:** 25 min.

8 bone-in pork loin chops (3/4 inch thick)
1 tablespoon vegetable oil
1/2 teaspoon salt
1/4 teaspoon pepper
2 cups apple cider *or* juice
3 tablespoons spicy brown mustard
3 medium red apples, sliced
1/2 cup sliced green onions
1/4 cup raisins
1/4 cup dried currants
2 tablespoons cornstarch
1/4 cup cold water

In a large skillet, brown pork chops in oil in batches on both sides. Sprinkle with salt and pepper. Return all chops to the skillet.

Combine cider and mustard; pour over meat. Bring to a boil; reduce heat. Cover and simmer for 13-18 minutes or until juices run clear. Remove chops to a serving platter; keep warm.

Add the apples, onions, raisins and currants to the skillet. Cover and cook over medium heat for 5-6 minutes or until apples are tender. Combine cornstarch and water until smooth; stir into apple mixture. Bring to a boil; cook and stir for 2 minutes or until thickened. Serve with pork chops. **Yield:** 8 servings.

Buttered Poppy Seed Noodles

(Pictured on page 224)

Prep/Total Time: 25 min.

1 package (16 ounces) egg noodles
1 medium onion, chopped
3 tablespoons butter
2 green onions, chopped
2 tablespoons poppy seeds
Salt and pepper to taste
1 tablespoon minced fresh parsley

Cook noodles according to package directions. Meanwhile, in a large heavy skillet, saute onion in butter until it begins to brown. Drain noodles; add to skillet. Cook and stir until noodles begin to brown.

Add the green onions, poppy seeds, salt and pepper; cook and stir 1 minute longer. Sprinkle with parsley. **Yield:** 8 servings.

Dutch Apple Salad

Prep: 30 min. + chilling

2 tablespoons all-purpose flour
1 tablespoon sugar
1 egg
1 cup milk
2 large Golden Delicious apples, chopped
2 large Red Delicious apples, chopped
1/2 cup finely chopped celery
1/2 cup seedless red grapes, quartered
1/2 cup chopped walnuts, toasted

In a small saucepan, combine flour and sugar. Whisk the egg and milk; stir into flour mixture until smooth. Bring to a boil over medium heat; cook and stir for 1-2 minutes or until thickened and bubbly. Transfer to a small bowl; cover and refrigerate until chilled.

Just before serving, combine the apples and celery in a large salad bowl. Drizzle with dressing; gently toss to coat. Sprinkle with grapes and walnuts. **Yield:** 8 servings.

Spice Cake

Prep: 25 min. **Bake:** 40 min. + cooling

1 cup raisins
1 cup boiling water
1/2 cup butter, softened
1/4 cup shortening
1-1/4 cups sugar
2 eggs
1 cup chunky applesauce
1 teaspoon vanilla extract
2-1/2 cups all-purpose flour
2 teaspoons ground cinnamon
1 teaspoon baking powder
1/2 teaspoon salt
1/2 teaspoon ground nutmeg
1/4 teaspoon baking soda
1/4 teaspoon ground ginger
1/2 cup chopped walnuts

Place raisins in a small bowl; cover with boiling water. Let stand for 5 minutes; drain and set aside.

In a large mixing bowl, cream the butter, shortening and sugar. Add eggs, one at a time, beating well after each. Beat in applesauce and vanilla. Combine the flour, cinnamon, baking powder, salt, nutmeg, baking soda and ginger; add to creamed mixture just until blended. Stir in walnuts and raisins.

Pour into a greased 11-in. x 7-in. x 2-in. baking dish. Bake at 350° for 40-45 minutes or until a toothpick inserted near the center comes out clean. Cool on a wire rack. Serve cake with Hard Sauce (recipe below). **Yield:** 8 servings.

Hard Sauce

Prep/Total Time: 15 min.

1 cup sugar
2 tablespoons all-purpose flour
1/4 teaspoon ground nutmeg
Dash ground allspice
1 cup cold water
2 tablespoons butter
1 teaspoon vanilla extract
1/4 to 1/2 teaspoon rum extract

In a small saucepan, combine the sugar, flour, nutmeg, allspice and water until smooth. Bring to a boil; cook and stir for 2 minutes or until thickened. Remove from the heat; stir in butter and extracts. Serve with spice cake. Refrigerate leftovers. **Yield:** about 1-1/3 cups.

Editors' Meals

Taste of Home is edited by 1,000 cooks across North America. Here, you'll "meet" some of them and see their family-favorite menus.

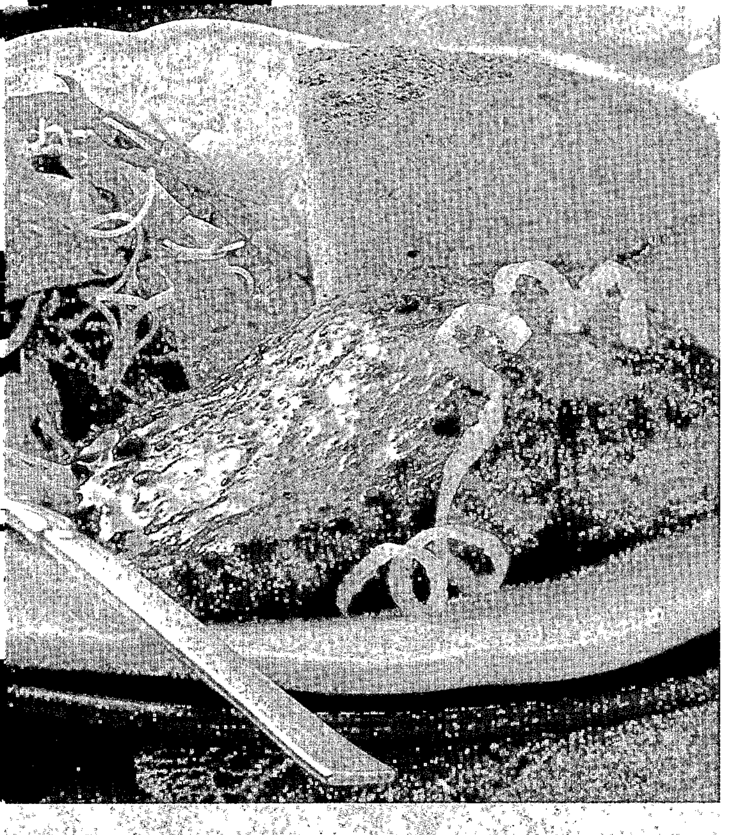

COOK'S BEST. Clockwise from top left: Meat-and-Potato Lovers' Menu (p. 238), Dressed-Up Salmon for Guests (p. 246), Crown Roast for Christmas (p. 230) and No-Fuss July 4th Cookout (p. 242).

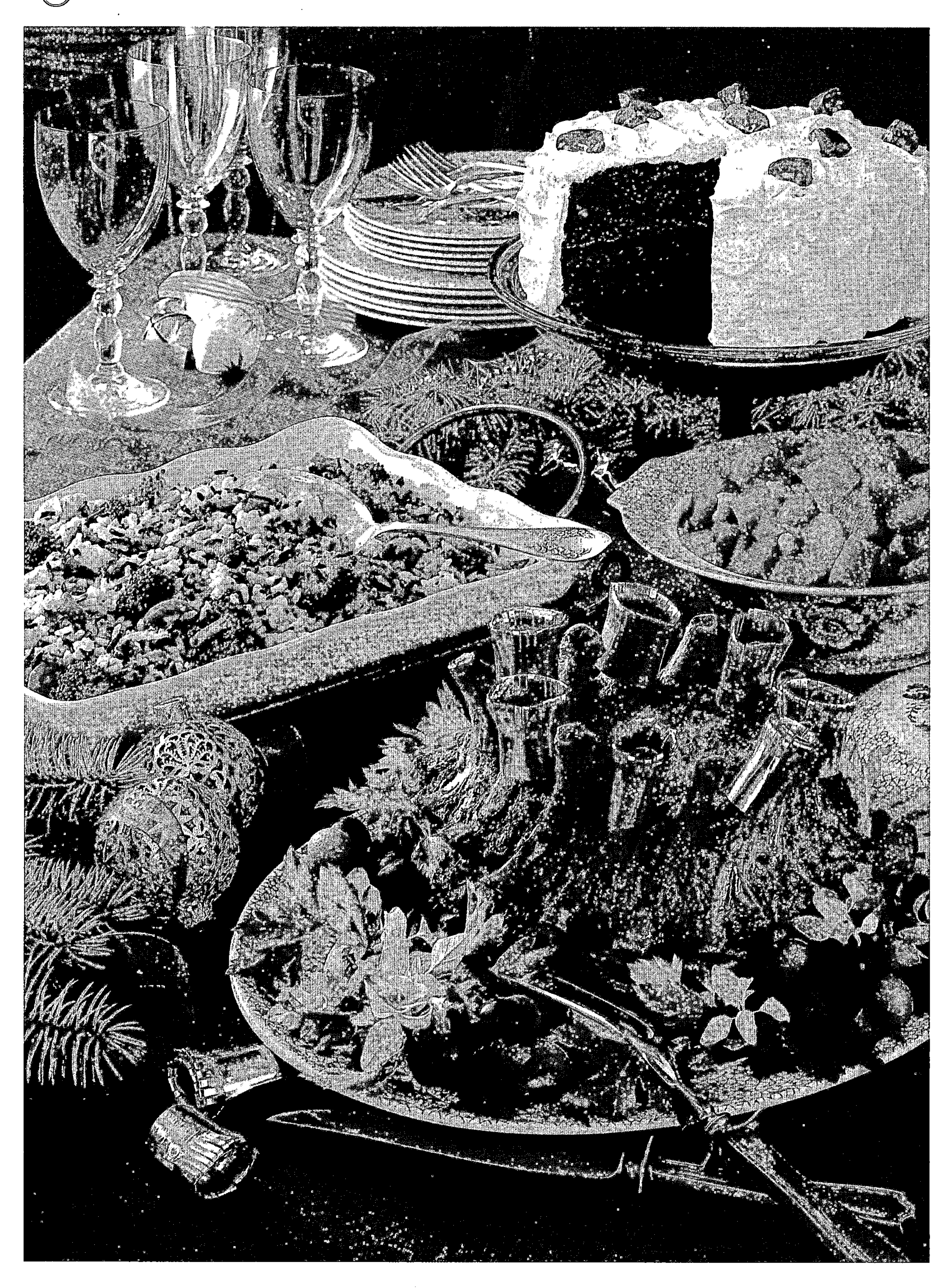

Crown Roast for Christmas

Elegant pork entree forms the centerpiece of this *Taste of Home* field editor's holiday meal.

By Dianne Bettin, Truman, Minnesota

I LOVE to feed people. And, even though our family has an extremely busy lifestyle, I don't mind taking time out to cook for family, friends and sometimes even strangers!

My husband, Doug, and I have a 350-sow hog operation and grain farm, and I have been volunteering with pork producers' organizations for almost 20 years.

So it's no surprise that pork stars on many of my menus. During the holidays and on other special occasions, I enjoy showcasing an elegant yet easy-to-prepare Crown Roast of Pork.

Some of my favorite side dishes to serve with the roast are Wild Rice Pilaf and Cashew-Peach Sweet Potatoes. Devil's Food Caramel Torte is a great finish!

A crown roast makes Christmas dinner extra special. I order the roast by how many ribs I want—figure on one per person. Just talk to your butcher or grocery store meat manager a few days ahead.

An herb rub accents the pork's flavor. Be sure to use a meat thermometer to guarantee that your roast is done to perfection (see tip below right).

I like to make the Wild Rice Pilaf a day ahead, which allows the rosemary and other flavors time to blend and makes mealtime preparations go smoother the day of the dinner. The pilaf reheats well in the microwave. I frequently take it to potlucks—the only bad thing about that is I rarely have any leftovers to bring home!

I ran across the Cashew-Peach Sweet Potatoes recipe while hunting for variations on the old standby sweet potato bake. I have a vegetable garden and found the recipe one particularly prolific year.

When I cook, I tend to focus on the main meal. The dessert oftentimes is an afterthought. That's why I was thrilled when my sister-in-law, Nicki, showed up at a gathering with scrumptious Devil's Food Caramel Torte. It was definitely one of those recipes I had to add to my collection.

I love food and confess that while most people travel to sightsee, I travel to eat! Even when I go to our famous Mall of America here in Minnesota, I spend more time eating than shopping. My favorite stop is a wonderful sushi bar.

As a member of the National Pork Board—an industry-nominated, government-appointed position—I help promote pork across the country and even overseas. I enjoy cooking new foods that I find on my travels, especially if they involve pork! Doug and our children—Tyler and Marisa—are good sports about trying everything I put in front of them.

Doug and I enjoy traveling and camping. Although the farm doesn't give him much free time, he manages to accompany me to some of my out-of-state meetings. We also enjoy time at the lake and around the campfire.

With our youngest off to college, our life will be very different now! I plan to fill some of the void with trying new recipes, especially tried-and-true ones from *Taste of Home*.

I hope you'll "pick pork" and enjoy this menu for your own Christmas dinner!

Dianne's Pork Pointers

- The old practice of overcooking pork is no longer necessary for food safety. Our improved product is best when left slightly pink in the center, cooked to an internal temperature of 160°.
- I remove my crown roast (and other cuts, too!) from the oven when it reaches 150° and let it rest for about 10 minutes before carving. This allows the temperature to continue to rise to 160° while sealing in the juices.
- New research from the USDA reveals that pork tenderloin contains only 2.98 grams of fat per 3-ounce cooked serving. A 3-ounce cooked skinless chicken breast has 3.03 grams of fat.

PICTURED AT LEFT: Crown Roast of Pork, Wild Rice Pilaf, Cashew-Peach Sweet Potatoes and Devil's Food Caramel Torte (recipes are on the next page).

Wild Rice Pilaf

Prep: 1 hour **Bake:** 25 min.

2 cans (14-1/2 ounces *each*) chicken broth
3/4 cup uncooked wild rice
1 cup uncooked long grain rice
1 large onion, chopped
2 medium carrots, halved lengthwise and sliced
1 garlic clove, minced
1/2 teaspoon dried rosemary, crushed
1/2 cup butter, cubed
3 cups fresh broccoli florets
1/4 teaspoon pepper

In a large saucepan, bring broth to a boil. Add wild rice; reduce heat. Cover and cook for 30 minutes. Add long grain rice; cook 20-25 minutes longer or until liquid is absorbed and rice is tender.

Meanwhile, in a large skillet, saute the onion, carrots, garlic and rosemary in butter until vegetables are tender. Stir in the rice, broccoli and pepper.

Transfer to a greased shallow 2-qt. baking dish. Cover and bake at 350° for 25-30 minutes or until broccoli is crisp-tender. Fluff with a fork before serving. **Yield:** 10 servings.

Crown Roast of Pork

Prep: 15 min. **Bake:** 3 hours + standing

1 tablespoon dried parsley flakes
1 tablespoon vegetable oil
1 teaspoon salt
1/2 teaspoon pepper
1 pork crown roast (14 ribs and about 8 pounds)
Foil *or* paper frills for rib ends

In a small bowl, combine the parsley, oil, salt and pepper; rub over roast. Place on a rack in a large shallow roasting pan. Cover rib ends with pieces of foil. Bake at 350° for 3 to 3-1/2 hours or until a meat thermometer reads 160°.

Transfer roast to a serving platter. Let stand for 10-15 minutes. Remove foil. Garnish rib ends with frills. Cut between ribs to serve. **Yield:** 14 servings.

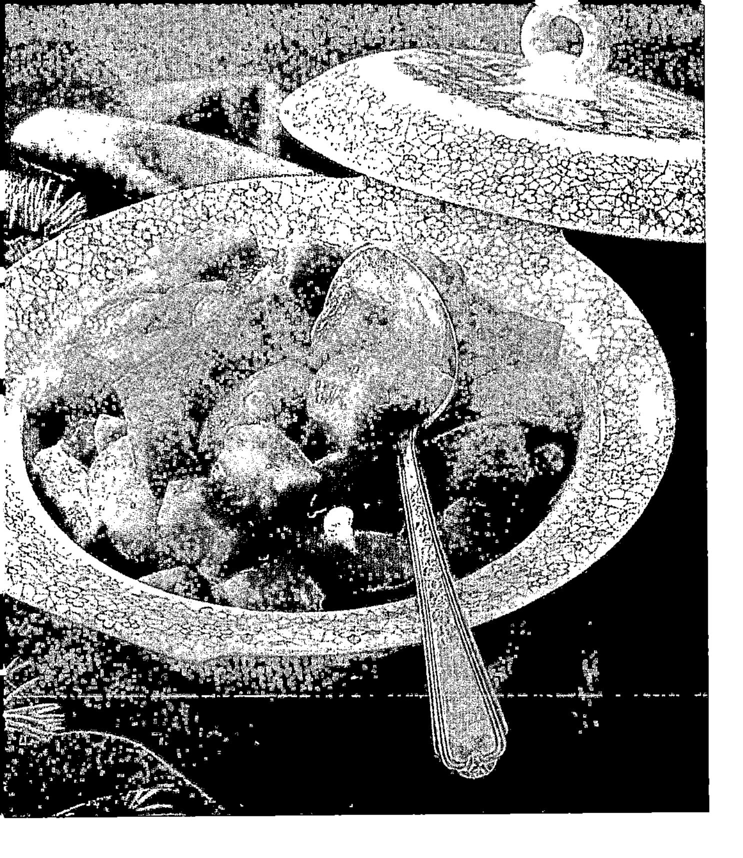

Cashew-Peach Sweet Potatoes

Prep: 45 min. **Bake:** 40 min.

6 medium sweet potatoes
1/2 cup packed brown sugar
1/3 cup coarsely chopped cashews
1/2 teaspoon salt
1/4 teaspoon ground ginger
1 can (15-1/4 ounces) sliced peaches, drained
3 tablespoons butter

Place sweet potatoes in a large saucepan or Dutch oven; cover with water. Bring to a boil. Reduce heat; cover and cook for 30-45 minutes or just until tender. Drain and cool slightly; peel and cut into cubes.

In a small bowl, combine brown sugar, cashews, salt and ginger. Place half of the sweet potatoes in an ungreased 11-in. x 7-in. x 2-in. baking dish; top with half of the peaches and brown sugar mixture. Repeat layers; dot with butter.

Cover and bake at 350° for 30 minutes. Uncover; bake 10 minutes longer or until bubbly and heated through. **Yield:** 10 servings.

Devil's Food Caramel Torte

Prep: 40 min. **Bake:** 25 min. + cooling

1 package (18-1/4 ounces) devil's food cake mix
1 cup buttermilk
1/2 cup vegetable oil
3 eggs
1 package (7 ounces) milk chocolate turtle candies, chopped, *divided*
1 tablespoon baking cocoa
1-1/2 cups heavy whipping cream
1/3 cup caramel ice cream topping
1 can (16 ounces) chocolate frosting
Additional milk chocolate turtle candies, broken, optional

Line two 9-in. round baking pans with waxed paper; grease the paper and set aside. In a large mixing bowl, combine the cake mix, buttermilk, oil and eggs. Beat on low speed for 30 seconds. Beat on medium for 2 minutes. Combine 1 cup candies and cocoa; fold into batter.

Pour into prepared pans. Bake at 350° for 25-30 minutes or until a toothpick inserted near the center comes out clean. Cool for 10 minutes before removing from pans to wire racks to cool completely. Remove waxed paper.

In a small mixing bowl, beat cream until it begins to thicken. Add caramel topping; beat until stiff peaks form. Fold in remaining candies.

Place one cake layer on a serving plate; spread with chocolate frosting. Top with remaining cake layer; frost top and sides of torte with cream mixture. Garnish with additional candies if desired. Refrigerate until serving. **Yield:** 12 servings.

Lovely Spring Feast of Lamb

This creative editor loves to write about food—and to share the secrets of her best-loved recipes.

By Millie Vickery, Lena, Illinois

WITH all the treasured recipes I've collected over the years, it's impossible to narrow them down to one favorite meal. But this dinner of Crusty Roast Leg of Lamb, Springtime Asparagus Medley, Cracked Pepper Salad Dressing, Banana Citrus Sorbet and Strawberry-Banana Angel Torte is a wonderful way to celebrate the spring season.

These five recipes are among hundreds I've included in my "Cooking with Millie" newspaper column over the years and put together in a cookbook with that same title. I love to try new recipes and adjust them to my taste...to share good recipes with friends...and to entertain.

Crusty Roast Leg of Lamb is a delightful entree for Easter. It would also be perfect for St. Patrick's Day, since lamb is a traditional favorite of the Irish. With its golden herb crust, this roast makes a beautiful presentation surrounded by apple slices and potatoes on the serving platter.

I grow lots and lots of mint in my little garden and like to serve lamb with mint jelly or mint sauce on the side. Fresh mint also makes a great garnish for any lamb dish, along with pear or apricot halves and red grapes.

For over 50 years, I cooked with joy for my late husband, Eugene "Vic." Now I live alone, but I still love to have family and guests for dinner. I like to be creative and have fun with food.

My mom, a wonderful cook, was always adding something for extra color and appeal. Her influence is evident in my eye-catching tossed salad, served with flavorful Cracked Pepper Salad Dressing. It is creamy, delicious and simple to mix up.

Often, I'll start a meal with a salad, then surprise guests with a sorbet refresher before the main course. Banana Citrus Sorbet is a recipe from my friend Adrianne St. George, who gave elegant parties.

A zippy sauce brings out the flavor of the vegetables in Springtime Asparagus Medley. Accented by blue cheese and crunchy almonds, this dish is delicious hot or cold.

When a recipe is easy but it looks like you fussed, it's a keeper. Strawberry-Banana Angel Torte, with its luscious creamy filling and fresh fruit, is one of these. Of course, you can make the angel food cake from scratch if you have the time and the egg whites. But buying the cake or baking it from a mix gets you off to a quick start.

Over the years, I've won awards in cooking contests and have given programs on garnishing. I've won many writing awards, too. Currently, I'm typing my husband's last book on the computer. He wrote 10 books after retiring from his practice, including *House Calls: Life of a Country Doctor*. Vic also wrote poems, including some fun ones about me and my cooking.

My life changed dramatically after I lost him. But I keep very busy. Recently, I sold the big Victorian home where we lived for decades and moved to an apartment. I still spend much of my time at our vacation home on nearby Apple Canyon Lake.

Whether here or there, you'll often find me in the kitchen, trying a new recipe or making an old favorite. I'm sure you and your family will enjoy the mouth-watering menu here!

Roasting Lamb

The leg of lamb you buy may have a thin, papery white membrane (the fell) covering it, which should be removed before roasting. Then trim the fat, which carries a strong taste that can overpower the delicate flavor of the meat. Leave just a few streaks of fat to provide moisture as the lamb cooks.

Place the lamb fat side up on the rack in a shallow roasting pan. Insert a meat thermometer into the thickest muscle, being careful not to let the thermometer rest on a bone or in fat.

PICTURED AT LEFT: Crusty Roast Leg of Lamb, Springtime Asparagus Medley, Cracked Pepper Salad Dressing, Banana Citrus Sorbet and Strawberry-Banana Angel Torte (recipes are on the next page).

Crusty Roast Leg of Lamb

Prep: 20 min. **Bake:** 2 hours + standing

1 boneless leg of lamb (4 to 5 pounds)
1 cup soft bread crumbs (about 2 slices)
2 tablespoons butter, melted
1/2 teaspoon herbes de Provence
Dash salt and pepper
1 large onion, finely chopped
1 can (14-1/2 ounces) chicken broth
2-1/2 pounds medium potatoes, peeled and cut into wedges
1 large tart apple, sliced

Place leg of lamb on a rack in a roasting pan. In a small bowl, combine the bread crumbs, butter and seasonings; spread over meat. Place onion in pan; pour broth over onion. Bake, uncovered, at 350° for 1 hour.

Add potatoes; bake 30 minutes longer. Add apple; bake 30 minutes longer or until potatoes are tender and meat reaches desired doneness (for medium-rare, a meat thermometer should read 145°; medium, 160°; well-done, 170°). Remove vegetables and apple and keep warm. Let roast stand for 10-15 minutes before slicing. **Yield:** 10 servings.

Springtime Asparagus Medley

Prep/Total Time: 25 min.

1 cup water
1-1/2 pounds fresh asparagus, trimmed and cut into 2-inch pieces
2 small tomatoes, cut into wedges
3 tablespoons cider vinegar
3/4 teaspoon Worcestershire sauce
1/3 cup sugar
1 tablespoon grated onion
1/2 teaspoon salt
1/2 teaspoon paprika
1/3 cup vegetable oil
1/3 cup sliced almonds, toasted
1/3 cup crumbled blue cheese, optional

In a large saucepan, bring water to a boil. Add asparagus; cover and cook for 3-5 minutes or until crisp-tender. Drain. Add tomatoes; cover and keep warm.

In a blender, combine vinegar, Worcestershire sauce, sugar, onion, salt and paprika; cover and process until smooth. While processing, gradually add oil in a steady stream. Pour over asparagus mixture and toss to coat. Transfer to a serving bowl; sprinkle with almonds and blue cheese if desired. Serve warm. **Yield:** 8-10 servings.

Cracked Pepper Salad Dressing

Prep: 15 min. + chilling

- **2 cups mayonnaise**
- **1/4 cup water**
- **1/4 cup milk**
- **1/4 cup buttermilk**
- **2 tablespoons grated Parmesan cheese**
- **1 tablespoon coarsely ground pepper**
- **2 teaspoons finely chopped green onion**
- **1 teaspoon lemon juice**
- **1/2 teaspoon garlic salt**
- **1/2 teaspoon garlic powder**

In a small bowl, whisk all ingredients until blended. Cover and chill for at least 1 hour. May be stored in the refrigerator for up to 2 weeks. **Yield:** 2-1/2 cups.

Banana Citrus Sorbet

(Pictured on page 234)

Prep: 10 min. + freezing

- **1/2 cup lemon juice**
- **3 medium ripe bananas, cut into chunks**
- **1-1/2 cups sugar**
- **2 cups cold water**
- **1-1/2 cups orange juice**

Place lemon juice and bananas in blender; cover and process until smooth. Add sugar; cover and process until blended. Transfer to a large bowl; stir in water and orange juice.

Fill cylinder of ice cream freezer two-thirds full; freeze according to manufacturer's directions. Refrigerate remaining mixture until ready to freeze. Transfer to a freezer container; freeze for 2-4 hours before serving. May be frozen for up to 1 month. **Yield:** 2-1/2 quarts.

Strawberry-Banana Angel Torte

Prep/Total Time: 20 min.

- **1 prepared angel food cake (8 inches)**
- **1/2 cup sour cream**
- **1/4 cup sugar**
- **1/4 cup pureed fresh strawberries**
- **3/4 cup sliced ripe bananas**
- **1/2 cup sliced fresh strawberries**
- **1 cup heavy whipping cream, whipped**
- **Halved fresh strawberries**

Split cake horizontally into three layers; place bottom layer on a serving plate. In a large bowl, combine the sour cream, sugar and pureed strawberries; fold in bananas and sliced strawberries. Fold in whipped cream.

Spread a third of the filling between each layer; spread remaining filling over top. Cover and refrigerate until serving. Garnish with the halved strawberries. **Yield:** 8-10 servings.

Meat-and-Potato Lovers' Menu

North meets South in this field editor's dinner of tender brisket, tasty sides and a delightful dessert.

By Darlis Wilfer, West Bend, Wisconsin

FOR THOSE who love to cook—and those who love to eat—I'm sharing a hearty meal that's great for a family gathering or a casual dinner party.

My popular menu features Beef Brisket with Mop Sauce, Comforting Potato Casserole, Tossed Salad with Carrot Dressing, Grandma's Honey Muffins and Poppy Seed Cake.

As you will see, I'm not a gourmet cook...I just like to put a different twist on food that people enjoy.

My husband, Chuck, and I have been married for over 50 years. He's a "meat-and-potatoes" guy, and I am fond of desserts. Three of our five children and their families live nearby, and we get together often for holidays and special occasions. I'm in my glory with a houseful of hungry folks to feed!

Beef Brisket with Mop Sauce was quick to gain everyone's approval. "You can make this again, Mom!" they agreed.

Before settling in Georgia, our son Mike and daughter-in-law Sharon lived in Texas. There, we discovered, mop sauce is traditionally prepared for Texas ranch-style barbecue in batches so large that it is brushed on the meat with a mop! We love the zesty flavor of the sauce and think you will, too.

Tossed Salad with Carrot Dressing always draws favorable comments. It tastes so fresh and is easy to mix up. We love going to the farmers market on Saturdays in summer and fall to buy fresh produce. I like to dress up homegrown lettuce with cut veggies, chow mein noodles, dried cranberries and shredded cabbage.

Comforting Potato Casserole was a dish I first tasted at a wedding reception, and I asked the caterer for the recipe. I make these snazzy potatoes often, as do our daughters, who appreciate that the recipe can be made ahead.

I fondly remember my Grandma Wheeler making her honey muffins. We'd eat them fresh from the oven...nice and warm. Getting the correct measurements for her recipe was a challenge because she didn't have them written down. She used "a pinch of this" and "a handful of that," and knew when there was enough flour because it "felt right."

Deciding which dessert recipe to add to my menu was difficult. I asked the family, and each one had a different suggestion. I chose old-fashioned Poppy Seed Cake. Tender and chock-full of poppy seeds, it has a yummy cream cheese frosting. I got the cake recipe from my longtime friend Mabel, who is of Finnish descent and bakes up a storm.

I'm from a German family in which the grandmothers and mothers took care to pass along cooking lore to their children. My mom, Lorraine, still remembers learning from her mom and grandmother. I'm now passing on those skills and heritage recipes to my children and grandchildren. I especially love to take an old recipe, tweak it to my liking and mix new with old. Usually, they're surprisingly compatible.

Browsing through cookbooks and *Taste of Home* gives me plenty of new ideas. What a privilege to have been asked to be one of the magazine's first field editors, in 1993.

I often think of the *Taste of Home* "family" of cooks and the many smiles they have brought to our table. For when my family and friends enjoy my cooking, they are also complimenting the many fellow subscribers who have shared their recipes.

My wish is that this meal brings smiles and good eating to you!

From-Scratch Frosting

It's a good idea to sift confectioners' sugar before using it to make homemade frosting. If there are any lumps in the sugar, there will likely be lumps in the frosting. In addition, if you will be piping frosting onto the cake, lumps in the frosting may clog your decorating tips.

PICTURED AT LEFT: Beef Brisket with Mop Sauce, Comforting Potato Casserole, Tossed Salad with Carrot Dressing, Grandma's Honey Muffins and Poppy Seed Cake (recipes are on the next page).

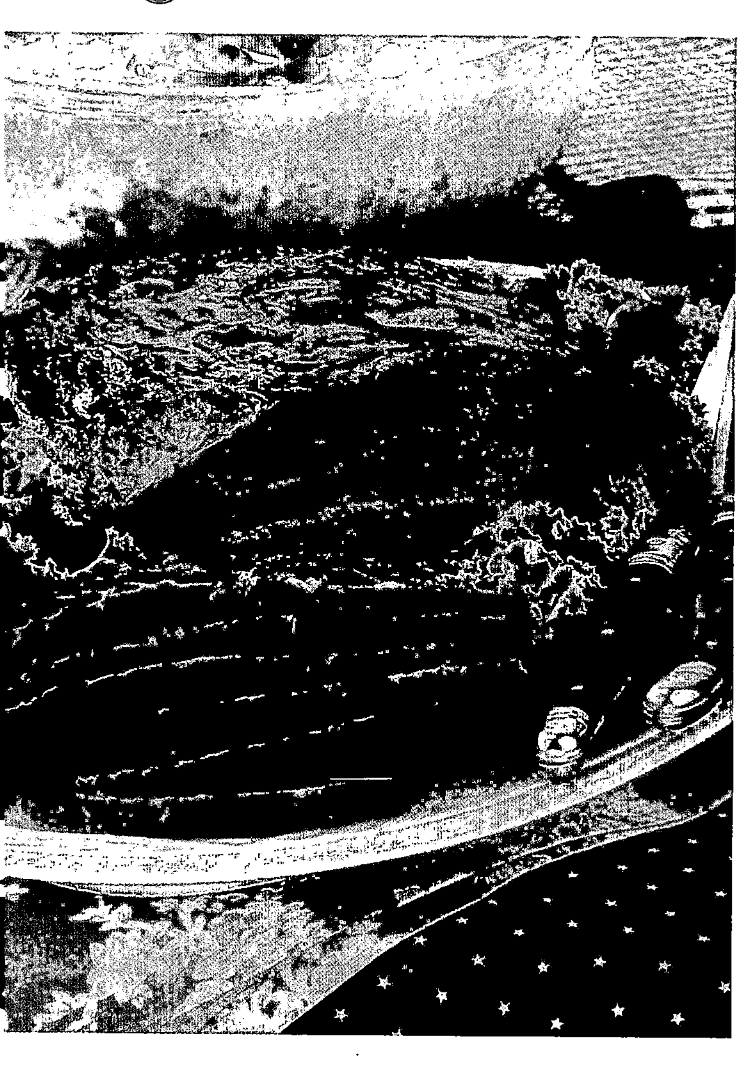

Beef Brisket with Mop Sauce

Prep: 20 min. **Bake:** 2 hours

- **1/2 cup water**
- **1/4 cup cider vinegar**
- **1/4 cup Worcestershire sauce**
- **1/4 cup ketchup**
- **1/4 cup dark corn syrup**
- **2 tablespoons vegetable oil**
- **2 tablespoons prepared mustard**
- **1 fresh beef brisket (3 pounds)**

In a large saucepan, combine the first seven ingredients. Bring to a boil, stirring constantly. Reduce heat; simmer for 5 minutes, stirring occasionally. Remove from the heat.

Place the brisket in a shallow roasting pan; pour sauce over the top. Cover and bake at 350° for 2 to 2-1/2 hours or until meat is tender. Let stand for 5 minutes. Thinly slice meat across the grain. **Yield:** 10-12 servings.

Editor's Note: This is a fresh beef brisket, not corned beef. The meat comes from the first cut of the brisket.

Comforting Potato Casserole

(Pictured on page 238)

Prep: 10 min. **Bake:** 1-1/2 hours

- **2 cans (10-3/4 ounces *each*) condensed cream of mushroom soup, undiluted**
- **2 cups (16 ounces) sour cream**
- **2 cups (8 ounces) shredded part-skim mozzarella cheese**
- **5 green onions, sliced**
- **1 package (32 ounces) frozen Southern-style hash brown potatoes, thawed**

In a large bowl, combine the soup, sour cream, cheese and onions; stir in potatoes until coated.

Transfer to a greased 2-1/2-qt. baking dish. Bake, uncovered, at 350° for 1-1/2 hours or until potatoes are tender. **Yield:** 10-12 servings.

Tossed Salad with Carrot Dressing

Prep/Total Time: 25 min.

- **3/4 cup red wine vinegar**
- **1 cup sugar**
- **2 celery ribs, cut into chunks**
- **1 small onion, cut into chunks**
- **1 small carrot, cut into chunks**
- **1/2 teaspoon salt**
- **3/4 cup vegetable oil**

SALAD:

- **8 cups spring mix salad greens**

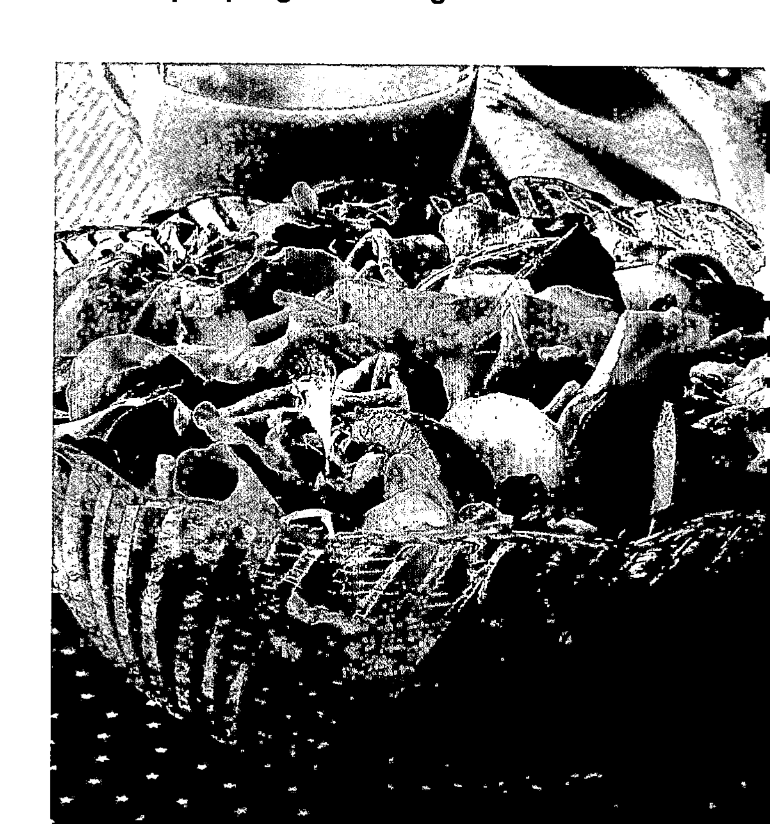

- 2 medium tomatoes, cut into wedges
- 1 medium cucumber, sliced
- 2 green onions, sliced
- 1/2 cup chow mein noodles
- 1/2 cup shredded cheddar cheese
- 1/2 cup dried cranberries

In a blender, combine the first six ingredients; cover and process until smooth. While processing, gradually add oil in a steady stream. Transfer to a bowl or small pitcher; cover and refrigerate until serving.

In a large bowl, combine the salad ingredients. Stir dressing and serve with salad. Refrigerate leftover dressing. **Yield:** 10 servings (3 cups dressing).

Grandma's Honey Muffins

Prep/Total Time: 30 min.

- **2 cups all-purpose flour**
- **1/2 cup sugar**
- **3 teaspoons baking powder**
- **1/2 teaspoon salt**
- **1 egg**
- **1 cup milk**
- **1/4 cup butter, melted**
- **1/4 cup honey**

In a bowl, combine the flour, sugar, baking powder and salt. In another bowl, whisk the egg, milk, butter and honey; stir into dry ingredients just until moistened.

Fill greased or paper-lined muffin cups three-fourths full. Bake at 400° for 15-18 minutes or until a toothpick comes out clean. Remove from pan to a wire rack. Serve warm. **Yield:** 1 dozen.

Poppy Seed Cake

Prep: 20 min. + standing **Bake:** 25 min. + cooling

- **1/3 cup poppy seeds**
- **1 cup milk**
- **4 egg whites**
- **3/4 cup shortening**
- **1-1/2 cups sugar**
- **1 teaspoon vanilla extract**
- **2 cups all-purpose flour**
- **2 teaspoons baking powder**

CREAM CHEESE FROSTING:

- **1 package (8 ounces) cream cheese, softened**
- **1/2 cup butter, softened**
- **1 teaspoon vanilla extract**
- **2 cups confectioners' sugar**

In a small bowl, soak poppy seeds in milk for 30 minutes. Place egg whites in a large mixing bowl; let stand at room temperature for 30 minutes.

In another large mixing bowl, cream the shortening, sugar and vanilla. Combine flour and baking powder; add to creamed mixture alternately with poppy seed mixture. Beat egg whites until soft peaks form; fold into batter.

Pour batter into a greased 13-in. x 9-in. x 2-in. baking dish. Bake at 375° for 25-30 minutes or until a toothpick inserted near the center comes out clean. Cool on a wire rack.

For frosting, in a small mixing bowl, beat the cream cheese, butter and vanilla until smooth. Gradually beat in confectioners' sugar. Spread over cake. Store in the refrigerator. **Yield:** 12-15 servings.

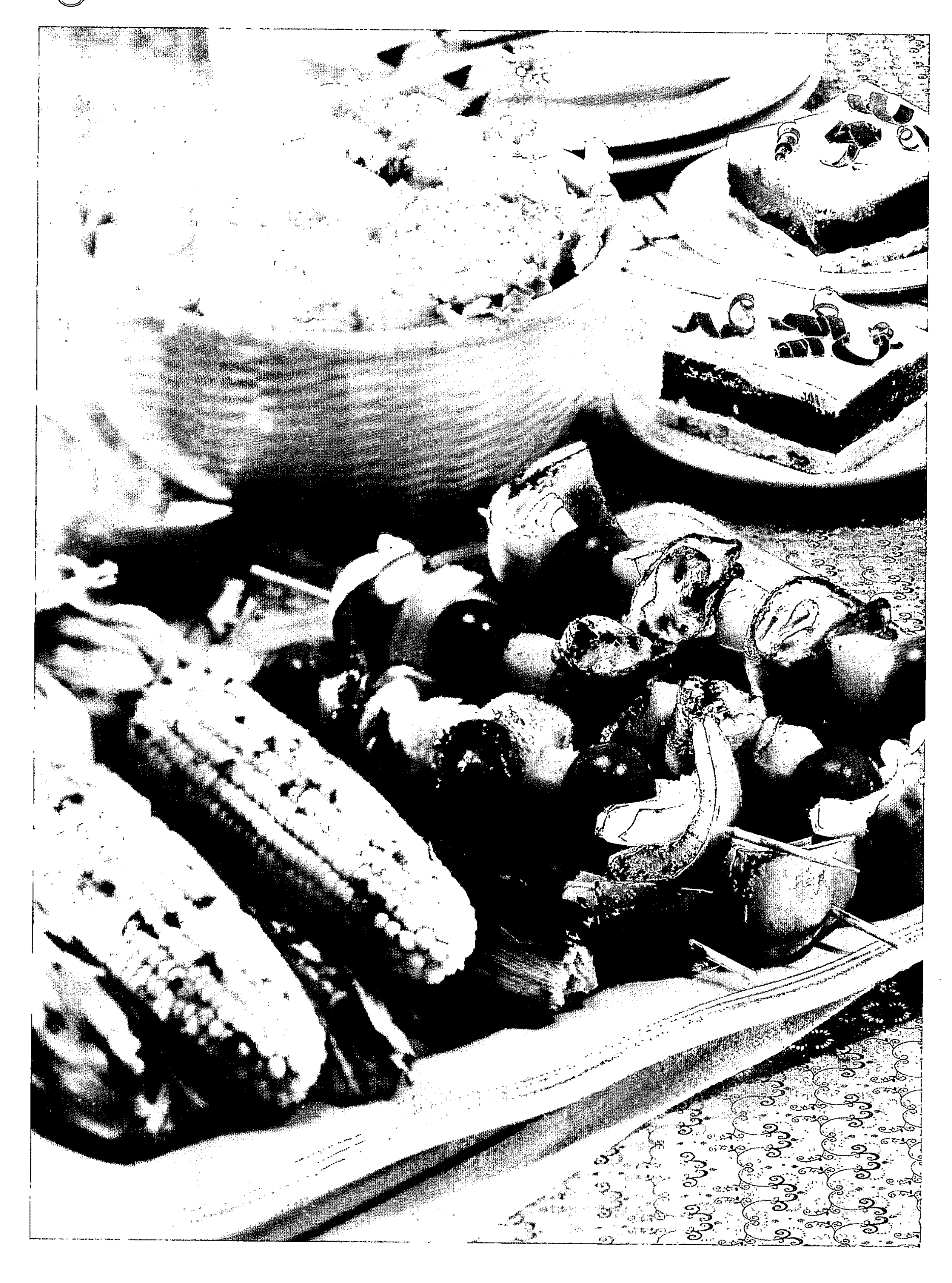

No-Fuss July 4th Cookout

This cool-headed editor escapes the heat of the kitchen during summer by relying on fabulous grilled fare.

By Angela Leinenbach, Mechanicsville, Virginia

AH, SUMMER! As kids, we longed for it because summer meant we were free from school. As adults, my husband, Stuart, and I crave the warm weather, the beauty of flowers blooming and the opportunity to get together with friends and family.

We always look forward to the Fourth of July holiday, as do our three children—Bethany, Thomas and Gabe—and our crazy English springer spaniel puppy, "Lady."

We meet at my in-laws' for a good old-fashioned cookout. On the menu are colorful Pineapple Chicken Kabobs, Dilly Potato Salad, Grilled Corn on the Cob and Black Forest Dream Dessert.

Grilling out means less time in a hot kitchen, less mess to clean up and more time for volleyball games, relaxation and fun!

Pineapple Chicken Kabobs resulted when I combined a couple of different recipes. The marinade does a terrific job of making the chicken and vegetables tasty and tender.

Our family absolutely loves this meal-on-a-stick, and it is also nice enough to serve when entertaining guests. I like to do all the prep work the night before. On the day of the cookout, it's a pleasure to have the main course all ready for the grill.

Is there anything better than fresh corn on the cob grilled to perfection? Last summer, my sister-in-law served it with a simple herb butter that made it even more delectable. Now, I don't want to eat my corn any other way!

To me, Fourth of July wouldn't be the same without potato salad. I've made Dilly Potato Salad over the years more times than I can count. It's definitely one of our best-liked hot-weather foods.

You probably have your own favorite potato salad, but I hope you will be adventurous and give this recipe a try. You'll be glad you did!

I do not have a large vegetable garden—just a few tomato plants—but I do like to grow a variety of herbs in a pot. It's not expensive, and I've been amazed at how fresh herbs add that perfect finishing touch to so many dishes.

Black Forest Dream Dessert will bring you raves...I promise! This rich, chilled dessert is a wonderful end to a summertime feast. The recipe makes a large panful, but I don't think you will have any complaints if there happen to be leftovers.

Summer is always exciting for us. One of the highlights is our town's annual Tomato Festival in July. We absolutely pride ourselves on growing the most delicious tomatoes in the area...and love finding tasty new ways to serve them.

Ripe tomatoes always bring back memories of when I was learning to cook. I can almost hear my grandmother as she taught me to make spaghetti sauce and meatballs, and my mother's laughter as we concocted recipes to try to win cooking contests.

Both were wonderful, loving women and fantastic cooks. Although they have passed away, I think of them so often when I am in the kitchen.

My grandmother bought me my first subscription to *Taste of Home* when the magazine first started. How thrilled she would have been to see this article and the accompanying recipes.

I am honored to be a *Taste of Home* field editor and to share recipes for my favorite meal. I hope you will enjoy it—on the Fourth or another fine summer day.

Selecting Sweet Corn

Look for corn that has fresh green, tightly closed husks with dark brown, dry (but not brittle) silk. The stem should be moist but not chalky, yellow or discolored.

Ears should have plump, tender, small kernels in tight rows up to the tip, and kernels should be firm enough to resist slight pressure. A fresh kernel will spurt "milk" if punctured.

PICTURED AT LEFT: Pineapple Chicken Kabobs, Dilly Potato Salad, Grilled Corn on the Cob and Black Forest Dream Dessert (recipes are on the next page).

Dilly Potato Salad

Prep: 40 min. + chilling

4 pounds red potatoes, halved
5 hard-cooked eggs
1 cup chopped dill pickles
1 small onion, chopped
1-1/2 cups mayonnaise
1 teaspoon celery seed
1/2 teaspoon salt
1/4 teaspoon pepper
Paprika

Place potatoes in a large kettle; cover with water. Bring to a boil. Reduce heat; cover and cook for 20-25 minutes or until tender. Drain and cool.

Cut potatoes into 3/4-in. cubes. Chop four eggs; slice remaining egg for garnish. In a large bowl, combine the potatoes, chopped eggs, pickles and onion.

In a small bowl, combine the mayonnaise, celery seed, salt and pepper. Pour over potato mixture and stir gently to coat. Sprinkle with paprika; garnish with sliced egg. Cover and refrigerate for at least 2 hours before serving. **Yield:** 12-14 servings.

Pineapple Chicken Kabobs

Prep: 30 min. + marinating **Grill:** 15 min.

2 cans (one 20 ounces, one 8 ounces) unsweetened pineapple chunks
1/3 cup Worcestershire sauce
8 boneless skinless chicken breast halves
1 package (1 pound) sliced bacon
2 large sweet onions
4 large green peppers
32 cherry tomatoes

Drain the pineapple, reserving juice; set pineapple aside. In a bowl, combine the Worcestershire sauce and reserved juice. Cut each piece of chicken into four strips. Pour 1 cup marinade into a large resealable plastic bag; add chicken. Seal bag and turn to coat; refrigerate for at least 1 hour. Cover and refrigerate remaining marinade for basting.

In a large skillet, cook bacon over medium heat until partially cooked but not crisp. Remove to paper towels to drain; cut in half widthwise. Cut each onion into 16 wedges; cut each pepper into eight pieces. Drain and discard marinade. Wrap a piece of bacon around each chicken strip.

On 16 metal or soaked wooden skewers, alternately thread the bacon-wrapped chicken, vegetables and pineapple. Grill, uncovered, over medium heat for 6-8 minutes on each side or until the juices run clear, basting frequently with the reserved marinade. **Yield:** 8 servings.

Grilled Corn on the Cob

Prep: 20 min. + soaking **Grill:** 25 min.

8 medium ears sweet corn
1/2 cup butter, softened
2 tablespoons minced fresh basil
2 tablespoons minced fresh parsley
1/2 teaspoon salt

Soak corn in cold water for 20 minutes. Meanwhile, in a small bowl, combine the butter, basil, parsley and salt. Carefully peel back cornhusks to within 1 in. of bottoms; remove silk. Spread butter mixture over corn.

Rewrap corn in husks and secure with kitchen string. Grill corn, covered, over medium heat for 25-30 minutes or until tender, turning occasionally. Cut string and peel back husks. **Yield:** 8 servings.

Black Forest Dream Dessert

Prep: 45 min. + chilling

1 cup all-purpose flour
2 tablespoons sugar
1/2 cup cold butter
1/2 cup flaked coconut
1/2 cup chopped walnuts, toasted
1 package (8 ounces) cream cheese, softened
1 cup confectioners' sugar
1 carton (8 ounces) frozen whipped topping, thawed, *divided*
1 can (21 ounces) cherry pie filling
1-1/2 cups semisweet chocolate chips
2-1/2 cups cold milk
2 packages (3.4 ounces *each*) instant vanilla pudding mix
Chocolate curls, optional

In a bowl, combine flour and sugar; cut in butter until crumbly. Stir in coconut and walnuts. Press into an ungreased 13-in. x 9-in. x 2-in. baking dish. Bake at 350° for 15-18 minutes or until lightly browned. Cool on a wire rack.

In a small mixing bowl, beat cream cheese until fluffy. Add confectioners' sugar; beat until smooth. Fold in 1 cup whipped topping. Spread over crust. Top with pie filling; cover and chill.

In a microwave-safe bowl, melt chocolate chips; stir until smooth. In a large bowl, whisk milk and pudding mixes for 2 minutes or until soft-set. Whisk a small amount of pudding into melted chocolate. Return all to the pudding, whisking constantly. Pour over cherry filling. Chill for 2 hours or until set.

Just before serving, spread remaining whipped topping over dessert. Garnish with chocolate curls if desired. **Yield:** 12 servings.

Dressed-Up Salmon for Guests

A flavorful marinade highlights the popular main course shared by this field editor and her husband.

By Krista Frank, Rhododendron, Oregon

WHEN we have company in the summer, I don't have to think twice about what to serve. Our favorite "impress-the-guests" meal is Double K Grilled Salmon, Herbed Onion Focaccia, Rice Noodle Salad and Summer Celebration Ice Cream Cake.

Each item is easy to prepare, and many parts of this menu can be made in advance, which is great for busy days.

My husband, Kevin, is a fly fisherman, and we love to cook his catch of salmon or steelhead. Double K Grilled Salmon is our "signature" dish.

After sampling it for the first time, one of Kevin's friends said he could eat this fish for the rest of his life and be perfectly happy. We are very proud of this homegrown recipe and playfully called it "Double K" for our two first names.

It came about when we looked for a spicy teriyaki-style marinade but couldn't find one we really liked. I checked cookbooks, taking note of ratios of sugar to vinegar and other ingredients used in marinades. We then combed our cupboards and drawers to see what we had in stock at home.

I started adding small amounts of all the ingredients until we came up with exactly what we wanted. I'm a big believer in not running to the store to buy one fancy ingredient I'll use only once. I always find a substitute, and if I can't, I just make something different.

Herbed Onion Focaccia is one of my old standbys that goes well with so many meals. I'd tried a bread recipe that called for just the onion. It was good but needed more oomph. Since I love basil and throw it into many dishes, I thought it might also spark up the bread.

Best served warm, its crust is crisp and salty. Inside, it's tender and pleasantly seasoned. The next day, it makes the best toast you've ever had! Over the years, I've given out this recipe more often than just about any other.

Although I make my Rice Noodle Salad more frequently in the summer, I serve it year-round. It's a fun and different side dish to take to a potluck. Sometimes, I add grilled or stir-fried chicken or beef to make it an all-in-one meal. I use its peanut dressing as a sauce in stir-fries, too.

I wanted to make my younger son an ice cream cake one year for his summer birthday, as he prefers ice cream to cake. He picked the ice cream flavors, and I used my brownies as a crust. Summer Celebration Ice Cream Cake has become his birthday tradition—we change the ice cream flavors depending on his favorites each particular year.

As an option, you can make individual servings of this refreshing dessert by baking the brownies in muffin tins.

My interest in cooking developed when Kevin and I were newlyweds with little extra money. I began watching a few local cooking shows on public TV with greater interest and checking out cookbooks and cooking magazines at the library. I also promised myself that each time I went to the store, I would buy one new ingredient or spice I hadn't tried before.

It was challenging to see how little I could spend at the grocery store and how far I could make the food go. I've come a long way as a cook as a result of experimenting while feeding my husband and our two sons, Michael and Jonathan.

My friends now call me for family-friendly meal ideas and guidance when they cook and bake—that makes me feel really great!

We live in a small community at the base of Mt. Hood, about 1 hour east of Portland. As a family, we like to hike mountain trails near our home. Many are historic wagon train routes. One trail not far from us takes you to a place where you can see rope burns in trees that were used as leverage in lowering wagons down a cliff.

Our area is wonderful for fly-fishing, camping and snowshoeing in the winter. In summer, we like to drive to the coast to go crabbing and spend time at the beach.

But no matter what the season, we love to share good food, like this favorite meal.

PICTURED AT LEFT: Double K Grilled Salmon, Herbed Onion Focaccia, Rice Noodle Salad and Summer Celebration Ice Cream Cake (recipes are on the next page).

Double K Grilled Salmon

Prep: 10 min. + marinating **Grill:** 20 min.

1/4 cup packed brown sugar
1/4 cup soy sauce
3 tablespoons unsweetened pineapple juice
3 tablespoons red wine vinegar
3 garlic cloves, minced
1 tablespoon lemon juice
1 teaspoon ground ginger
1/2 teaspoon pepper
1/2 teaspoon hot pepper sauce
1 salmon fillet (2 pounds)

In a small bowl, combine the first nine ingredients. Pour 3/4 cup into a large resealable plastic bag; add salmon. Seal bag and turn to coat; refrigerate for 1 hour, turning occasionally. Set aside remaining marinade for basting.

Coat grill rack with nonstick cooking spray before starting the grill. Drain and discard marinade. Place salmon skin side down on rack. Grill, covered, over medium heat for 5 minutes. Brush with reserved marinade. Grill 15-20 minutes longer or until fish flakes easily with a fork. **Yield:** 8 servings.

Herbed Onion Focaccia

Prep: 30 min. + rising **Bake:** 20 min.

1 cup water (70° to 80°)
1/3 cup finely chopped onion
1 tablespoon sugar
1-1/2 teaspoons salt
1 teaspoon grated Parmesan cheese
1/2 teaspoon garlic powder
1/2 teaspoon dried basil
1/2 teaspoon dill weed
1/2 teaspoon pepper
3 cups all-purpose flour
2 teaspoons active dry yeast

TOPPING:

1 tablespoon olive oil
1/2 teaspoon grated Parmesan cheese
1/2 teaspoon dried parsley flakes
1/4 teaspoon salt
1/8 teaspoon pepper

In bread machine pan, place the first 11 ingredients in order suggested by manufacturer. Select dough setting (check dough after 5 minutes of mixing; add 1 to 2 tablespoons of water or flour if needed).

When cycle is completed, turn dough onto a greased baking sheet and punch down (dough will be sticky). With lightly oiled hands, pat dough into a 9-in. circle. Brush with oil; sprinkle with Parmesan cheese, parsley, salt and pepper. Cover and let rise in a warm place until doubled, about 45 minutes.

Bake at 400° for 18-20 minutes or until golden brown. Cut into wedges and serve warm. **Yield:** 1 loaf (1-1/2 pounds).

Rice Noodle Salad

Prep/Total Time: 25 min.

1 package (8.8 ounces) thin rice noodles
2 cups fresh spinach, cut into strips
1 large carrot, shredded
1/2 cup pineapple tidbits
1/4 cup minced fresh cilantro
1 green onion, chopped
SESAME PEANUT DRESSING:
1/4 cup unsalted peanuts
1/4 cup water
1/4 cup lime juice
2 tablespoons soy sauce
1 tablespoon brown sugar
1 tablespoon vegetable oil
1 teaspoon sesame oil
1/2 teaspoon ground ginger
1/4 teaspoon crushed red pepper flakes

Cook noodles according to package directions. Meanwhile, in a large salad bowl, combine the spinach, carrot, pineapple, cilantro and onion.

In a blender, combine the dressing ingredients; cover and process until blended. Drain noodles and rinse in cold water; drain well. Add to spinach mixture. Drizzle with dressing and toss to coat. Serve immediately. **Yield:** 8-10 servings.

Summer Celebration Ice Cream Cake

(Also pictured on back cover)

Prep: 15 min. **Bake:** 20 min. + freezing

1 cup sugar
3 tablespoons butter, melted
3 tablespoons orange yogurt
1 egg
1 teaspoon grated orange peel
1 teaspoon vanilla extract
3/4 cup all-purpose flour
1/3 cup baking cocoa
1 cup (6 ounces) semisweet chocolate chips
1-3/4 quarts vanilla ice cream, softened
4 to 6 squares (1 ounce *each*) semisweet chocolate
1 tablespoon shortening
Mixed fresh berries

Line an 8-in. square baking dish with foil and grease the foil; set aside. In a large bowl, combine the sugar, butter, yogurt, egg, orange peel and vanilla until blended. Combine flour and cocoa; stir into sugar mixture. Add chocolate chips.

Spread into prepared dish. Bake at 325° for 20-25 minutes or until a toothpick inserted near the center comes out with moist crumbs. Cool on a wire rack.

Spread the ice cream over cake. Cover and freeze for 3 hours or until firm.

Remove from the freezer 10 minutes before serving. In a microwave-safe bowl, melt chocolate and shortening; stir until smooth. Using foil, lift dessert out of dish; gently peel off foil. Cut into squares. Garnish with berries and drizzle with chocolate. **Yield:** 9 servings.

Easy and Cheesy Supper

A few shortcuts don't compromise the rich, satisfying flavor of this editor's delightful dinner.

By Becky Ruff, Monona, Iowa

ALTHOUGH I love to cook, I don't always have the time for long, drawn-out preparations. This lasagna menu is one that doesn't tie me up for hours and hours in the kitchen but is applauded by everyone.

My Hot Cheddar Mushroom Spread, Golden Squash Soup, Simple Sausage Lasagna, Broccoli with Mustard Sauce and Caramel-Pecan Cheesecake Pie are tasty and satisfying yet easy to fix.

Several years ago, I got together with a group of friends from high school for the weekend, and each of us brought our favorite appetizer to snack on. One of the women made Hot Cheddar Mushroom Spread, and we couldn't get enough of it.

She happily shared the recipe for this rich and chunky spread, which I've made many times since for various family holiday get-togethers.

Golden Squash Soup made me change my tune about a wonderful autumn vegetable. My early memories of squash include sitting at the table when I was little while my mom tried to coax me into eating it. Even the addition of butter and brown sugar on top wouldn't do the trick.

So I had never been a squash-eater until I found this recipe in a newspaper a number of years ago. It sounded interesting, so I tried it. The mellow squash flavor mixed with the zip of onions and cheddar cheese made me a squash convert.

I think this soup is especially good for fall and winter meals. If I'm in a hurry, I microwave the squash to speed up the cooking time.

Simple Sausage Lasagna is a dish I "developed" after seeing a recipe in a church cookbook that did not call for precooking the noodles. I have loved lasagna since I can remember, but the traditional preparation was too time-consuming after getting home from work.

Now, even when I have enough time, I still don't cook the noodles first for lasagna. My kids, Tony and Katrina, love this convenient version, and it makes a great dish for family get-togethers or anytime. So, my feeling is, why fix anything else?

Easy but not ordinary, my Broccoli with Mustard Sauce has a unique blend of flavors. Tony loves this sauce on a variety of vegetables, and the only "trouble" I've ever had with this recipe was not making enough to please everyone.

An impressive treat with a pretty layered look, Caramel-Pecan Cheesecake Pie is irresistible and perfect for special meals. I've taken it to potlucks, and I always leave with an empty pie plate. Use your favorite homemade pie crust if you wish, or a ready-made frozen crust works well if time is short.

When putting together a yummy dessert like this pie, I think back to when I was about 3 years old and prepared "cakes" for my father to enjoy when he came home from work. I'm afraid, however, that the enjoyment was mainly mine since my cakes consisted of water mixed with every spice I could find in my mom's cupboard!

But Dad would always smack his lips at this concoction and praise my cooking abilities. As I got older, my parents let me keep experimenting in the kitchen—and I have retained my love for cooking. It's also safe to say that my culinary skills have dramatically improved since then!

Besides making meals for my family and working in the dietary department at a residential care facility, I look for other food-related opportunities as well. I've helped neighbors and relatives prepare food for large parties, and I've prepared and served food at church functions for up to 200 people.

A resident of northeast Iowa for most of my life, I can't imagine living anywhere else. Most people think of Iowa as being flat, but in this corner of the state, it's very hilly with scenic bluffs overlooking the Mississippi River. The fall season brings tourists (leaf-lookers) to this part of the state to take in the spectacular autumn colors.

It's also a time of year when a harvest of rich flavors—like those in this favorite meal of mine—are appreciated. Hope you'll give my recipes a try!

PICTURED AT LEFT: Hot Cheddar Mushroom Spread, Golden Squash Soup, Simple Sausage Lasagna, Broccoli with Mustard Sauce and Caramel-Pecan Cheesecake Pie (recipes are on the next page).

Hot Cheddar Mushroom Spread

Prep/Total Time: 25 min.

2 cups mayonnaise
2 cups (8 ounces) shredded cheddar cheese
2/3 cup grated Parmesan cheese
4 cans (4-1/2 ounces *each*) sliced mushrooms, drained
1 envelope ranch salad dressing mix
Minced fresh parsley
Assorted crackers

In a large bowl, combine the mayonnaise, cheeses, mushrooms and dressing mix. Spread into a greased 9-in. pie plate.

Bake, uncovered, at 350° for 20-25 minutes or until cheese is melted. Sprinkle with parsley. Serve with crackers. **Yield:** 3 cups.

Editor's Note: Reduced-fat or fat-free mayonnaise is not recommended for this recipe.

Golden Squash Soup

(Pictured on page 250)

Prep/Total Time: 30 min.

5 medium leeks (white portion only), sliced
2 tablespoons butter
1-1/2 pounds butternut squash, peeled, seeded and cubed (about 4 cups)
4 cups chicken broth
1/4 teaspoon dried thyme
1/4 teaspoon pepper
1-3/4 cups shredded cheddar cheese
1/4 cup sour cream
2 tablespoons thinly sliced green onion

In a large saucepan, saute leeks in butter until tender. Stir in the squash, broth, thyme and pepper. Bring to a boil. Reduce heat; cover and simmer for 10-15 minutes or until squash is tender. Cool slightly.

In a blender, cover and process squash mixture in small batches until smooth; return to the pan. Bring to a boil. Reduce heat to low. Add cheese; stir until soup is heated through and cheese is melted. Garnish with sour cream and onion. **Yield:** 6 servings.

Simple Sausage Lasagna

Prep: 20 min. **Bake:** 55 min. + standing

1 pound bulk pork sausage
1 jar (26 ounces) spaghetti sauce
1/2 cup water
2 eggs, beaten
1 carton (24 ounces) cottage cheese
1/3 cup grated Parmesan cheese
1 to 2 tablespoons dried parsley flakes
1/2 teaspoon *each* garlic powder, pepper, dried basil and oregano
9 uncooked lasagna noodles
3 cups (12 ounces) shredded part-skim mozzarella cheese

In a large skillet, cook sausage over medium heat until no longer pink; drain. Stir in spaghetti sauce and wa-

ter. Simmer, uncovered, for 10 minutes. Meanwhile, in a large bowl, combine the eggs, cottage cheese, Parmesan cheese, parsley and seasonings.

Spread 1/2 cup meat sauce into a greased 13-in. x 9-in. x 2-in. baking dish. Layer with three noodles, a third of the cheese mixture, meat sauce and mozzarella cheese. Repeat layers twice.

Cover and bake at 375° for 45 minutes. Uncover; bake 10 minutes longer or until noodles are tender. Let stand for 15 minutes before serving. **Yield:** 12 servings.

Broccoli with Mustard Sauce

Prep/Total Time: 25 min.

2 pounds fresh broccoli florets, cauliflowerets *or* sliced carrots
1/2 cup mayonnaise
1/3 cup milk
1/4 cup grated Parmesan cheese
1/4 cup shredded Swiss cheese
2 teaspoons lemon juice
2 teaspoons prepared mustard
Salt and pepper to taste

Place broccoli in a steamer basket; place in a large saucepan over 1 in. of water. Bring to a boil; cover and steam for 5-8 minutes or until crisp-tender.

Meanwhile, in a small microwave-safe bowl, combine the remaining ingredients. Cover and microwave at 50% power for 2 minutes or until heated through, stirring every 30 seconds. Drain broccoli. Serve with sauce. **Yield:** 4 servings.

Editor's Note: This recipe was tested in a 1,100-watt microwave.

Caramel-Pecan Cheesecake Pie

Prep: 15 min. **Bake:** 35 min. + chilling

1 package (8 ounces) cream cheese, softened
1/2 cup sugar
4 eggs
1 teaspoon vanilla extract
1 unbaked pastry shell (9 inches)
1-1/4 cups chopped pecans
1 cup caramel ice cream topping

In a small mixing bowl, beat the cream cheese, sugar, 1 egg and vanilla until smooth. Spread into pastry shell; sprinkle with pecans.

In a small bowl, whisk remaining eggs; gradually whisk in caramel ice cream topping until blended. Pour over pecans.

Bake at 375° for 35-40 minutes or until lightly browned (loosely cover with foil after 20 minutes if pie browns too quickly). Cool on a wire rack for 1 hour. Refrigerate for 4 hours or overnight before slicing. Refrigerate leftovers. **Yield:** 6-8 servings.

Meals in Minutes

Home-style cooking in a snap—it's as easy as relying on the fast and family-pleasing menus here. You'll find 12 complete dinners you can get on the table in a mere 30 minutes...or less!

MAKE IT SNAPPY: Clockwise from upper left: Catch Up to Mealtime with Seafood (p. 260), Combo Features Sub, Side and Smoothie (p. 264), Suppertime Goes South of the Border (p. 258) and Sizzling Steak Dinner Is Speedy (p. 270).

Mouth-Watering Menu Stars Saucy Chops

IN SPRING when the weather's turning warm, the kitchen is the last place you want to be. So spring into action and breeze through these quick dishes.

Ready in 30 minutes, the menu our Test Kitchen put together here features recipes from three cooks who give "fast food" a flavorful new meaning.

Tammy Messing of Ruth, Michigan shares Skillet Barbecued Pork Chops. "Between keeping up with our three children and helping my carpenter husband, Jeff, build our new home, I don't have time to stand around the stove," she notes.

"On days I volunteer at church or shuttle between after-school activities, I'm glad this dinner comes together in one skillet. The sauce makes the pork chops so moist and tender."

From Kansas City, Missouri, Elizabeth Freise sent the recipe for her Broccoli Garbanzo Salad, a fun and different side dish.

"This salad goes well with Mexican and Italian food," she says. "It's a nice change of pace from lettuce—and doesn't require much prep work.

"To make it a main dish, just add broiled chicken strips, your favorite cheese and extra veggies," she suggests.

In Arkansaw, Wisconsin, Brenda Drier serves Chocolate Rice Dessert as a fast finish to a meal. "Even with calves to feed and chores to finish, I can plan this quick treat for supper," confirms the dairy farm wife. "It's so yummy, it's worth it!"

Skillet Barbecued Pork Chops

Prep: 10 min. **Cook:** 15 min.

- **4 boneless pork loin chops (1/2 inch thick)**
- **1 teaspoon seasoned salt**
- **1 tablespoon butter**
- **1 medium onion, chopped**
- **1/2 cup water**
- **1/2 cup packed brown sugar**
- **1 cup honey barbecue sauce**
- **1 tablespoon Worcestershire sauce**
- **2 teaspoons cornstarch**
- **1 tablespoon cold water**

Sprinkle pork chops with seasoned salt. In a large skillet, brown chops on both sides in butter over medium-high heat. Remove chops. In the drippings, saute onion until golden brown. Add the water, brown sugar, barbecue sauce and Worcestershire sauce. Return chops to the skillet. Bring to a boil. Reduce heat; cover and simmer for 15 minutes or until meat juices run clear. Remove chops; keep warm.

Combine cornstarch and cold water until smooth; stir into skillet. Bring to a boil; cook and stir for 2 minutes or until thickened. Serve sauce over pork. **Yield:** 4 servings.

Broccoli Garbanzo Salad

Prep: 10 min. + chilling

- **1 package (12 ounces) broccoli florets**
- **1 can (15 ounces) garbanzo beans *or* chickpeas, rinsed and drained**
- **1 cup pitted ripe olives, drained**
- **1/3 to 1/2 cup Italian salad dressing**

Place broccoli in a steamer basket over 1 in. of boiling water in a saucepan. Cover and steam for 5-6 minutes or until tender. Rinse with cold water; drain and place in a bowl. Add the garbanzo beans and olives. Drizzle with dressing; toss to coat. Refrigerate until serving. **Yield:** 4-6 servings.

Chocolate Rice Dessert

Prep/Total Time: 10 min.

- **4-1/2 teaspoons sugar**
- **1 teaspoon ground cinnamon**
- **1/8 teaspoon salt**
- **3 cups cooked long grain rice, warmed**
- **1/4 cup hot fudge sauce *or* chocolate ice cream topping**
- **1/4 cup whipped topping**
- **4 maraschino cherries**

In a bowl, combine the sugar, cinnamon and salt. Add rice and mix well. Divide among four dessert dishes or parfait glasses. Top with hot fudge sauce. Dollop with whipped topping and garnish with maraschino cherries. **Yield:** 4 servings.

Suppertime Goes South of the Border

HECTIC LIFESTYLES require simple solutions when it comes to getting food on the table for your busy family. The home economists in our Test Kitchen put together this speedy, satisfying meal using three reader favorites. Featuring a spiced-up main dish, citrusy side and delectable dessert, this complete dinner is sure to become a fast favorite at your home, too!

"We often enjoy Mexican Pork Chops over rice to catch every drop of the flavorful sauce," says field editor Nancy Negvesky of Somerville, New Jersey. "You can use mild, medium or hot salsa, depending on your preference. If the pork chops are too spicy for you, just eliminate the cumin-chili powder rub."

"I enjoyed a delicious broccoli dish in Mexico a few years ago and tried making it in my own kitchen. I think my Broccoli with Lemon Sauce tastes very much like what I had at the restaurant," writes Nancy Larkin, a field editor in Maitland, Florida.

A quick, homemade chocolate sauce is layered with sweet berries and tender pound cake for Individual Strawberry Trifles, shared by Karen Scaglione of Nanuet, New York. "I sometimes garnish the desserts with confectioners' sugar," Karen notes.

Mexican Pork Chops

Prep/Total Time: 25 min.

✓ **Uses less fat, sugar or salt. Includes Nutrition Facts and Diabetic Exchanges.**

- **1 teaspoon ground cumin**
- **1 teaspoon chili powder**
- **4 boneless pork loin chops (4 ounces *each* and 1/2 inch thick)**
- **1 tablespoon vegetable oil**
- **1-1/4 cups salsa**
- **1 teaspoon baking cocoa**
- **1/8 teaspoon ground cinnamon**
- **2 tablespoons minced fresh cilantro**
- **1 green onion, chopped**

Combine cumin and chili powder; rub over both sides of pork. In a large skillet, brown pork chops in oil on both sides over medium heat.

In a small bowl, combine the salsa, cocoa and cinnamon; pour over pork. Bring sauce to a boil. Reduce heat; simmer, uncovered, for 8-10 minutes or until meat is tender, turning chops once and stirring sauce occasionally. Sprinkle with cilantro and green onion. **Yield:** 4 servings.

Nutrition Facts: 1 pork chop with 1/3 cup sauce equals 213 calories, 10 g fat (3 g saturated fat), 55 mg cholesterol, 390 mg sodium, 4 g carbohydrate, 1 g fiber, 22 g protein. **Diabetic Exchanges:** 3 lean meat, 1 vegetable, 1/2 fat.

Broccoli with Lemon Sauce

Prep/Total Time: 20 min.

- **1 bunch broccoli, cut into spears**
- **1/2 cup sour cream**
- **3 to 4 tablespoons milk**
- **1 teaspoon lemon juice**
- **1/2 teaspoon grated lemon peel**

Place broccoli in a steamer basket; place in a large saucepan over 1 in. of water. Bring to a boil; cover and steam for 5-7 minutes or until crisp-tender.

Meanwhile, in a small microwave-safe bowl, combine the sour cream, milk, lemon juice and peel. Microwave, uncovered, at 50% power for 1-1/2 minutes or until heated through, stirring every 30 seconds. Serve with broccoli. **Yield:** 4 servings.

Editor's Note: This recipe was tested in a 1,100-watt microwave.

Individual Strawberry Trifles

Prep/Total Time: 20 min.

- **1/2 cup semisweet chocolate chips**
- **1/2 cup heavy whipping cream**
- **2 tablespoons orange juice**
- **2 cups sliced fresh strawberries**
- **4 slices pound cake, cubed**

In a small saucepan, melt the chocolate chips with the heavy whipping cream over low heat; stir until smooth. Remove from the heat; stir in the orange juice. Cool to room temperature.

In four dessert glasses or bowls, layer the strawberries, pound cake cubes and chocolate mixture. **Yield:** 4 servings.

Catch Up to Mealtime With Seafood

RUNNING BEHIND? Then you need a quick-fix meal that's ready to serve when your family walks in the door. You won't have to "fish" for compliments with this mouth-watering menu, which is on the table in just 30 minutes and full of great flavors, too.

Ruth Hayward's refreshing Strawberry-Bacon Spinach Salad is sweet and crunchy with a tangy dressing. "I made this recipe for our prayer group, and everyone enjoyed it," Ruth writes from Lake Charles, Louisiana.

"An onion soup and sour cream mixture really adds zip to Busy-Day Baked Fish," says Beverly Krueger of Yamhill, Oregon. "Your family would never guess that it's so quick and easy to prepare."

Field editor Marian Platt of Sequim, Washington shares her luscious Coffee Whip Dessert, a smooth and creamy finale fit for everyday meals or even special occasions...and it takes just minutes to fix!

Strawberry-Bacon Spinach Salad

Prep/Total Time 15 min.

1 package (6 ounces) fresh baby spinach
1 pint fresh strawberries, sliced
8 bacon strips, cooked and crumbled
1/4 cup chopped red onion
1/4 cup chopped walnuts
1 cup mayonnaise
1/2 cup sugar
1/4 cup raspberry vinegar

In a salad bowl, combine the spinach, strawberries, bacon, onion and walnuts. In a small bowl or pitcher, combine the mayonnaise, sugar and vinegar. Serve with salad. **Yield:** 6-8 servings.

Busy-Day Baked Fish

Prep/Total Time: 30 min.

1 cup (8 ounces) sour cream
2 tablespoons onion soup mix
1-1/2 cups seasoned bread crumbs
2-1/2 pounds fresh *or* frozen fish fillets, thawed
1/4 cup butter, melted
1/3 cup shredded Parmesan cheese

In a shallow bowl, combine sour cream and soup mix. Place bread crumbs in another shallow bowl. Cut fish into serving-size pieces; coat with the sour cream mixture, then roll in the crumbs.

Place in two greased 13-in. x 9-in. x 2-in. baking dishes. Drizzle with butter. Bake, uncovered, at 425° for 12 minutes. Sprinkle with Parmesan cheese; bake 2-6 minutes longer or until fish flakes easily with a fork. **Yield:** 6-8 servings.

Coffee Whip Dessert

Prep/Total Time: 30 min.

1 cup water
2 tablespoons instant coffee granules
6-1/2 cups miniature marshmallows
1 cup heavy whipping cream
Whipped cream and additional instant coffee granules, optional

In a large saucepan, bring water to a boil. Remove from the heat; stir in coffee. Add marshmallows; cook for 5-6 minutes over low heat until marshmallows are melted, stirring occasionally. Pour into a large bowl; cover and refrigerate until slightly thickened.

In a small mixing bowl, beat the heavy whipping cream until soft peaks form; fold into the marshmallow mixture. Spoon into individual dessert dishes. Garnish with whipped cream and additional coffee granules if desired. **Yield:** 8 servings.

Coffee Clues

Do you like to keep containers of instant coffee in your pantry to use in recipes, such as Coffee Whip Dessert (above)? Keep the following storage tips in mind:

Unopened containers of instant coffee may be stored at room temperature for up to a year. Once the package has been opened, the coffee loses its flavor quickly. Store opened packages in the refrigerator for up to 3 weeks.

Chicken and Veggies Please Families

YOU JUST GOT HOME, the clock's ticking and the rest of the family will be home soon. Don't worry! You won't disappoint them with this appealing, nutritious and quick menu.

Apricot Honey Chicken from Kathy Hawkins of Gurnee, Illinois boasts a sweet, fruity sauce. There's no need to heat up the oven for this flavorful entree, so you can even enjoy it during the summer.

"I saw something similar to Favorite Herbed Potatoes prepared on TV and decided to make my own version using the herbs I had on hand," says Naomi Olson of Hamilton, Michigan. "To cut prep time, I substituted canned potatoes for fresh ones, and we like it just as much."

Gingered Cranberry-Carrot Slaw gives this meal a nice tang with cranberries and a sweet-tart pineapple dressing. Genise Krause of Sturgeon Bay, Wisconsin shared the refreshing recipe.

Apricot Honey Chicken

Prep/Total Time: 25 min.

✓ **Uses less fat, sugar or salt. Includes Nutrition Facts and Diabetic Exchanges.**

4 boneless skinless chicken breast halves (5 ounces *each*)
1 tablespoon canola oil
3 tablespoons apricot preserves
2 tablespoons orange juice
4 teaspoons honey

In a large skillet, cook chicken in oil over medium heat for 7-9 minutes on each side or until juices run clear. Combine the preserves, orange juice and honey; pour over chicken. Cook for 2 minutes or until heated through. **Yield:** 4 servings.

Nutrition Facts: 1 chicken breast half equals 243 calories, 7 g fat (1 g saturated fat), 78 mg cholesterol, 74 mg sodium, 16 g carbohydrate, trace fiber, 29 g protein. **Diabetic Exchanges:** 4 very lean meat, 1 starch, 1 fat.

Favorite Herbed Potatoes

Prep/Total Time: 20 min.

✓ **Uses less fat, sugar or salt. Includes Nutrition Facts and Diabetic Exchanges.**

2 cans (15 ounces *each*) whole potatoes, drained and halved lengthwise
1 tablespoon butter
1 tablespoon olive oil
1-1/2 teaspoons dried parsley flakes
1/2 teaspoon dried basil
1/2 teaspoon dried thyme
1/2 teaspoon rubbed sage
1/2 teaspoon dried rosemary, crushed
1/2 teaspoon garlic powder
1/4 teaspoon paprika
1/4 teaspoon pepper

In a large skillet over medium heat, cook potatoes in butter and oil for 6-8 minutes or until browned. Add the remaining ingredients; cook and stir for 2-3 minutes or until well coated. **Yield:** 4 servings.

Nutrition Facts: 3/4 cup equals 152 calories, 7 g fat (2 g saturated fat), 8 mg cholesterol, 491 mg sodium, 22 g carbohydrate, 3 g fiber, 3 g protein. **Diabetic Exchanges:** 1-1/2 starch, 1 fat.

Gingered Cranberry-Carrot Slaw

Prep/Total Time: 10 min.

✓ **Uses less fat, sugar or salt. Includes Nutrition Facts and Diabetic Exchanges.**

1/2 cup unsweetened pineapple juice
2 tablespoons honey
1 tablespoon lemon juice
1 teaspoon cider vinegar
3 cups shredded carrots
1/2 cup dried cranberries
1 to 2 tablespoons minced fresh gingerroot

In a large bowl, whisk the unsweetened pineapple juice, honey, lemon juice and cider vinegar until smooth. Add the carrots, dried cranberries and ginger; toss to coat. Cover and refrigerate slaw until serving. **Yield:** 4 servings.

Nutrition Facts: 3/4 cup equals 133 calories, trace fat (trace saturated fat), 0 cholesterol, 30 mg sodium, 35 g carbohydrate, 3 g fiber, 1 g protein. **Diabetic Exchanges:** 1-1/2 starch, 1 vegetable.

Combo Features Sub, Side and Smoothie

FOR A REFRESHING summertime (or anytime) meal, try this perky menu compiled by our Test Kitchen. You can put some of your garden harvest to good use with these reader recipes, plus have supper ready in a snap.

"I came up with Hot Italian Ham Subs after tasting a delicious deli sandwich while traveling," says Leann Hillmer of Sylvan Grove, Kansas. Tomato and basil add color and flavor to her hearty recipe.

Jennifer Yoder writes from Amberg, Wisconsin, "I love fresh vegetables, especially zucchini. My mother-in-law's recipe for Zucchini Mozzarella Medley is a wonderful way to fix it."

"Strawberry Orange Smoothies came out of one of my experiments," writes Deb MacNeil, Westminster, Colorado. "If you make this recipe ahead of time, place it in a pitcher in the fridge and stir it for an additional 10-15 seconds before serving."

Strawberry Orange Smoothies

Prep/Total Time: 15 min.

✓ **Uses less fat, sugar or salt. Includes Nutrition Facts.**

3 cups orange juice
1 carton (8 ounces) strawberry yogurt
6 fresh strawberries
1 can (8 ounces) crushed pineapple, drained
1 medium firm banana, cut into chunks

Place half of each ingredient in a blender; cover and process until blended. Pour into chilled glasses. Repeat with remaining ingredients. Serve immediately. **Yield:** 6 servings.

Nutrition Facts: 1 cup equals 144 calories, 1 g fat (trace saturated fat), 2 mg cholesterol, 21 mg sodium, 33 g carbohydrate, 1 g fiber, 2 g protein.

Shake Up Smoothies

Feel free to experiment with whatever ingredients you may have on hand when making smoothies. For example, try substituting yogurt of a different flavor, another kind of fruit juice or different fresh, frozen or canned fruits.

Hot Italian Ham Subs

Prep/Total Time: 20 min.

1/2 cup olive oil
1/4 cup red wine vinegar
1 tablespoon Dijon mustard
1 teaspoon sugar
1/2 teaspoon pepper
1/2 teaspoon minced fresh basil
1/4 teaspoon salt
4 submarine buns, split
12 slices deli ham
12 slices tomato
12 slices part-skim mozzarella cheese
12 fresh basil leaves
4 slices sweet onion, halved
8 slices provolone cheese

In a jar with a tight-fitting lid, combine the first seven ingredients; shake well. Generously drizzle dressing over cut sides of buns.

On bun bottoms, layer the ham, tomato and mozzarella cheese. On bun tops, layer the basil leaves, onion and provolone cheese. Place on a baking sheet.

Broil 6-8 in. from the heat for 2-3 minutes or until cheese is melted. Place bun tops over bottoms; serve immediately. **Yield:** 4 servings.

Zucchini Mozzarella Medley

Prep/Total Time: 25 min.

3 cups sliced zucchini
1 cup sliced onion
1 garlic clove, minced
1/2 teaspoon dried basil
1/4 teaspoon dried oregano
1/8 teaspoon salt
3 tablespoons butter
1 medium tomato, cut into 12 wedges
1 cup (4 ounces) shredded part-skim mozzarella cheese

In a large skillet, saute the zucchini, onion, garlic and seasonings in butter until crisp-tender.

Gently stir in tomato wedges; sprinkle with cheese. Remove from the heat; cover and let stand for 1-2 minutes or until cheese is melted. **Yield:** 4 servings.

Have a Fiesta With Mexican Spread

PRESSED for time? It's tempting to stop at a fast food restaurant, but this zippy menu is better for you...and you'll have it ready in just 30 minutes.

For the entree, enjoy Tex-Mex Pork Chops from JoAnn Dalrymple of Claremore, Oklahoma. Complete the meal with Chili Cheddar Penne, shared by Aaron Werner of Madison, Wisconsin, and Zucchini Bean Salad from Carol Waugh, Bellingham, Washington.

Tex-Mex Pork Chops

Prep/Total Time: 20 min.

✓ Uses less fat, sugar or salt. Includes Nutrition Facts and Diabetic Exchanges.

Butter-flavored nonstick cooking spray
1 small onion, chopped
6 boneless pork loin chops (5 ounces *each*)
1 cup salsa
1 can (4 ounces) chopped green chilies
1/2 teaspoon ground cumin
1/4 teaspoon pepper

In a large skillet coated with butter-flavored spray, saute onion until tender. Add pork chops; cook over medium heat for 5-6 minutes on each side or until juices run clear.

Combine the salsa, chilies, cumin and pepper; pour over pork. Bring to a boil. Reduce heat; cover and simmer until heated through. **Yield:** 6 servings.

Nutrition Facts: 1 pork chop equals 223 calories, 8 g fat (3 g saturated fat), 68 mg cholesterol, 433 mg sodium, 9 g carbohydrate, 5 g fiber, 32 g protein. **Diabetic Exchanges:** 3 lean meat, 1 vegetable, 1 fat.

Chili Cheddar Penne

Prep/Total Time: 25 min.

1-1/3 cups uncooked penne pasta
4 teaspoons butter
4 teaspoons all-purpose flour
1 cup milk
2 cups (8 ounces) shredded cheddar cheese
4 teaspoons taco seasoning
1/4 teaspoon salt
2/3 cup frozen corn, thawed
2/3 cup chopped fresh tomatoes
1 can (4 ounces) chopped green chilies, drained
Sliced avocado, optional

Cook pasta according to package directions. Meanwhile, in a large saucepan, melt butter over medium heat. Stir in flour until smooth; gradually add milk. Bring to a boil; cook and stir for 2 minutes or until thickened.

Reduce heat to medium. Stir in the cheddar cheese, taco seasoning and salt. Cook and stir for 2-3 minutes or until the cheese is melted. Drain the pasta; stir into the cheese sauce. Cook and stir for 3 minutes or until heated through.

Stir in the corn, tomatoes and green chilies just until combined. Garnish with sliced avocado if desired. **Yield:** 6 servings.

Zucchini Bean Salad

Prep/Total Time: 30 min.

✓ Uses less fat, sugar or salt. Includes Nutrition Facts and Diabetic Exchanges.

1 cup cut fresh green beans
1 can (16 ounces) kidney beans, rinsed and drained
1-1/2 cups thinly sliced halved zucchini
1 medium green pepper, julienned
3 green onions, thinly sliced
3 tablespoons cider vinegar
2 tablespoons canola oil
3/4 teaspoon sugar
3/4 teaspoon seasoned salt
1/4 teaspoon pepper

Place green beans in a small saucepan and cover with water. Bring to a boil; cover and cook for 8-10 minutes or until crisp-tender. Drain and rinse in cold water.

In a large bowl, combine the green beans, kidney beans, zucchini, green pepper and onions. In a small bowl, whisk the vinegar, oil, sugar, seasoned salt and pepper. Pour over bean mixture; toss to coat. Cover and chill until serving. **Yield:** 6 servings.

Nutrition Facts: 3/4 cup equals 127 calories, 5 g fat (trace saturated fat), 0 cholesterol, 315 mg sodium, 17 g carbohydrate, 5 g fiber, 6 g protein. **Diabetic Exchanges:** 1 starch, 1 vegetable, 1 fat.

Beef-and-Biscuit Supper Is Satisfying

IN THE MARKET for some quick and simple dinner recipes? You'll soon be sold on Melody Smaller's approach to cooking.

"My days stay full running a discount general store in the front of our home here in Fowler, Colorado," Melody notes. And after closing her business for the day, she wastes no time shopping for something fast, filling and flavorful to whip up for supper.

Quick dishes, such as those Melody shares here, are just the ticket for time-conscious cooks. Like all the menus served up in this chapter, her meal is on the table in 30 minutes or less.

"I created these speedy Veggie Beef Patties using ingredients I had on hand," she explains. "They make a nice entree for a dinner or potluck. Or, you can serve them with barbecue sauce on a bun for a casual lunch or cookout.

"Zucchini Lettuce Salad is a great way to use up an abundant crop of squash, and shredding the veggies is a breeze with a food processor. Topped with bottled dressing and cheese, this colorful side is a flavorful addition to any meal."

To round out the dinner, Melody stirs up Paprika Cheese Biscuits. "My husband, Tom, loves their cheddar flavor," Melody says. "They're so tender, I eat them as a snack."

As for Melody's customers, they sometimes leave the store with more than they bargained for. She's always happy to pass along a labor-saving recipe, free of charge.

"To lighten up the burgers, substitute ground turkey for beef," Melody suggests. "You could also stuff them with a vegetable-rice mix or mushrooms and cheese."

To shake up her salad, try homemade vinegar and oil dressing. Add chopped tomatoes, pimientos or red peppers to boost the color.

Keen on Zucchini

To choose the freshest zucchini, look for a firm, heavy squash with a moist stem end and shiny skin. Generally, smaller squash are sweeter and more tender than larger ones.

One medium (1/3 pound) zucchini yields about 1-1/2 cups shredded or 2 cups sliced zucchini. Store zucchini in a plastic bag in the refrigerator crisper for up to 4 days. Leave it unwashed until you're ready to use it.

Veggie Beef Patties

Prep/Total Time: 25 min.

1/4 cup grated carrot
1/4 cup finely chopped onion
1/4 cup finely chopped green pepper
1 pound ground beef
Salt and pepper to taste

In a bowl, combine the carrot, onion and green pepper. Crumble beef over mixture and mix well. Shape into four patties. Season with salt and pepper. Pan-fry over medium heat until meat is no longer pink. **Yield:** 4 servings.

Zucchini Lettuce Salad

Prep/Total Time: 10 min.

2 cups shredded leaf lettuce
1/2 cup shredded zucchini
1/2 cup sliced ripe olives
1/4 cup chopped red onion
1/2 cup Italian salad dressing
1/4 cup shredded Parmesan cheese

In a serving bowl, combine the lettuce, zucchini, olives and onion. Drizzle with dressing; sprinkle with Parmesan cheese. Serve immediately. **Yield:** 4 servings.

Paprika Cheese Biscuits

Prep/Total Time: 20 min.

2-1/4 cups biscuit/baking mix
1/2 cup shredded cheddar cheese
2/3 cup milk
1 tablespoon butter, melted
1/2 teaspoon paprika

In a bowl, combine the biscuit mix and cheese. With a fork, stir in milk just until moistened. Turn onto a floured surface; knead 10 times. Roll dough to 1/2-in. thickness; cut with a 2-1/2-in. biscuit cutter. Place on an ungreased baking sheet. Brush with butter; sprinkle with paprika. Bake at 450° for 8-10 minutes or until golden brown. **Yield:** 8 biscuits.

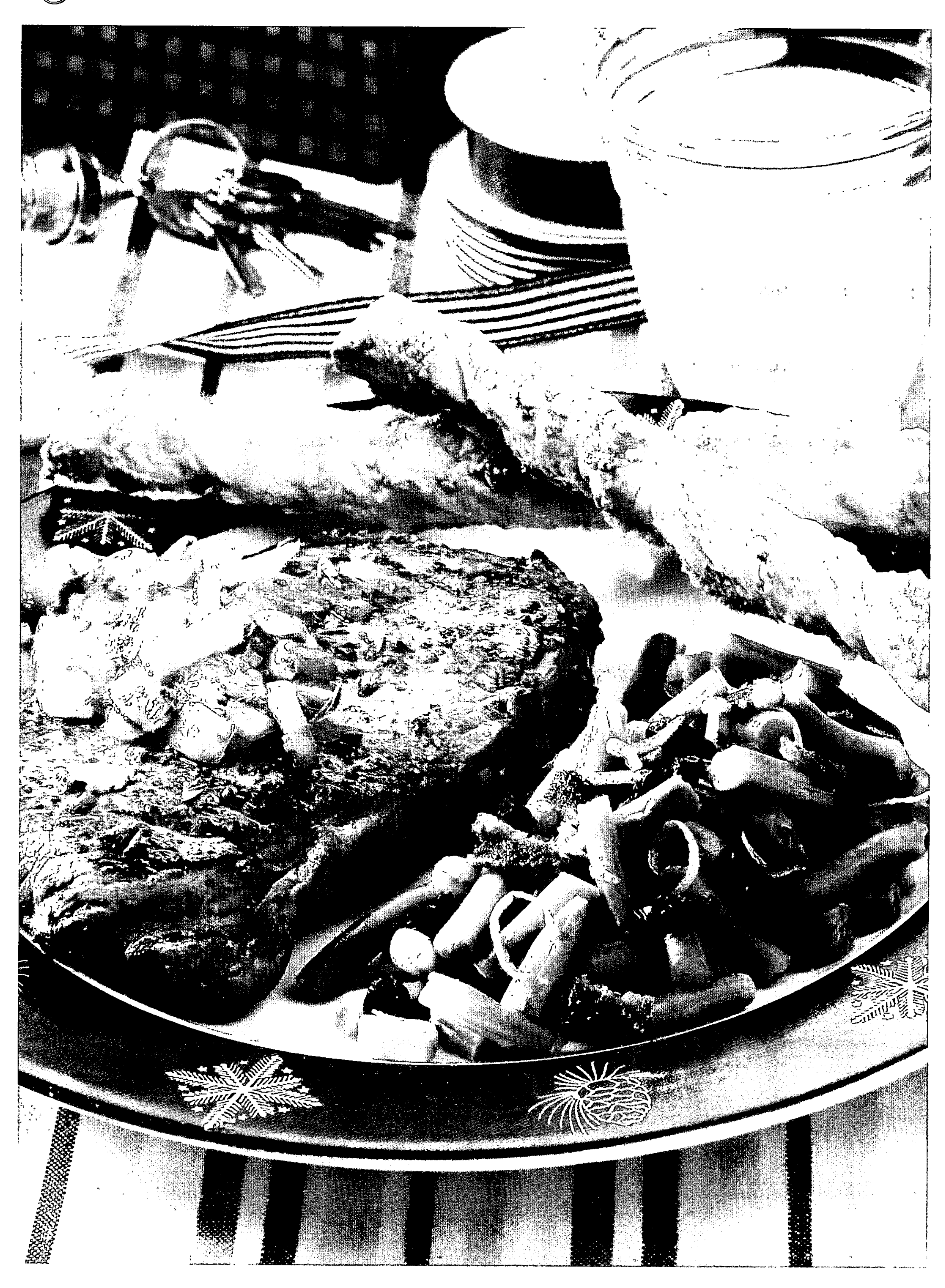

Sizzling Steak Dinner Is Speedy

DURING the Christmas season, busy days of cookie baking, shopping and gift wrapping leave precious little time for dinner preparation. This simple but satisfying meal, consisting of three reader-favorite dishes, can be on the table in just 30 minutes...and still win rave reviews from the whole family!

"Saucy Skillet Steaks couldn't be easier to make," relates Karen Haen of Sturgeon Bay, Wisconsin. "Though I prefer these juicy rib eyes, I have also used this versatile recipe for everything from chicken breasts and fish to veal and hamburgers. We love the tasty onion topping."

"To make my Holiday Green Beans festive red and green for Christmastime, I add dried cranberries," notes field editor Darlene Brenden from her kitchen in Salem, Oregon. "A touch of sweet honey complements the cranberries' tartness and the tangy hint of citrus."

Serve the quick-to-fix Cheddar Bread Twists as a side dish or as an appetizer with your favorite dip. Tracy Travers of Fairhaven, Massachusetts shared the no-fuss recipe, which relies on convenient frozen puff pastry and just four other ingredients.

Saucy Skillet Steaks

Prep/Total Time: 30 min.

4 beef rib eye steaks (3/4 inch thick)
1 large onion, chopped
4 garlic cloves, minced
1/4 cup butter, cubed
2 tablespoons Dijon mustard
Salt and pepper to taste
1/3 cup beef broth
1 tablespoon minced fresh parsley

In a large nonstick skillet, brown the beef rib eye steaks over medium-high heat for 1-2 minutes on each side. Remove and keep warm. In the same skillet, saute the onion and garlic in butter until tender, stirring to loosen browned bits.

Brush the steaks with mustard; sprinkle with salt and pepper. Return to the pan. Stir in broth. Cook for 5-7 minutes on each side or until meat reaches desired doneness (for medium-rare, a meat thermometer should read 145°; medium, 160°; well-done, 170°). Spoon the onion mixture over steaks; sprinkle with parsley. **Yield:** 4 servings.

Holiday Green Beans

Prep/Total Time: 15 min.

1 package (16 ounces) frozen cut green beans
1 teaspoon grated orange peel
1/2 cup dried cranberries
1/2 cup real bacon bits
2 tablespoons honey

Cook the green beans according to package directions, adding the orange peel during cooking; drain. Add the cranberries, bacon and honey; toss to combine. **Yield:** 4 servings.

Cheddar Bread Twists

Prep/Total Time: 25 min.

1 sheet frozen puff pastry, thawed
1 egg white
1 tablespoon cold water
1/2 cup shredded cheddar cheese
Dash salt

Place the puff pastry on a greased baking sheet. In a small bowl, beat the egg white and water; brush over pastry. Sprinkle with cheese and salt.

Cut into ten 1-in. strips; twist each strip three times. Bake at 400° for 10-13 minutes or until golden brown. **Yield:** 10 breadsticks.

About Puff Pastry

Puff pastry dough is a rich dough made by placing chilled butter between layers of pastry dough, then rolling it out, folding it into thirds and allowing it to rest. This process is repeated six to eight times, producing a pastry with many layers of dough and butter.

Commercially prepared puff pastry dough is the perfect time-saving choice for recipes such as Cheddar Bread Twists (above). It also can be used in recipes that call for homemade puff pastry dough, such as croissants.

Skillet Dinner Fits Families

ALL THE TIME in the world is a luxury Patty Burk seldom has. But that doesn't keep her from inviting guests to her table all the way from Japan.

"So far, my husband, Tom, our children and I have been the host family to six international students," she reports from Nanaimo, British Columbia. Luckily, much of what she needs for a fast, no-fuss supper is as close as her kitchen cupboard.

Like all the dishes served up in this chapter, the recipes Patty shares here are on the table in 30 minutes or less.

"Perfect for guests or family meals, Salsa Chicken is full of color and flavor. The simple sauce is tasty over pork chops, too," Patty adds. "Some mornings, I put the ingredients in the slow cooker. That way, this delicious dinner is ready to enjoy when we walk in the door that evening.

"My quick rice side dish gets a zippy twist from lemon-pepper seasoning. It complements the chicken and other Mexican dishes nicely."

For a fun-looking dessert or a refreshing treat, Patty relies on limes. "I cut the fruit in half and remove the pulp," she describes.

"Then my children, Emily, Jon and Ethan, fill the lime cups with a variety of ice cream and sherbet flavors. They even request these sundaes for birthday parties. Kids love them!

"We always manage to have dinner as a family," Patty says. "Table talk ranges from the kids' Celtic dance classes and Tom's floor hockey team to my volunteer work and our visiting students' progress with English."

Inevitably, Patty's quick dishes elicit one hearty international response—"Yum!"

Salsa Chicken

Prep/Total Time: 15 min.

1-1/2 pounds boneless skinless chicken breasts, cut into thin strips
1 tablespoon vegetable oil
1 medium green pepper, cut into thin strips
2 cups salsa

In a large skillet, saute chicken in oil for 3 minutes. Add the green pepper; cook for 3 minutes or until peppers are crisp-tender. Stir in salsa; bring to a boil. Reduce heat; simmer, uncovered, for 3 minutes or until chicken juices run clear. **Yield:** 4 servings.

Lemon-Pepper Rice

Prep/Total Time: 10 min.

2 cups water
1 to 2 teaspoons lemon-pepper seasoning
2 cups uncooked instant rice

In a saucepan, bring the water and lemon-pepper seasoning to a boil. Stir in the rice. Remove from the heat; cover and let the rice stand for 5 minutes. Fluff with a fork. **Yield:** 4 servings.

Lime Sundaes

Prep: 20 min. + freezing

4 large limes
1 cup lime sherbet
1 cup vanilla ice cream
Fresh mint, optional

Cut limes in half lengthwise; scoop out pulp and save for another use. Fill each lime half with 2 tablespoons of sherbet and ice cream. Return to the freezer until serving. Garnish with mint if desired. **Yield:** 4 servings.

Sundae Fun

Try making the yummy sundaes at left using scooped-out oranges or lemons rather than limes. Or, at Christmastime, fill each lime with raspberry sherbet for a festive red-and-green color combination.

To dress up your sundaes a bit more, add a dollop of whipped cream or whipped topping and a maraschino cherry to each one.

Dig Right into Chunky Stew

FAST FOOD is the answer when Elizabeth Freise of Kansas City, Missouri gets extra busy. And without question, her favorite kind comes from her very own kitchen.

"I work second shift and arrive home fairly late from my job at a college library," Elizabeth explains. Rather than reading through complex recipes, she'd prefer to check out speedy dishes that shorten her meal-making efforts.

Quick menus, such as the one Elizabeth shares here, are guaranteed best-sellers among busy cooks. Like all of the meals presented in this chapter, hers is ready to eat in 30 minutes or less.

"Loaded with fully cooked sausage, pre-chopped vegetables and instant rice, my Spicy Cajun Stew practically makes itself. It's my all-time favorite skillet dinner," she notes.

"Sometimes, I put the ingredients for the stew in my slow cooker, switch it on and go. When I get home from work, I can sit down to dinner...or carry it to a potluck with friends.

"My Green Bean Salad is ready on the spot, too. It has only a few on-hand fixings to toss together. With a hint of garlic, it stands up beside a highly flavored main dish. Plus, the tangy lemon dressing refreshes the palate."

As she dives into dessert, Elizabeth admits she feels like a kid again. "I invented Mandarin Fluff as an easy grown-up version of the frozen cream-filled pops we all enjoyed as children," she says.

"The mandarin oranges and whipped topping recreate the tangy orange and ice cream flavors of those yummy pops. And one bite cools the heat from the rest of the meal."

Since putting together dinner has become so easy, she has ample time for other pet projects and pastimes, Elizabeth relates. "The time I end up saving in the kitchen, I use for flower gardening, creative writing, crafting and walking with 'Marlo,' my energetic German shepherd."

In the stew, feel free to substitute turkey sausage for kielbasa or collard greens for spinach, Elizabeth suggests. "To complete the meal, serve it with corn bread, rolls or toast rounds."

For another change of taste, replace the green beans in Elizabeth's salad with sliced avocado or blanched zucchini. Your favoite bottled Italian is an optional dressing.

"If you like, you can make your Mandarin Fluff dessert even fruitier by adding cubes of cantaloupe, cherries and grapes," Elizabeth recommends. "I've also discovered that custard pudding is a tasty stand-in for frozen whipped topping."

Spicy Cajun Stew

Prep/Total Time: 20 min.

1 package (16 ounces) fully cooked kielbasa *or* Polish sausage, cut into 1/4-inch slices
2 cans (10 ounces *each*) diced tomatoes and green chilies, undrained
1 can (14-1/2 ounces) chicken broth
1 package (10 ounces) frozen chopped spinach, thawed and drained
1/2 to 3/4 cup uncooked instant rice

In a large skillet, saute sausage until lightly browned; drain. Add tomatoes and broth. Bring to a boil. Stir in spinach. Return to a boil; cook for 2 minutes. Stir in the rice. Cover and remove from the heat. Let stand for 5 minutes. Stir with a fork. **Yield:** 5 servings.

Green Bean Salad

Prep/Total Time: 15 min.

1/2 cup water
3 cups frozen cut green beans
3 to 4 tablespoons lemon juice
1 garlic clove, minced
1/2 teaspoon salt
Dash pepper
2 tablespoons olive oil
2 large tomatoes, cut into wedges

In a saucepan, bring water to a boil; add beans. Return to a boil; cook for 5 minutes (the beans will be crisp-tender). Immediately rinse beans in cold water.

For the dressing, in a small bowl, combine the lemon juice, garlic, salt and pepper; whisk in the olive oil. Place the beans and tomatoes on salad plates; drizzle with the dressing. **Yield:** 5 servings.

Mandarin Fluff

Prep/Total Time: 5 min.

2 cans (15 ounces *each*) mandarin oranges, well drained
1 carton (8 ounces) frozen whipped topping, thawed
1/2 cup chopped pecans

Place oranges in a bowl. Fold in whipped topping. Just before serving, fold in nuts. **Yield:** 5-6 servings.

Sandwich Supper Starts on the Grill

PLAYING beat the clock daily at dinner? You can get ahead of the game in no time flat while serving something simple and deliciously satisfying.

Ready in 30 minutes, the menu our Test Kitchen put together here features recipes from three cooks eager to get you up to speed in the kitchen.

Marlene Wiczek of Little Falls, Minnesota shares hearty Sirloin Steak Sandwiches. "My husband, Bill, and I raise black Angus on our hobby farm. So beef is big on our menu," she notes.

"I keep several of these easy steak sandwiches in the freezer. When family or friends stop over, I heat them through, add a side of baked beans or fruit salad, and we have an instant picnic."

From Orrville, Ohio, reader Krista Musser sent a no-fuss spud recipe that bowls over her husband and son. "My Swiss Potato Soup is bubbling on the stovetop faster than Scott and our son, Donald, can ask what's for dinner," she laughs.

"Sometimes I start simmering the ingredients in the slow cooker in the morning. Then, when I get home from work, I add the cheese and bacon."

In Lapeer, Michigan, Heidi Wilcox pays attention to the little things—whether tending to the children at her in-home day care or sweetening up supper for her own two kids.

"Since I usually have all the ingredients on hand, Pound Cake Cobbler is as easy as pie to assemble," Heidi relates. "If you like, change the canned filling to another flavor, such as raspberry or strawberry."

Sirloin Steak Sandwiches

Prep/Total Time: 20 min.

1 boneless beef sirloin steak (1 pound)
4 onion rolls, split
1/4 cup mayonnaise
2 to 4 tablespoons prepared mustard
4 teaspoons prepared horseradish
4 slices Swiss cheese

Grill steak, uncovered, over medium heat for 5-8 minutes on each side or until meat reaches desired doneness (for medium-rare, a meat thermometer should read 145°; medium, 160°; well-done, 170°). Spread cut side of roll tops with mayonnaise, mustard and horseradish. Slice steak diagonally; place on roll bottoms. Top with cheese and roll tops. **Yield:** 4 servings.

Swiss Potato Soup

Prep/Total Time: 30 min.

5 bacon strips, diced
1 medium onion, chopped
2 cups water
4 medium potatoes, peeled and cubed
1-1/2 teaspoons salt
1/8 teaspoon pepper
1/3 cup all-purpose flour
2 cups milk
1 cup (4 ounces) shredded Swiss cheese

In a saucepan, cook bacon until crisp; remove to paper towels with a slotted spoon. Drain, reserving 1 tablespoon drippings.

Saute the onion in the bacon drippings until tender. Add the water, potatoes, salt and pepper. Bring to a boil. Reduce the heat; simmer, uncovered, for 12 minutes or until the potatoes are tender.

Combine flour and milk until smooth; gradually stir into potato mixture. Bring to a boil; cook and stir for 2 minutes or until thickened and bubbly. Remove from the heat; stir in cheese until melted. Garnish with bacon. **Yield:** 4 servings.

Pound Cake Cobbler

Prep/Total Time: 15 min.

1 frozen pound cake (10-3/4 ounces), thawed
1 can (21 ounces) cherry *or* blueberry pie filling
1/3 cup water
1/2 teaspoon almond extract
Whipped topping
2 tablespoons sliced almonds, toasted

Cut the pound cake into 1-in. cubes. Place in a microwave-safe 1-1/2-qt. dish or 9-in. pie plate. In a small bowl, combine the pie filling, water and almond extract; spoon over the cake cubes. Cover and microwave on high for 3-5 minutes or until heated through. Spoon onto dessert plates; garnish with whipped topping and almonds. **Yield:** 4 servings.

Editor's Note: This recipe was tested in an 850-watt microwave.

They'll Gobble Up Turkey Meal

HER PLATE is as full as it's ever been, says Jean Komlos of Plymouth, Michigan. But that doesn't mean she spends every day in the kitchen.

"If I'm not working on an afghan, I'm busy with my card club or walking a frisky terrier or sheltie as our neighborhood dog-sitter," says this active grandmother. "I can't see taking time to decide what to cook when there are other fun things to do."

Quick menus, such as the one Jean shares here, are fast friends to cooks of any age. Her meal is set for the table in 30 minutes or less.

Broiled Turkey Tenderloins are perfect for weekday meals or special dinners with her grandchildren, Jean notes. "The simple sauce gets its tang from citrus, a little kick from cayenne and subtle sweetness from molasses. It keeps well in the refrigerator, so it can be made ahead. I often fix extra to use later with other meats.

"My Cheesy Chive Potatoes are a speedy side dish that complements most any entree," she continues. "Feta cheese adds a rich zesty flavor. A neighbor supplies me with garden-fresh chives, but you can use frozen or dried chives for more convenience."

For dessert, Jean puts together a fuss-free parfait. "Cherry yogurt, chocolate syrup and whipped topping blend into a yummy fluff. It's lovely with the alternating layers of fruit," she shares.

Broiled Turkey Tenderloins

Prep/Total Time: 20 min.

3/4 cup orange juice concentrate
1/3 cup molasses
1/4 cup ketchup
3 tablespoons prepared mustard
2 tablespoons soy sauce
1/2 teaspoon garlic powder
1/4 teaspoon cayenne pepper
1/8 teaspoon ground cumin
3 turkey breast tenderloins (about 1-1/2 pounds), cut lengthwise in halves

In a saucepan, whisk the first eight ingredients; bring to a boil. Set aside 3/4 cup for serving. Brush tenderloins on both sides with remaining sauce.

Broil 6 in. from the heat for 5 minutes, basting once. Turn; broil 5-8 minutes longer or until juices run clear. Serve with reserved sauce. **Yield:** 6 servings.

Cheesy Chive Potatoes

Prep/Total Time: 20 min.

6 medium potatoes, peeled and cubed
1/2 cup milk
1/2 cup crumbled feta cheese
1 tablespoon butter
1/2 teaspoon salt
1/8 teaspoon pepper
2 tablespoons minced chives

Place potatoes in a large saucepan and cover with water. Bring to a boil. Reduce heat; cover and boil for 10-15 minutes or until tender.

Drain and add the milk, feta cheese, butter, salt and pepper; mash. Stir in chives. **Yield:** 6 servings.

Cherry Chocolate Dessert

Prep/Total Time: 10 min.

1 carton (8 ounces) cherry yogurt
1/2 cup chocolate syrup
1 carton (8 ounces) frozen whipped topping, thawed
2 cups sliced strawberries *or* bananas
Mint sprigs

In a large bowl, combine yogurt and chocolate syrup; fold in whipped topping. Set aside 12 strawberry slices. In six dessert dishes, place half of the remaining fruit; top with half of the yogurt mixture. Repeat layers. Garnish with reserved strawberries and mint sprigs. **Yield:** 6 servings.

Feta Cheese Facts

Feta cheese is a white, salty, semi-firm cheese. It was traditionally made from sheep or goat's milk, but today it is also made using cow's milk.

After feta cheese is formed in a special mold, it is sliced into large pieces, salted and soaked in brine. Although feta cheese is mostly associated with Greek cooking, "feta" comes from the Italian word "fette," meaning "slice of food."

Enjoy the distinctive flavor of feta in Cheesy Chive Potatoes (page 278). Or try it tossed with your favorite green salad, stuffed in a pita bread sandwich or sprinkled on a pizza.

Meals on a Budget

Want to save money on groceries? Just rely on the wallet-friendly menus here. Each is low-cost but high in family-pleasing flavor.

ECONOMICAL EATING. Clockwise from upper left: Feed Your Family for $1.40 a Plate! (p. 284), Feed Your Family for $1.90 a Plate! (p. 282), Feed Your Family for $1.69 a Plate! (p. 288) and Feed Your Family for $1.96 a Plate! (p. 286).

Feed Your Family For $1.90 a Plate!

IF REAL spareribs don't always fit in your budget, try this lip-smacking—and economical—substitute along with two satisfying sides.

Janice Porterfield of Atlanta, Texas turns chicken thighs into Chicken Spareribs with a zippy barbecue-style sauce. You might have to make extras!

"I always take my Cottage Cheese Yeast Rolls to potlucks, where they disappear quickly," writes Angie Merriam of Springfield, Ohio. "These rolls are nice and light, so they go well with many dishes."

Using fresh spinach instead of frozen really enhances the flavor of classic Creamed Spinach, shared by Ann Van Dyk of Wrightstown, Wisconsin.

Chicken Spareribs

Prep: 5 min. **Cook:** 30 min.

8 bone-in chicken thighs
2 tablespoons vegetable oil
1 cup water
2/3 cup packed brown sugar
2/3 cup soy sauce
1/2 cup apple juice
1/4 cup ketchup
2 tablespoons cider vinegar
2 garlic cloves, minced
1 teaspoon crushed red pepper flakes
1/2 teaspoon ground ginger
2 tablespoons cornstarch
2 tablespoons cold water

In a Dutch oven, brown chicken in oil in batches on both sides; drain. Return all of the chicken to the pan.

In a bowl, combine the water, brown sugar, soy sauce, apple juice, ketchup, vinegar, garlic, pepper flakes and ginger; pour over chicken. Bring to a boil. Reduce heat; cover and simmer for 20 minutes or until chicken juices run clear.

Remove chicken to a platter and keep warm. In a small bowl, combine cornstarch and cold water until smooth; stir into cooking juices. Bring to a boil; cook and stir for 2 minutes or until thickened. Spoon over chicken. **Yield:** 4 servings.

Cottage Cheese Yeast Rolls

Prep: 30 min. + rising **Bake:** 10 min. per batch

2 packages (1/4 ounce *each*) active dry yeast
1/2 cup warm water (110° to 115°)
2 cups cottage cheese
2 eggs
1/4 cup sugar
2 teaspoons salt
1/2 teaspoon baking soda
4-1/2 cups all-purpose flour

In a large mixing bowl, dissolve yeast in warm water. In a small saucepan, heat cottage cheese to 110°-115°. Add the cottage cheese, eggs, sugar, salt, baking soda and 2 cups flour to yeast mixture; beat until smooth. Stir in enough remaining flour to form a firm dough (dough will be sticky).

Turn onto a floured surface; knead until smooth and elastic, about 6-8 minutes. Place in a greased bowl, turning once to grease top. Cover and let rise in a warm place until doubled, about 1 hour.

Punch the dough down. Turn onto a lightly floured surface; divide the dough into 30 pieces. Shape each piece into a roll. Place 2 in. apart on greased baking sheets. Cover rolls and let rise until doubled, about 30 minutes. Bake at 350° for 10-12 minutes or until golden brown. Remove from baking sheets to wire racks. **Yield:** 2-1/2 dozen.

Creamed Spinach

Prep/Total Time: 25 min.

3/4 pound fresh spinach, torn
2 tablespoons olive oil
6 tablespoons butter, cubed
1/4 cup chopped onion
1/4 cup all-purpose flour
1/2 teaspoon salt
1/8 teaspoon ground nutmeg
1-1/2 cups milk

In a Dutch oven, cook spinach in oil for 3 minutes or until wilted. Transfer to a cutting board; chop. Melt butter in the Dutch oven. Add onion; saute for 2 minutes or until crisp-tender.

Stir in flour, salt and nutmeg until combined. Gradually whisk in milk until blended. Bring to a boil; cook and stir 2 minutes or until thickened. Add chopped spinach. Reduce heat to low; cook, uncovered, for 5 minutes or until heated through. **Yield:** 4 servings.

Feed Your Family For $1.40 a Plate!

THIS MONEY-SAVING MENU centers on Curry-Berry Turkey Salad, a wholesome medley from Neva Arthur of New Berlin, Wisconsin. Jumbo Banana-Carrot Muffins shared by Julye Byrd, Azle, Texas, wonderfully complement that cool main dish. For a complete meal, add the treat from Ruby Williams of Bogalusa, Louisiana—Cinnamon-Chocolate Snackin' Cake.

Curry-Berry Turkey Salad

Prep/Total Time: 25 min.

✓ Uses less fat, sugar or salt. Includes Nutrition Facts and Diabetic Exchanges.

2 cups cubed cooked turkey breast
2 celery ribs, sliced
1/4 cup chopped red onion
2 tablespoons golden raisins
1/2 cup mayonnaise
2 teaspoons sugar
2 teaspoons lime juice
3/4 teaspoon curry powder
1/2 teaspoon grated lime peel
1/4 teaspoon salt
3 cups sliced fresh strawberries

In a large bowl, combine the turkey, celery, onion and raisins. In a small bowl, combine mayonnaise, sugar, lime juice, curry, lime peel and salt. Pour over turkey mixture; toss gently to coat. Just before serving, gently stir in strawberries. **Yield:** 6 servings.

Nutrition Facts: 1 cup (prepared with fat-free mayonnaise) equals 127 calories, 2 g fat (trace saturated fat), 42 mg cholesterol, 303 mg sodium, 14 g carbohydrate, 3 g fiber, 15 g protein. **Diabetic Exchanges:** 2 very lean meat, 1 fruit.

Jumbo Banana-Carrot Muffins

Prep: 25 min. **Bake:** 25 min.

1-1/2 cups all-purpose flour
3/4 cup sugar
1 teaspoon baking powder
1 teaspoon baking soda
1/2 teaspoon salt
1/2 teaspoon ground cinnamon
1/4 teaspoon ground nutmeg

2 eggs, *separated*
1 tablespoon honey
1/4 teaspoon grated orange peel
2 medium ripe bananas, mashed
1 cup shredded carrots
1/2 cup unsweetened applesauce

In a large bowl, combine the first seven ingredients. In a small mixing bowl, beat egg yolks until light and lemon-colored. Beat in honey and orange peel. Fold in the bananas, carrots and applesauce. Stir into dry ingredients just until moistened.

In another small mixing bowl, beat egg whites on high speed until stiff peaks form; fold into batter a third at a time.

Fill greased or paper-lined muffin cups two-thirds full. Bake at 350° for 25-30 minutes or until a toothpick comes out clean. Cool for 5 minutes before removing from pan to a wire rack. **Yield:** 9 jumbo muffins.

Editor's Note: This recipe contains no oil or butter.

Cinnamon-Chocolate Snackin' Cake

Prep: 30 min. **Bake:** 35 min. + cooling

2 cups all-purpose flour
2 cups sugar
1 teaspoon baking soda
1 teaspoon ground cinnamon
1/2 teaspoon salt
1-1/2 cups miniature marshmallows
3 tablespoons baking cocoa
1 cup cola
1/2 cup vegetable oil
1/2 cup butter, cubed
2 eggs, beaten
1/2 cup buttermilk
1 teaspoon vanilla extract

FROSTING:

3 tablespoons baking cocoa
6 tablespoons cola
1/2 cup butter, cubed
3-3/4 cups confectioners' sugar
1 cup chopped pecans, *divided*

In a large bowl, sift together flour, sugar, baking soda, cinnamon and salt. Stir in marshmallows; set aside.

In a small saucepan, bring cocoa, cola, oil and butter to a boil. Pour over flour mixture; mix well. Stir in eggs, buttermilk and vanilla. Pour into a greased 13-in. x 9-in. x 2-in. baking dish. Bake at 350° for 35-40 minutes or until a toothpick inserted near the center comes out clean (do not overbake). Cool on a wire rack.

For frosting, in a large saucepan, bring the cocoa, cola and butter to a boil. Remove from the heat. Stir in confectioners' sugar and 3/4 cup nuts until blended. Spread over cake; sprinkle with remaining nuts. Cool until frosting is set before cutting. **Yield:** 15 servings.

Editor's Note: Diet cola is not recommended for this recipe.

Feed Your Family For $1.96 a Plate!

THIS INVITING MENU featuring reader-favorite recipes starts off with After-Holiday Ham on Biscuits. Billie George from Saskatoon, Saskatchewan puts her extra ham and hard-cooked eggs to good use by making this creamy, comforting main dish.

A light, lemony dressing drapes the well-seasoned Vinaigrette Asparagus Salad shared by Linda Lacek of Winter Park, Florida. And from Amy Voights of Brodhead, Wisconsin, refreshing Orange Slush makes a delightful accompaniment to just about any meal.

After-Holiday Ham on Biscuits

Prep/Total Time: 30 min.

1 cup all-purpose flour
2 teaspoons baking powder
1/4 teaspoon salt
3 tablespoons cold butter
1/2 cup milk
CREAM SAUCE:
1 cup cubed fully cooked ham
1/4 cup chopped onion
3 tablespoons butter
1/2 teaspoon chicken bouillon granules
1/2 teaspoon Worcestershire sauce
1/8 teaspoon pepper
3 tablespoons all-purpose flour
1-3/4 cups milk
3 hard-cooked eggs, chopped
1 tablespoon minced fresh parsley

In a bowl, combine the flour, baking powder and salt. Cut in butter until mixture resembles coarse crumbs. Stir in milk just until moistened. Turn onto a lightly floured surface; knead 8-10 times.

Pat or roll out to 1/2-in. thickness; cut with a floured 2-1/2-in. biscuit cutter. Place 2 in. apart on a greased baking sheet. Bake at 425° for 10-12 minutes or until golden brown.

Meanwhile, in a large skillet, saute ham and onion in butter for 3-4 minutes or until onion is crisp-tender. Stir in bouillon, Worcestershire and pepper. Combine flour and milk until smooth; gradually stir into pan. Bring to a boil; cook and stir for 2 minutes or until thickened. Gently stir in eggs and parsley.

Split warm biscuits in half horizontally; top with ham mixture. **Yield:** 4 servings.

Vinaigrette Asparagus Salad

Prep: 15 min. + chilling

✓ **Uses less fat, sugar or salt. Includes Nutrition Facts and Diabetic Exchanges.**

2 tablespoons lemon juice
2 tablespoons olive oil
1 teaspoon minced fresh tarragon *or* 1/4 teaspoon dried tarragon
1 garlic clove, minced
1/2 teaspoon Dijon mustard
1/4 teaspoon pepper
Dash salt
1 pound fresh asparagus, trimmed

For vinaigrette, in a jar with a tight-fitting lid, combine the first seven ingredients; shake well. Cover and refrigerate for at least 1 hour.

In a large skillet, bring 1/2 in. of water to a boil. Add asparagus; cover and boil for 3 minutes. Drain and immediately place asparagus in ice water. Drain and pat dry. Place in a serving bowl; cover and refrigerate.

Just before serving, shake vinaigrette and drizzle over asparagus. **Yield:** 4 servings.

Nutrition Facts: 1 serving equals 77 calories, 7 g fat (1 g saturated fat), 0 cholesterol, 59 mg sodium, 3 g carbohydrate, 1 g fiber, 2 g protein. **Diabetic Exchanges:** 1 vegetable, 1 fat.

Orange Slush

Prep/Total Time: 15 min.

1 cup water
1 can (6 ounces) frozen orange juice concentrate
1 carton (6 ounces) vanilla yogurt
1/2 cup sugar
1/2 cup cold milk
1/2 teaspoon vanilla extract
10 to 12 ice cubes

In a blender, combine water, juice concentrate, yogurt, sugar, milk and vanilla; cover and process until smooth. While processing, add a few ice cubes at a time until mixture achieves desired thickness. Pour into chilled glasses; serve immediately. **Yield:** 4 servings.

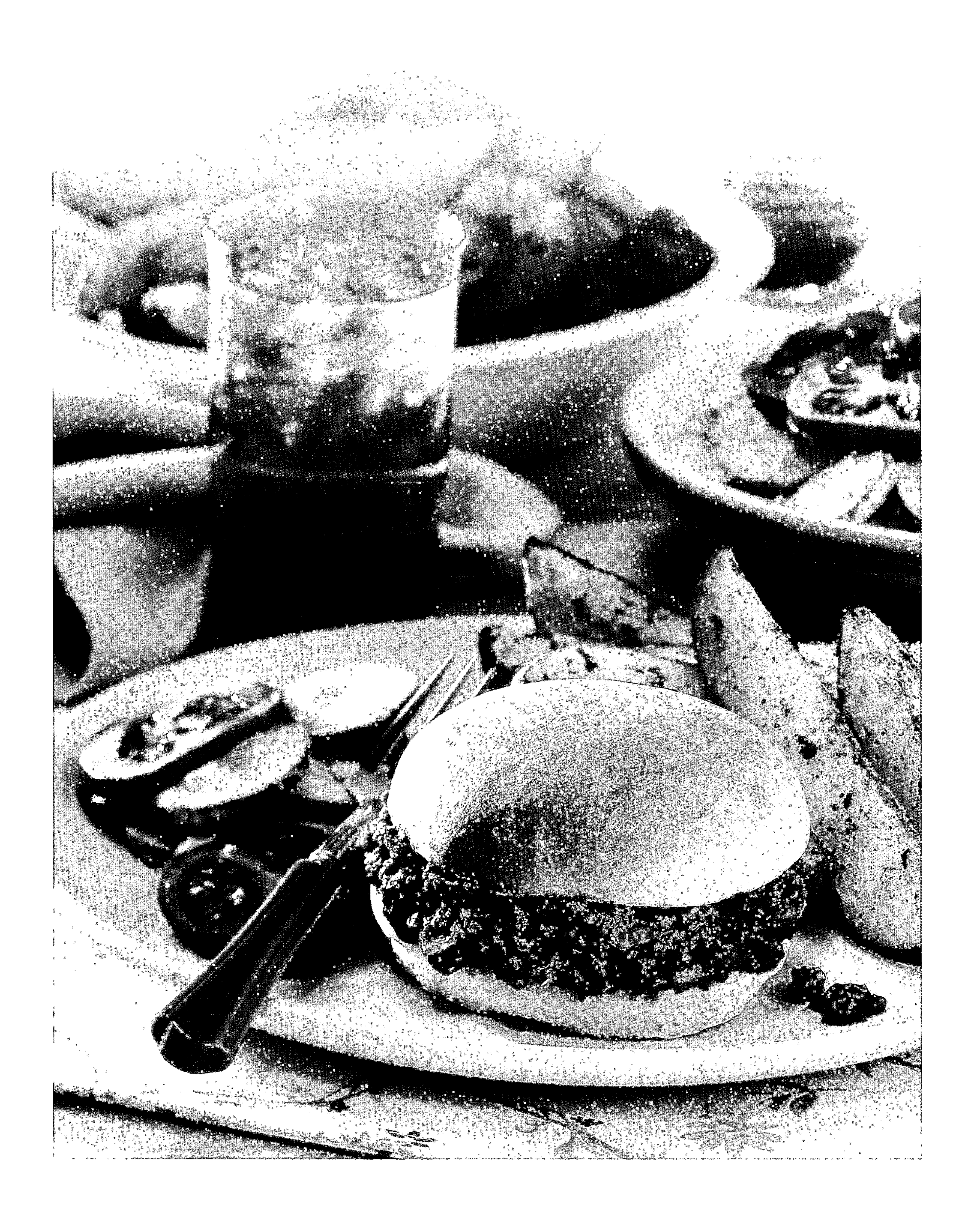

Feed Your Family For $1.69 a Plate!

IN SUMMERTIME, the livin' is easy. So simplify meals during that laid-back season by preparing the all-time-favorite sloppy joes here. They're sure to be popular with the entire family.

You'll find that casual dining can be cost-effective as well as satisfying. You can feed your family these saucy beef sandwiches, along with a garden-fresh veggie side dish and spicy spuds, for just $1.69 a plate. Now that's good eating!

"My grandchildren love my Family-Pleasing Sloppy Joes," Patricia Ringle writes from Edgar, Wisconsin. "I like this recipe because it can be made ahead of time and can also be put in the slow cooker. I've found it freezes well, too."

"German Cukes and Tomatoes really sparks up a meal and goes fairly well with everything from grilled steak to tuna salad," says field editor Karen Ann Bland of Gove, Kansas. "I like to serve it in individual, lettuce-lined glass bowls with a sour cream star piped on top and a sprinkling of dill."

From Keego Harbor, Michigan, Autumn McNamara writes, "When my husband and I barbecue with friends, we always bring Spicy Potato Wedges, and everyone loves them. They're a favorite accompaniment to almost anything we cook on the grill."

Family-Pleasing Sloppy Joes

Prep: 10 min. **Cook:** 45 min.

2 pounds ground beef
1 large onion, chopped
1-1/4 cups ketchup
1/2 cup water
1 tablespoon brown sugar
1 tablespoon white vinegar
1/2 teaspoon salt
1/2 teaspoon ground mustard
1/2 teaspoon chili powder
1/4 teaspoon ground allspice
8 sandwich buns, split

In a Dutch oven, cook the beef and onion over medium heat until the meat is no longer pink; drain. Add the ketchup, water, brown sugar, vinegar, salt, mustard, chili powder and allspice. Bring to a boil.

Reduce heat; simmer, uncovered, for 35-40 minutes or until heated through.

Spoon about 1/2 cup meat mixture onto each bun. **Yield:** 8 servings.

German Cukes and Tomatoes

Prep/Total Time: 20 min.

✓ **Uses less fat, sugar or salt. Includes Nutrition Facts and Diabetic Exchanges.**

2 medium cucumbers, thinly sliced
4 green onions, thinly sliced
2 tablespoons minced fresh parsley
1/4 cup sour cream
2 tablespoons snipped fresh dill
1 tablespoon cider vinegar
1/2 teaspoon salt
1/4 teaspoon prepared mustard
1/8 teaspoon pepper
1 tablespoon milk, optional
3 small tomatoes, sliced

In a large bowl, combine the cucumbers, onions and parsley. In a small bowl, combine the sour cream, dill, vinegar, salt, mustard and pepper. Stir in milk if a thinner dressing is desired. Pour over cucumber mixture and toss to coat. Divide among eight salad plates; top with tomatoes. **Yield:** 8 servings.

Nutrition Facts: 3/4 cup equals 37 calories, 1 g fat (1 g saturated fat), 5 mg cholesterol, 158 mg sodium, 5 g carbohydrate, 1 g fiber, 1 g protein. **Diabetic Exchanges:** 1 vegetable, 1/2 fat.

Spicy Potato Wedges

Prep: 15 min. **Bake:** 30 min.

1/4 cup vegetable oil
1 tablespoon chili powder
2 teaspoons onion powder
2 teaspoons garlic salt
1 teaspoon sugar
1 teaspoon paprika
3/4 teaspoon salt
1/4 to 1/2 teaspoon cayenne pepper
3-1/2 pounds large red potatoes, cut into wedges

In a large bowl, combine the first eight ingredients; add potatoes and toss to coat. Arrange in a single layer on greased baking sheets. Bake at 400° for 30-35 minutes or until potatoes are tender and golden brown, turning once. **Yield:** 8 servings.

Feed Your Family For $1.72 a Plate!

BREAK a boring mealtime routine with terrific Beef Barley Soup from Louise Laplante, Hanmer, Ontario. For super sides, try Tomato Zucchini Salad from Suzanne Kesel of Cohocton, New York and Yeast Corn Bread Loaf from Fred Barnsdale, Pahokee, Florida.

Beef Barley Soup

Prep: 20 min. **Cook:** 1 hour

2 pounds beef stew meat, cut into 1-inch pieces
1 tablespoon vegetable oil
5 cups water
4 celery ribs, chopped
4 medium carrots, chopped
1 large onion, chopped
1 can (14-1/2 ounces) diced tomatoes, undrained
2 tablespoons tomato paste
4 teaspoons beef bouillon granules
1 teaspoon *each* dried oregano, thyme, basil and parsley flakes
1/2 teaspoon salt
1/4 teaspoon pepper
1 cup quick-cooking barley

In a Dutch oven, brown meat in oil on all sides; drain. Add the water, celery, carrots, onion, tomatoes, tomato paste, bouillon and seasonings. Bring to a boil. Reduce heat; cover and simmer for 50 minutes.

Stir in the barley; cover and simmer 10-15 minutes longer or until barley is tender. **Yield:** 8 servings (about 2 quarts).

Tomato Zucchini Salad

Prep/Total Time: 20 min.

✓ **Uses less fat, sugar or salt. Includes Nutrition Facts and Diabetic Exchanges.**

2 cups water
4 small zucchini, thinly sliced
1/8 teaspoon salt
2 small tomatoes, cut into wedges
2 slices red onion, separated into rings

DRESSING:

3 tablespoons olive oil
1 tablespoon balsamic vinegar
1 tablespoon minced fresh tarragon *or* 1 teaspoon dried tarragon
1 tablespoon Dijon mustard
1/2 teaspoon salt
1/2 teaspoon hot pepper sauce
1 garlic clove, minced
1 tablespoon minced fresh parsley

In a large saucepan, bring water to a boil. Add zucchini; cover and boil for 2-3 minutes. Drain and immediately place zucchini in ice water. Drain and pat dry; sprinkle with salt. In a large bowl, combine the zucchini, tomatoes and onion.

In a jar with a tight-fitting lid, combine the oil, vinegar, tarragon, mustard, salt, hot pepper sauce and garlic; shake well. Pour over vegetables and gently toss to coat. Sprinkle with parsley. **Yield:** 8 servings.

Nutrition Facts: 3/4 cup equals 63 calories, 5 g fat (1 g saturated fat), 0 cholesterol, 238 mg sodium, 4 g carbohydrate, 1 g fiber, 1 g protein. **Diabetic Exchanges:** 1 vegetable, 1 fat.

Yeast Corn Bread Loaf

Prep: 20 min. + rising **Bake:** 35 min. + cooling

1 package (1/4 ounce) active dry yeast
1-1/4 cups warm water (110° to 115°), *divided*
1 cup yellow cornmeal
1/4 cup nonfat dry milk powder
3 tablespoons butter, softened
2 tablespoons sugar
1-1/2 teaspoons salt
2-1/4 to 2-3/4 cups all-purpose flour

In a large mixing bowl, dissolve yeast in 1/4 cup warm water. Add cornmeal, milk powder, butter, sugar, salt, remaining water and 1-1/4 cups flour. Beat until smooth. Stir in enough remaining flour to form a soft dough. Turn onto a floured surface; knead until smooth and elastic, about 6-8 minutes. Place in a greased bowl, turning once to grease top. Cover and let rise in a warm place until doubled, about 1 hour.

Punch dough down. Shape into a loaf. Place in a greased 9-in. x 5-in. x 3-in. loaf pan. Cover and let rise until doubled, about 30 minutes. Bake at 375° for 35-40 minutes or until golden brown. Remove from pan to a wire rack to cool. **Yield:** 1 loaf.

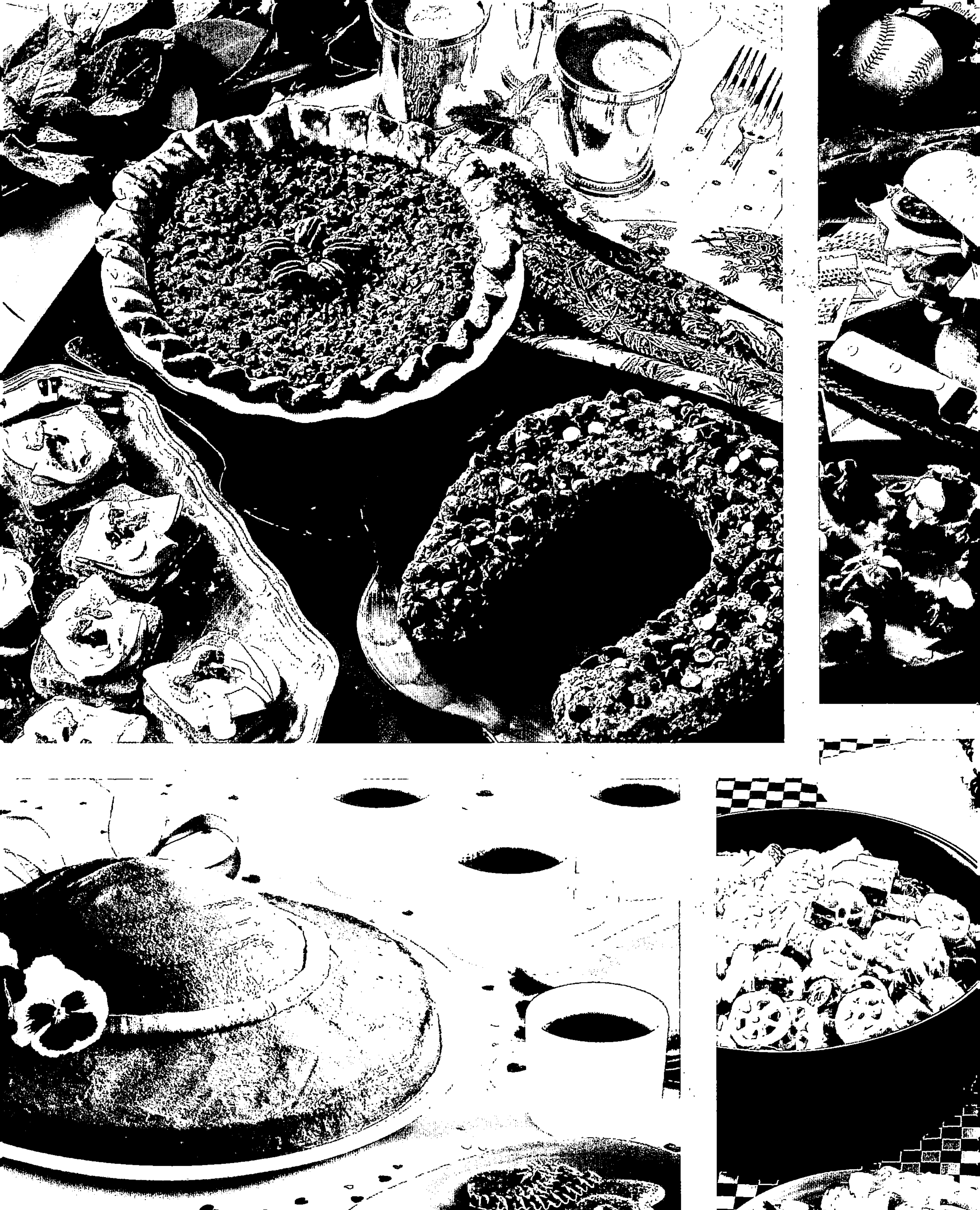

Getting in the Theme of Things

Have some fun with your food! With these festive party recipes from creative cooks, you can whip up delightfully themed menus.

PLAYFUL FARE. Clockwise from top left: Derby Treats Are a Sure Bet! (p. 296), Have a Ball with Sporty Spread (p. 298), You 'Auto' Try This! (p. 302) and Revel in Red Hat Theme (p. 300).

'Brunch with Rudolph' Shines

By Laurel Leslie, Sonora, California

FOR A GET-TOGETHER at Christmastime, my husband and I had fun hosting "Brunch with Rudolph."

We all enjoyed Clam Fondue in a Bread Bowl, Herbed Dip for Veggies, Warm Potato Salad, Green Chili Egg Puff and Praline Cheesecake.

Clam Fondue in a Bread Bowl

Prep: 25 min. **Bake:** 40 min.

3 cans (6-1/2 ounces *each*) minced clams
2 packages (8 ounces *each*) cream cheese, cubed
1 tablespoon minced fresh parsley
2 teaspoons lemon juice
2 teaspoons Worcestershire sauce
2 teaspoons minced chives
1/2 teaspoon salt
6 drops hot pepper sauce
2 round loaves (1 pound *each*) sourdough bread

Drain two cans of clams and discard liquid. Drain remaining can, reserving the liquid. In a food processor, combine clams, cream cheese, parsley, lemon juice, Worcestershire, chives, salt, hot pepper sauce and reserved clam juice; cover and process until smooth.

Cut the top fourth off one loaf of bread; carefully hollow out loaf, leaving a 1-in. shell. Set removed bread aside. Fill shell with clam mixture; replace top. Wrap tightly in heavy-duty foil; place on a baking sheet. Bake at 350° for 40-45 minutes.

Meanwhile, cut reserved bread and the second loaf into cubes. Wrap in heavy-duty foil; place in oven during the last 15 minutes of baking. Unwrap bowl and remove top. Serve with cubes. **Yield:** 3-1/2 cups.

Herbed Dip for Veggies

Prep: 15 min. + chilling

1 cup mayonnaise
1/2 cup sour cream
1 tablespoon minced chives
1 tablespoon grated onion
1 tablespoon capers, drained
1/2 teaspoon salt
1/2 teaspoon lemon juice
1/2 teaspoon Worcestershire sauce
1/4 teaspoon minced fresh parsley
1/4 teaspoon paprika
1/8 teaspoon curry powder
Dash garlic salt
Assorted fresh vegetables

In a bowl, combine the first 12 ingredients. Cover and refrigerate for 1 hour. Serve with vegetables. Refrigerate leftovers. **Yield:** 1-1/2 cups.

Warm Potato Salad

Prep: 25 min. **Cook:** 15 min.

2 pounds small red potatoes
1/4 pound fresh snow peas, halved
1/2 pound sliced fresh mushrooms
1/2 cup diced fully cooked ham
1 medium green pepper, julienned
1 celery rib, diced
3 tablespoons chopped green onions
2 tablespoons minced fresh parsley

DRESSING:

1/2 cup olive oil
3 tablespoons white wine vinegar
2 tablespoons lemon juice
1 teaspoon dried oregano
1/2 teaspoon salt
1/2 teaspoon pepper
1 tablespoon minced fresh parsley
4 red lettuce leaves

Place potatoes in a large saucepan; cover with water. Bring to a boil. Reduce heat; cover and cook for 10 minutes. Add peas; cook, uncovered, 1 minute longer or until potatoes are tender. Drain. When cool enough to handle, quarter potatoes into a large bowl. Add mushrooms, ham, green pepper, celery, onions and parsley.

In a small bowl, combine the oil, vinegar, lemon juice, oregano, salt and pepper. Pour over potato mixture and toss gently. Sprinkle with parsley. Serve warm in a lettuce-lined bowl. **Yield:** 7 servings.

Green Chili Egg Puff

Prep: 15 min. **Bake:** 35 min.

10 eggs
1/2 cup all-purpose flour
1 teaspoon baking powder
1/2 teaspoon salt
4 cups (16 ounces) shredded Monterey Jack cheese

2 cups (16 ounces) small-curd cottage cheese
1 can (4 ounces) chopped green chilies

In a large mixing bowl, beat eggs on medium-high speed for 3 minutes or until light and lemon-colored. Combine the flour, baking powder and salt; gradually add to eggs. Stir in the cheeses and chilies.

Pour into a greased 13-in. x 9-in. x 2-in. baking dish. Bake, uncovered, at 350° for 35-40 minutes or until a knife inserted near the center comes out clean. Let stand for 5 minutes before serving. **Yield:** 12 servings.

Praline Cheesecake

Prep: 20 min. **Bake:** 55 min. + chilling

1-1/2 cups crushed vanilla wafers (about 50)
1/4 cup sugar
1/4 cup butter, melted
16 whole vanilla wafers

FILLING:

3 packages (8 ounces *each*) cream cheese, softened
1 cup sugar
1/2 cup sour cream
1 teaspoon vanilla extract
3 eggs, lightly beaten

TOPPING:

25 caramels
2 tablespoons milk
1/2 cup chopped pecans, toasted

In a small bowl, combine crumbs, sugar and butter. Press onto bottom of a greased 9-in. springform pan. Stand whole wafers around edge of pan, pressing lightly into crumbs; set aside. In a large mixing bowl, beat cream cheese, sugar, sour cream and vanilla until smooth. Add eggs; beat on low speed just until combined. Pour into crust. Place pan on a baking sheet.

Bake at 325° for 55-60 minutes or until center is almost set. Cool on a wire rack for 10 minutes. Carefully run a knife around edge of pan to loosen; cool 1 hour longer. Refrigerate overnight.

Place caramels and milk in a microwave-safe bowl. Microwave, uncovered, on high for 1 minute; stir until smooth. Remove sides of springform pan. Drizzle caramel mixture over cheesecake; sprinkle with pecans. Refrigerate leftovers. **Yield:** 12 servings.

Editor's Note: This recipe was tested in a 1,100-watt microwave.

Derby Treats Are a Sure Bet!

IF YOU can't make it to Louisville this May, you might want to host a fun-filled Kentucky Derby party of your own. Race-day treats from Annette Grahl and Emily Baldwin will help you set the scene.

Annette (left) owns Scottwood Bed & Breakfast in Midway, Kentucky, located about an hour's drive away from the famous Churchill Downs racetrack. "As you might expect, we get plenty of business during the weekend of the race," she says.

"I serve our guests pecan-topped Brie, grits with shrimp, fresh asparagus dishes and Mini Hot Browns, my version of a traditional Derby specialty.

"The celebrated Hot Brown open-faced sandwich was dreamed up by the chef of the Brown Hotel in Louisville in the 1920s," Annette relates. "I make a smaller version, with juicy turkey slices and crispy bacon stacked on toasted rye bread and topped with a rich cheese sauce.

"Another Derby cornerstone, the mint julep, normally has a healthy dose of Kentucky bourbon," says Annette. "But I concocted an alcohol-free recipe so all ages can enjoy the fresh mint, tangy lemon and ginger ale fizz. Served in chilled stainless steel glasses, Mock Mint Julep makes for pleasant sipping."

On race day, the Grahls' bed-and-breakfast guests cast their votes for the winning horse. "When we return from the race, the champion 'bettors' are presented with bourbon chocolates," adds Annette.

Raised in Kentucky, Emily Baldwin (above) moved

to Fort Collins, Colorado and discovered, "My friends here had never been to a Kentucky Derby party. So I hosted one for them.

"All of the race-day 'spectators' were delighted with my Southern-style spread, which included barbecue chicken, biscuits, potato salad and baked beans."

For an exciting finish, Emily made Kentucky Pecan Pie, a lusciously dense dessert. Its smooth chocolate filling and toasted pecans are definitely a winning combination.

"I also served a Pretzel Horseshoe," she relates. "Youngsters at the party loved its fun horseshoe shape, and the crunchy, sweet 'n' salty taste is popular with people of all ages.

"Waiting for the big race on TV, the kids had a grand time playing 'pin the tail on the horse,' and we adults placed our bets. The winners roped in both 100 Grand candy bars and horseshoe magnets."

Mini Hot Browns

Prep/Total Time: 30 min.

1 teaspoon chicken bouillon granules
1/4 cup boiling water
3 tablespoons butter
2 tablespoons all-purpose flour
3/4 cup half-and-half cream
1 cup (4 ounces) shredded Swiss cheese
18 slices snack rye bread
6 ounces sliced deli turkey
1 small onion, thinly sliced and separated into rings
5 bacon strips, cooked and crumbled
2 tablespoons minced fresh parsley

In a small bowl, dissolve bouillon in water; set aside. In a small saucepan, melt butter over medium heat. Stir in flour until smooth; stir in cream and bouillon. Bring to a boil; cook and stir for 1-2 minutes or until thickened. Stir in cheese until melted. Remove from the heat.

Place bread slices on two baking sheets. Layer each with turkey, onion and cheese sauce. Sprinkle with bacon. Bake at 350° for 10-12 minutes or until heated through. Sprinkle with parsley. **Yield:** 1-1/2 dozen.

Mock Mint Julep

Prep: 15 min. + standing

2 cups cold water
1-1/2 cups sugar
3/4 cup lemon juice
6 mint sprigs
5 cups ice cubes
2-1/2 cups ginger ale, chilled
Lemon slices and additional mint, optional

In a bowl, combine the water, sugar, lemon juice and mint; stir well. Let stand for at least 45 minutes.

Strain and discard mint. Place ice cubes in two 2-qt. pitchers; add half of the lemon mixture and ginger ale to each. Garnish with lemon and mint if desired. **Yield:** 13 servings (about 3 quarts).

Kentucky Pecan Pie

Prep: 10 min. **Bake:** 35 min. + cooling

3 eggs, lightly beaten
1 cup light corn syrup
1/2 cup packed brown sugar
1/2 teaspoon vanilla extract
1/4 teaspoon salt
1 cup chopped pecans
1 cup (6 ounces) semisweet chocolate chips
1 unbaked pastry shell (9 inches)

In a small bowl, whisk the eggs, corn syrup, brown sugar, vanilla and salt. Stir in pecans and chocolate chips. Pour into pastry shell. Cover edges with foil.

Bake at 350° for 25 minutes. Remove foil; bake 10-15 minutes longer or until a knife inserted near the center comes out clean. Cool on a wire rack. Refrigerate leftovers. **Yield:** 6-8 servings.

Pretzel Horseshoes

Prep: 20 min. + cooling

1 package (16 ounces) large marshmallows
3 tablespoons butter, cubed
4 cups crushed pretzels
1 cup semisweet chocolate chips
1/2 cup butterscotch chips

Place two 5-in.-diameter bowls upside down in the center of two 9-in. round pans; coat bowls and pans with nonstick cooking spray.

In a large microwave-safe bowl, melt marshmallows and butter on high for 1 minute. Stir in pretzels. Quickly add chips; stir gently until coated. Divide mixture in half. Press one portion evenly into one prepared pan. Shape into a horseshoe by forming mixture around bowl, leaving a 4-in. opening at one end. Repeat with remaining mixture. Cool for 15 minutes.

Remove bowls. Run a knife around edges of pans to loosen. Invert each onto a serving platter. Reshape horseshoes by gently spreading sides apart to elongate if necessary. Cool completely. **Yield:** 2 horseshoes (14 servings each).

ANGELS

Have a Ball with Sporty Spread

By Cathy Runyon, Allendale, Michigan

NOWHERE, in my experience, does baseball fever strike with more intensity than among us reporters at the *Advance Newspaper*. Our newsroom crew especially loves rooting on our local baseball team, the West Michigan Whitecaps. To celebrate the Whitecaps' opening day, I treated my co-workers to a "ballpark buffet" lunch.

One friend joked that I'd probably have only beans and hot dogs. Naturally, I threw a curveball and served Home Run Slugger Subs. I trimmed long French bread to make the hoagies look like bats—stuffed with cold cuts, veggies and cheese.

Bases Loaded Nachos also were a big hit. They're great to munch, offering popular Southwest flavor.

We washed it all down with Lost-in-the-Sun Punch. Scoops of sherbet floated in the bowl like high flies to left field (the gang chuckled at this description).

For dessert, I made Curveball Cupcakes. After frosting the cupcakes white, I used red decorating frosting to "stitch the seams."

Pinstriped cotton fabric was perfect for a tablecloth. I put a piece of parchment paper cut to look like home plate at its center. On home plate, I criss-crossed two bat-shaped subs on a clear cutting board. Next to the sandwiches, I set a real fielder's glove holding one of my baseball cupcakes.

Home Run Slugger Sub

Prep/Total Time: 15 min.

- **1 French bread baguette (1 pound and 20 inches long)**
- **1/4 pound thinly sliced fully cooked ham**
- **1/4 pound thinly sliced bologna**
- **1/4 pound thinly sliced hard salami**
- **4 romaine leaves**
- **6 slices Swiss cheese**
- **6 slices Colby cheese**
- **1 medium tomato, sliced**

With a sharp knife, cut one end of the baguette in the shape of a baseball bat handle. Slice the loaf in half lengthwise.

On the bottom half, layer the ham, bologna, salami, romaine, cheeses and tomato slices. Replace top. Secure with toothpicks if necessary. Cut into slices. **Yield:** 8 servings.

Bases Loaded Nachos

Prep/Total Time: 30 min.

- **1 cup refried beans, warmed**
- **3 cups tortilla chips**
- **1 cup shredded lettuce**
- **1 cup (4 ounces) finely shredded cheddar cheese**
- **1 large tomato, chopped**
- **1 can (3.8 ounces) sliced ripe olives, drained**
- **1/4 cup sour cream**

Spread the refried beans over the tortilla chips, about 1 teaspoon on each. Place chips on a serving platter. Sprinkle with lettuce, cheddar cheese, tomato and ripe olives. Top with a dollop of sour cream. Serve immediately. **Yield:** 8 servings.

Lost-in-the-Sun Punch

Prep/Total Time: 5 min.

- **1 carton (64 ounces) orange juice, chilled**
- **1 bottle (2 liters) lemonade, chilled**
- **6 cans (12 ounces *each*) cream soda, chilled**
- **1 pint pineapple sherbet**

In a large punch bowl, combine the orange juice, lemonade and soda. Top with scoops of sherbet. Serve immediately. **Yield:** 16-20 servings (4 quarts).

Curveball Cupcakes

Prep: 50 min. **Bake:** 20 min. + cooling

- **1 package (18-1/4 ounces) yellow cake mix**
- **1 can (16 ounces) vanilla frosting**
- **1 tube red decorating frosting**

Prepare cake batter according to package directions. Fill paper-lined muffin cups two-thirds full. Bake at 350° for 20-22 minutes or until a toothpick comes out clean. Cool for 10 minutes before removing from pans to wire racks to cool completely.

Frost cupcakes with vanilla frosting. Use red frosting to pipe stitch marks to resemble baseballs. **Yield:** 2 dozen.

Revel in Red Hat Theme

THE FUN-LOVING LADIES of the Red Hat Society know how to host lively parties. Just consider the crimson-colored recipes here from creative *Taste of Home* readers.

"There are 12 in our group, the 'Southern Charms' Red Hatters," relates Patsy Snyder (above left) of Rural Hall, North Carolina. "We had a get-together at my cousin's cottage, nestled on the edge of the Yadkin River. Wearing our finest purple outfits and red hats, we enjoyed a 'Red Hattitude' luncheon on a spacious wrap-around porch.

"The ladies couldn't get over our special dessert. They 'oohed' and 'aahed' at Red Hat Cake—a moist, golden cake decorated with bright red frosting, colorful flowers and a wispy purple bow."

Red Chapeau Sugar Cookies also caught attention, Patsy notes. "The cutouts were sprinkled with red sugar and frosted with purple sashes and bows."

From Newfield, New York, Margaret Blomquist organized a luncheon for the "Red Petunias," featuring chicken and Red Potato Salad.

"My sister, Teeny, hid the salad in a fancy red hat," says Margaret. "At mealtime, she pulled the top off of the hat, and we dug right in!"

Red Hat Cake

Prep: 1 hour **Bake:** 20 min. + cooling

1 cup butter, softened
2-1/4 cups sugar
5 eggs
3 teaspoons vanilla extract
3-1/2 cups all-purpose flour
2 teaspoons baking powder

1/2 teaspoon salt
1 cup milk

FROSTING:

1 cup shortening
1/4 cup butter, softened
7-1/2 cups confectioners' sugar
1/2 cup half-and-half cream
3 tablespoons red food coloring
3 teaspoons vanilla extract
1/2 teaspoon salt

In a large mixing bowl, cream butter and sugar. Add eggs, one at a time, beating well after each addition. Beat in vanilla. Combine the flour, baking powder and salt; add to creamed mixture alternately with milk.

Line a 14-in. round cake pan or deep-dish pizza pan with waxed paper; lightly coat with nonstick cooking spray. Pour 1-1/2 cups batter into a greased 2-qt. round baking dish; pour remaining batter into prepared cake pan or pizza pan.

Bake at 350° for 20-30 minutes or until a toothpick inserted near the center comes out clean. Cool for 10 minutes before removing from pans to wire racks to cool completely.

For frosting, in a large mixing bowl, cream the shortening, butter and confectioners' sugar until light and fluffy. Add cream, food coloring, vanilla and salt; beat until smooth.

Place larger cake on a serving plate; spread with frosting. Top with smaller cake; frost. Decorate as desired. Save any remaining frosting for another use. **Yield:** 16-20 servings.

Red Chapeau Sugar Cookies

Prep: 45 min. + chilling
Bake: 10 min. per batch + cooling

✓ **Uses less fat, sugar or salt. Includes Nutrition Facts and Diabetic Exchanges.**

1 cup butter, softened
2 cups packed light brown sugar
2 eggs
2 teaspoons lemon extract
4-1/2 cups all-purpose flour
1 teaspoon baking soda
1/2 teaspoon cream of tartar
Red colored sugar *or* decorating frosting, optional

In a large mixing bowl, cream the butter and brown sugar. Beat in the eggs and lemon extract. Combine the flour, baking soda and cream of tartar; gradually add to the creamed mixture. Cover and refrigerate dough for 1 hour or until easy to handle.

Divide the dough into fourths. On a lightly floured surface, roll one portion to 1/8-in. thickness. Cut with a floured 6-in. hat-shaped cookie cutter. Place 1 in. apart on ungreased baking sheets. Repeat with the remaining dough.

Sprinkle with red colored sugar if desired. Bake at 350° for 6-8 minutes or until edges begin to brown. Remove to wire racks to cool. Decorate with frosting if desired. **Yield:** 5 dozen.

Nutrition Facts: 1 cookie equals 92 calories, 3 g fat (2 g saturated fat), 15 mg cholesterol, 57 mg sodium, 14 g carbohydrate, trace fiber, 1 g protein. **Diabetic Exchanges:** 1 starch, 1/2 fat.

Editor's Note: Hat-shaped cookie cutters are available from *www.ShopTasteofHome.com.*

Red Potato Salad

Prep: 40 min. + chilling

5 pounds medium red potatoes, halved
5 hard-cooked eggs, chopped
1 celery rib, finely chopped
1/2 medium onion, finely chopped
1-1/2 cups mayonnaise
1/4 cup sweet pickle relish
3 tablespoons sugar
2 tablespoons dried parsley flakes
2 teaspoons prepared mustard
1 teaspoon salt
1 teaspoon cider vinegar
1/8 teaspoon pepper

Place potatoes in a large kettle; cover with water. Bring to a boil. Reduce heat; cover and cook for 20-25 minutes or until tender. Drain and cool. Cut potatoes into 3/4-in. cubes.

In a large bowl, combine the potatoes, eggs, celery and onion. In a small bowl, combine the remaining ingredients. Pour over potato mixture and stir gently to coat. Cover and refrigerate for 6 hours or overnight. **Yield:** 17 servings (3/4 cup each).

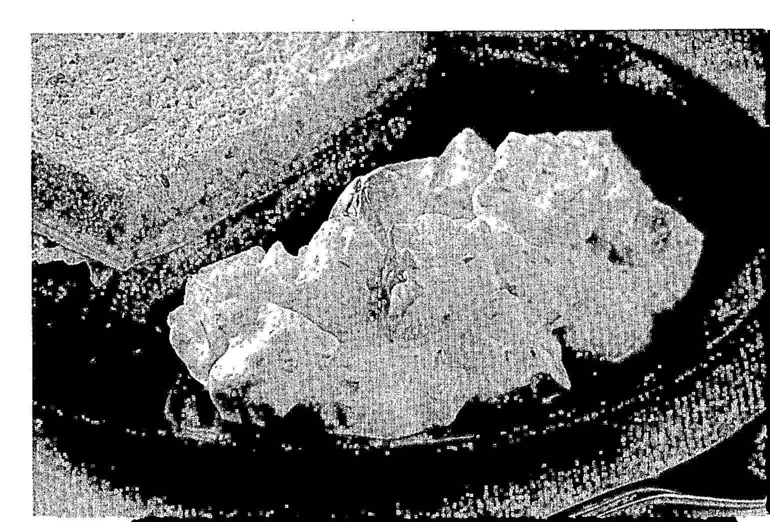

Go!
RCR

You 'Auto' Try This!

By Amber Kimmich, Powhatan, Virginia

MY SON, Wyatt, was born during the NASCAR racing season, so I had no problem choosing a race-car theme for his first birthday. Relatives and friends at the party agreed that the menu and motif were fun for racing fans of all ages.

On "race day," we lined the driveway with checkered flags. As guests arrived, we gave each child a goodie bag filled with toy cars and candy cars and trucks.

Pit-Stop Stromboli was the main course for our race-day meal. This hearty, all-time champ features a three-cheese filling and plenty of pepperoni slices baked into a homemade bread dough roll.

Another first-place finisher was my Cruisin' Crostini. These tasty rounds are similar to bruschetta, but much easier. People always seem to savor the pretty red tomatoes and melted cheese.

Guests peeled rubber getting to the table to munch on Champion Chicken Puffs, too. These tender bites are made with hassle-free refrigerated crescent rolls and a flavorful chicken and cream cheese filling.

Complementing the meal was Wheely-Good Pasta Salad. Red pepper pieces accent this yummy side dish, and the pasta wheels really "drive" the theme.

Everyone got fired up when they saw my Racetrack Cake. This eye-catching treat is really fun to make...and eat!

I baked two 13- by 9-inch cakes (one white, one chocolate), placed them side by side to form a large rectangle and frosted the whole thing white.

For the track and pit stop area, I used gray icing outlined with yellow frosting. Edible green glitter designated grass areas. (If you wish, add a blue gel lake inside the track like the one at the Daytona 500.) The cake's black-and-white checkered sides were made with a star tip.

Our young guests roared with delight when they saw the miniature race cars cruising the track on the cake, with one about to pass the checkered flags.

Bread Flour Facts

The best flour for bread-machine recipes (such as Pit-Stop Stromboli above right), bread flour is made from high-gluten hard wheat flour that gives strength and elasticity to yeast doughs. Bread flour can also be used for other yeast-raised doughs, strudel, puff pastry and pasta where strength and elasticity are desired.

Pit-Stop Stromboli

Prep: 20 min. + rising **Bake:** 30 min.

3/4 cup plus 2 tablespoons water (70° to 80°)
3 tablespoons olive oil
2 tablespoons warm milk (70° to 80°)
3 cups bread flour
1 teaspoon salt
1/2 teaspoon sugar
1-1/2 teaspoons active dry yeast
2 slices part-skim mozzarella cheese, cut into strips
2 slices Colby-Monterey Jack cheese, cut into strips
2 slices provolone cheese, cut into strips
28 slices pepperoni

In bread machine pan, place the first seven ingredients in order suggested by manufacturer. Select dough setting (check dough after 5 minutes of mixing; add 1 to 2 tablespoons of water or flour if needed).

When cycle is completed, turn dough onto a lightly floured surface. Roll into a 16-in. x 10-in. rectangle. Cover with plastic wrap; let rest for 10 minutes.

Arrange cheese and pepperoni evenly over dough to within 1/2 in. of edges. Roll up jelly-roll style, starting with a long side. Pinch seam to seal; tuck ends under. Place seam side down on a greased baking sheet. Cover; let rise in a warm place for 30 minutes.

Bake at 400° for 30-35 minutes or until cheese is bubbly. Remove to a wire rack. Cut into slices; serve warm. **Yield:** 12 servings.

Cruisin' Crostini

Prep/Total Time: 20 min.

24 slices French bread (1/4 inch thick)
1/4 cup olive oil
1/2 teaspoon garlic powder
6 slices part-skim mozzarella cheese
1-1/2 cups chopped seeded tomatoes

Place bread slices on ungreased baking sheets. Combine oil and garlic powder; brush over bread. Cut each slice of cheese into four pieces. Top each slice of bread with tomatoes and a piece of cheese. Broil 3-4 in. from the heat for 2-3 minutes or until cheese is melted. **Yield:** 2 dozen.

Champion Chicken Puffs

(Pictured on page 302)

Prep/Total Time: 30 min.

4 ounces cream cheese, softened
1/2 teaspoon garlic powder
1/2 cup shredded cooked chicken
2 tubes (8 ounces *each*) refrigerated crescent rolls

In a small bowl, beat cream cheese and garlic powder until smooth. Stir in chicken.

Unroll crescent dough; separate into 16 triangles. Cut each triangle in half lengthwise, forming two triangles. Place 1 teaspoon of chicken mixture in the center of each. Fold each short side over filling; press sides to seal and roll up.

Place 1 in. apart on greased baking sheets. Bake at 375° for 12-14 minutes or until golden brown. Serve warm. **Yield:** 32 appetizers.

Wheely-Good Pasta Salad

(Pictured on page 302)

Prep/Total Time: 25 min.

1 package (16 ounces) wagon wheel pasta
8 ounces cheddar cheese, cut into small cubes
1 medium sweet red pepper, diced
1 can (3.8 ounces) sliced ripe olives, drained
2 teaspoons minced fresh oregano
1 bottle (16 ounces) creamy Parmesan Romano salad dressing

Cook pasta according to package directions; drain and rinse in cold water. In a large serving bowl, combine the pasta, cheese, red pepper, olives and oregano. Drizzle with dressing and toss to coat. Cover and refrigerate until serving. **Yield:** 12 servings.

Racetrack Cake

Prep: 3 hours **Bake:** 30 min. + cooling

1 package (18-1/4 ounces) white cake mix
1 package (18-1/4 ounces) chocolate cake mix
10 cups buttercream frosting
Black, yellow and red food coloring
Green edible glitter
4 miniature cars
2 miniature checkered flags

Prepare and bake each cake according to package directions, using greased 13-in. x 9-in. x 2-in. baking pans. Cool for 10 minutes before inverting onto wire racks to cool completely. Transfer cakes to a covered board and position side-by-side.

Frost top of cakes with 5-1/3 cups of frosting. Tint 1-1/2 cups frosting black. Cut a small hole in the corner of a pastry or plastic bag; fill with black frosting. Outline edge of cake. Using a #17 star tip, pipe a checkered pattern on sides of cake with 1 cup white frosting and remaining black frosting.

Tint 2/3 cup frosting gray; create an oval racetrack in middle of cake. Tint 3/4 cup frosting yellow; pipe lines around track and infield. Tint 1/2 cup frosting red; pipe lettering on corners of cake. For grass, sprinkle green glitter on infield. Position cars and checkered flags. **Yield:** 24-30 servings.

Substitutions & Equivalents

Equivalent Measures

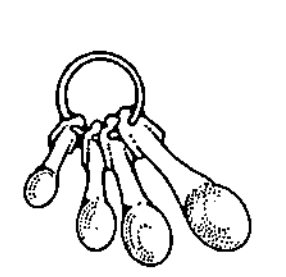

3 teaspoons	=	1 tablespoon	**16 tablespoons**	=	1 cup
4 tablespoons	=	1/4 cup	**2 cups**	=	1 pint
5-1/3 tablespoons	=	1/3 cup	**4 cups**	=	1 quart
8 tablespoons	=	1/2 cup	**4 quarts**	=	1 gallon

Food Equivalents

Grains

Macaroni	1 cup (3-1/2 ounces) uncooked	=	2-1/2 cups cooked
Noodles, Medium	3 cups (4 ounces) uncooked	=	4 cups cooked
Popcorn	1/3 to 1/2 cup unpopped	=	8 cups popped
Rice, Long Grain	1 cup uncooked	=	3 cups cooked
Rice, Quick-Cooking	1 cup uncooked	=	2 cups cooked
Spaghetti	8 ounces uncooked	=	4 cups cooked

Crumbs

Bread	1 slice	=	3/4 cup soft crumbs, 1/4 cup fine dry crumbs
Graham Crackers	7 squares	=	1/2 cup finely crushed
Buttery Round Crackers	12 crackers	=	1/2 cup finely crushed
Saltine Crackers	14 crackers	=	1/2 cup finely crushed

Fruits

Bananas	1 medium	=	1/3 cup mashed
Lemons	1 medium	=	3 tablespoons juice, 2 teaspoons grated peel
Limes	1 medium	=	2 tablespoons juice, 1-1/2 teaspoons grated peel
Oranges	1 medium	=	1/4 to 1/3 cup juice, 4 teaspoons grated peel

Vegetables

Cabbage	1 head	=	5 cups shredded	**Green Pepper**	1 large	=	1 cup chopped
Carrots	1 pound	=	3 cups shredded	**Mushrooms**	1/2 pound	=	3 cups sliced
Celery	1 rib	=	1/2 cup chopped	**Onions**	1 medium	=	1/2 cup chopped
Corn	1 ear fresh	=	2/3 cup kernels	**Potatoes**	3 medium	=	2 cups cubed

Nuts

Almonds	1 pound	=	3 cups chopped	**Pecan Halves**	1 pound	=	4-1/2 cups chopped
Ground Nuts	3-3/4 ounces	=	1 cup	**Walnuts**	1 pound	=	3-3/4 cups chopped

Easy Substitutions

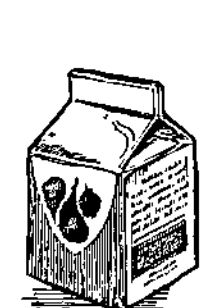

When you need...		*Use...*
Baking Powder	1 teaspoon	1/2 teaspoon cream of tartar + 1/4 teaspoon baking soda
Buttermilk	1 cup	1 tablespoon lemon juice *or* vinegar + enough milk to measure 1 cup (let stand 5 minutes before using)
Cornstarch	1 tablespoon	2 tablespoons all-purpose flour
Honey	1 cup	1-1/4 cups sugar + 1/4 cup water
Half-and-Half Cream	1 cup	1 tablespoon melted butter + enough whole milk to measure 1 cup
Onion	1 small, chopped (1/3 cup)	1 teaspoon onion powder *or* 1 tablespoon dried minced onion
Tomato Juice	1 cup	1/2 cup tomato sauce + 1/2 cup water
Tomato Sauce	2 cups	3/4 cup tomato paste + 1 cup water
Unsweetened Chocolate	1 square (1 ounce)	3 tablespoons baking cocoa + 1 tablespoon shortening *or* oil
Whole Milk	1 cup	1/2 cup evaporated milk + 1/2 cup water

Cooking Terms

HERE'S a quick reference for some of the cooking terms used in *Taste of Home* recipes:

Baste—To moisten food with melted butter, pan drippings, marinades or other liquid to add more flavor and juiciness.

Beat—A rapid movement to combine ingredients using a fork, spoon, wire whisk or electric mixer.

Blend—To combine ingredients until *just* mixed.

Boil—To heat liquids until bubbles form that cannot be "stirred down." In the case of water, the temperature will reach 212°.

Bone—To remove all meat from the bone before cooking.

Cream—To beat ingredients together to a smooth consistency, usually in the case of butter and sugar for baking.

Dash—A small amount of seasoning, less than 1/8 teaspoon. If using a shaker, a dash would comprise a quick flip of the container.

Dredge—To coat foods with flour or other dry ingredients. Most often done with pot roasts and stew meat before browning.

Fold—To incorporate several ingredients by careful and gentle turning with a spatula. Used generally with beaten egg whites or whipped cream when mixing into the rest of the ingredients to keep the batter light.

Julienne—To cut foods into long thin strips much like matchsticks. Used most often for salads and stir-fry dishes.

Mince—To cut into very fine pieces. Used often for garlic or fresh herbs.

Parboil—To cook partially, usually used in the case of chicken, sausages and vegetables.

Partially Set—Describes the consistency of gelatin after it has been chilled for a small amount of time. Mixture should resemble the consistency of egg whites.

Puree—To process foods to a smooth mixture. Can be prepared in an electric blender, food processor, food mill or sieve.

Saute—To fry quickly in a small amount of fat, stirring almost constantly. Most often done with onions, mushrooms and other chopped vegetables.

Score—To cut slits partway through the outer surface of foods. Often used with ham or flank steak.

Stir-Fry—To cook meats and/or vegetables with a constant stirring motion in a small amount of oil in a wok or skillet over high heat.

General Recipe Index

This handy index lists every recipe by food category, major ingredient and/or cooking method, so you can easily locate recipes to suit your needs.

✓ Recipe includes Nutrition Facts and Diabetic Exchanges.

✓ Recipe includes Nutrition Facts and Diabetic Exchanges.

✓Recipe includes Nutrition Facts and Diabetic Exchanges.

✓ Recipe includes Nutrition Facts and Diabetic Exchanges.

✓ Recipe includes Nutrition Facts and Diabetic Exchanges.

✓ Recipe includes Nutrition Facts and Diabetic Exchanges.

✓ Recipe includes Nutrition Facts and Diabetic Exchanges.

✓ Recipe includes Nutrition Facts and Diabetic Exchanges.

✓ Recipe includes Nutrition Facts and Diabetic Exchanges.

✓ Recipe includes Nutrition Facts and Diabetic Exchanges.

Alphabetical Recipe Index

This handy index lists every recipe in alphabetical order so you can easily find your favorites.

A

B

C

✓ Recipe includes Nutrition Facts and Diabetic Exchanges.

D

E

F

G

H

✓ Recipe includes Nutrition Facts and Diabetic Exchanges.

I

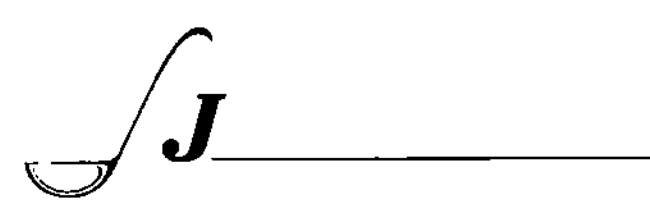

J

K

L

M

N

O

P

R

✓ Recipe includes Nutrition Facts and Diabetic Exchanges.

S

T

V

W

Y

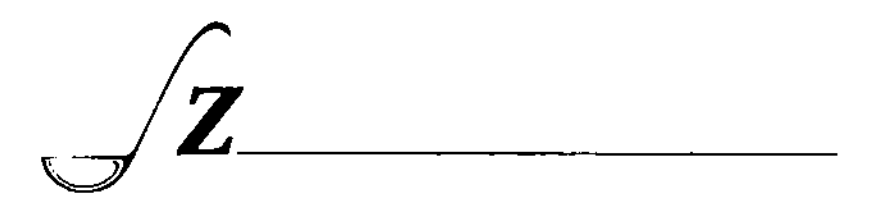

Z

✓ Recipe includes Nutrition Facts and Diabetic Exchanges.